Bioinformatics, Proteomics and Genomics

Bioinformatics, Proteomics and Genomics

Editor: Charles Malkoff

www.callistoreference.com

Callisto Reference,
118-35 Queens Blvd., Suite 400,
Forest Hills, NY 11375, USA

Visit us on the World Wide Web at:
www.callistoreference.com

ISBN: 978-1-63239-804-8 (Hardback)

The publisher's policy is to use permanent paper from mills that operate a sustainable forestry policy. Furthermore, the publisher ensures that the text paper and cover boards used have met acceptable environmental accreditation standards.

Trademark Notice: Registered trademark of products or corporate names are used only for explanation and identification without intent to infringe.

Printed in the United States of America.

Cataloging-in-publication Data

Bioinformatics, proteomics and genomics / edited by Charles Malkoff.
    p. cm.
Includes bibliographical references and index.
ISBN 978-1-63239-804-8
1. Bioinformatics. 2. Proteomics. 3. Genomics. 4. Protein-protein interactions. 5. Computational biology.
6. Molecular biology. 7. Molecular genetics. I. Malkoff, Charles.
QH324.2 .B56 2017
572.8--dc23

Table of Contents

Preface...IX

Chapter 1   **Exploring the Human Plasma Proteome for Humoral Mediators of Remote Ischemic Preconditioning - A Word of Caution**..1
Erik Helgeland, Lars Ertesvåg Breivik, Marc Vaudel, Øyvind Sverre Svendsen, Hilde Garberg, Jan Erik Nordrehaug, Frode Steingrimsen Berven, Anne Kristine Jonassen

Chapter 2   **New Candidate Biomarkers in the Female Genital Tract to Evaluate Microbicide Toxicity**..10
Scott Fields, Benben Song, Bareza Rasoul, Julie Fong, Melissa G. Works, Kenneth Shew, Ying Yiu, Jon Mirsalis, Annalisa D'Andrea

Chapter 3   **MitProNet: A Knowledgebase and Analysis Platform of Proteome, Interactome and Diseases for Mammalian Mitochondria**..............................22
Jiabin Wang, Jian Yang, Song Mao, Xiaoqiang Chai, Yuling Hu, Xugang Hou, Yiheng Tang, Cheng Bi, Xiao Li

Chapter 4   **Relating Diseases by Integrating Gene Associations and Information Flow through Protein Interaction Network**..39
Mehdi Bagheri Hamaneh, Yi-Kuo Yu

Chapter 5   **Prediction of Interactions between Viral and Host Proteins using Supervised Machine Learning Methods**..53
Ranjan Kumar Barman, Sudipto Saha, Santasabuj Das

Chapter 6   **A Network-Based Classification Model for Deriving Novel Drug-Disease Associations and Assessing their Molecular Actions**.................................63
Min Oh, Jaegyoon Ahn, Youngmi Yoon

Chapter 7   **Identification of Potential Serum Proteomic Biomarkers for Clear Cell Renal Cell Carcinoma**...75
Juan Yang, Jin Yang, Yan Gao, Lingyu Zhao, Liying Liu, Yannan Qin, Xiaofei Wang, Tusheng Song, Chen Huang

Chapter 8   **Structural and Functional Characterization of Cleavage and Inactivation of Human Serine Protease Inhibitors by the Bacterial SPATE Protease EspPα from Enterohemorrhagic _E. coli_**...84
André Weiss, Hanna Joerss, Jens Brockmeyer

Chapter 9   **Systems Level Analysis and Identification of Pathways and Networks Associated with Liver Fibrosis**...96
Mohamed Diwan M. AbdulHameed, Gregory J. Tawa, Kamal Kumar, Danielle L. Ippolito, John A. Lewis, Jonathan D. Stallings, Anders Wallqvist

Chapter 10   **Computational Approaches for Predicting Biomedical Research Collaborations**..110
Qing Zhang, Hong Yu

Chapter 11   **Identification of the PLK2-Dependent Phosphopeptidome by Quantitative Proteomics**..123
Cinzia Franchin, Luca Cesaro, Lorenzo A. Pinna, Giorgio Arrigoni, Mauro Salvi

Chapter 12   **Proteomic-Coupled-Network Analysis of T877A-Androgen Receptor Interactomes can Predict Clinical Prostate Cancer Outcomes between White (Non- Hispanic) and African-American Groups**..134
Naif Zaman, Paresa N. Giannopoulos, Shafinaz Chowdhury, Eric Bonneil, Pierre Thibault, Edwin Wang, Mark Trifiro, Miltiadis Paliouras

Chapter 13   **Adaptive Firefly Algorithm: Parameter Analysis and its Application**..........................148
Ngaam J. Cheung, Xue-Ming Ding, Hong-Bin Shen

Chapter 14   **Protein Interaction Networks Reveal Novel Autism Risk Genes within GWAS Statistical Noise**..160
Catarina Correia, Guiomar Oliveira, Astrid M. Vicente

Chapter 15   **Community Structure Detection for Overlapping Modules through Mathematical Programming in Protein Interaction Networks**...........................171
Laura Bennett, Aristotelis Kittas, Songsong Liu, Lazaros G. Papageorgiou, Sophia Tsoka

Chapter 16   **Identification of Kernel Proteins Associated with the Resistance to *Fusarium* Head Blight in Winter Wheat (*Triticum aestivum* L.)**.....................................186
Dawid Perlikowski, Halina Wiśniewska, Tomasz Góral, Michał Kwiatek, Maciej Majka, Arkadiusz Kosmala

Chapter 17   **Crystal Structure of Human Protein N-Terminal Glutamine Amidohydrolase, an Initial Component of the N-End Rule Pathway**..197
Mi Seul Park, Eduard Bitto, Kyung Rok Kim, Craig A. Bingman, Mitchell D. Miller, Hyun-Jung Kim, Byung Woo Han, George N. Phillips Jr

Chapter 18   **GroupRank: Rank Candidate Genes in PPI Network by Differentially Expressed Gene Groups**..205
Qing Wang, Siyi Zhang, Shichao Pang, Menghuan Zhang, Bo Wang, Qi Liu, Jing Li

Chapter 19   **Significant Low Prevalence of Antibodies Reacting with Simian Virus 40 Mimotopes in Serum Samples from Patients Affected by Inflammatory Neurologic Diseases, Including Multiple Sclerosis**...212
Elisa Mazzoni, Silvia Pietrobon, Irene Masini, John Charles Rotondo, Mauro Gentile, Enrico Fainardi, Ilaria Casetta, Massimiliano Castellazzi, Enrico Granieri, Maria Luisa Caniati, Maria Rosaria Tola, Giovanni Guerra, Fernanda Martini, Mauro Tognon

Chapter 20   **Proteome Folding Kinetics is Limited by Protein Halflife**.. 219
Taisong Zou, Nickolas Williams, S. Banu Ozkan, Kingshuk Ghosh

Permissions

List of Contributors

Index

Preface

Proteins have been studied for many decades. However, the study of proteins was revolutionized in the late twentieth century. Technological advances in the field of computational modeling has paved way for the observation of protein behavior and the corresponding process of formation and transformation. This book on bioinformatics, proteomics and genomics discusses the recent developments that have taken place in these fields. The various sub-fields of bioinformatics, proteomics and genomics along with technological progress that have future implications are glanced at in this book. It brings forth some of the most innovative concepts and elucidates the unexplored aspects of these fields. This text includes contribution of experts and scientists which will provide innovative insights to readers. This book will be useful for students, experts and professionals in the fields of molecular biology, genetics and proteomics. The extensive content of this book provides the readers with a thorough understanding of the subject.

This book has been an outcome of determined endeavour from a group of educationists in the field. The primary objective was to involve a broad spectrum of professionals from diverse cultural background involved in the field for developing new researches. The book not only targets students but also scholars pursuing higher research for further enhancement of the theoretical and practical applications of the subject.

It was an honour to edit such a profound book and also a challenging task to compile and examine all the relevant data for accuracy and originality. I wish to acknowledge the efforts of the contributors for submitting such brilliant and diverse chapters in the field and for endlessly working for the completion of the book. Last, but not the least; I thank my family for being a constant source of support in all my research endeavours.

Editor

Exploring the Human Plasma Proteome for Humoral Mediators of Remote Ischemic Preconditioning - A Word of Caution

Erik Helgeland[1], Lars Ertesvåg Breivik[1], Marc Vaudel[1], Øyvind Sverre Svendsen[3], Hilde Garberg[1], Jan Erik Nordrehaug[2], Frode Steingrimsen Berven[1], Anne Kristine Jonassen[1]*

1 Department of Biomedicine, Faculty of Medicine and Dentistry, University of Bergen, Bergen, Norway, 2 Department of Clinical Science, Faculty of Medicine and Dentistry, University of Bergen, Bergen, Norway, 3 Department of Anaesthesia and Surgical Services, Haukeland University Hospital, Bergen, Norway

Abstract

Despite major advances in early revascularization techniques, cardiovascular diseases are still the leading cause of death worldwide, and myocardial infarctions contribute heavily to this. Over the past decades, it has become apparent that reperfusion of blood to a previously ischemic area of the heart causes damage in and of itself, and that this ischemia reperfusion induced injury can be reduced by up to 50% by mechanical manipulation of the blood flow to the heart. The recent discovery of remote ischemic preconditioning (RIPC) provides a non-invasive approach of inducing this cardioprotection at a distance. Finding its endogenous mediators and their operative mode is an important step toward increasing the ischemic tolerance. The release of humoral factor(s) upon RIPC was recently demonstrated and several candidate proteins were published as possible mediators of the cardioprotection. Before clinical applicability, these potential biomarkers and their efficiency must be validated, a task made challenging by the large heterogeneity in reported data and results. Here, in an attempt to reproduce and provide more experimental data on these mediators, we conducted an unbiased in-depth analysis of the human plasma proteome before and after RIPC. From the 68 protein markers reported in the literature, only 28 could be mapped to manually reviewed (Swiss-Prot) protein sequences. 23 of them were monitored in our untargeted experiment. However, their significant regulation could not be reproducibly estimated. In fact, among the 394 plasma proteins we accurately quantified, no significant regulation could be confidently and reproducibly assessed. This indicates that it is difficult to both monitor and reproduce published data from experiments exploring for RIPC induced plasma proteomic regulations, and suggests that further work should be directed towards small humoral factors. To simplify this task, we made our proteomic dataset available via ProteomeXchange, where scientists can mine for novel potential targets.

Editor: Yiru Guo, University of Louisville, United States of America

Funding: This study was funded by the F. Mohn foundation, the Grieg Foundation, Bergen Heart Foundation, Lærdal Foundation for Acute Medicine and Bergen Medical Research Foundation. Erik Helgeland was supported by the University of Bergen. Lars Breivik was supported by the Norwegian Council on Cardiovascular disease. The funders had no role in study design, data collection and analysis, decision to publish, or preparation of the manuscript.

Competing Interests: The authors have declared that no competing interests exist.

* Email: anne.jonassen@biomed.uib.no

Introduction

Remote ischemic preconditioning (RIPC) is an emerging treatment for reducing ischemia reperfusion injury (IRI) in the heart. Several proof-of-concept studies and small randomized controlled trials have demonstrated that the human heart is amenable to RIPC [1–5]. Recently, it was also reported that 3 cycles of 5 min upper arm ischemia substantially reduced myocardial injury after coronary artery bypass graft in large pools of patients. This was demonstrated by a significant decrease in cardiac troponin I (cTnI) release, significantly improving prognosis with a reduction in mortality in the group receiving RIPC compared to controls [6].

The signaling pathway of RIPC in the human heart is just starting to be uncovered [7], but how the RIPC stimulus is transferred from the arm to the heart remains unclear. Compelling preclinical evidence suggests communication via one or more unknown humoral factors: First, it seems that a period of reperfusion after the RIPC stimulus is required for protection, suggesting that wash-out of blood borne factors and transport to the site of protection is involved [8–10]. Secondly, it was demonstrated that effluent from preconditioned hearts could transfer the protection to naïve recipient hearts and that the protection is mediated via small, unknown hydrophobic factors of protein nature between 3.5 and 15–30 kDa [11–16]. Moreover, the fact that this humoral factor is effective at a remote location after dilution in blood or perfusion fluid, hints at a large concentration change which should be detectable by modern proteomic techniques.

In fact, several proteins were recently found to be regulated after RIPC, paving the way for potential use of these cardioprotective compounds in the clinic. Notably, Hepponstall et al. conducted an ambitious study where 806 differentially expressed peptides were identified after RIPC. Among them, 133 could be mapped to 48 protein sequences [17]. In addition, 2D gel experiments reported 33 regulated spots with 6 identifiable proteins. Surprisingly, only one of the given protein accession numbers could be found in both the mass spectrometry and the 2D gel result sets. Subsequently, Pang et al. found 14 proteins to be differentially regulated by RIPC and validated these findings by Western blotting [18]. Notably, only one of the accessions reported, Gelsolin (UniProt accession number P06396), could be mapped to those published by Hepponstall et al. Additionally, albeit with different sequences, both studies underlined the regulation of Apolipoprotein A-I. However, Pang et al. found this protein to be down-regulated, while Hepponstall et al. found it down-regulated using gels and up-regulated when applying a gel free technique in the same study. The latter result is in concordance with the study of Hibert et al. showing a 30% up-regulation of Apolipoprotein A-I, postulating it to be the principal factor behind RIPC mediated cardioprotection [19]. Later on, however, Hilbert et al. published another report where this protein appeared not to be regulated, but seven other related proteins presented a Mann-Whitney test p-value <0.05 [20]. Notably, no protein passed a more stringent 0.01 threshold and all proteins showed moderate regulations (between 0.58:1 and 1.2:1), consistently below the two fold change regulation level generally used in biology. The results of Davidson et al. suggested SDF-1α to be the main mediator of RIPC, presumably communicating the cardioprotection via SDF-1α/CXCR4 signaling [21]. Importantly, Przyklenk [22] criticized the latter study for its limitations, as the plasma levels of SDF-1α should have been monitored both before and after the RIPC stimulus, thereby failing to validate the factor as a mediator of RIPC.

Discovery studies "often significantly overestimate their findings" as claimed by Gosho et al. [23]. Consequently, prior to clinical testing, protein markers must be validated, preferably using targeted mass spectrometry based proteomics [24]. For that, quantitative assays are built in order to target and quantify the compounds of interest. Setting up such an assay requires experimental data on the digestion, peptide elution, ionization, and fragmentation profile of the targeted protein. This information is, however, not available from any of the mentioned mass spectrometry datasets. Notably, they are not available in public repositories, despite this being standard publication guidelines in proteomics [25,26]. In fact, the peptide information is not available and in the case of Hepponstall et al., not even the estimated protein ratios are reported.

The disagreement in the literature, even in studies from the same group or within the same study, together with the lack of provided information, show the striking need for a transparent proteomic dataset stringently monitoring the proteomic changes induced by RIPC. Here, we present a quantitative in depth sequencing of the plasma proteome before and after RIPC. Six healthy donors underwent RIPC according to the protocol used in consensus with the literature and whose efficiency has been long established. In contrast to previously reported studies, the digested plasma samples were multiplexed using isobaric tags and fractionated, thereby limiting inter-sample artifacts and dramatically increasing sample coverage. The protein identification and quantification was achieved using open source software applying stringent quality criteria to avoid reporting artifact regulations. The acquired raw data was deposited to the ProteomeXchange

consortium [27] together with identification results, and can thereby freely be accessed, inspected and even reprocessed to better plan further experiments [28].

Materials and Methods

This study was approved by the regional ethics committee for medical and health research in Western Norway (REK 2010/1642-1). Written informed consent was obtained from all participants and the study conformed to the principles in the declaration of Helsinki. Six healthy adult male donors aged 29 ± 2 years old (mean $\pm$ SD), not on any medication, underwent the RIPC protocol which consisted of 3 cycles of 5 min upper arm ischemia alternating with 5 min reperfusion (Figure 1). All subjects rested upright on a bench for 15 min before ischemia was induced by inflating a blood pressure cuff to 200 mmHg. Venous blood samples were collected from the ipsilateral arm after 14 min of rest in the beginning, and at 1 and 4 min into each reperfusion period. Blood samples were collected in K_2 EDTA-coated tubes (Vacutainer, BD). Samples were centrifuged at 1,500 rcf at 4°C for 15 min within 30 min after sampling. 1.5 ml plasma was transferred to Eppendorf tubes before a second centrifugation at 15,000 rcf at 4°C for 15 min to remove any cell contaminants or cell fragments. ~90% of the clear supernatant was removed using a 1 ml pipette and stored at −80°C. The whole process from sampling to freezing of each individual sample took less than 75 min. Samples from 1 and 4 min of each of the reperfusion periods were pooled, resulting in 12 samples: a sample before (control) and after RIPC for each of the six individuals.

Chemicals

Trypsin was purchased from Promega. N-octyl-β-D-glycopyranoside (NOG), acetonitrile (ACN), formic acid (FA), ammonium formate and water were purchased from Sigma-Aldrich. Water and ACN were of HPLC quality.

Abundant protein depletion and concentration

20 μl of plasma from each sample was depleted using a human Multiple Affinity Removal System (MARS Hu-14) 4.6 mm×50 mm LC column (Agilent Technologies) according to the protocol provided by the supplier, using a Dionex 3000-series LC system. The MARS column depletes the plasma of albumin, IgG, antitrypsin, IgA, transferrin, haptoglobin, fibrinogen, alpha-2-macroglobulin, alpha-1-acid glycoprotein, IgM, apolipoprotein AI, apolipoprotein AII, complement C3 and transthyretin. The protein depleted plasma samples were concentrated using 3 kDa ultracentrifugation filters (Amicon Ultra-4, Millipore, Bedford, MA) pre-treated with 0.1% NOG.

Protein digestion and iTRAQ labeling

The entire depleted protein sample was reduced, cysteine blocked, trypsin digested (1:20, trypsin:protein, w/w), iTRAQ labeled (114, 115, 116 and 117) and combined according to the protocol using the chemicals provided (AB Sciex). The iTRAQ 4-Plex kit allowed us to multiplex two conditions from two donors per kit, resulting in three parallel experiments as detailed in Table 1. Both conditions (before and after RIPC) followed the exact same downstream workflow for every individual.

Mix-mode fractionation

iTRAQ labeled peptides were fractionated into 60 fractions using a mixed-mode (MM) reversed phase anion exchange (RP-AX) Sielc Promix column as described by Philips et al. [29] (MP-10.250.0530, 1.0×250 mm, 5 μm, 300 Å, Sielc Technologies,

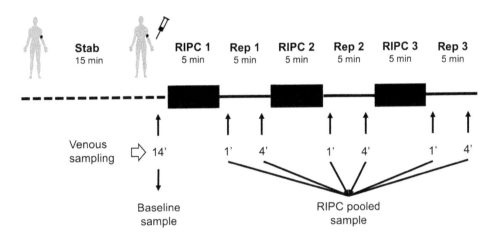

Figure 1. Experimental protocol. A peripheral venous catheter was inserted in the cubital fossa of subject for blood sampling. The subject rested for 14 min reclined on a bench before the baseline sample was drawn. The blood pressure cuff was inflated to 200 mmHg for 5 min before being released. Blood samples were drawn at 1 and 4 min into reperfusion from the ipsilateral arm. Blood samples were centrifuged to collect plasma which was stored at −80°C. Before analysis, all six reperfusion samples were pooled for each subject.

Prospect Heights, Illinois) coupled to an Agilent 1260 series LC system (Agilent Technologies, Palo Alto, CA). The iTRAQ labeled peptides were reconstituted in 20 mM ammonium formate, 3% ACN (buffer A) and loaded on the column in 85% buffer A for 10 minutes at a flowrate of 50 µl/min. The peptides were eluted from the column increasing the contents of buffer B (2 mM ammonium formate, 80% ACN, pH 3.0), from 15% to 60% in 35 minutes and further to 100% buffer over 10 minutes. Buffer B was held constant for 5 minutes before the column was equilibrated for 10 minutes in 85% buffer A. The fractions from the first 10 minutes of the gradient were discarded.

LC-MS/MS analyses

Fifty fractions from the MM RP-AX separation from each sample were analyzed on an LTQ-Orbitrap Velos Pro (Thermo Scientific) coupled to a Dionex Ultimate NCR-3000RS LC system. The fractions were dissolved in 1% FA and trapped on the pre-column (Dionex, Acclaim PepMap 100, 2 cm×75 µm i.d, 3 µm C18 beads) in buffer A (2% ACN, 0.1% FA) at a flowrate of 5 µl/min for 5 minutes before separation by reverse phase chromatography (Dionex, Acclaim PepMap 100, 15 cm×75 µm i.d., 3 µm C18 beads) at a flow of 280 nL/min. The fractions were run on three nano LC gradients: The first fifteen fractions were run on a LC gradient consisting of a gradient starting at 5% buffer B (90%ACN, 0.1% FA) ramping to 12% buffer B over 55 minutes (5–60 min), the gradient was subsequently ramped to 30% buffer B in 30 minutes (60–90 min), increased to 90% B in 10 minutes (90–100 min), held for 5 minutes (100–105 min) followed by ramping to 5% buffer B for 3 minutes (105–108) and equilibration

of the column in 12 minutes (108–120); fractions 16–35 were separated on the following LC gradient: 0–5.0 minutes 5% buffer B, 5.0–5.5 minutes 8% buffer B, 5.5–60 minutes 20% buffer, 60–90 minutes 35% buffer B; the last fractions (36–50) were separated using the following gradient: 0–5.0 minutes 5% buffer B, 5.0–5.5 minutes 8% buffer B, 5.5–90 minutes 40% buffer. The last part of the nano LC gradient is similar for all three gradients.

The mass spectrometer was operated in data-dependent-acquisition (DDA) mode to automatically switch between full scan MS and MS/MS acquisition. The instrument was controlled by Tune 2.6.0 and Xcalibur 2.1. Survey full scan MS spectra (from m/z 300 to 2,000) were acquired in the Orbitrap with resolution R = 60,000 at m/z 400 (after accumulation to a target value of 1E6 in the linear ion trap with maximum allowed ion accumulation time of 500 ms). The 7 most intense eluting peptides above an ion threshold of 1,000 counts and charge states 2 or higher, were sequentially isolated in the high-pressure linear ion trap to a target value of 5E5 at a maximum allowed accumulation time of 1,000 ms, and isolation width maintained at 2 Da. Fragmentation in the Higher-Energy Collision Dissociation (HCD) cell was performed with a normalized collision energy of 40%, and activation time of 0.1 ms. Fragments were detected in the Orbitrap at a resolution of 7,500 with first mass fixed at m/z 100.

Data analysis

All RAW data were transformed into mgf peak lists using the ProteoWizard software [30] package version 2.2.2954. The obtained peak lists were searched with OMSSA [31] version 2.1.9 and X!Tandem [32] Cyclone 2013.2.01.1 using SearchGUI

Table 1. Repartition on every iTRAQ channel of the samples at baseline and after RIPC for the six donors.

iTRAQ Channel	Experiment 1	Experiment 2	Experiment 3
114	Donor 1 baseline	Donor 3 RIPC	Donor 5 baseline
115	Donor 1 RIPC	Donor 3 baseline	Donor 5 RIPC
116	Donor 2 baseline	Donor 4 RIPC	Donor 6 baseline
117	Donor 2 RIPC	Donor 4 baseline	Donor 6 RIPC

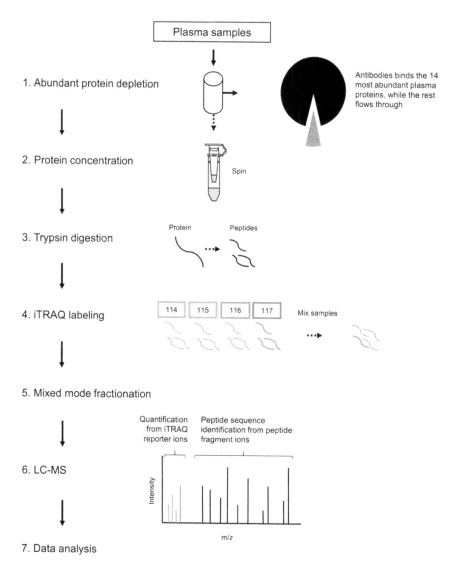

Figure 2. Analysis workflow. Plasma samples were depleted by a MARS Hu-14 column and subsequently concentrated by 3 kDa ultracentrifugation filters. Next, samples were reduced, cysteine blocked and trypsin digested before iTRAQ labeling. The iTRAQ labeled peptides were fractioned into 60 fractions using a mixed-mode reverse phase anion exchanger. Finally, fractions were analyzed on an LTQ-Orbitrap Velos Pro connected to a Dionex Ultimate NCR-3000RS LC system.

[33] version 1.12.2. Peak lists were searched against a concatenated target/decoy [34] version of the human complement of the UniProtKB/Swiss-Prot database [35] (downloaded on September 2012). The decoy sequences were created by reversing the target sequences in SearchGUI. Search settings were as follows: Trypsin with a maximum of 2 missed cleavages; 10 ppm as MS, 0.6 Da as MS/MS tolerances, respectively; fixed modifications: methylthio of Cys (+45.987721 Da) and iTRAQ on Lys and peptide N-term (+144.105918 Da); and variable modifications: oxidation of Met (+ 15.994915 Da) and iTRAQ on Tyr (+144.105918 Da). All other OMSSA or X!Tandem settings were kept at the default values set in SearchGUI. Peptides and proteins were inferred from the search engine results using PeptideShaker (http://peptide-shaker.googlecode.com) [36]. Peptide to Spectrum Matches (PSMs), peptides and proteins were validated at a stringent 1% FDR estimated using the decoy hits. The mass spectrometry proteomics data have been deposited to the ProteomeXchange Consortium(http://proteomecentral.proteomexchange.org) [27] via the

PRIDE partner repository [37,38] with the dataset identifier PXD000605 and DOI 10.6019/PXD000605.

For every validated protein, the iTRAQ reporter ions were extracted from spectra of validated PSMs and deisotoped using the isotope abundance matrix [39]. Intensities were normalized using the median intensity in order to limit the ratio deviation [40] and peptide and protein ratios were estimated using maximum likelihood estimators [41]. Ratios were log2 converted and normalized to the median to avoid inter-sample bias. Only those proteins presenting two or more validated and quantified peptides were retained for the quantitative analysis. Standard contaminants as well as all proteins with known affinity to the antibodies in the MARS column were excluded from downstream statistical analysis. A paired two-sided t-test was conducted on protein ratios using a p-value threshold of 0.01. Subsequently, in order to distinguish RIPC specific regulations from random biological variability, an asymmetric normal distribution was drawn from the background ratios (calibrated on the median and ±34.1 percentiles) and only those proteins having a probability <1% to be derived from the background were considered confidently regulated. Finally, protein abundance indexes were estimated using spectral counting where the spectral counting index is simply the number of validated PSMs divided by the protein molecular weight [42].

Results

The plasma proteome from six healthy donors was compared at baseline and after RIPC as illustrated in Figure 1 and 2. In total, 727 proteins were confidently identified (<1% FDR). Among them, 409 proteins could be accurately quantified by two or more unique peptides. Of these, 393 remained after exclusion of known contaminant and depletion-affected proteins, and are listed in Table S1. These 393 accurately quantified plasma proteins showed accurate and reproducible stability: 388 (98.7%) presented a fold change between 0.91:1 and 1.1:1, which is in accordance with the known technical variability of iTRAQ quantification reported in the literature [43]. In order to assess the relevance of the regulations, the quantified proteins were classified into four categories (as detailed in Table S1): (A) proteins with low variability among donors (t-test passed) *and* confident regulation to the background (no protein); (B) low donor variability *but no* confident regulation to the background (4 proteins); (C) high donor variability (t-test failed) but confident regulation to the background (42 proteins); and (D) high donor variability (t-test failed) and no confident regulation to the background (347 proteins). These proteins are plotted in Figure 3 and 4. Notably, no protein was validated by both statistical tests (category A).

Four proteins had a significant p-value (<0.01) between the baseline and RIPC samples for the six subjects examined (category B; Kallistatin, Complement C2, Lactoylglutathione lyase (GLO1), and Cysteine-rich secretory protein 3 (CRISP-3))(Table S1). The fold changes for these, however, were very low (<10% change). We considered changes below 20% to be within what can be explained by the technical variability of the analytical approach. This was corroborated by the statistical analysis, showing no confident regulation to the background. Interestingly, all highly regulated proteins were detected in a much lower abundance (see Figure 4) and failed the t-test (p-value>0.01), suggesting that these cannot be trusted for clinical application.

In contrast, high ratios were systematically found for contaminants and proteins affected by depletion. Remarkably, the remaining amount of proteins targeted by the MARS hu-14 depletion column was consistently less abundant in RIPC samples, hinting at a systematic artifact in the depletion procedure. Among them, fibronectin, previously reported to be regulated after RIPC has been identified in the bound fraction after MARS hu-14 depletion [44]. Fibronectin was confidently regulated to the background (ratios of 0.62:1) in close agreement with Apolipoprotein A-I (ratio of 0.61:1) and Haptoglobin (ratio of 0.63:1), both also targeted by the MARS column. In conclusion, we cannot rule out that the difference in relative abundance measured for these proteins are not simply protocol related. On the other hand, our ability to quantify these potential experimental artifacts demonstrates the reliability of the quantification procedure and rules out the hypothesis that regulations were systematically lost along the workflow.

Discussion

RIPC of the upper arm is an easy, practical and non-invasive way of inducing protection from ischemia reperfusion induced injury (IRI) in the heart, and the remote ischemic conditioning

stimulus can even be applied during or immediately after an ischemic insult (as reviewed in [45–47]). The protocol used in the present study was chosen for its proven clinical efficiency: It was demonstrated to reduce the release of troponins (marker for cardiac damage) [3–5] and improve survival [6] after coronary artery bypass graft surgery. Shimizu et al. used a human RIPC protocol of 4×5 min, sampled venous blood before and after the RIPC intervention, and demonstrated that both whole plasma and plasma dialysate <15 kDa could reduce IRI in the isolated rabbit heart [15]. This protocol is comparable, but not identical, to the one used for the proteomic studies mentioned above, ensuring comparability between the results.

Table S2 gives a generated list of all biomarkers found in the literature, in addition to the ones identified in this study, including the measured ratios and t-test results. As detailed in the introduction, Hepponstall et al. [17] report 53 regulated proteins after RIPC. The provided accession numbers were mapped to UniProt entries using the Picr service of the European Bioinformatics Institute [48], leading to mapping of only 14 Swiss-Prot sequences – all other sequences having low or no evidence. 11 of them were confidently quantified in our study (see Table S2). Only Fibronectin (t-test passed) and Haptoglobin (t-test failed) were found to be regulated. However, the regulation of these factors may be experimental artifacts as discussed earlier.

From the 14 proteins reported as regulated by Pang et al. [18], 13 could be mapped to Swiss-Prot accessions, and 12 were accurately quantified in our experiment. Among these, only Apolipoprotein A-I was found to be regulated (0.61:1 ratio). However, as detailed already, it is one of the targets of the depletion procedure. As a result, it is impossible to validate whether this controversial protein is actually regulated by RIPC or whether the up- and down-regulations reported in the literature may be due to experimental artifacts.

Furthermore, we did not quantify SDF-1α, a purported signaling molecule of RIPC [21,22]. Nor did we identify a single of its peptides. This protein is a good example of the challenges posed when identifying RIPC mediators, as peptide based protein quantification heavily relies on the ability to detect at least two peptides, which can become defying for small proteins. It is important to note that most of the reported potential mediators of RIPC mediated cardioprotection do not meet the mass expectation (<30 kDa). More experimental data are, thus, necessary to assess the role of SDF-1α in RIPC.

In our study, we identified 4 proteins (category B) to be significantly regulated according to the t-test, but showed no confident regulation to the background; Kallistatin, Complement C2, Lactoylglutathione lyase (GLO1), and Cysteine-rich secretory protein 3 (CRISP-3) (Table S1). Despite claiming an overall negative study outcome, as no factors were released in abundance and confidently regulated to the background, it is perhaps timely to consider whether RIPC might be mediated by a consortium of slightly regulated humoral factors, acting on one or several signalling networks, which in concert induce protection. Interestingly, we identify 3 proteins (kallistatin, kallikrein, and kininogen-1) that might inter-relate in the kallikrein-kinin signalling pathway. Of further interest, and adding to the complexity, Complement 2 as part of the complement system may also exerts cross-talk with proteins of the kinin-generating systems. The 4 category B proteins will be discussed further below.

Kallistatin, in particular, proves to be a very interesting candidate in terms of cytoprotective capabilities (UniProt accession P29622, 48 kDa, p-value 0.002, fold-change 1.04). Chao and co-workers identified kallistatin as a tissue kallikrein binding protein (KBP) and a unique serine proteinase inhibitor (serpin) [49]. Later,

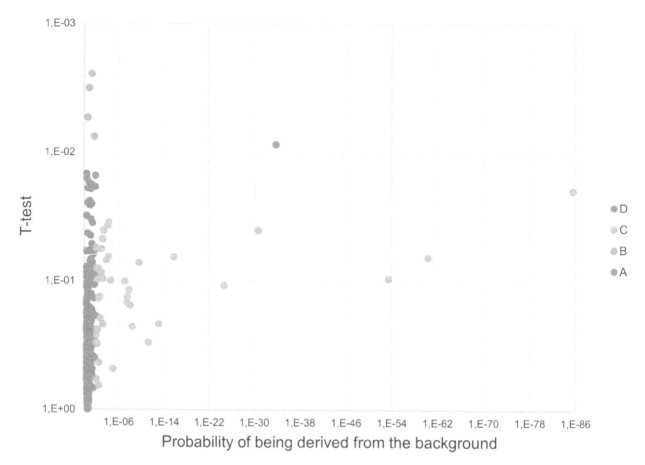

Figure 3. Volcano plot. The significance of the relative regulation of 394 proteins was inspected using (1) a paired two-sided t-test (y axis) and (2) by estimating the probability for the regulation derived from the background (x axis). Proteins are clustered into four categories based on the statistical test passing the threshold (see text for details).

kallistatin has been ascribed many other functions unrelated to its interaction with tissue kallikrein, including lowering blood pressure, vasodilatation, preventing cardiac remodelling and offering protection against cardiovascular injuries by preventing apoptosis, oxidative stress, and inflammation [50–54]. Moreover, the effects of kallistatin seems to be mediated via pro-survival PI3K/Akt/NO dependent signalling, and is postulated to be activated by a yet unidentified kallistatin specific cell surface receptor or binding protein [52]. Kallistatin may also act as an inhibitor of kallikrein [49], and we found kallikrein to be slightly down-regulated in our data (UniProt accession P03952, category D, 71 kDa, p-value 0.014, fold change 0.96). Kallikrein may be activated by lowered plasma pH [55] due to flow restriction imposed by the ischemic conditioning cycles, which in turn can reduce kinin breakdown, enhance bradykinin (BK) formation [56–58], and inhibit BK degradation [59]. We also observed that the MK-RPPGFSPFR-SS peptide located at amino-acids 381 to 389 of Kininogen-1 (UniProt accession P01042, catgory D, 72 kDa) was recorded with a 10 times increase in the number of spectra when compared to the median of the peptides for this protein. The peptide may be a product of Kininogen-1 degradation. Plasma and tissue kallikreins converts kiniogens to produce vasoactive kinin peptides, such as bradykinin and lys-bradykinin. BK is known to exert anti-ischemic effects and for being a possible mediator of ischemic preconditioning, although the peptide presence could not directly be related to RIPC in our experiment (ratio of 1.02:1, p-value of 0.43). But kinin receptors were

previously shown to be influenced by RIPC [60], so monitoring kininogen degradation products might be a promising approach elucidating the RIPC mechanism.

The complement system, which is part of the innate humoral immune system, was slightly up-regulated (UniProt accession P06681, 83 kDa, p-value 0.003, fold-change 1.02). Weissman et al. [61] demonstrated that complement components are deposited in ischemia reperfused myocardium. In addition, animal models of IR in other organ systems like the gut, kidney, and skeletal muscle indicate that the complement system is a key mediator of IRI [62–67]. However, the precise mechanism of complement activation in ischemic tissue has not been clearly elucidated due to the lack of appropriate experimental models, restricted knowledge of the molecular processes causing complement activation during hypoxia in cells, and how it exerts cross-talk between different complement activation pathways [68]. Despite the fact that complement activation during IR is associated with cellular injuries; it is intriguing that the complement system exerts cross-talk with proteins of the kinin generating systems [69].

Lactoylglutathione lyase (also known as glyoxalase I or GLO1) was slightly down-regulated after RIPC (UniProt accession Q04760, 23 kDa, p-value 0.005, fold-change 0.92). Glyoxalase 1 (GLO1) in combination with glyoxalase 2 and the co-factor gluthatione constitute the glyoxalase system, which is responsible for the detoxification of methylglyoxal (MG) [70]. MGs are highly reactive metabolites of glucose degradation pathway, protein and fatty acid metabolism. MG itself is cytotoxic and pro-apoptotic.

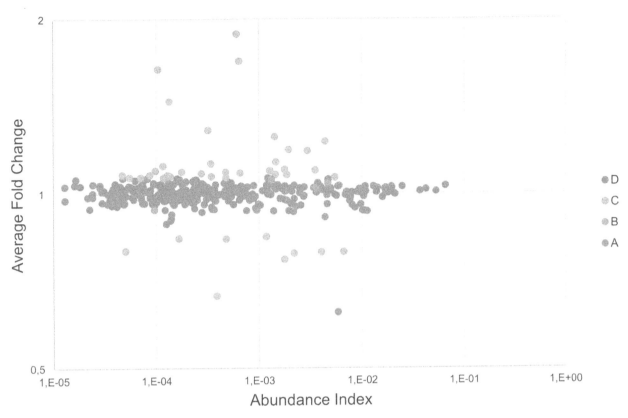

Figure 4. Protein regulation vs. abundance index. The protein regulation is plotted against the abundance index and every protein is classified according to the result of the statistical tests.

GLO1 might be a key factor for detoxifying MG and protecting organs against IR injury [71], and may also prevent hyperglyce-mia-induced diabetic complications [72,73]. Despite a potential protective role of GLO1, it appears down regulated in our study, and could, thus, be unrelated to the cardioprotective properties of RIPC mediating humoral factors.

Our data also identified glycoprotein human cysteine-rich secretory protein 3 (CRISP-3) as slightly up-regulated (UniProt accession P54108, 28 kDa, p-value 0.007, fold-change 1.06). CRISP-3 is believed to play a role in innate immunity. High levels of human CRISP-3 was found in plasma bound to α_1-1B-glycoprotein (A1BG-like) (a plasma protein of unknown function) [74]. The A1BG–CRISP-3 complex is thought to inhibit the toxic effect of snake venom metalloproteinases or myotoxins. Udby et al. suggests that the A1BG–CRISP-3 complex displays a similar function in protecting the circulation from a potentially harmful effect of free CRISP-3, although the overall function of CRIPS-3 is unclear [74]. Cardiac related effects of CRISP-3 has not been described in the literature as of yet. We also identified the proposed A1BG binding partner in our data (P04217; category D; 54 kDa; p-value 0.16; fold change 1.02).

It is crucial to consider that even with the best analytical approaches available today, there is a limit to how many proteins that can be identified and quantified in plasma. Notably, isoforms, posttranslational modifications and degraded proteins are a vast field of investigation and future experiments might, thus, be directed at other targets in blood. Moreover, we decided to pool all the reperfusion samples for each individual, making sure all potential relevant proteins were present in the sample. However, pooling of the samples may have masked possible time-dependent

RIPC induced protein/mediator alterations, in addition to averaging the relative abundance of potential candidate proteins. It also reduces the ability to monitor changes occurring after a single conditioning cycle only. Future work might, thus, improve the time resolution of the experiment.

In addition to technical analytical factors as described here, other aspects of the RIPC procedure such as the number of cycles and stimulus site should be considered when mining for blood borne humoral factors. Loukogeogakis et al. elegantly demon-strated a dose-response effect with regards to both site and number of cycles, when exploring for the protective effect of RIPC on endothelial IRI of the arm. The maximum protective effect was obtained with 3 cycles of IR of the arm and at least 2 cycles of the leg [75]. This study is complemented by the study of Hong et al., suggesting that the protective effects of RIPC by lower limb ischemia are greater than those induced by upper limb ischemia [76]. The reason that fewer cycles of the lower limb induced protection may possibly be due to a larger lower limb mass leading to greater release of humoral factors. However, according to a very recent systematic review and meta-analysis [77], it is clear that the optimal RIPC stimulus has not been demonstrated, and it is unknown whether upper or lower limb ischemia is superior. Furthermore, the most favorable timing/duration of the stimulus is unclear. Once the optimal stimulus algorithm, site and timing for the RIPC procedure and more certain indications of the RIPC mediated humoral factor(s) mediating protection are established, it will be timely to compare the healthy plasma RIPC proteome to that of diseased patient populations. Furthermore, comparison of age-matched young and old, male and females might also delineate possible and important age and sex differences.

In conclusion, using shotgun quantification techniques, the detectable portion of the plasma proteomes of six healthy adult males was stable after remote ischemic preconditioning, and in contrast to the literature, no significant changes were identified, other than those that appeared to be related to contaminants or the analytical process itself. Four proteins did, however, show low donor variability (passed the t-test) *but was not* confidently regulated to the background. One way of moving forward could be to increase the workflow sensitivity, with regards to time or biological targets, *e.g.* by selecting on hydrophobicity and size.

Supporting Information

Table S1   The complete table of experimental data from all three iTRAQ experiments. All proteins we identified were classified into four categories based on two statistical tests: (A) proteins with low variability among donors (t-test passed) *and* confident regulation to the background (no protein); (B) low donor variability *but no* confident regulation to the background (4 proteins); (C) high donor variability (t-test failed) but confident regulation to the background (42 proteins); and (D) high donor variability (t-test failed) and no confident regulation to the background (347 proteins). (XLSX)

Table S2   A list of all biomarkers found in the literature, in addition to the ones identified in this study. This includes their measured ratios and t-test results. (XLSX)

Acknowledgments

The EBI is acknowledged for the quality of the products they make available to the community, notably Picr and PRIDE which were central for this study. The PRIDE team is acknowledged for efficient handling of our data.

Author Contributions

Conceived and designed the experiments: AKJ LB EH JEN FSB. Performed the experiments: AKJ LB EH MV ØSS HKG. Analyzed the data: EH MV HKG FSB. Contributed reagents/materials/analysis tools: AKJ LB EH MV ØSS FSB JEN. Wrote the paper: AKJ EH LB MV FSB.

References

1. Botker HE, Kharbanda R, Schmidt MR, Bottcher M, Kaltoft AK, et al. (2010) Remote ischaemic conditioning before hospital admission, as a complement to angioplasty, and effect on myocardial salvage in patients with acute myocardial infarction: a randomised trial. Lancet 375: 727–734.

2. Cheung MM, Kharbanda RK, Konstantinov IE, Shimizu M, Frndova H, et al. (2006) Randomized controlled trial of the effects of remote ischemic preconditioning on children undergoing cardiac surgery: first clinical application in humans. J Am Coll Cardiol 47: 2277–2282.

3. Hausenloy DJ, Mwamure PK, Venugopal V, Harris J, Barnard M, et al. (2007) Effect of remote ischaemic preconditioning on myocardial injury in patients undergoing coronary artery bypass graft surgery: a randomised controlled trial. Lancet 370: 575–579.

4. Thielmann M, Kottenberg E, Boengler K, Raffelsieper C, Neuhaeuser M, et al. (2010) Remote ischemic preconditioning reduces myocardial injury after coronary artery bypass surgery with crystalloid cardioplegic arrest. Basic Res Cardiol 105: 657–664.

5. Venugopal V, Hausenloy DJ, Ludman A, Di Salvo C, Kolvekar S, et al. (2009) Remote ischaemic preconditioning reduces myocardial injury in patients undergoing cardiac surgery with cold-blood cardioplegia: a randomised controlled trial. Heart 95: 1567–1571.

6. Thielmann M, Kottenberg E, Kleinbongard P, Wendt D, Gedik N, et al. (2013) Cardioprotective and prognostic effects of remote ischaemic preconditioning in patients undergoing coronary artery bypass surgery: a single-centre randomised, double-blind, controlled trial. Lancet 382: 597–604.

7. Heusch G, Musiolik J, Kottenberg E, Peters J, Jakob H, et al. (2012) STAT5 activation and cardioprotection by remote ischemic preconditioning in humans: short communication. Circ Res 110: 111–115.

8. Gho BC, Schoemaker RG, van den Doel MA, Duncker DJ, Verdouw PD (1996) Myocardial protection by brief ischemia in noncardiac tissue. Circulation 94: 2193–2200.

9. McClanahan TB, Nao BS, Wolke LJ, Martin BJ, T E., Mertz aKPG (1993) Brief renal occlusion and reperfusion induces myocardial infarct size in rabbits. FASEB J 7.

10. Weinbrenner C, Nelles M, Herzog N, Sarvary L, Strasser RH (2002) Remote preconditioning by infrarenal occlusion of the aorta protects the heart from infarction: a newly identified non-neuronal but PKC-dependent pathway. Cardiovasc Res 55: 590–601.

11. Breivik L, Helgeland E, Aarnes EK, Mrdalj J, Jonassen AK (2010) Remote postconditioning by humoral factors in effluent from ischemic preconditioned rat hearts is mediated via PI3K/Akt-dependent cell-survival signaling at reperfusion. Basic Res Cardiol 106: 135–145.

12. Dickson EW, Lorbar M, Porcaro WA, Fenton RA, Reinhardt CP, et al. (1999) Rabbit heart can be "preconditioned" via transfer of coronary effluent. Am J Physiol 277: H2451–2457.

13. Serejo FC, Rodrigues LF, Jr., da Silva Tavares KC, de Carvalho AC, Nascimento JH (2007) Cardioprotective properties of humoral factors released from rat hearts subject to ischemic preconditioning. J Cardiovasc Pharmacol 49: 214–220.

14. Dickson EW, Blehar DJ, Carraway RE, Heard SO, Steinberg G, et al. (2001) Naloxone blocks transferred preconditioning in isolated rabbit hearts. J Mol Cell Cardiol 33: 1751–1756.

15. Shimizu M, Tropak M, Diaz RJ, Suto F, Surendra H, et al. (2009) Transient limb ischaemia remotely preconditions through a humoral mechanism acting directly on the myocardium: evidence suggesting cross-species protection. Clin Sci (Lond) 117: 191–200.

16. Lim SY, Yellon DM, Hausenloy DJ (2010) The neural and humoral pathways in remote limb ischemic preconditioning. Basic Res Cardiol 105: 651–655.

17. Hepponstall M, Ignjatovic V, Binos S, Monagle P, Jones B, et al. (2012) Remote ischemic preconditioning (RIPC) modifies plasma proteome in humans. PLoS One 7: e48284.

18. Pang T, Zhao Y, Zhang NR, Jin SQ, Pan SQ (2013) Transient limb ischemia alters serum protein expression in healthy volunteers: complement C3 and vitronectin may be involved in organ protection induced by remote ischemic preconditioning. Oxid Med Cell Longev 2013: 859056.

19. Hibert P, Prunier-Mirebeau D, Beseme O, Chwastyniak M, Tamareille S, et al. (2013) Apolipoprotein a-I is a potential mediator of remote ischemic preconditioning. PLoS One 8: e77211.

20. Hibert P, Prunier-Mirebeau D, Beseme O, Chwastyniak M, Tamareille S, et al. (2014) Modifications in rat plasma proteome after remote ischemic preconditioning (RIPC) stimulus: identification by a SELDI-TOF-MS approach. PLoS One 9: e85669.

21. Davidson SM, Selvaraj P, He D, Boi-Doku C, Yellon RL, et al. (2013) Remote ischaemic preconditioning involves signalling through the SDF-1alpha/CXCR4 signalling axis. Basic Res Cardiol 108: 377.

22. Przyklenk K (2013) 'Going out on a limb': SDF-1alpha/CXCR4 signaling as a mechanism of remote ischemic preconditioning? Basic Res Cardiol 108: 382.

23. Gosho M, Nagashima K, Sato Y (2012) Study designs and statistical analyses for biomarker research. Sensors (Basel) 12: 8966–8986.

24. Aebersold R, Burlingame AL, Bradshaw RA (2013) Western blots versus selected reaction monitoring assays: time to turn the tables? Mol Cell Proteomics 12: 2381–2382.

25. Kinsinger CR, Apffel J, Baker M, Bian X, Borchers CH, et al. (2012) Recommendations for mass spectrometry data quality metrics for open access data (corollary to the Amsterdam Principles). J Proteome Res 11: 1412–1419.

26. Martens L, Nesvizhskii AI, Hermjakob H, Adamski M, Omenn GS, et al. (2005) Do we want our data raw? Including binary mass spectrometry data in public proteomics data repositories. Proteomics 5: 3501–3505.

27. Vizcaino JA, Deutsch EW, Wang R, Csordas A, Reisinger F, et al. (2014) ProteomeXchange provides globally coordinated proteomics data submission and dissemination. Nat Biotech 32: 223–226.

28. Barsnes H, Martens L (2013) Crowdsourcing in proteomics: public resources lead to better experiments. Amino Acids 44: 1129–1137.

29. Phillips HL, Williamson JC, van Elburg KA, Snijders AP, Wright PC, et al. (2010) Shotgun proteome analysis utilising mixed mode (reversed phase-anion exchange chromatography) in conjunction with reversed phase liquid chromatography mass spectrometry analysis. Proteomics 10: 2950–2960.

30. Kessner D, Chambers M, Burke R, Agus D, Mallick P (2008) ProteoWizard: open source software for rapid proteomics tools development. Bioinformatics 24: 2534–2536.

31. Geer LY, Markey SP, Kowalak JA, Wagner L, Xu M, et al. (2004) Open mass spectrometry search algorithm. J Proteome Res 3: 958–964.

32. Craig R, Beavis RC (2004) TANDEM: matching proteins with tandem mass spectra. Bioinformatics 20: 1466–1467.

33. Vaudel M, Barsnes H, Berven FS, Sickmann A, Martens L (2011) SearchGUI: An open-source graphical user interface for simultaneous OMSSA and X!Tandem searches. Proteomics 11: 996–999.

34. Elias JE, Gygi SP (2007) Target-decoy search strategy for increased confidence in large-scale protein identifications by mass spectrometry. Nat Methods 4: 207–214.

35. Apweiler R, Bairoch A, Wu CH, Barker WC, Boeckmann B, et al. (2004) UniProt: the Universal Protein knowledgebase. Nucleic Acids Res 32: D115–119.

36. Barsnes H, Vaudel M, Colaert N, Helsens K, Sickmann A, et al. (2011) compomics-utilities: an open-source Java library for computational proteomics. BMC Bioinformatics 12: 70.

37. Vizcaino JA, Cote RG, Csordas A, Dianes JA, Fabregat A, et al. (2013) The PRoteomics IDEntifications (PRIDE) database and associated tools: status in 2013. Nucleic Acids Res 41: D1063–1069.

38. Martens L, Hermjakob H, Jones P, Adamski M, Taylor C, et al. (2005) PRIDE: the proteomics identifications database. Proteomics 5: 3537–3545.

39. Vaudel M, Sickmann A, Martens L (2010) Peptide and protein quantification: a map of the minefield. Proteomics 10: 650–670.

40. Vaudel M, Sickmann A, Martens L (2013) Introduction to opportunities and pitfalls in functional mass spectrometry based proteomics. Biochim Biophys Acta.

41. Burkhart JM, Vaudel M, Zahedi RP, Martens L, Sickmann A (2011) iTRAQ protein quantification: a quality-controlled workflow. Proteomics 11: 1125–1134.

42. Powell DW, Weaver CM, Jennings JL, McAfee KJ, He Y, et al. (2004) Cluster analysis of mass spectrometry data reveals a novel component of SAGA. Mol Cell Biol 24: 7249–7259.

43. Burkhart JM, Vaudel M, Gambaryan S, Radau S, Walter U, et al. (2012) The first comprehensive and quantitative analysis of human platelet protein composition allows the comparative analysis of structural and functional pathways. Blood 120: e73–82.

44. Yadav AK, Bhardwaj G, Basak T, Kumar D, Ahmad S, et al. (2011) A systematic analysis of eluted fraction of plasma post immunoaffinity depletion: implications in biomarker discovery. PLoS One 6: e24442.

45. Hausenloy DJ (2013) Cardioprotection techniques: preconditioning, postconditioning and remote con-ditioning (basic science). Curr Pharm Des 19: 4544–4563.

46. Heusch G (2013) Cardioprotection: chances and challenges of its translation to the clinic. Lancet 381: 166–175.

47. Przyklenk K (2013) Reduction of Myocardial Infarct Size with Ischemic "Conditioning": Physiologic and Technical Considerations. Anesth Analg.

48. Cote RG, Jones P, Martens L, Kerrien S, Reisinger F, et al. (2007) The Protein Identifier Cross-Referencing (PICR) service: reconciling protein identifiers across multiple source databases. BMC Bioinformatics 8: 401.

49. Zhou GX, Chao L, Chao J (1992) Kallistatin: a novel human tissue kallikrein inhibitor. Purification, characterization, and reactive center sequence. J Biol Chem 267: 25873–25880.

50. Chao J, Stallone JN, Liang YM, Chen LM, Wang DZ, et al. (1997) Kallistatin is a potent new vasodilator. J Clin Invest 100: 11–17.

51. Gao L, Yin H, S. Smith R J, Chao L, Chao J (2008) Role of kallistatin in prevention of cardiac remodeling after chronic myocardial infarction. Lab Invest 88: 1157–1166.

52. Chao J, Yin H, Yao YY, Shen B, Smith RS, Jr., et al. (2006) Novel role of kallistatin in protection against myocardial ischemia-reperfusion injury by preventing apoptosis and inflammation. Hum Gene Ther 17: 1201–1213.

53. Shen B, Gao L, Hsu YT, Bledsoe G, Hagiwara M, et al. (2010) Kallistatin attenuates endothelial apoptosis through inhibition of oxidative stress and activation of Akt-eNOS signaling. Am J Physiol Heart Circ Physiol 299: H1419–1427.

54. Liu Y, Bledsoe G, Hagiwara M, Shen B, Chao L, et al. (2012) Depletion of endogenous kallistatin exacerbates renal and cardiovascular oxidative stress, inflammation, and organ remodeling. Am J Physiol Renal Physiol 303: F1230–1238.

55. Renaux JL, Thomas M, Crost T, Loughraieb N, Vantard G (1999) Activation of the kallikrein-kinin system in hemodialysis: role of membrane electronegativity, blood dilution, and pH. Kidney Int 55: 1097–1103.

56. Dray A, Perkins M (1993) Bradykinin and inflammatory pain. Trends Neurosci 16: 99–104.

57. Dietze GJ, Wicklmayr M, Rett K, Jacob S, Henriksen EJ (1996) Potential role of bradykinin in forearm muscle metabolism in humans. Diabetes 45 Suppl 1: S110–114.

58. Boix F, Rosenborg L, Hilgenfeldt U, Knardahl S (2002) Contraction-related factors affect the concentration of a kallidin-like peptide in rat muscle tissue. J Physiol 544: 127–136.

59. Edery H, Lewis GP (1962) Inhibition of plasma kininase activity at slightly acid ph. Br J Pharmacol Chemother 19: 299–305.

60. Saxena P, Shaw OM, Misso NL, Naran A, Shehatha J, et al. (2011) Remote ischemic preconditioning stimulus decreases the expression of kinin receptors in human neutrophils. J Surg Res 171: 311–316.

61. Weisman HF, Bartow T, Leppo MK, Marsh HC, Jr., Carson GR, et al. (1990) Soluble human complement receptor type 1: in vivo inhibitor of complement suppressing post-ischemic myocardial inflammation and necrosis. Science 249: 146–151.

62. Diepenhorst GM, van Gulik TM, Hack CE (2009) Complement-mediated ischemia-reperfusion injury: lessons learned from animal and clinical studies. Ann Surg 249: 889–899.

63. Pemberton M, Anderson G, Vetvicka V, Justus DE, Ross GD (1993) Microvascular effects of complement blockade with soluble recombinant CR1 on ischemia/reperfusion injury of skeletal muscle. J Immunol 150: 5104–5113.

64. Wada K, Montalto MC, Stahl GL (2001) Inhibition of complement C5 reduces local and remote organ injury after intestinal ischemia/reperfusion in the rat. Gastroenterology 120: 126–133.

65. Zhao H, Montalto MC, Pfeiffer KJ, Hao L, Stahl GL (2002) Murine model of gastrointestinal ischemia associated with complement-dependent injury. J Appl Physiol (1985) 93: 338–345.

66. Stahl GL, Xu Y, Hao L, Miller M, Buras JA, et al. (2003) Role for the alternative complement pathway in ischemia/reperfusion injury. Am J Pathol 162: 449–455.

67. Karpel-Massler G, Fleming SD, Kirschfink M, Tsokos GC (2003) Human C1 esterase inhibitor attenuates murine mesenteric ischemia/reperfusion induced local organ injury. J Surg Res 115: 247–256.

68. Schwaeble WJ, Lynch NJ, Clark JE, Marber M, Samani NJ, et al. (2011) Targeting of mannan-binding lectin-associated serine protease-2 confers protection from myocardial and gastrointestinal ischemia/reperfusion injury. Proc Natl Acad Sci U S A 108: 7523–7528.

69. Bossi F, Peerschke EI, Ghebrehiwet B, Tedesco F (2011) Cross-talk between the complement and the kinin system in vascular permeability. Immunol Lett 140: 7–13.

70. Engelbrecht B, Stratmann B, Hess C, Tschoepe D, Gawlowski T (2013) Impact of GLO1 knock down on GLUT4 trafficking and glucose uptake in L6 myoblasts. PLoS One 8: e65195.

71. Inagi R, Kumagai T, Fujita T, Nangaku M (2010) The role of glyoxalase system in renal hypoxia. Adv Exp Med Biol 662: 49–55.

72. Ahmed U, Dobler D, Larkin SJ, Rabbani N, Thornalley PJ (2008) Reversal of hyperglycemia-induced angiogenesis deficit of human endothelial cells by overexpression of glyoxalase 1 in vitro. Ann N Y Acad Sci 1126: 262–264.

73. Wautier JL, Schmidt AM (2004) Protein glycation: a firm link to endothelial cell dysfunction. Circ Res 95: 233–238.

74. Udby L, Sorensen OE, Pass J, Johnsen AH, Behrendt N, et al. (2004) Cysteine-rich secretory protein 3 is a ligand of alpha1B-glycoprotein in human plasma. Biochemistry 43: 12877–12886.

75. Loukogeorgakis SP, Williams R, Panagiotidou AT, Kolvekar SK, Donald A, et al. (2007) Transient limb ischemia induces remote preconditioning and remote postconditioning in humans by a K(ATP)-channel dependent mechanism. Circulation 116: 1386–1395.

76. Hong DM, Jeon Y, Lee CS, Kim HJ, Lee JM, et al. (2012) Effects of remote ischemic preconditioning with postconditioning in patients undergoing off-pump coronary artery bypass surgery–randomized controlled trial. Circ J 76: 884–890.

77. The Remote Preconditioning Trialists G, Healy DA, Khan WA, Wong CS, Moloney MC, et al. (2014) Remote preconditioning and major clinical complications following adult cardiovascular surgery: Systematic review and meta-analysis. Int J Cardiol.

New Candidate Biomarkers in the Female Genital Tract to Evaluate Microbicide Toxicity

Scott Fields⁹, **Benben Song**⁹, **Bareza Rasoul, Julie Fong, Melissa G. Works, Kenneth Shew, Ying Yiu, Jon Mirsalis, Annalisa D'Andrea***

Biosciences Division, SRI International, Menlo Park, California, United States of America

Abstract

Vaginal microbicides hold great promise for the prevention of viral diseases like HIV, but the failure of several microbicide candidates in clinical trials has raised important questions regarding the parameters to be evaluated to determine in vivo efficacy in humans. Clinical trials of the candidate microbicides nonoxynol-9 (N9) and cellulose sulfate revealed an increase in HIV infection, vaginal inflammation, and recruitment of HIV susceptible lymphocytes, highlighting the need to identify biomarkers that can accurately predict microbicide toxicity early in preclinical development and in human trials. We used quantitative proteomics and RT-PCR approaches in mice and rabbits to identify protein changes in vaginal fluid and tissue in response to treatment with N9 or benzalkonium chloride (BZK). We compared changes generated with N9 and BZK treatment to the changes generated in response to tenofovir gel, a candidate microbicide that holds promise as a safe and effective microbicide. Both compounds down regulated mucin 5 subtype B, and peptidoglycan recognition protein 1 in vaginal tissue; however, mucosal brush samples also showed upregulation of plasma proteins fibrinogen, plasminogen, apolipoprotein A-1, and apolipoprotein C-1, which may be a response to the erosive nature of N9 and BZK. Additional proteins down-regulated in vaginal tissue by N9 or BZK treatment include CD166 antigen, olfactomedin-4, and anterior gradient protein 2 homolog. We also observed increases in the expression of C-C chemokines CCL3, CCL5, and CCL7 in response to treatment. There was concordance in expression level changes for several of these proteins using both the mouse and rabbit models. Using a human vaginal epithelial cell line, the expression of mucin 5 subtype B and olfactomedin-4 were down-regulated in response to N9, suggesting these markers could apply to humans. These data identifies new proteins that after further validation could become part of a panel of biomarkers to effectively evaluate microbicide toxicity.

Editor: J. Gerardo Garcia-Lerma, Centers for Disease Control and Prevention, United States of America

Funding: This project has been funded in whole with Federal funds from the National Institute of Allergy and Infectious Diseases, National Institutes of Health, Department of Health and Human Services, under Contract No. HHSN272200700043C. JM and AD received the funding to support the work. The funder approved the manuscript for submission.

Competing Interests: The authors have declared that no competing interests exist.

* Email: annalisa.dandrea@sri.com

⁹ These authors contributed equally to this work.

Introduction

Topical microbicides have been proposed as agents to prevent the transmission of HIV by creating chemical, biological, and/or physical barriers to infection, or by blocking or inactivating the virus at the mucosal surface where infection can occur. An ideal microbicide would need to demonstrate both protection against HIV infection and low toxicity after repeated use. Although several candidate microbicides initially appeared promising in preclinical safety studies, they later proved to be ineffective in clinical trials [1–13]. In some cases, they actually increased the risk of infection, e.g. cellulose sulfate [12]. Similarly, nonoxynol-9 (N9), a contraceptive spermicide that has previously been shown to be safe in preclinical and phase I studies, generated disappointing clinical data as a protective microbicide [11,13]. In fact, repetitive use of N9 resulted in genital irritation/inflammation and increased risk of acquiring HIV [11]. The limited success of putative microbicides in clinical trials demonstrates a need for better parameters to predict the safety of candidates undergoing preclinical development. One approach is to develop a robust series of biomarkers capable of predicting cellular and molecular changes occurring in the vaginal mucosa/epithelium during microbicide treatment. Such markers could have utility both in preclinical development and, eventually, in clinical development as well.

The current preferred pre-clinical model for assessment of microbicide safety is the rabbit vaginal irritation (RVI) model [14–20]. The assay requires euthanizing all study animals, endpoints of the RVI are mostly histological, and in the recent years, the limitations of this model in detecting potential toxicity, have clearly demonstrated that additional parameters must be included in the evaluation of new candidate microbicides. Recent studies using this model have identified changes in inflammatory cytokines in vaginal lavage fluids in response to compounds with toxic characteristics [18,19]. Although the significance of these cytokines has not been fully clarified, they may play a role in creating an environment more susceptible to pathogen infection. For example, the presence of elevated levels of pro-inflammatory factors may increase the proliferation of immune cells in the vaginal tissue parenchyma or increase migration of HIV susceptible immune cells to the vaginal tract.

Although the RVI model has been used for many years to evaluate potential toxic effects of microbicide candidates, it is not without shortcomings. The human and rabbit vaginal tissues show significant structural differences (stratified squamous versus columnar epithelium, respectively) [21,22] that may be responsible for different susceptibilities to treatment. Moreover, the rabbit lacks cyclic reproductive stages and cervical mucus production, all factors that may affect the environment in the vagina [23].

More recently, mouse vaginal irritation (MVI) models have been developed and used to evaluate microbicide safety [14,15,24–28]. Advantages of the MVI include a well characterized immune system and a comprehensive protein database that makes it amenable to proteomics studies. In addition, it is significantly less expensive than the RVI model; uses a smaller animal that requires less test articles for safety testing; is not a USDA-regulated species; and provides toxicological results for cross-species comparison, making it an attractive, alternative model for pre-clinical evaluation of vaginal microbicides.

Here, we have used both the RVI and MVI models to identify biomarkers that could be incorporated in the safety evaluations of vaginal anti-HIV microbicides. For our studies we selected to use two known vaginal irritants [18,29], the spermicide N9 and the antiseptic benzalkonium chloride (BZK) as models of toxic compounds for vaginal administration, and we identified vaginal proteins that were altered following application of these compounds.

Materials and Methods

Reagents

Nonoxynol-9 (N9) was purchased from Spectrum Chemicals and Laboratory Products (Gardena, CA); tenofovir was purchased from AK Scientific (Union City, CA); benzalkonium chloride (BZK) and high viscosity sodium carboxymethylcellulose (CMC) were purchased from Sigma Aldrich (Saint Louis, MO); mouse and rabbit apolipoprotein A-I, apolipoprotein C-1, CD166, and rabbit IL-8 ELISA kits were purchased from EIAab (Wuhan, China); mouse fibrinogen ELISA kit was purchased from Abcam (Cambridge, MA); mouse plasminogen ELISA kit was purchased from GenWay (San Diego, CA); mouse IL-8 ELISA kit was purchased from MyBioSource (San Diego, CA); rabbit plasminogen ELISA kit was purchased from Alpco (Salem, NH); rabbit fibrinogen ELISA kit was purchased from Molecular Innovations (Novi, MI); medroxyprogesteron acetate was purchased from Pharmacia (New York, NY); CellTiter-Glo was purchased from Promega (Madison, WI); RNeasy kits were purchased from Qiagen (Valencia, CA); rabbit anti-human olfactomedin 4 (OLFM-4) polyclonal antibody was purchased from Abcam (Cambridge, MA); rabbit anti-human mucin 5B polyclonal antibody was purchased from Bioss USA Antibodies (Woburn, MA); rabbit anti-mouse CD166 was purchased from GeneTex (Irvine, CA); rabbit anti-mouse β-actin was purchased from Abcam (Cambridge, MA); IRDye 800 Goat anti-rabbit IgG was purchased from Li-Cor (Lincoln, NE); mouse anti-human β-actin monoclonal antibody was purchased from Cell Signaling Technology (Danvers, MA); donkey anti-mouse IgG antibody conjugated to IRDye 680 LT and goat anti-rabbit IgG antibody conjugated to IRDye 800CW were purchased from LI-COR biosciences (Lincoln, NE). Precast 4–12% Bis-Tris SDS-PAGE, 3–8% Tris-acetate SDS-PAGE gels, and PDVF transfer membranes

were purchased from Life Technologies (Grand Island, NY); oligonucleotides were purchased from Integrated DNA Technologies (Coralville, IA).

Microbicide preparation

8% N9, and 1% tenofovir were prepared in 2% sodium carboxymethylcellulose, and 2% BZK was prepared in ultrapure water. Vehicle, N9, and BZK microbicides were prepared no more than three days prior to use and were stored at 4°C. Prior to dosing, microbicides were allowed to reach ambient temperature. Buffergel, was obtained from ReProtect Inc. in its clinical formulation, and was stored at room temperature until use.

Cell lines and challenge with N9

Vk2 (E6/E7) cells (CRL-2616) were obtained from American Type Culture Collection (Rockville, MD). Cells were grown in keratinocyte-serum free medium (K-SFM) containing 5 ng/ml recombinant epidermal growth factor and 50 μg/ml bovine pituitary extract all purchased from Invitrogen Corporation (Grand Island, NY). A maximum tolerated dose of N9 was determined by serially diluting the microbicide from 1% to 0.001% in cell culture medium and each dilution added to individual wells of an 80–90% confluent monolayer of Vk2 Cells in 96-well plates. Cell viability was performed using CellTiter-Glo.

Quantitative RT-PCR

Frozen vaginas were minced and placed in RNAlater (Life Technologies, Grand Island, NY) and stored at −20°C until processed. Total RNA was isolated (RNeasy kit) from vaginal tissues of 3 animals per treatment group using Ambion *mir*Vana Kit (Life Technologies, Grand Island, NY). cDNA was prepared using Superscript RTII (Life Technologies, Grand Island, NY). Quantitative RT-PCR was conducted using the LightCycler 480 RT-PCR machine and SYBR green I master mix (Roche, Indianapolis, IN). RT-PCR oligos used are shown in **Table S1**. The threshold cycle for each test gene and glyceraldehyde 3-phosphate dehydrogenase (GADPH) was determined and an average of threshold cycles was calculated for each test gene. Each test gene's average threshold cycle was normalized to the GADPH average threshold cycle and expression relative to GADPH was reported as $2^{\text{Ct" GAPDH} - \text{Ct gene}}$ (where "Ct" indicates cycle threshold).

Immunoblot analysis of mouse and human proteins

Mouse proteins were extracted from vaginal tissues using SDS extraction buffer (50 mM Tris pH 7.5, 0.5% SDS, Halt Protease inhibitor (Pierce, Rockford, IL) for 1.5 hours on ice after vortexing with 0.5 mm glass beads (Next Advance, Averill Park, NY) five times with 1 min vortexing and 1 min rests. Protein concentration was measured using a Nanodrop spectrophotometer (Thermo Scientific, Rockford, IL) and BCA assay (Pierce, Rockford, IL). Protein lysates (50 μg) were resolved on a 10% SDS-Page gel. The gel was transferred (wet transfer at 300 mA) to a nitrocellulose membrane and then blocked with 5% non-fat dry milk in 1 × PBS. For the detection of CD166 antigen in mouse samples, the primary antibody was applied in blocking buffer overnight at 4°C; Secondary antibody was used incubated at room temperature for 1 hour. Following extensive washing, the membranes were scanned using the LI-COR Odyssey imaging system.

To detect OLFM-4 and β-actin, Vk2 cell extracts were resolved on 4–12% Bis-Tris SDS-PAGE gels and mucin 5B was resolved on 3–8% Tris-acetate SDS-PAGE gels f. Gels were transferred to PDVF membranes and probed with the appropriate primary and secondary antibodies and detected using the LI-COR Odyssey imaging system.

Animals

All animal experiments were reviewed and approved by the Institutional Animal Care and Use Committee (IACUC) at SRI International and were performed in accordance with relevant guidelines and regulations in a facility accredited by the Association for the Assessment and Accreditation of Laboratory Animal Care International (AAALAC).

Mouse vaginal irritation studies

Female CD1 mice were purchased from Harlan Laboratories (Livermore, CA) and were housed in hanging polypropylene cages with hardwood chip bedding; Purina rodent chow #5002 and reverse osmosis purified water were provided *ad libitum*. All mice (7–9 weeks of age) were hormonally synchronized with two subcutaneous doses of 3 mg of medroxyprogesterone acetate on days −8 and −1. The mice were randomized into treatment groups consisting of between 6 and 12 animals per group per study and were dosed daily for 11 days with 50 μl of the microbicide candidates using a luer-lock syringe attached to a 20-gauge applicator inserted 0.8 cm into the vagina. Mouse vaginal brush collection (7 hours after dosing) was done using MicroBrush X (MicroBrush International, Grafton, WI). Brushes were pre-wetted in saline and inserted approximately 0.8 cm into the vaginal cavity. The brushes were turned 90 degrees clockwise and counterclockwise three times, removed, and placed in saline containing 1× protease inhibitor cocktail (Thermo Scientific). Any visual presence of blood was noted but not scored. Vaginal brush samples were collected on days 7 and 10. For the first two studies, a single brush sample per mouse per collection day was obtained, but on the last two studies, three brushes were used per mouse per collection day to maximize the yield of sample recovery. Each individual brush was placed into 0.1 ml of 1× PBS pH 7.6 plus 1× protease inhibitors (Thermo Scientific). The brushes were vortexed in the collection buffer and in the cases where multiple brushes were used, the samples were pooled, mixed, aliquoted, and snap frozen in dry ice/ethanol prior to storage at −80°C. Vaginas were harvested on the final day of the study and snap frozen using dry ice/ethanol baths prior to storage at −80°C.

Rabbit vaginal irritation studies

Female white New Zealand White rabbits (20–28 weeks of age) were obtained from Harlan Laboratories; and were housed individually in stainless steel cages; Teklad Rabbit Diet and reverse osmosis purified water were provided *ad libitum*. Animals were randomized into treatment groups consisting of 6 animals per group per study and one day prior to dosing the vaginal cavities were washed with 1 ml sterile 0.9% saline solution. The rabbits were dosed daily for 10 days with 1 ml of the microbicide candidates using a syringe attached to a lightly curved 6 inch stainless steel ball-tipped cannula that was inserted approximately 5–6 inches into the vagina. On days 7 and 10 (collection 7 hours after last dose), sterile cotton tipped swabs were used to collect vaginal cavity material. Three cotton swabs were used per rabbit per collection day. Each individual swab was placed into 0.5 ml of 1× PBS pH 7.6 plus 1× protease inhibitors (Thermo Scientific). The swabs were vortexed in the collection buffer and all three collection tubes per rabbit were pooled, mixed, aliquoted, and snap frozen in liquid nitrogen prior to storage at −80°C.

ELISA assays

Vaginal brush samples were thawed and tested using the individual kit's instructions. Briefly, each dilution of the standard was run in triplicate and samples were run in duplicate. The OD settings for each ELISA were set by the manufacturer. Values above or below the standard curve are reported as having a value equal to the highest or lowest value in the standard for the kit. Plates were scanned using a Spectramax 2 plate reader (Molecular Devices, Sunnyvale, CA). Generation of standard curves, determination of unknown protein concentrations, and statistical calculations were accomplished using GraphPad Prism software.

Vaginal tissue protein extraction for proteomic analysis

Vaginal tissue samples were thawed on ice, weighted, and washed with PBS buffer. For tissue protein extraction, approximately 100 mg vaginal tissue was minced into 2 mm pieces in 1 ml lysis buffer (50 mM Tris, pH 7.4, 0.5% SDS, EDTA-free protease inhibitor cocktail). Approximately 0.2 ml glass beads (0.5 mm, Sigma-Aldrich, St. Louis, MO) were added to each sample. Subsequently, samples were homogenized 5 times with 1 min for each time by a Bullet-Blender (Next Advance, Averill Park, NY) at 4°C and then left on ice for 1 hour. Debris was removed by centrifugation at 12,000 g for 15 min at 4°C, and supernatants were stored at −80°C for proteomics sample preparation. Total protein of the samples was measured using a NanoDrop 2000 (Thermo Scientific, USA).

Proteomics sample preparation and TMT labeling

Filter-aided sample preparation (FASP) using 10k cutoff filters, developed by Mann [30], was applied to purify and digest both brush samples and tissue protein lysates. Briefly, 600 μL brush sample or 200 μL tissue protein lysate was transferred into a 1.5 mL Microcon YM-10 centrifugal unit (Millipore, Billerica, MA, USA). Protein reduction, alkylation, and tryptic digestion were performed step by step in the centrifugal unit. After overnight tryptic digestion at 37°C, the peptides were eluted twice with 150 μL 50 mM ammonium bicarbonate. The total protein or peptide concentration in each step was measured using NanoDrop 2000. The eluted peptides were dried by vacuum centrifugation and then resuspended in 50 mM Tetraethylammonium bromide (TEAB) for Tandem Mass Tag (TMT) labeling.

50 μg digested peptides from each tissue sample and 10 μg peptides from each brush sample were incubated with an amine-reactive 6-plex TMT tag (Thermo Scientific, USA) for 2 hours at room temperature. In mouse studies 1 and 2, the digested peptides from brush samples from the same group of animals were pooled together for TMT labeling due to low peptide content. Reactions were quenched by adding 8 μL 5% hydroxylamine and incubated at room temperature for 15 min. The labeled peptides were then combined together and excess TMT tags were removed by a 3×8 mm SCX trap column then desalted by a 3×8 mm C18 RP trap column (Bruker-Michrom, CA, USA). Purified labeled peptides were dried by vacuum centrifugation and then resuspended in 0.1% formic acid (FA) for Multidimensional Protein Identification Technology (MudPIT) analysis.

Nano-LC-MS/MS

Each TMT labeled sample was separated using a nano-LC system (Agilent 1200, Palo Alto, CA, USA) and analyzed with an LTQ-XL Orbitrap ETD (Thermo Fisher Scientific, San Jose, CA, USA) equipped with a nano-ESI source. Full MS spectra were acquired in positive mode over a 350–1800 *m/z* range, followed by four CID (collision induced dissociation) and HCD (higher-energy

Table 1. Summary of proteins whose levels significantly altered upon N9 or BZK treatment determined by proteomic analysis in mouse vaginal brush and tissue samples.

Vaginal brush samples (day7)	Gene name	N†	N9		BZK		Tenofovir
			avg*	std	avg	std	
Apolipoprotein A-1	APOA1	4	3.80	2.46	8.22	3.73	0.7
Apolipoprotein C-1	APOC1	4	7.01	3.91	13.36	8.02	0.22
Fibrinogen, alpha polypeptide isoform 2	FGA	4	4.30	3.00	6.60	2.96	1.23
Plasminogen	PLG	4	7.81	7.94	8.44	6.28	0.77
Heat shock cognate 71 kDa protein	HSPA8	2	3.62	0.02	3.87	0.22	0.83
Corticosteroid-binding globulin	SERPINA6	3	11.17	10.06	17.08	14.67	0.35
Peptidoglycan recognition protein 1††	PGLYRP-1	3	0.09	0.07	0.10	0.10	3
Mucin 5, subtype B, tracheobronchial	Muc5B	2	0.06	0.01	0.04	0.03	1.07
Destrin	DSTN	3	0.28	0.12	0.27	0.14	1.64
Carbonyl reductase 3	CAR3	2	0.29	0.21	0.28	0.21	1.48

Vaginal tissue samples	Gene name	N†	N9		BZK		Tenofovir
			avg	std	avg	std	
Peptidoglycan recognition protein 1	PGLYRP-1	2	0.39	0.07	0.43	0.16	ND
CD166 antigen	CD166	2	0.47	0.07	0.4	0.01	ND
Mucin 5, subtype B, tracheobronchial	Mucin 5B	3	0.43	0.05	0.30	0.15	0.66
Olfactomedin-4	OLFM-4	3	0.49	0.35	0.39	0.31	0.99
Anterior gradient protein 2 homolog	AGR2	3	0.41	0.15	0.31	0.26	0.94
Calcium-activated chloride channel regulator 1	CLCA2	2	0.39	0.16	0.26	0.02	0.62
Protein-glutamine gamma-glutamyltransferase 2	TGM2	3	0.55	0.35	0.58	0.28	0.66

*Average signal compared to vehicle controls which are set to a default value of 1,
†number of studies where protein was identified (for both N9 and BZK), ND=not detected,
††proteins shown in bold are identified in both brush and tissue samples.

collisional dissociation) events on the four most intense ions selected from the full MS spectrum and using a dynamic exclusion time of 30 s [31]. Four CID scans (maximum inject time 100 ms, minimum signal threshold 500 counts, collision energy 35%, activation time 30 ms, isolation width 1.0 m/z) were used for peptide identification and four corresponding HCD scans (maximum inject time 300 ms, minimum signal threshold 500 counts, collision energy 45%, activation time 30 ms, isolation width 1.0 m/z) were used for quantitation.

Database Search

Acquired tandem mass spectra were searched against the European Bioinformatics Institute International Protein Index mouse protein database (version 3.73). A decoy database containing the reverse sequence of all the proteins was appended to estimate false discovery rate (FDR). The search was performed using SEQUEST algorithm incorporated in Proteome Discoverer 1.3.0.339 (Thermo Finnigan, CA, USA). The precursor mass accuracy was limited to 13 ppm and fragment ion mass tolerance was set at 1.1 Da. Fully tryptic enzyme specificity and up to two missed cleavages were allowed. Fixed modifications included carbamidomethylation on cysteines and variable modifications included oxidation on methionines and TMT adduction to peptide N-termini, lysines, and tyrosines. Peptide quantitation was also performed in Proteome Discoverer 1.3.0.339 in the same workflow. A TMT 6-plex quantitation method was used for HCD-based quantitation. Mass tolerance was set at 150 ppm for reporter TMT tags. The intensity of each peptide was normalized to protein median intensity before calculating the ratio of different tags from the same peptides. Protein quantitation was calculated based on the data of each quantified peptide.

Statistical analysis

For ELISAs and RT-PCR assays, Student's T-tests were conducted with Welch's correction for compound groups versus vehicle controls. P values ≤0.05 are considered.

For proteomics analysis, FDR calculated through a decoy database search was set as 0.05 for both protein identification and quantitation. For each study, standard deviation was calculated for the protein level alteration in every group.

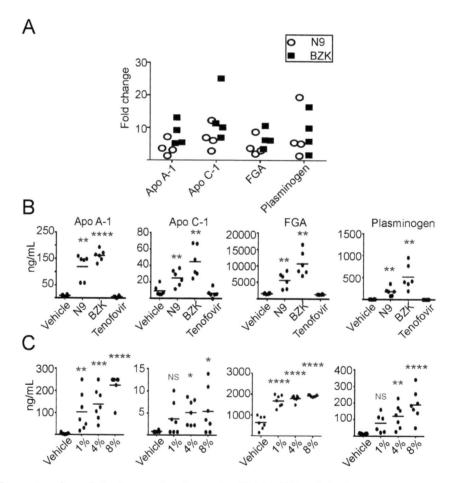

Figure 1. Quantitative proteomic analysis of mouse brush samples. (A) LC-MS/MS analysis of mouse proteins in vaginal brush samples that are affected by treatment with N9 (open, black circles) or BZK (closed squares). Shown are proteins that are increased compared to vehicle controls. The following proteins are depicted in the graph: fibrinogen alpha polypeptide isoform 2 (FGA), plasminogen (PLG), apolipoprotein A-1 (Apo A-1), and apolipoprotein C-1 (Apo C-1). Each data point represents the average of three replicates per individual experimental study. Only proteins shown to change in four studies are represented. (B) ELISA analysis of mouse proteins in vaginal brush samples that are increased in response to treatment with 8% N9 or 2% BZK. Statistical analysis was performed using a two-tailed Student's T-test of N9 and BZK samples compared to vehicle controls (**** = p≤0.0001, ** = p≤0.01). (C) ELISA analysis of mouse proteins in vaginal brush samples that are increased in response to treatment with vehicle, 1% N9, 4% N9, or 8% N9. Statistical analysis was performed using a two-tailed student's t-test of N9 and BZK samples compared to vehicle controls (**** = p≤0.0001, *** = p≤0.001, ** = p≤0.01, * = p≤0.05, NS = not significant).

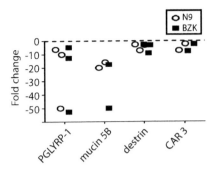

Figure 2. LC MS/MS proteomic analysis of mouse proteins in vaginal brush samples that are down-regulated in response to treatment with N9 (open, black circles) or BZK (closed squares). Shown are proteins that were decreased relative to vehicle controls. The following proteins are shown: peptidoglycan recognition protein 1 (PGLYRP-1), mucin 5 subtype B (mucin 5B), destrin, and carbonyl reductase 3 (CAR3). Each data point represents the average of three replicates per individual experimental study. Only proteins shown to change in a minimum of two studies are represented.

Results

Proteomic analysis of mouse vaginal brush and tissue samples reveal changes in several proteins after treatment with N9 and BZK

To identify protein changes induced by vaginal microbicide candidates with an undesirable safety profile, mice were treated with either vehicle, 8% N9, 2% BZK, or 1% tenofovir (a non-toxic microbicide) [32] for 10 consecutive days (as per standard preclinical evaluations in mice and rabbits [23]), with mucosal samples and tissue samples collected at day 7 and 10 for analysis by 2D LC-MS/MS, respectively.

Proteins from vaginal brushes. Changes in proteins present in mucosal secretions upon treatment with toxic compounds can be exploited to define biomarkers of toxicity. In this study, mucosal secretions were collected by brush sampling. Proteins that showed a consistent >2-fold change relative to vehicle control in at least two studies were selected for further analysis. A total of 1017 proteins were identified with a false discovery rate (FDR) cutoff of 0.05. After 7 days exposure to toxic microbicides, a total of 10 proteins were selected for further analysis (**Table 1**). Six proteins were increased upon exposure to N9 and BZK, while 4 proteins decreased. Notably the increased proteins are thought to be serum derived, probably as a result of exudates due to the erosive nature of the microbicides. These consisted of apolipoprotein A-1 (Apo A-1; range change 2 to 7-fold for N9 and 5 to 13-fold for BZK), apolipoprotein C-1 (Apo C-1; range change 3 to 12-fold and 3 to 12-fold for BZK), fibrinogen alpha polypeptide isoform 2 (FGA; range change 3 to 9-fold and 2 to 8-fold for BZK), plasminogen (PLG; range change 2 to 19-fold and 2 to 8-fold for BZK), heat shock cognate 71 kDa protein (HSPA8; range change 3 to 4-fold for both N9 and BZK), and Corticosteroid-binding globulin (SERPINA6; range change 2 to 21-fold and 3 to 31-fold for BZK), respectively (**Figure 1A** and **Table 1**). In contrast the following four proteins were decreased in vaginal brush samples (**Table 1** and **Figure 2**): peptidoglycan recognition protein 1 (PGLYRP-1; range change 6 to 50-fold for N9 and 5 to 52–fold for BZK), mucin 5 subtype B (mucin 5B, range change 16 to 20-fold for N9 and 20 to 50-fold for BZK), destrin (DSTN, range change 2.5 to 7-fold for N9 and 3 to 10-fold for BZK), and carbonyl reductase 3 (CAR3, range change 2.5 to 7-

fold for N9 and 2 to 10-fold for BZK). These proteins are produced by vaginal tissue and are thought to confer protection against microorganisms.

Tenofovir treatment had no significant effect on the expression levels of those proteins, with the exception of Apo C-1 and PGLYRP-1, which were decreased 4-fold and increased 3-fold in vaginal brush samples, respectively. A summary of their known or putative biological functions of the proteins affected by N9 and BZK treatment is listed in **Table S2**.

Proteins that were consistently upregulated in four replicate proteomic experiments (**Figure 1A**) were then confirmed by ELISA analysis. Upon exposure to N9, Apo A-1 was increased an average of 12-fold over vehicle (ranging from a 4 to 29-fold increase) or tenofovir control and plasminogen was elevated an average of 27-fold (ranging from an 18 to 71-fold increase). Both Apo C-1 and fibrinogen alpha polypeptide isoform 2 (FGA) showed a lesser increase, averaging ~3-fold each (ranging from a 0.53 decrease to a 7.4-fold increase and a 1.55 to 6 fold-increase, respectively) (**Figure 1B**). Levels of these proteins were even higher following treatment with BZK when compared with vehicle/tenofovir control treatment. Apo A-1 showed a 17-fold increase (range 9–36-fold); plasminogen increased 76-fold (range 15 to 191-fold), Apo C-1 increased 5-fold (range 1 to 13-fold), and FGA showed a 7-fold increases, (range 4 to 12-fold), respectively. The biological function of these proteins is summarized in **Table S2**. The identified proteins are all serum proteins possibly present in the brush sample due to the tissue damaging effect of N9 and BZK, which in few circumstances caused bleeding in the vaginal cavity (not shown). Blood was predominantly found in brush samples in mice and rabbits treated with BZK, but was only rarely observed in animals (mice or rabbits) treated with N9, emphasizing that the presence of blood could not be used as a predictor to detect toxic effects of test articles. To cover a range of concentrations of N9 that were representative of the dose used in clinical trials (3.5%) and to reduce the disruption of the cervical epithelium, we examined the levels of Apo A-1, Apo C-1, FGA, and plasminogen using lower concentrations of N9 (1%, 4%, and 8%). With the exception of Apo C-1, N9 exhibited a dose dependent increase of each marker in vaginal brush samples. Apo A-1 and FGA showed a marked increase with as little 1% N9 and all 4 proteins were elevated with 4% N9 which increased further with 8% N9 (**Figure 1C**). These data suggest that these proteins could be incorporated into a rapid and sensitive set of biomarkers for the evaluation of microbicide toxicity. In addition, one experiment was conducted with a clinical trial formulation of Buffergel, a microbicide that was safe but failed to protect women against HIV. Buffergel treatment, like tenofovir did not demonstrate an increase in these proteins when tested by ELISA (**Figure S1**).

Proteins from Vaginal Tissues. The use of proteomics for identification of proteins in vaginal fluids revealed several proteins altered by treatment with N9 and BZK; however, focusing only on vaginal secretions limits the potential pool of markers that could be utilized for evaluating microbicide toxicity. Therefore, we proceeded to analyze the changes induced by N9 and BZK in vaginal tissues. Proteomic analysis of vaginal tissue lysates revealed an average of 1,585 unique proteins. Of these, five proteins were found to be consistently down-regulated (average of 2- to 3-fold) (**Table 1** and **Figure 3A**), including PGLYRP-1 (range 2 to 3-fold for both N9 and BZK), CD166 (range 2 to 2.5-fold for N9 and 2.5 to 3-fold for BZK), mucin 5B (range 2 to 2.5-fold for N9 and 2 to 6-fold for BZK), olfactomedin-4 (OLFM-4) (range 1 to 4-fold for N9 and 1 to 5-fold for BZK), and anterior gradient protein 2

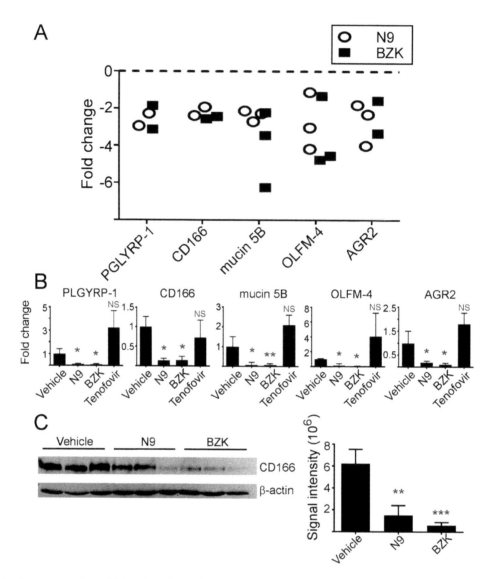

Figure 3. Quantitative proteomic and RT-PCR analysis of mouse tissue samples. (A) LC MS/MS proteomic analysis of mouse vaginal tissue samples down-regulated in response to treatment with N9 (open, black circles) or BZK (black squares). The following proteins are depicted in the graphs: peptidoglycan recognition protein 1 (PGLYRP-1), CD166 antigen (CD166), mucin 5 subtype B (mucin 5B), olfactomedin-4 (OLFM-4), and anterior gradient protein 2 homolog (AGR2). Each data point represents the average of three replicates per individual experimental study. Only proteins shown to change in a minimum of two studies are represented. (B) RT-PCR analysis of mouse mRNA expression. Relative expression levels are normalized to Glyceraldehyde 3-phosphate dehydrogenase (GADPH). Statistical analysis was performed using a two-tailed Student's T-test of N9, BZK, and tenofovir samples compared to vehicle controls (*** = p≤0.001, * = p≤0.05, NS = not significant). (C) Immunoblot of mouse CD166 antigen. Three mice per condition were analyzed. The CD166 signal was normalized to the β-actin signal for quantitation. Statistical analysis was performed using a two-tailed student's T-test of N9 and BZK samples compared to vehicle controls (*** = p≤0.001, ** = p≤0.01).

homolog (AGR2) (range 2 to 4-fold for N9 and 1 to 15-fold for BZK.

The down-regulation was confirmed by RT-PCR analysis (**Figure 3B**) and CD166 was further confirmed by immunoblot (**Figure 3C**). Changes in mRNA expression levels could be detected for these proteins following treatment with as little as 1% N9 gel (**Figure S1B**). Tenofovir treatment had minimal effect (less than two-fold) on these protein levels in vaginal tissues, suggesting that the changes observed were restricted to compounds with toxic activities. Among all proteins identified from vaginal brushes and tissues, PGLYRP-1 and mucin 5B were the only two proteins affected by treatment in both set of samples. In summary, these results highlight the limited number of proteins in the vaginal tissues that change upon exposure to N9 and BZK, potentially

identifying new candidate biomarkers to evaluate safety of candidate microbicides.

Treatment with N9 and BZK up-regulates the expression of several chemokine genes

One of the limitations of our proteomic approach is the difficulty in reliably detecting proteins of low molecular weight (e.g. chemokines). Previously we demonstrated that the chemokine CCL2 (MCP-1) was up-regulated in rabbit brush and tissue samples when treated with N9 and BZK [18], and other groups have shown CCL2 and CCL5 (RANTES) to be up-regulated in the MVI model in response to treatment with N9 [33]. To determine if the vaginal levels of additional chemokines were affected by treatment with N9 or BZK, we selected chemokines that had been shown to be upregulated during a preliminary microarray analysis (not shown) and performed RT-PCR for CCL3, CCL5 and CCL7 on rabbit vaginal tissue samples. Our analysis showed upregulated expression of CCL3 and CCL7 in response to N9 or BZK, but not in response to treatment with tenofovir or vehicle (**Figure 4**). While CCL5 was increased upon exposure to N9, as previously reported [33], its expression was not affected by BZK, indicating that CCL5 is not consistently upregulated by all compounds with toxic potential.

Treatment with N9 and BZK induces similar changes in rabbit brush samples

Although the RVI model is considered the preferred model for preclinical evaluation of vaginal microbicides, its many limitations have prompted investigators to develop other animal models. The MVI model has been used extensively to examine the efficacy of potential topical microbicides in the prevention of HSV-2 infection [34] and several studies have provided evidence for the MVI model to be a valuable tool for preclinical assessment of toxicity associated with exposure to candidate microbicides [33,35]. To provide further evidence in support of using the MVI model, we performed a proteomic analysis to determine whether proteins affected by N9 and BZK treatment in mice were consistent across species. For this evaluation, rabbits were treated with the same concentrations of N9 and BZK used in the mouse studies, and brush samples were analyzed. Our proteomic analysis from N9 or BZK treated samples showed a concordance between mouse and rabbit with upregulation of FGA, Apo A-1, Apo C-1, plasminogen, and corticosteroid binding protein (**Table 2**). ELISA assays confirmed the changes observed by proteomics analysis for FGA, Apo A-1, and plasminogen in brush samples (**Figure 5**), with no significant effect observed upon treatment with tenofovir. Apo C-1 showed a similar trend although the changes were not statistically significant (not shown). The lack of reagents specific for rabbit proteins impeded the testing of additional markers. In addition, the proteomics analysis also confirmed the down-regulatory effect of N9 and BZK on destrin and proteins from the mucin family; in rabbit we detected down-regulation of mucin 1 rather than mucin 5B. Some of the other proteins affected by N9 and BZK in MVI model (such as CAR3, and heat shock cognate 71 kDa protein) were not identified in our proteomic analysis in the rabbit study, and we see two possible explanations. In the first case, the anatomy of rabbits may play a critical role as the bladder empties into the vagina, therefore diluting all proteins or test articles present in the vagina. As a consequence, the residence time of the test articles can be significantly impacted, therefore masking its real impact on tissues. Alternatively, some of the proteins identified in mice did not appear represented in the rabbit proteomic database and therefore would not be detected using our proteomic approaches.

N9 treatment of a human vaginal epithelial cell line up-regulates the mRNA expression levels of proinflammatory proteins while reducing the expression levels of mucin 5B, and OLFM-4

Due to the difficulty in obtaining human vaginal tissue from patients treated with N9 or BZK, we tested the effects of N9 on cultured human vaginal epithelial cells. For these studies we selected human Vk2 (E6/E7) cells, a vaginal epithelial cell line, frequently used for vaginal studies [36–38]. Expression of OLFM-4, mucin 5B and PGLYRP-1 in untreated cells was evaluated using RT-PCR and immunoblots. VK2 cells had detectable levels of mRNA and protein for OLFM-4 and mucin 5B mRNA (**Figure S2**) but not PGLYRP-1 (data not shown). To determine whether N9 may down-regulate the expression of these genes in Vk2 cells, cultures were treated with N9 and transcript changes were monitored following treatment. As was previously demonstrated [38], N9 is highly toxic to Vk2 cells. We used 0.001% N9 as 80% of the cells remained viable after 24 hours at this concentration (**Figure S3**). Expression of OLFM-4 was highly reduced by 24 hours (14.6-fold, ranging from 12.5 to 25-fold), while mucin 5B, although already significantly reduced at 24 hours (2.5 fold, ranging from 2 to 3-fold) showed a total decrease of 12.4-fold (ranging from 9 to 14-fold) at 48 hours (**Figure 6, right panel**). As previously reported [38], N9 increased the expression of the pro-inflammatory marker cox2 10-fold (ranging from 8 to 15-fold) within 6 hours of treatment (**Figure 6, left panel**) compared to vehicle control cells. We also found that IL-8 was increased approximately 3-fold (ranging from 2 to 4-fold). The data obtained from cell lines suggests that selected proteins identified in proteomic assays in mice and rabbits may in fact serve as predictive biomarker for human studies.

Discussion

Monitoring changes in the vagina that can efficiently and reliably predict the safety of candidate microbicides is an important step during preclinical and clinical studies of putative new microbicides. For preclinical evaluations, use of the RVI model has long been the primary model to assess the toxicity of vaginal products. This model focuses on histopathological observations and has been repeatedly shown to be less than optimal for preclinical assessments as it is not a good structural model for the human vagina due to the differences in the epithelium (squamous in human versus columnar in rabbits) and position of the urethra inside the vagina. In addition, the study required euthanizing ~30–35 rabbits for each microbicide evaluated. The unfortunate findings during human clinical trials with N9, and cellulose sulfate (CS), both of which enhanced HIV acquisition [12,13], have clearly pointed out the need for more comprehensive and meaningful testing.

Although considerable effort has been directed toward the discovery of biomarkers that can be linked to microbicide-induced cervicovaginal inflammation [5,26,39,40], currently there is no FDA approved panel of biomarkers that can be used to demonstrate that changes in specific proteins will increase susceptibility to HIV infection. Identifying proinflammatory markers is a rational approach, and we and others [4,18,33] have reported the presence of inflammatory mediators and chemokines

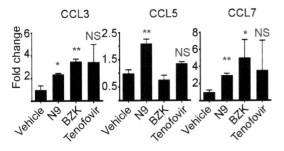

Figure 4. RT-PCR analysis of mouse CCL3, CCL5, and CCL7 gene expression. Relative expression levels are normalized to Glyceraldehyde 3-phosphate dehydrogenase (GADPH). Statistical analysis was performed using a two-tailed Student's T-test of N9, BZK, and Tenofovir samples (3 replicates) compared to vehicle controls (*** = p≤0.001, ** = p≤0.01, * = p≤0.05, NS = not significant).

Table 2. Alteration of protein levels in rabbit vaginal brush samples (day 7) after microbicide treatment.

Vaginal brush samples (day7)	Gene name	Treatment Group*								
		N9			BZK			Tenofovir		
		avg*	std	N†	avg	std	N†	avg	std	N
Apolipoprotein A-1††	APOA1	17.1	12.5	3	44.2	16.3	3	0.90	0.92	3
Apolipoprotein C-1	APOC1	7.4	3.5	3	10.4	4.2	3	0.84	0.14	3
Fibrinogen, alpha polypeptide isoform 2	FGA	7.9	0.38	3	29.8	25.5	3	0.61	0.70	3
Plasminogen	PLG	5.2	1.2	3	16.5	4.1	3	0.76	0.38	3
Corticosteroid-binding globulin	SERPINA6	7.7	3.60	2	4.0	ND	1	0.72	0.11	2
Peptidoglycan recognition protein 1	PGLYRP1	2.0	1.07	3	2.9	1.3	2	0.27	0.09	2
Mucin 1	MUC1	0.86	0.61	3	0.29	0.2	3	1.30	0.93	3
Destrin	DSTN	0.54	0.28	3	1.0	0.2	3	1.21	0.26	3

* Average signal compared to vehicle controls which are set to a default value of 1,
† number of rabbits where signal was detected in the study,
†† proteins shown in bold are identified in both MVI and RVI models.

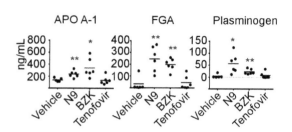

Figure 5. ELISA analysis of rabbit proteins in vaginal brush samples increased in response to treatment with N9 or BZK. Statistical analysis were performed using a two-tailed Student's T-test of N9 and BZK samples compared to vehicle controls (** = p≤0.01, * = p≤ 0.05, NS = not significant).

that could be responsible for the recruitment of monocytes and T cells in the vaginal milieu. Although the presence of chemokine attractants such as CCL2, CCL3, CCL5, MIP-2, and IL-1β, provides a potential rationale for the increased HIV infection rate observed during clinical trials [13], our data suggests that focusing only on proinflammatory proteins provides an incomplete picture. Considering the complexity of the microenvironment of vagina and the presence of different cytokines under various physiological conditions [40], an attempt to predict microbicide toxicity using cytokines and chemokines alone may not be sufficient. Inclusion of additional biomarkers that can detect changes in the epithelium barrier and local mucosal immunity may provide a more accurate indicator of potential safety issues with a product.

In this study we have used an untargeted LC-MS/MS based quantitative proteomic approach to identify and compare the qualitative and quantitative changes induced by N9 and BZK in both mice and rabbits. Our studies identified a panel of proteins that are altered during treatment with N9 and BZK in the mucosal vaginal environment. Using the MVI model, we identified six proteins (Apo A-1, Apo C-1, FGA, plasminogen, destrin, CAR3), whose levels were increased only in the vaginal brush samples (**Figure 1 and 2**). The elevated levels of these proteins are consistent with traces of blood in collected brush samples from some animals (mice and rabbits) treated with N9 and BZK. In addition, three proteins (CD166, OLFM-4, AGR2) were decreased only in the vaginal tissues and two proteins (mucin 5B and PGLYRP-1) were down-regulated in both brush and vaginal tissue samples. These are found on the surface of vaginal epithelia and

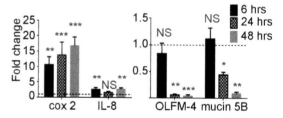

Figure 6. mRNA expression of cyclooxygenase-2 (cox2), Interleukin 8 (IL-8), mucin 5B, and olfactomedin-4 (OLFM-4) in Vκ2 cells treated with N9 for 6, 24, or 48 hours. Shown is the average of two replicate experiments, each sample performed in duplicate for each time point. Also shown is the relative expression levels of N9 treated cells compared to untreated cells, which was set to an expression level of 1 (dashed line). Statistical analysis was performed using a two-tailed Student's T-test of N9 samples compared to untreated controls at each time point (*** = p≤0.001, ** = p≤0.01, * = p≤0.05, NS = not significant).

probably shed into the mucosa. Using ELISA assays we also identified two additional chemokines, CCL3 and CCL7, and confirmed the presence of two previously known chemokines, CCL5 [33] and our previously reported CCL2 [18], as additional proteins that could be added to a potential biomarker panel. By proteomics analysis we didn't consistently detect changes in cytokines or chemokines known to be altered following treatment with N9, fact that could be due to the loss of small size of the proteins during sample preparation procedure. Therefore, we performed RT-PCR to evaluate changes on those small proteins.

Among the down-regulated proteins, OLFM-4, PGLYRP-1 and mucin 5B, have been reported to play important roles in protection against microbial infection [41–46]. OLFM-4 is known to be a suppressor of proinflammatory responses in gut epithelium [45], and it has been implicated in maintenance of persistent *Helicobacter pylori* infection where the bacteria use OLFM-4 to suppress the inflammatory response by the intestinal epithelium [45]. Loss of OLFM-4 results in rapid clearance of *H. pylori* due to an aggressive host inflammatory response that includes the release of CCL3, CCL5, and CCL7 [45], chemokines that were identified in this screen. The localized production of these chemokines may be responsible for the recruitment of HIV susceptible monocytes and T cells to sites of vaginal damage induced by the toxic microbicide compounds [18,33]. Additionally, epithelial cells have been shown to produce chemokines as well as express chemokine receptors [47,48], which play a role in the migration of immune modulating cells. OLFM-4 may play a similar role in the vagina by suppressing an inflammatory response to vaginal microflora and maintaining proper microbe and tissue homeostasis in the vaginal cavity.

PGLYRP-1 is known to be highly bactericidal [49,50] and like OLFM-4, also appears to play a role in suppressing inflammation

[51]. Thus, PGLYRP-1 could play a dual role in maintaining normal microflora and targeted killing of pathogenic bacteria. Mucin 5B has been shown to be required for respiratory tract health, and *muc5b*$^{-/-}$ mice exhibit severe morbidity and mortality due to an inability to clear routine debris (e.g., hair), leading to persistent bacterial infection of the lungs [44]. Additionally, human mucin 5B has demonstrated anti-HIV properties *in vitro*, presumably because of interactions between mucin 5B and the glycoproteins found in the viral envelope [52]. AGR2, one of the proteins that we found to be down-regulated in vaginal tissues, has been implicated in the production of mucus [53,54], so the reduction in AGR2 may have affected levels of mucin 5B protein. Thus mucin 5B may also play two roles in the vagina by maintaining the proper bacterial flora as observed in airway epithelium and providing defense against viral pathogens like HIV.

Given the diversity of the vaginal microflora, there must be a delicate balance between allowing for persistent growth of normal bacterial flora and suppression of an inflammatory response to remove them. N9 is known to alter the normal microflora of the vagina, including a marked decrease in several lactobacillus species with a subsequent increase in other potentially harmful bacterial species [55] that could induce an inflammatory response. We propose a model (**Figure 7**) in which these anti-inflammatory/protective proteins provide a stout epithelial barrier that support a balanced microbiota in the vagina; treatment with compounds such as N9, by inducing inflammation, causes damage to the vaginal epithelium, reduces the levels of these proteins leading to recruitment and dangerous exposure of HIV susceptible cells at the sites of tissue damage.

Supporting Information

Figure S1   Quantitative proteomic and RT-PCR analysis of mouse brush and tissue samples from mice treated with vehicle, buffergel, Tenofovir, and/or a range of N9 concentrations. (A) ELISA analysis of day 7 mouse vaginal brush samples isolated from animals treated with either vehicle, Buffergel, or 1% Tenofovir. (B) mRNA expression analysis of CD166 antigen (CD166), peptidoglycan recognition protein 1 (PGLYRP-1), olfactomedin-4 (OLFM-4), mucin 5 subtype B (mucin 5B), and anterior gradient protein 2 homolog (AGR2) isolated from day 10 mouse vaginal tissues treated with vehicle, 1%, 4%, or 8% N9, Buffergel, or 1% Tenofovir. Statistical analysis was performed using a two-tailed Student's T-test of N9 samples compared to untreated controls at each time point (*** = p≤0.001, ** = p≤0.01, * = p≤0.05, NS = not significant). (EPS)

Figure S2   Analysis of mucin 5B expression in Vk2 cells. mRNA (left) and protein expression (right) of mucin 5B and olfactomedin-4 (OLFM-4) in untreated Vκ2 cells. Shown are two replicate cDNA samples amplified using gene-specific RT-PCR oligonucleotides, and duplicate protein samples resolved by SDS-PAGE, then blotted to PDVF membranes and stained with anti-mucin 5B, anti-OLFM-4, or anti-actin antibodies. Glyceraldehyde 3-phosphate dehydrogenase (GAPDH) is included as a loading control for the RT-PCR. (EPS)

Figure S3   Viability of Vk2 cells treated with 0.001% N9 for 6, 24, and 48 hrs. Viability testing was performed using CellTiter Glo which measures ATP levels in the cells. (EPS)

Normal vaginal epithelium

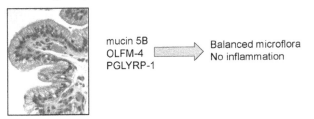

Vaginal epithelium after N9 treatment

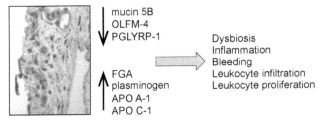

Figure 7. Model for the protective roles of mucin 5B, olfactomedin-4 (OLFM-4), and peptidoglycan recognition protein 1 (PGLYRP-1) in vaginal epithelium. In normal epithelium (top) these proteins maintain a balanced microflora and suppress inflammation. However, when exposed to toxic microbicides such as N9 (bottom) the epithelium becomes inflamed causing bleeding and release of serum proteins into the vaginal cavity. In addition, the microflora is altered leading to dysbiosis which could boost the inflammation response. The result is an influx and expansion of leukocytes that may be susceptible to HIV.

Table S1   List of RT-PCR primers.
(DOCX)

Table S2   List of biomarker candidates and their biological function.
(DOCX)

Acknowledgments

We would like to thank Dr. Paul Stein for his efforts in reviewing the manuscript and Dr. Hao Zhang for his guidance. We would also like to thank SRI International's Laboratory Animal Management Department for help with animal husbandry and help with carrying out the animal studies.

Author Contributions

Conceived and designed the experiments: SF BS AD. Performed the experiments: SF BS BR JF MGW KS YY. Analyzed the data: SF BS BR AD. Wrote the paper: SF BS JM AD.

References

1. Catalone B, Kish-Catalone T, Neely E, Budgeon L, Ferguson M, et al. (2005) Comparative safety evaluation of the candidate vaginal microbicide C31G. Antimicrob Agents Chemother 49: 1509–1520.
2. Catalone BJ, Miller SR, Ferguson ML, Malamud D, Kish-Catalone T, et al. (2005) Toxicity, inflammation, and anti-human immunodeficiency virus type 1 activity following exposure to chemical moieties of C31G. Biomed Pharmacother 59: 430–437.
3. Cohen J (2008) AIDS research. Microbicide fails to protect against HIV. Science 319: 1026–1027.
4. Cone R, Hoen T, Wong X, Abusuwwa R, Anderson D, et al. (2006) Vaginal microbicides: detecting toxicities in vivo that paradoxically increase pathogen transmission. BMC Infectious Diseases 6: 90.
5. Fichorova R, Tucker L, Anderson D (2001) The molecular basis of nonoxynol-9-induced vaginal inflammation and its possible relevance to human immunodeficiency virus type 1 transmission. J Infect Dis 184: 418–428.
6. Honey K (2007) Microbicide trial screeches to a halt. J Clin Invest 117: 1116.
7. Kreiss J, Ngugi E, Holmes K, Ndinya-Achola J, Waiyaki P, et al. (1992) Efficacy of nonoxynol 9 contraceptive sponge use in preventing heterosexual acquisition of HIV in Nairobi prostitutes. JAMA 268: 477–482.
8. Mayer KH, Peipert J, Fleming T, Fullem A, Moench T, et al. (2001) Safety and tolerability of BufferGel, a novel vaginal microbicide, in women in the United States. Clin Infect Dis 32: 476–482.
9. Roddy RE, Zekeng L, Ryan KA, Tamoufe U, Weir SS, et al. (1998) A controlled trial of nonoxynol 9 film to reduce male-to-female transmission of sexually transmitted diseases. N Engl J Med 339: 504–510.
10. Skoler-Karpoff S, Ramjee G, Ahmed K, Altini L, Plagianos MG, et al. (2008) Efficacy of Carraguard for prevention of HIV infection in women in South Africa: a randomised, double-blind, placebo-controlled trial. The Lancet 372: 1977–1987.
11. Van Damme L, Chandeying V, Ramjee G, Rees H, Sirivongrangson P, et al. (2000) Safety of multiple daily applications of COL-1492, a nonoxynol-9 vaginal gel, among female sex workers. COL-1492 Phase II Study Group. AIDS 14: 85–88.
12. Van Damme L, Govinden R, Mirembe FM, Guédou F, Solomon S, et al. (2008) Lack of Effectiveness of Cellulose Sulfate Gel for the Prevention of Vaginal HIV Transmission. New England Journal of Medicine 359: 463–472.
13. Van Damme L, Ramjee G, Alary M, Vuylsteke B, Chandeying V, et al. (2002) Effectiveness of COL-1492, a nonoxynol-9 vaginal gel, on HIV-1 transmission in female sex workers: a randomised controlled trial. Lancet 360: 971–977.
14. Kish-Catalone TM, Lu W, Gallo RC, DeVico AL (2006) Preclinical evaluation of synthetic -2 RANTES as a candidate vaginal microbicide to target CCR5. Antimicrob Agents Chemother 50: 1497–1509.
15. Zeitlin L, Hoen TE, Achilles SL, Hegarty TA, Jerse AE, et al. (2001) Tests of Buffergel for contraception and prevention of sexually transmitted diseases in animal models. Sex Transm Dis 28: 417–423.
16. Kaminsky M, Szivos MM, Brown KR, Willigan DA (1985) Comparison of the sensitivity of the vaginal mucous membranes of the albino rabbit and laboratory rat to nonoxynol-9. Food Chem Toxicol 23: 705–708.
17. Tien D, Schnaare RL, Kang F, Cohl G, McCormick TJ, et al. (2005) In vitro and in vivo characterization of a potential universal placebo designed for use in vaginal microbicide clinical trials. AIDS Res Hum Retroviruses 21: 845–853.
18. Alt C, Harrison T, Dousman L, Fujita N, Shew K, et al. (2009) Increased CCL2 expression and macrophage/monocyte migration during microbicide-induced vaginal irritation. Curr HIV Res 7: 639–649.
19. Fichorova RN, Bajpai M, Chandra N, Hsiu JG, Spangler M, et al. (2004) Interleukin (IL)-1, IL-6, and IL-8 predict mucosal toxicity of vaginal microbicidal contraceptives. Biol Reprod 71: 761–769.
20. Anderson RA, Feathergill KA, Diao XH, Cooper MD, Kirkpatrick R, et al. (2002) Preclinical evaluation of sodium cellulose sulfate (Ushercell) as a contraceptive antimicrobial agent. J Androl 23: 426–438.
21. Kurita T (2010) Developmental origin of vaginal epithelium. Differentiation 80: 99–105.
22. Barberini F, Correr S, De Santis F, Motta PM (1991) The epithelium of the rabbit vagina: a microtopographical study by light, transmission and scanning electron microscopy. Arch Histol Cytol 54: 365–378.
23. Costin GE, Raabe HA, Priston R, Evans E, Curren RD (2011) Vaginal irritation models: the current status of available alternative and in vitro tests. Altern Lab Anim 39: 317–337.
24. Achilles SL, Shete PB, Whaley KJ, Moench TR, Cone RA (2002) Microbicide efficacy and toxicity tests in a mouse model for vaginal transmission of Chlamydia trachomatis. Sex Transm Dis 29: 655–664.
25. Cone RA, Hoen T, Wong X, Abusuwwa R, Anderson DJ, et al. (2006) Vaginal microbicides: detecting toxicities in vivo that paradoxically increase pathogen transmission. BMC Infect Dis 6: 90.
26. Catalone BJ, Kish-Catalone TM, Budgeon LR, Neely EB, Ferguson M, et al. (2004) Mouse model of cervicovaginal toxicity and inflammation for preclinical evaluation of topical vaginal microbicides. Antimicrob Agents Chemother 48: 1837–1847.
27. Milligan G, Dudley K, Bourne N, Reece A, Stanberry L (2002) Entry of inflammatory cells into the mouse vagina following application of candidate microbicides: comparison of detergent-based and sulfated polymer-based agents. Sex Transm Dis 29: 597–605.
28. Achilles S, Shete P, Whaley K, Moench T, Cone R (2002) Microbicide efficacy and toxicity tests in a mouse model for vaginal transmission of Chlamydia trachomatis. Sex Transm Dis 29: 655–664.
29. Patton DL, Kidder GG, Sweeney YC, Rabe LK, Hillier SL (1999) Effects of multiple applications of benzalkonium chloride and nonoxynol 9 on the vaginal epithelium in the pigtailed macaque (Macaca nemestrina). Am J Obstet Gynecol 180: 1080–1087.
30. Wisniewski JR, Zougman A, Nagaraj N, Mann M (2009) Universal sample preparation method for proteome analysis. Nat Methods 6: 359–362.
31. Liu Y, Parman T, Schneider B, Song B, Galande AK, et al. Serum biomarkers reveal long-term cardiac injury in isoproterenol-treated African green monkeys. J Proteome Res 12: 1830–1837.
32. Redd AD, Mullis CE, Wendel SK, Sheward D, Martens C, et al. (2014) Limited HIV-1 superinfection in seroconverters from the CAPRISA 004 Microbicide Trial. J Clin Microbiol 52: 844–848.
33. Galen BT, Martin AP, Hazrati E, Garin A, Guzman E, et al. (2007) A comprehensive murine model to evaluate topical vaginal microbicides: mucosal inflammation and susceptibility to genital herpes as surrogate markers of safety. J Infect Dis 195: 1332–1339.
34. Roy S, Gourde P, Piret J, Desormeaux A, Lamontagne J, et al. (2001) Thermoreversible gel formulations containing sodium lauryl sulfate or n-Lauroylsarcosine as potential topical microbicides against sexually transmitted diseases. Antimicrob Agents Chemother 45: 1671–1681.
35. Catalone B, Kish-Catalone T, Budgeon L, Neely E, Ferguson M, et al. (2004) Mouse model of cervicovaginal toxicity and inflammation for preclinical evaluation of topical vaginal microbicides. Antimicrob Agents Chemother 48: 1837–1847.
36. Fichorova RN, Yamamoto HS, Delaney ML, Onderdonk AB, Doncel GF (2011) Novel vaginal microflora colonization model providing new insight into microbicide mechanism of action. MBio 2: e00168-00111.
37. Wagner RD, Johnson SJ (2012) Probiotic lactobacillus and estrogen effects on vaginal epithelial gene expression responses to Candida albicans. J Biomed Sci 19: 58.
38. Zalenskaya IA, Cerocchi OG, Joseph T, Donaghay MA, Schriver SD, et al. (2011) Increased COX-2 Expression in Human Vaginal Epithelial Cells Exposed to Nonoxynol-9, a Vaginal Contraceptive Microbicide that Failed to Protect Women from HIV-1 Infection. American Journal of Reproductive Immunology 65: 569–577.
39. Fichorova R, Bajpai M, Chandra N, Hsiu J, Spangler M, et al. (2004) Interleukin (IL)-1, IL-6, and IL-8 predict mucosal toxicity of vaginal microbicidal contraceptives. Biol Reprod 71: 761–769.
40. Fichorova RN (2004) Guiding the vaginal microbicide trials with biomarkers of inflammation. J Acquir Immune Defic Syndr 37 Suppl 3: S184–193.
41. Osanai A, Sashinami H, Asano K, Li S-J, Hu D-L, et al. (2011) Mouse Peptidoglycan Recognition Protein PGLYRP-1 Plays a Role in the Host Innate Immune Response against Listeria monocytogenes Infection. Infection and Immunity 79: 858–866.
42. Liu C, Gelius E, Liu G, Steiner H, Dziarski R (2000) Mammalian Peptidoglycan Recognition Protein Binds Peptidoglycan with High Affinity, Is Expressed in Neutrophils, and Inhibits Bacterial Growth. Journal of Biological Chemistry 275: 24490–24499.
43. Liu C, Xu Z, Gupta D, Dziarski R (2001) Peptidoglycan Recognition Proteins: A Novel Family of Four Human Innate Immunity Pattern Recognition Molecules. Journal of Biological Chemistry 276: 34686–34694.

44. Roy MG, Livraghi-Butrico A, Fletcher AA, McElwee MM, Evans SE, et al. (2014) Muc5b is required for airway defence. Nature 505: 412–416.

45. Liu W, Yan M, Liu Y, Wang R, Li C, et al. (2010) Olfactomedin 4 down-regulates innate immunity against Helicobacter pylori infection.

46. Liu W, Yan M, Sugui JA, Li H, Xu C, et al. (2013) Olfm4 deletion enhances defense against Staphylococcus aureus in chronic granulomatous disease. The Journal of Clinical Investigation 123: 3751–3755.

47. Mulayim N, Palter SF, Kayisli UA, Senturk L, Arici A (2003) Chemokine receptor expression in human endometrium. Biol Reprod 68: 1491–1495.

48. Patterson BK, Landay A, Andersson J, Brown C, Behbahani H, et al. (1998) Repertoire of chemokine receptor expression in the female genital tract: implications for human immunodeficiency virus transmission. Am J Pathol 153: 481–490.

49. Kashyap DR, Wang M, Liu LH, Boons GJ, Gupta D, et al. (2011) Peptidoglycan recognition proteins kill bacteria by activating protein-sensing two-component systems. Nat Med 17: 676–683.

50. Lu X, Wang M, Qi J, Wang H, Li X, et al. (2006) Peptidoglycan recognition proteins are a new class of human bactericidal proteins. J Biol Chem 281: 5895–5907.

51. Saha S, Qi J, Wang S, Wang M, Li X, et al. (2009) PGLYRP-2 and Nod2 are both required for peptidoglycan-induced arthritis and local inflammation. Cell Host Microbe 5: 137–150.

52. Habte H, de Beer C, Lotz Z, Roux P, Mall A (2010) Anti-HIV-1 activity of salivary MUC5B and MUC7 mucins from HIV patients with different CD4 counts. Virology Journal 7: 269.

53. Park S-W, Zhen G, Verhaeghe C, Nakagami Y, Nguyenvu LT, et al. (2009) The protein disulfide isomerase AGR2 is essential for production of intestinal mucus. Proceedings of the National Academy of Sciences 106: 6950–6955.

54. Zhou Y, Shapiro M, Dong Q, Louahed J, Weiss C, et al. (2002) A calcium-activated chloride channel blocker inhibits goblet cell metaplasia and mucus overproduction. Novartis Found Symp 248: 150–165; discussion 165–170, 277-182.

55. Ravel J, Gajer P, Fu L, Mauck CK, Koenig SS, et al. (2012) Twice-daily application of HIV microbicides alter the vaginal microbiota. MBio 3.

MitProNet: A Knowledgebase and Analysis Platform of Proteome, Interactome and Diseases for Mammalian Mitochondria

Jiabin Wang[⁹], **Jian Yang**[⁹], **Song Mao, Xiaoqiang Chai, Yuling Hu, Xugang Hou, Yiheng Tang, Cheng Bi, Xiao Li***

College of Life Sciences, Sichuan University, Ministry of Education Key Laboratory for Bio-resource and Eco-environment, Sichuan Key Laboratory of Molecular Biology and Biotechnology, Chengdu, People's Republic of China

Abstract

Mitochondrion plays a central role in diverse biological processes in most eukaryotes, and its dysfunctions are critically involved in a large number of diseases and the aging process. A systematic identification of mitochondrial proteomes and characterization of functional linkages among mitochondrial proteins are fundamental in understanding the mechanisms underlying biological functions and human diseases associated with mitochondria. Here we present a database MitProNet which provides a comprehensive knowledgebase for mitochondrial proteome, interactome and human diseases. First an inventory of mammalian mitochondrial proteins was compiled by widely collecting proteomic datasets, and the proteins were classified by machine learning to achieve a high-confidence list of mitochondrial proteins. The current version of MitProNet covers 1124 high-confidence proteins, and the remainders were further classified as middle- or low-confidence. An organelle-specific network of functional linkages among mitochondrial proteins was then generated by integrating genomic features encoded by a wide range of datasets including genomic context, gene expression profiles, protein-protein interactions, functional similarity and metabolic pathways. The functional-linkage network should be a valuable resource for the study of biological functions of mitochondrial proteins and human mitochondrial diseases. Furthermore, we utilized the network to predict candidate genes for mitochondrial diseases using prioritization algorithms. All proteins, functional linkages and disease candidate genes in MitProNet were annotated according to the information collected from their original sources including GO, GEO, OMIM, KEGG, MIPS, HPRD and so on. MitProNet features a user-friendly graphic visualization interface to present functional analysis of linkage networks. As an up-to-date database and analysis platform, MitProNet should be particularly helpful in comprehensive studies of complicated biological mechanisms underlying mitochondrial functions and human mitochondrial diseases. MitProNet is freely accessible at http://bio.scu.edu.cn:8085/MitProNet.

Editor: Miguel A. Andrade-Navarro, Johannes-Gutenberg University of Mainz, Germany

Funding: This work was supported partially by National Natural Science Foundation of China (Grant No. 61001149) and the National Science and Technology Major Project of the Ministry of Science and Technology of China (Grant No. 2012ZX10005001-010). The funders had no role in study design, data collection and analysis, decision to publish, or preparation of the manuscript.

Competing Interests: The authors have declared that no competing interests exist.

* Email: lix@scu.edu.cn

⁹ These authors contributed equally to this work.

Introduction

Almost all eukaryotic organisms possess mitochondria as their essential cellular components that function as the center of energy production, metabolism, signaling, apoptosis and cell growth [1]. Mitochondrial dysfunctions are known to be associated with a broad spectrum of metabolic and age-related diseases in humans, including diabetes mellitus, several cancer types, cardiovascular disorders, and neurodegenerative diseases such as Alzheimer's and Parkinson's disease [2–6]. Since these mitochondria-related diseases are caused by multigenic factors and have complex clinical phenotypes, they still remain to be poorly understood and difficult to develop medical therapy. In mammals, it is estimated that the mitochondrion is composed of about 1500 distinct proteins, the vast majority of which (above 99%) are nuclear-encoded except for thirteen polypeptides of the respiratory chain that are encoded in the mitochondrial genome (mtDNA) [7,8].

In order to understand better the roles mitochondria play in human health and disease, our priority is to define and characterize the mitochondrial proteome [9]. In the past few years, many research communities have made great efforts to identify mitochondrial proteins using different approaches, including genetics, proteomics and bioinformatics methods. In particular, mass spectrometry-based technologies exhibit the capability of high-throughput proteins identification, and have been widely utilized to define and characterize the mammalian mitochondrial proteome, which resulted in the publication of various proteomics data sets. Meanwhile, many web-accessible databases, such as MitoP2 [10], MitoProteome [11], MitoMiner [8], MitoRes [12], MiGenes [13] and MitoCarta [14], were developed to store the mitochondrial protein data that were curated manually from the biochemical literatures or collected from the large-scale proteomic studies. Among these, some performed the bioinformatics methods to improve the confidence and the coverage of mitochondrial proteomes [14].

Despite these significant successes in identifying mitochondrial proteins, the high complexity of the current data sets coupled with the tissue and development heterogeneity of mitochondrial proteins [15] are a major challenge to their use in understanding of the mammalian mitochondrial proteome and discovering susceptible genes in complex mitochondrial diseases. Firstly, a lack of common standards hinders us from defining the comprehensive and accurate mitochondrial proteome. By combining various experimental datasets from the proteomic studies, an integrative analysis showed that about 7300 proteins were identified as mitochondrial, which significantly excesses the estimated size of the mammalian mitochondrial proteome. The large number of proteins reveals the presence of false discovery in large-scale proteomic studies. This is mainly due to the purified mitochondria are often contaminated by other non-mitochondrial organelles such as microsomes and cytoskeletons whose proteins are falsely identified as mitochondrial [7]. Secondly, with the rapidly increasing number of newly discovered mitochondrial proteins, a critical task beyond protein identification is to annotate cellular functions for newly-identified mitochondrial proteins and to associate their functional roles with human mitochondrial disorders. The investigation [14] on MitoCarta which may represent the largest comprehensive compendium of mammalian mitochondrial proteins to date indicated that about a quarter of proteins in the inventory were not annotated to a biological process in terms of Gene Ontology (GO) annotation [16]. If we expand to the whole mitochondrial proteome, a greater number of mitochondrial proteins will remain to be uncharacterized.

With the increase in the availability of genomic and proteomic data, computational approaches have been proposed for inferring the biological function of mitochondrial proteins, prioritizing and predicting candidate genes susceptible to mitochondrial disorders. Many computational approaches follow the idea termed 'guilt-by-association' that the function of one protein could be transferred from another protein with known function relying on their biological relationship [17]. The large-scale genomic and proteomic datasets allow us to measure quantitatively the biological relationship between two genes, including gene expression profiling, protein-protein interactions, phylogenetic profiling, and synthetic genetic analysis and so on. For example, using phylogenetic profiling analysis across hundreds of species, Pagliarini *et al.* identified 19 novel factors that are involved in the assembly of complex I of the mitochondrial respiratory chain [14]. More recently, the biological relationships among a set of genes/proteins can be represented as a network such as gene co-expression network, transcription regulation network and protein interaction network, which provides us a global perspective of understanding mitochondrial biology and disease at a systems level [18–20]. Nevertheless, most of those studies on mitochondria used only individual data source or data type, which led to insufficient coverage of the mitochondrial proteome and thus potentially limited their predictive ability.

A reasonable alternative would be to utilize the functional linkage network (FLN) integrated from heterogeneous datasets generated from successful efforts on larger scale assembly. The integration of complementary knowledge from heterogeneous sources is essential to understand the system as a whole and obtain well populated networks. Comparing with the networks derived from individual data type, the FLNs are denser and less biased

towards a kind of particular evidence. Many successes have been achieved in predicting gene functions and prioritizing disease genes through utilizing the FLN-based scheme. Although several FLN databases have been distributed, such as STRING [21], Reactome [22] and BioGRID [23], there are very few FLN databases that are designed specifically for mitochondria.

To address the issue of single data set or type, Franke et al. [24] constructed a functional linkage network (FLN) by integrating multiple types of genome-wide data, and utilized the FLN for disease gene prioritization. However, it is speculated that the performance of this FLN was highly dependent on Gene Ontology (GO) annotations, and as a result, the predictions tended to be biased towards well-characterized genes, and thus limit capacity on inferences. In another study, Linghu et al. [25] integrated multiple genome-wide features to construct an evidence-weighted FLN, and used a neighborhood-weighting decision rule for disease gene prioritization successfully. Nevertheless, while specialized in mitochondrion, a specific FLN among proteins in this organelle using a combination of multiple types of data focusing its message exclusively on functional associations among mitochondrial proteins, would deliver superior performance. To date, only two databases specialized for mitochondrial protein interactions are public available, Mitointeractome [26] and InterMitoBase [27]. Mitointeractome is a representative interaction database for mitochondria which includes predicted protein-protein interactions (PPIs) based on structural and homologous information. InterMitoBase contains well-annotated PPIs between mitochondrial and mitochondrial/non-mitochondrial proteins integrated from a wide range of resources. However, the both of databases cover only PPI information, which is not sufficient for characterizing functional associations among mitochondrial proteins. Therefore, it is necessary to construct a database covering the entire FLN that characterizes the global functional associations among mitochondrial proteins.

In this study, we performed a machine-learning classifier to integrate mitochondrial proteins from 23 proteomic datasets for compiling an inventory of mammalian mitochondrial proteins. Comparing with other datasets, the list of mitochondrial proteins comprising 1124 proteins reveals a larger coverage and better accuracy. A mitochondria-specific FLN was constructed by integrating 15 heterogeneous genomic and proteomic datasets, resulting in 32,951 weighted functional linkages among 1072 mitochondrial proteins. Furthermore, the mitochondria-specific FLN was utilized to identify and prioritize candidate genes for typical mitochondrial diseases. The results show the inventory of mitochondrial proteins and the FLN among mitochondrial proteins should be valuable resources in comprehensive studies of complicated biological mechanisms underlying mitochondrial functions and human mitochondrial diseases.

Results and Discussion

General procedure

The overall procedure (Figure 1) included three steps. The first step was to compile an inventory of mammalian mitochondrial proteins by means of collection from various proteomic experimental datasets and several publicly-available databases. In the second step, a FLN among mitochondrial proteins was constructed through integrating functional features from heterogeneous 'omic' data sources. Finally, the FLN was then used to identify and prioritize candidate genes for mitochondrial diseases.

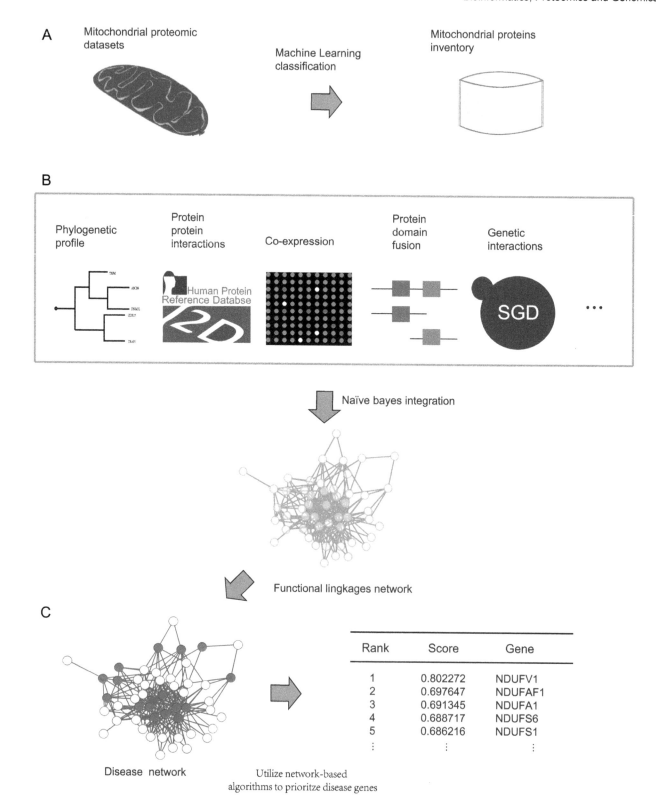

Figure 1. A flowchart depicting the work. (A) Step 1: obtaining a mitochondrial proteins inventory utilizing machine learning classification. (B) Step 2: constructing the FLN by integrating 11 genomic features including protein-protein interaction, domain-domain interaction, shared domains, genomic context, genetic interaction, phenotypic semantic similarity, co-expression, GO semantic similarity, protein expression profiles, disease involvement and operon based on the Naïve bayes model. (C) Step 3: ranking the disease candidate genes utilizing the FLN and a network-based algorithm. The table on the right shows the ranking scores of the top 5 candidate genes for mitochondrial complex I deficiency.

An inventory of mammalian mitochondrial proteins

Although Pagliarini *et al.* presented the most comprehensive mammalian mitochondrial proteome (the MitoCarta database) with nearly 1100 proteins and estimated that their compendium covers more than 85% of the mitochondrial proteome [14], Meisinger *et al.* speculated that they may underestimate the size of the mammalian mitochondrial proteome and that the total number of mammalian genes for mitochondrial proteins could approach 1500 [28]. Considering the limitation of the databases, as a first step, we needed to compile an inventory of mammalian mitochondrial proteins that covers as many proteins as possible in the organelle. Thus, we made an extensive collection of mammalian mitochondrial proteins identified experimentally.

Despite various proteomics-scale experiments successfully identified mitochondrial proteins, a combined experimental datasets from these proteomic studies showed that about 7300 proteins were identified as mitochondrial proteins, which significantly exceeded the estimated size of the mammalian mitochondrial proteome. The large number of proteins reveals the presence of false discovery in large-scale proteomic studies. The previous investigation revealed that there is a high conservation among mammalian mitochondrial proteomes [8], hence it is a complement to compile a comprehensive inventory of mitochondrial proteins by integrating the proteomic datasets from a wide range of mammalian mitochondria. Here we collected 23 proteomic datasets from three model mammals including human (*H. sapiens*), mouse (*M. musculus*) and rat (*R. norvegicus*) for the integration (Table 1). To reduce false discovery, moreover, we

performed a machine-learning classifier to integrate mitochondrial proteins.

We used weka, a software that collecting a set of machine learning algorithms for data mining tasks [51], to integrate mitochondrial proteomic datasets. As a first step of machine learning, a gold standard positive (GSP) set and gold standard negative (GSN) set were constructed. Based on the test set, various machine-learning classifiers including AdaBoostM1, Id3, J48, Logistic, MultiClassClassifier, MultilayerPerceptron, NaiveBayes and RandomForest were trained. We assessed the prediction performance by 10-fold cross-validation, showing that the AdaBoostM1 classifier [52] achieved the best, prediction with a high sensitivity of 0.93 (Table S1). The AdaBoost classifier was then applied to identify mitochondrial proteins form 23 proteomic datasets, which resulted in 1109 proteins as positives, 550 of which were the known mitochondrial proteins in the GSP set. There were 15 proteins defined in the GSP were falsely classified as non-mitochondrial proteins. To achieve a comprehensive database of mitochondrial proteins, the high-confidence list was curated manually to include these proteins. As a result, we created an inventory of high-confidence mammalian mitochondrial proteins that includes 1124 mitochondrial proteins (Table S2), which consists of 1109 proteins predicted by the AdaBoostM1 classifier as well as 15 missing proteins from the GSP set. In order to utilize sufficiently the proteomic resources, we further classified the remaining about 6100 proteins as middle-confidence or low-confidence using a simple voting policy. The voting policy was described as follows: a protein was classified as middle-confidence

Table 1. Integrated mitochondrial proteomic datasets for an inventory of mammalian mitochondrial proteins.

Datasets	Species	Number of Proteins	Tissue/organ/cell	Method
Calvo S et al. [29]	H. sapiens	1048		Prediction
Taylor SW et al. [30]	H. sapiens	600	Heart	MS
Rezaul K et al. [31]	H. sapiens	656	T leukemia cells	MS
Xie J et al. [32]	H. sapiens	180	Immortalized lymphoblastoid cell lines	2-GE
Ozawa T et al. [33]	M. musculus	48	Cell line BNL1ME (liver)	GFP
Mootha VK et al. [34]	M. musculus	462	Brain, heart, kidney, and liver	MS
Jin J et al. [35]	M. musculus	781	Dopaminergic cells	MS
Kislinger T et al. [36]	M. musculus	1872	Brain, heart, kidney, liver, lung, and placenta	MS
Da Cruz S et al. [37]	M. musculus	97	Liver	MS
Johnson DT et al. [15]	R. norvegicus	292	Brain, liver, heart, and kidney	MS
Forner F et al. [38]	R. norvegicus	503	Muscle, heart, and liver	MS
Reifschneider NH et al. [39]	R. norvegicus	110	Kidney, Liver, Heart, Skeletal Muscle and Brain	BN
Palmfeldt J et al. [40]	H. sapiens	2591	Skin fibroblast	MS
Lefort N et al. [41]	H. sapiens	892	Skeletal muscle	MS
Bousette N et al. [42]	M. musculus	2087	Heart	MS
Fang X et al. [43]	M. musculus	2165	Brain	MS
Zhang J et al. [44]	M. musculus	916	Heart	MS
Deng WJ et al. [45]	R. norvegicus	624	Liver	MS
Wu L et al. [46]	H. sapiens	1149	T leukemia cells	MS
Catherman AD et al. [47]	H.sapiens	1326	H1299 cells	MS
Hansen J et al. [48]	H.sapiens	2138	human lymphoblastoid cells	MS
Chappell NP et al. [49]	H.sapiens	1523	Epithelial ovarian cancer cell	MS
Chen X et al. [50]	R. norvegicus	1215	rat INS-1 cells	MS

MS, mass spectrometry. 2-GE, two-dimensional gel electrophoretic. GFP, green fluorescent protein. BN, blue-native.

if it is included in MitoP2 or MitoCarta dataset, or was identified from more than five proteomic experiments, while the remaining were low-confidence. The high-confidence mitochondrial proteins were strongly supported by the 23 datasets, which may represent the most common proteins in mitochondria. Some other proteins however may intermittently bind to the surface of mitochondria, making it hard to discover by mass spectrometry, thus may fall into the middle-confidence or even low-confidence category. Nevertheless, by integrating sufficient datasets from various experimental conditions, the risk for the latter case will drop a lot. Considering the fact that some proteins may expressed under certain circumstances or special tissues, the information for tissue/organ origin of a protein was retained for researchers' judgments on our web pages. The 1124 high-confidence proteins as well as the 1159 middle-confidence proteins together made up the MitoCom dataset.

To evaluate the quality of MitoCom, a comparison between MitoCom (high-confidence proteins) and two mitochondrial databases, MitoPred [53] and MitoCarta, was carried out by using the MitoP2 dataset as the reference set. As shown in table 2, the high-confidence proteins in MitoCom showed considerable overlap with MitoPred and MitoCarta, meanwhile it retained a wider coverage, greater sensitivity and lower false discovery rate, which can reduce the "noise" in high-throughput mammalian mitochondrial protein identification effectively. The venn diagram (figure 2) between these three datasets and the middle-confidence proteins showed that the high-confidence proteins had about 74% overlap with MitoCarta and MitoPred, while keeping 288 proteins that identified uniquely by this work. The high-confidence proteins in MitoCom extended the mitochondrial proteome while the middle-confidence proteins can be a clue for a more complete mitochondrial proteome. Thus, our inventory of mammalian mitochondrial proteins would be more comprehensive and accurate in comparison to other databases, which enables it to be a powerful tool for mitochondrial proteome studies.

Functional linkages among mitochondrial proteins

With the rapidly increasing number of discovered mitochondrial proteins, a critical task beyond protein identification is to annotate cellular functions for newly-identified mitochondrial proteins and to associate their functional roles with human mitochondrial disorders. We have pursued these goals by integrating genomic features from heterogeneous data sources to build quantitative functional links among mitochondrial proteins. Since a single data source usually reflects only one type of functional association between proteins (genes), and its coverage is relatively limited, functional associations from multiple data sources should be jointed to achieve larger coverage and better accuracy.

In the previous step, we have built an inventory of 1124 mammalian mitochondrial proteins. This yielded 631688 potential mitochondrial protein-protein functional linkages. To validate these protein pairs, we systematically combined 11 genomic features about 15 datasets (Table 3) using machine learning algorithm.

The integrated features were shown as follows:

- **Protein-protein interaction (PPI).** Protein-protein interactions are fundamental to all biological processes. The interacting proteins should have closely functional association.

- **Domain-domain interactions.** Proteins perform their biological functions often through domains as units. Thus two proteins may have similar function if they contain domains with capability of interacting.

- **Shared domains.** As well known, domain is the functional unit in protein. Hence, proteins possess the same set of domains should have similar function.

- **Genomic context.** Genomic context including phylogenetic profiles and Rosetta Stone can be powerful evidence for functional linkages between genes. Gene pair that has similar phylogenetic profile or appears in a gene fusion event tends to be functionally associated [69,70].

- **GO Semantic Similarity.** Gene ontology defines a gene function with a hierarchical structure in three dimensions including cellular component, molecular function and biological process. Two genes with terms that share the same parent far from root should be functional associated [24]. Thus, the GO semantic similarity can be used to measure function association between genes.

- **Genetic interaction.** Genetic interactions, such as synthetic lethal and synthetic growth, infer those involved genes have strong correlation. These correlations are also evidences of functional associations.

- **Phenotypic semantic similarity.** Genes leading to similar phenotypes should have functional linkages, as similar phenotypes may need similar substances or involve similar processes.

- **Gene co-expression.** Genes encoding proteins that are involved in the same process are expected to be simultaneously expressed in time and space [71]. Therefore, genes with similar expression patterns should have related function. To profile gene expression, four microarray datasets were selected. GSE1133 and GSE4726 interrogate the expression of the vast majority of protein-encoding human and mouse gene that can give us a global view on gene expression profile at the genome scale, while GSE4330 and GSE6210 studied the influence of mutant in PGC1α and PGC1β, both of which are transcriptional coactivator that potently stimulates mitochondrial biogenesis and respiration of cells, focusing on mitochondrial-specific genes.

- **Proteomic profiles.** Similar to gene co-expression profile, proteomic profile may lead to better understanding of mitochondrial feature at protein level. Thomas Kislinger et

Table 2. Quality comparison of MitoCom with other mitochondrial databases.

Database	Number	Sensitivity	False discovery rate
MitoCom*	1109	97.34%	11.30%
MitoCarta	1013	86.10%	13.70%
MitoPred	910	50.10%	14.80%

*Just the high-confidence proteins.

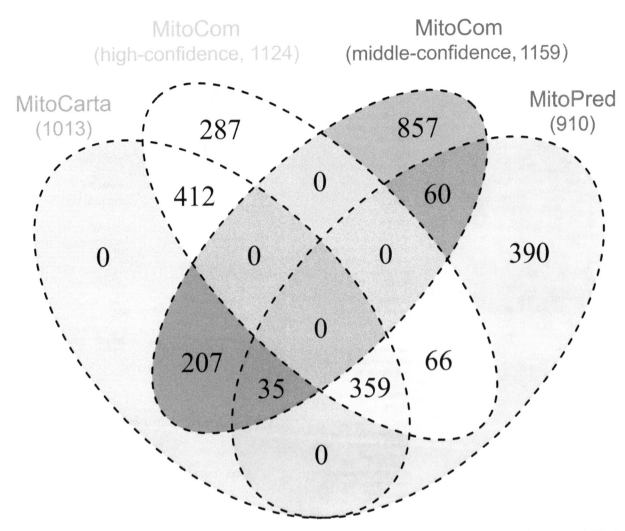

Figure 2. Venn diagram of the four datasets: MitoCom (high-confidence), MitoCom (middle-confidence), MitoCarta and MitoPred.

al [36] examined the protein content of four organellar compartments in six mouse organs, which could be a valuable resource. We extracted the mitochondrial-specific proteomic profile from this dataset.

- **Diseases involvement.** Genes annotated in the same disease tend to have functional associations.
- **Operon.** Based on the endosymbiotic theory, mitochondrion may evolve from an ancestor of *Rickettsia prowazekii*, which shares a lot of homological genes with mitochondrial genome [72]. As a functional unit, operon contains a series of genes that involved in same biological process. Therefore, mitochondrial genes whose homologies appear in the same operon in *Rickettsia prowazekii* should be an evidence for functional associations.

To implement the machine learning algorithm, a GSP and a GSN were first constructed (see materials and methods). Based on the well-defined GSP and GSN, we investigated the coverage of each genomic feature, revealing that several datasets had very low coverage (<20%). Only five datasets including GO semantic similarity, gene co-expression, proteomics profiles and phenotypic semantic similarity covered over 20% on the GSP and GSN (Table S3). For integrating these datasets, we used a naïve Bayes classifier [73,74] owing to its two advantages. First, it can integrate

heterogeneous kinds of evidence and tolerate missing data among them. Second, it is simple but highly efficient to tackle data in a large scale with short time consumption.

As a prerequisite for using naïve Bayes classifier, all the datasets should be conditionally independent. We assessed the statistical independence between each pair of datasets with coverage more than 20% by calculating the PCC. As shown in Table S4, these datasets are relatively independent with the maximum PCC is only 0.217. Following the naïve Bayes theorem, a likelihood ratio (LR) corresponding to a specific biological evidence could be used to measure the predictive power or confidence degree. Thus we measured the power of individual datasets to infer functional linkages by using the naïve Bayes model. Each dataset was divided into several bins, and then the LR for each bin was calculated according to the GSP and the GSN. As shown in Figure S1, all the 15 datasets were clearly correlated with LRs and all the datasets had one or more bins with LR>1, which suggested that the 15 datasets can be used to infer functional linkages between genes.

To evaluate the performances of individual dataset model and integrated model, we carried out five-fold cross-validation and drew the receiver operating characteristic (ROC) curve (Figure 3A). The figure showed that the integrated model had the largest area under ROC curves (AUC), demonstrating the superiority of data integration. The results also suggested that

Table 3. Functional features for mammalian mitochondrial FLN construction.

Functional features	Data sets	Description	Scale	Data source
Protein-protein interaction		Protein-protein interaction.	Genome-scale	HPRD [54], I2d [55]
Domain-domain interaction		Protein pairs have interacting protein domains.	Genome-scale	3did [56]
Shared domains		Proteins pairs sharing same protein domains.	Genome-scale	Interpro [57]
Genomic context	Rosetta Stone	Gene fusion events.	Genome-scale	Prolinks [58]
	Phylogenetic profiles	Phylogenetic Profiles [59] of 1086 genes among 600 species.(Table S6)	Genome-scale	NCBI, KEGG [60]
Genetic interaction		Mutations in two genes produce a phenotype that is greatly different from each mutation's individual effects.	Genome-scale	Saccharomyces Genome Database [61]
Phenotypic semantic similarity		Sementic simlilarity of mouse phenotypic terms.	Genome-scale	Mammalian Phenotype Browser [62]
Co-expression	GSE1133 [63]	Gene expression profile of the vast majority of protein-encoding human and mouse genes in 79 human and 61 mouse tissues.	Genome-scale	GEO [64]
	GSE4726 [65]	A quantitative and comprehensive atlas of gene expression in mouse development.	Genome-scale	GEO
	GSE4330 [29]	Microarray time-course of mouse myotubes transduced with the transcriptional co-activator PGC-1α, which is known to induce mitochondrial biogenesis in muscle cells.	Mitochondria-specific	GEO
	GSE6210 [66]	Gene expression profile in liver tissue and quadriceps muscle in mice between control and the PCG-1β mutant, a transcriptional coactivator that potently stimulates mitochondrial biogenesis and respiration of cells.	Mitochondria-specific	GEO
GO semantic similarity		GO Sementic similarity of genes sharing the same biological process terms	Genome-scale	The Gene Ontology [16]
Protein expression profiles		Mitochondrial protein profiles of protein-coding genes in heart, brain, liver, kidney and lung.	Mitochondria-specific	Results of Thomas Kislinger et al [36]

Table 3. Cont.

Functional features	Data sets	Description	Scale	Data source
Disease involvement		A pair of genes that annotated in the same disease.	Mitochondria-specific	OMIM [67]
Operon		Operon data of *Rickettsia prowazekii.*	Mitochondria-specific	Database of prOkaryotic OpeRons [68]

individual data models have limited capability to correctly identify functional linkages between genes. Most of individual dataset models including gene co-expression model and proteomic profile model have similar performances with an AUC around 0.6, much lower than the integrated model. The rest datasets except for the GO semantic similarity model showed no difference to the reference line, indicating their inefficiency. A clear exception was GO semantic similarity model, which had an AUC of 0.772, a little lower than the integrated data. The GSP and GSN were derived from prior knowledge, which will introduce in bias when estimating the GO semantic similarity model that was also derived from prior knowledge. If we use this model to predict novel function linkage, the prediction ability is limited. Therefore, we can conclude that data integrating approach is the best when try to predict novel functional linkages.

Furthermore, we classified the 15 datasets as genomic-scale and mitochondria-specific according to dataset source and data scale. A dataset was considered as mitochondria-specific if the dataset was generated from an experiment was aimed at mitochondrial study, like GSE4330, GSE6210 and proteomic profile, If a dataset

contains information only derived from the mitochondrial proteome, such as diseases involvement, operon and GO semantic similarity, it was also considered as mitochondria-specific. As shown in Figure 3B, the integrated mitochondria-specific model had a larger AUC than the integrated genome-scale model, which indicated that the mitochondria-specific dataset was more powerful to construct FLN.

After data integration, each protein pair has been attached a LR score. A cutoff of LR was determined afterward, which representing as an indicator of whether a protein pair is functional associated (that is, yes if the composite LR is above the LR cutoff, no if not). We used the ratio of true positive (TP) to false positive (FP) to measure the prediction accuracy, and plotted the TP/FP ratio as a function of LR cutoff (Figure 4). We found that there is an apparent positive correlation between the TP/FP ratio and LR cutoff, but the sensitivity decreases monotonically and the FLN scale shrinks simultaneously with the increase of LR cutoff. A composite LR cutoff of 2.5 was selected where the TP/FP ratio was 1, which means that we can achieve 50% prediction accuracy at this resolution. Based on this LR cutoff, the resulting FLN is

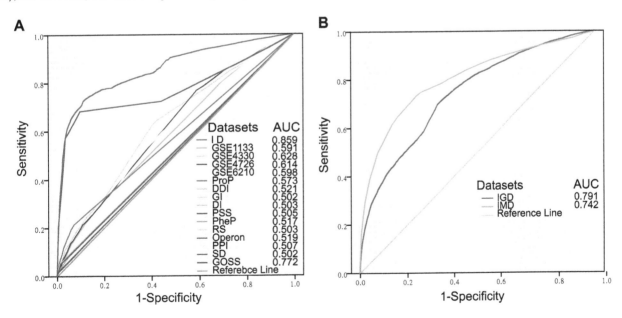

Figure 3. ROC curves for evaluating the performances of various data sources using cross-validations. (A) ROC curves and AUC of individual dataset and integrated dataset. The data sources are highlighted in different colors. (B) ROC curves and AUC of mitochondrial-specific (green) and genome-scale (blue) datasets. ID: Integrated datasets; ProP: Protein expression profiles; DDI: Domain-Domian Interaction; GI: Genetic Interaction; DI: Disease Involvement; PSS: Phenotypic Semantic Similarity; PheP: Phylogenetic Profiles; RS: Rosetta Stone; PPI: Protein-Protein Interaction; SD: Shared Domains; GOSS: GO Semantic Similarity; IGD: Integrated Genomic-scale Datasets; IMG: Integrated Mitochondrial-specific Datasets; ROC: receiver operating characteristic; AUC: area under ROC curves.

comprised of 1072 proteins (covering approximately 71% of the mitochondrial proteome) and 32951 weighted functional linkages (Table 4), the average number of functional linked neighbors per protein is 61. The mitochondria-specific FLN owns such high coverage and linkage density, which is essential to the successful utilization of the FLN for disease gene prediction and prioritization.

Disease candidate gene prioritization

With the FLN, we aimed at using the information to prioritize candidates for mitochondrial diseases. The utility of FLN for disease candidates prioritization based on the assumption that genes underlying the same or related diseases tend to be functionally related [69]. Based on this assumption, FLNs have been successfully used to identify novel disease genes in recent studies [74–76]. Meanwhile, many network-based methods have been developed to prioritize candidates, for example, random walk, neighborhood-based and diffusion kernel methods. These methods mostly locate the known disease genes in network as "seeds" first, and then score the associated neighborhoods of these seeds by specific algorithm, and finally candidates are prioritized based on the scores of candidates.

In this work, four network-based methods were chosen for disease candidate prioritization. The average adjacency ranking (AAR) rule has been successfully used by Guan Y et al. to predict novel pathway components [74]. PageRank with Priors (PRP), K-step Markov (KSM) and Heat Kernel Diffusion Ranking (HKDR) methods were also used to prioritize disease candidates based on PPI networks [75]. Goncalves et al analyzed the performance of the four methods, indicating their applicability in prioritizing disease candidates [76].

Despite the impacts of ranking approaches, FLN should outperform the single source networks for the reason that multiple evidence increases coverage/density and reduces bias toward individual sources [76]. We evaluated the effectiveness of the four ranking algorithms utilizing the FLN and two single source networks including PPI network and co-expression network to prioritize candidates, both of which were derived from single data source. Furthermore, because the ranking algorithms are also susceptible to the network scale and density, the FLN was expanded into a scale-larger network named the FLNhm by including the middle-confidence mitochondrial genes and their functional linkages (the LR cutoff wasn't used). We downloaded the disease data from the OMIM database, and extracted those that have at least two OMIM-annotated disease genes present in the networks for identifying disease candidates. Owing to the scale difference, different sets of mitochondrial diseases and disease genes were analyzed when utilizing the four networks respectively. Using known disease-associated genes as "seeds", Leave-one-out cross-validation tests were conducted. ROC curves were plotted to visualize the performance with AUC values as quantitative measures.

For the reason that algorithms performance differently with the parameter set and the scale of network different, different test parameter sets were empirically selected to decide the best algorithm and its optimal parameter set for each network. (see materials and methods).We decided the optimal parameters of the algorithms on each network based on the AUC (Table S5). Figure 5 showed ROC curves of the four algorithms with optimal parameters on the four networks. The HKDR, PRP and KSM algorithms outperformed neighborhood algorithm AAR, which indicated that the three algorithms utilizing the whole topology information were superior to algorithms utilizing local topology information. It may be the result of that the algorithms that utilize the whole topology can compensate for missing links by exploiting higher order neighborhoods and path redundancies [76]. HKDR and PRP algorithms performed best respectively on the FLN and the FLNhm. KSM had a poor performance compared with PRP and HKDR on FLN and FLNhm, but outperformed the two algorithms on the PPI network and the co-expression network, suggest that KSM algorithm was better in compensating for missing links than HKDR and PRP algorithm when being utilized in single source networks.

Furthermore, we also observed that the performances of the four algorithms dropped orderly and significantly in FLN, FLNhm, PPI network and co-expression network. As a single source network, the PPI network and co-expression network were supported to be less informative with limited coverage and large number of false positive linkages. Therefore, PPI network and co-expression network performed worse than FLN and FLNhm as expected. The FLNhm, which was denser and with bigger coverage than FLN, but performed worse than FLN, indicated that topology also play an important role in the performance of network. Being the best performance of cross-validation, HKDR algorithm with its optimal parameter (n = 3) on the FLN were chosen to rank candidates of mitochondrial diseases.

Mitochondrial complex I deficiency: a case study

Mitochondrial complex I deficiency, the most common cause of mitochondrial disorders (accounts for ~30% cases of respiratory

Table 4. Descriptions and parameters of four networks.

	Description	Number of Nodes	Number of Edges	Average number of neighbors	Density
FLN	FLN among the proteins with high confidence	1072	32951	61.476	0.057
FLNhm	FLN among the proteins with high or middle confidence	1992	1983036	1991.000	1
PPI network	Protein-protein interactions network derived from HPRD and I2D	1322	9049	12.850	0.01
Co-expression network	Co-expression network derived from microarray experiment GSE1133	1684	1417186	1683.000	1

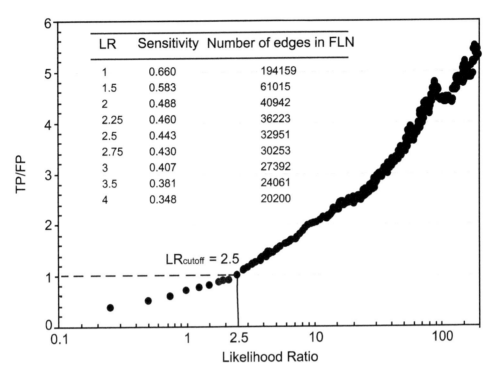

LR	Sensitivity	Number of edges in FLN
1	0.660	194159
1.5	0.583	61015
2	0.488	40942
2.25	0.460	36223
2.5	0.443	32951
2.75	0.430	30253
3	0.407	27392
3.5	0.381	24061
4	0.348	20200

Figure 4. TP/FP ratios vs. LR cutoff, and corresponding sensitivity. TP: True Positive; FP: False Positive. Sensitivity = TP/(TP+FN).

chain deficiency in humans) [77], causes a wide range of clinical disorders, ranging from lethal neonatal disease to adult-onset neurodegenerative disorders. Phenotypes include macrocephaly with progressive leukodystrophy, nonspecific encephalopathy, hypertrophic cardiomyopathy, myopathy, liver disease, Leigh syndrome, Leber hereditary optic neuropathy, and some forms of Parkinson disease. It shows extreme genetic heterogeneity. Up to now, mutations in 17 genes encoding mitochondrial complex I subunits have been described in the OMIM database. However, these 17 genes account for disease in only a minority of mitochondrial complex I patients. Since mitochondrial complex I has at least 45 subunits [78,79], mutations in any of the other approximately 30 supernumerary subunit genes could potentially cause mitochondrial complex I deficiency, even mutations in other genes functionally associated with mitochondrial complex I subunits are also possible causes. Here, heat diffusion was applied to rank and screen promising candidates of mitochondrial complex I deficiency based on linkage with known disease genes, then we assessed the ability of prioritization to identify unknown causes.

Fifteen of these disease causing genes are present in our function linkage network. The importance of each gene in the function linkage network relative to mitochondrial complex I deficiency was ranked using these 15 genes as seeds. We investigated the top 15 candidates (Table 5), almost all of which could be associated with mechanisms of mitochondrial complex I deficiency (Figure 6). In the top three, the *NADH dehydrogenase 1 beta subcomplex, 8, 19 kDa (NDUFB8)* is known to encode a subunit of mitochondrial complex I [79,80]. Haack *et al.* found mutations in *NDUFB8* result in decreased activity and amount of mitochondrial complex I [81]. And the *cytochrome c oxidase subunit Vb (COX5B)*, known to cooperate with mitochondrial complex I in respiratory electron transport chain, is a terminal enzyme of the mitochondrial respiratory chain [82]. *Electron-transfer-flavoprotein, alpha polypeptide (ETFA)*, in the third place, shuttles electrons between primary flavoprotein dehydrogenases and the membrane-bound

electron transfer flavoprotein ubiquinone oxidoreductase [83]. Mutations in *ETFA* are causative for multiple acyl-CoA dehydrogenase deficiency, and result in decreased activity of mitochondrial complexes I [84,85]. It is worth noting that the *NADH dehydrogenase Fe-S protein 3, 30 kDa (NDUFS3)*, ranked 4th, encodes one of the iron-sulfur protein components of mitochondrial NADH: ubiquinone oxidoreductase (complex I) [79,80]. Benit *et al.* found mutations in *NDUFS3* related to isolated mitochondrial complex I deficiency by using a combination of denaturing high performance liquid chromatography and sequence analysis [86]. Haack *et al.* also reported pathogenic mutations in *NDUFS3* caused isolated mitochondrial complex I deficiency by combining unbiased exome analysis, sequential filter, and functional investigation [81]. The *NADH dehydrogenase 1 beta subcomplex, 7, 18 kDa (NDUFB7)*, ranked 14th, encodes a subunit of mitochondrial complex I [79], Triepels *et al.* found pathogenic mutations in *NDUFB7* in the patients of mitochondrial complex I deficiency [87].

Despite continued progress in our understanding of the molecular basis of mitochondrial complex I deficiency, the genetic defect remains elusive in many cases. With the application of the function linkage network, potential pathogenic causes could be ranked and prioritized. Furthermore, top ranked candidates could guide the design of new disease-genes association studies and offer clues for new treatment strategies.

Database and web server

We constructed a database named MitoProNet for storing our results including mammalian mitochondrial proteins, the FLN and human disease information. MitoProNet is an object-relational database implemented by mysql accessible via a user-friendly web interface written in JSP.

The main contents of MitoProNet are demonstrated in Figure 7 including proteome section, disease section and FLN among proteins or genes, which could be accessed by browsing or

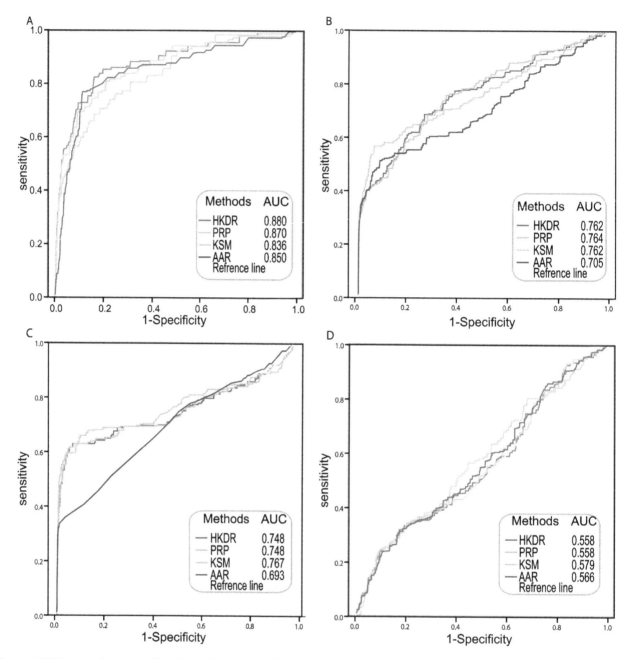

Figure 5. ROC curves for evaluating the performances of four networks on disease-gene prioritization. (A) The ROC curve for FLN. (B) The ROC curve for FLNhm. (C) The ROC curve for PPI network. (D) The ROC curve for co-expression network. AAR: Average Adjacency Ranking; PRP: PageRank with Priors; KSM: K-Step Markov; HKDR: Heat Kernel Diffusion Ranking; FLN: Functional Linkage Network among high-confidence mitochondrial proteins; FLNhm: Functional Linkage Network among high-confidence and middle-confidence mitochondrial proteins; PPIN: Protein-Protein Interaction Network; CEN: Co-Expression Network.

searching in MitProNet. Users can browse proteome data and disease data by clicking the proteome interface and the disease interface. The proteome interface provides comprehensive data of mammalian mitochondrial proteins that were identified experimentally. Results could be displayed orderly according to experiment, confidence level or organisms. The disease interface provides comprehensive information about typical mitochondrial diseases, including description, known disease genes, top ranking disease candidates ranked in our study, as well as functional linkages network among these genes. Users can also click the name of a protein of interest, the results include description of the

protein and its annotation information will be displayed via HTML pages. Moreover, a local functional linkages network can be visualized online as a scalable vector graphics (SVG) file, which provides the means for a fast visual evaluation of the protein's functional association with other proteins. The search interface also allows users to source the proteins or diseases of interest conveniently by using a variety of keywords include gene IDs, gene symbols, protein IDs and OMIM IDs. And Figure 8 showed a case of browsing and searching in MitProNet. All these data presented in MitProNet can be downloaded freely through our download interface.

Table 5. The 30 top-ranking genes for mitochondrial complex I deficiency.

Ranking	Score	GeneID	Symbol	Description
1	0.802272	4723	NDUFV1	NADH dehydrogenase flavoprotein 1, 51 kDa
2	0.697647	51103	NDUFAF1	NADH dehydrogenase 1 alpha subcomplex, assembly factor 1
3	0.691345	4694	NDUFA1	NADH dehydrogenase 1 alpha subcomplex, 1, 7.5 kDa
4	0.688717	4726	NDUFS6	NADH dehydrogenase Fe-S protein 6, 13 kDa
5	0.686216	4719	NDUFS1	NADH dehydrogenase Fe-S protein 1, 75 kDa
6	0.685317	4720	NDUFS2	NADH dehydrogenase Fe-S protein 2, 49 kDa
7	0.68423	4709	NDUFB3	NADH dehydrogenase 1 beta subcomplex, 3, 12 kDa
8	0.681527	4729	NDUFV2	NADH dehydrogenase flavoprotein 2, 24 kDa
9	0.676788	4724	NDUFS4	NADH dehydrogenase Fe-S protein 4, 18 kDa
10	0.65894	79133	C20orf7	chromosome 20 open reading frame 7
11	0.656693	126328	NDUFA11	NADH dehydrogenase 1 alpha subcomplex, 11, 14.7 kDa
12	0.656337	91942	NDUFAF2	NADH dehydrogenase 1 alpha subcomplex, assembly factor 2
13	0.656292	55572	FOXRED1	FAD-dependent oxidoreductase domain containing 1
14	0.656166	25915	NDUFAF3	NADH dehydrogenase 1 alpha subcomplex, assembly factor 3
15	0.656105	80224	NUBPL	nucleotide binding protein-like
16	0.115148	4714	NDUFB8	NADH dehydrogenase 1 beta subcomplex, 8, 19 kDa
17	0.109928	1329	COX5B	cytochrome c oxidase subunit Vb
18	0.090152	2108	ETFA	electron-transfer-flavoprotein, alpha polypeptide
19	0.087915	4722	NDUFS3	NADH dehydrogenase Fe-S protein 3, 30 kDa
20	0.083753	6390	SDHB	succinate dehydrogenase complex, subunit B, iron sulfur (Ip)
21	0.078834	1743	DLST	dihydrolipoamide S-succinyltransferase (E2 component of 2-oxo-glutarate complex)
22	0.070645	54205	CYCS	cytochrome c, somatic
23	0.068273	509	ATP5C1	ATP synthase, H+ transporting, mitochondrial F1 complex, gamma polypeptide 1
24	0.067436	506	ATP5B	ATP synthase, H+ transporting, mitochondrial F1 complex, beta polypeptide
25	0.06552	1345	COX6C	cytochrome c oxidase subunit VIc
26	0.061017	25828	TXN2	thioredoxin 2
27	0.060686	6391	SDHC	succinate dehydrogenase complex, subunit C, integral membrane protein, 15 kDa
28	0.060526	50	ACO2	aconitase 2, mitochondrial
29	0.060351	4713	NDUFB7	NADH dehydrogenase 1 beta subcomplex, 7, 18 kDa
30	0.058394	740	MRPL49	mitochondrial ribosomal protein L49

Conclusions

In our work, we carried out a comprehensive mammalian mitochondrial proteomic study through a three-step approach. We compiled an extensive inventory of mammalian mitochondrial proteins by combining 23 genomic-scale datasets. Our inventory showed considerable overlap with MitoPred and MotoCarta, the two best existing mitochondrial databases, but held greater sensitivity and lower false discovery rate. The high-confidence proteins along with the middle-confidence proteins provide a narrowed scope of candidates for mitochondrial proteins with relatively high possibility. We also constructed a comprehensive and high quality weighted FLN among mitochondrial proteins through integrating 15 heterogeneous functional features. With the comprehensive features integrated, the FLN is less biased towards single evidence and can be more accurate and with higher coverage. The high coverage and linkage density is essential to the successful utilization of the FLN for disease gene prediction and prioritization. Thus the FLN we presented can provide valuable resource for researches on mammalian mitochondrial proteomics. One important utility of the FLN is for mitochondrial disease genes predicting and prioritizing. The top-ranking candidates for the mitochondrial diseases reported in this work represent the highly possible risk genes for the specific disease, which provide a narrowed spectrum of suspects for these important human diseases and will promote the disease-genes association studies and offer clues for new treatment strategies. Moreover, with the identification of new disease genes, these results can be further integrated into our framework for better disease gene predictions. Furthermore, a web-based database MitProNet was also implemented. Researchers can easily locate a gene of interest and analyze those tightly associated genes. The visualization of local FLN around the gene can be a rapid and convenient approach to inspect the relationship of those associated genes. The disease related network present an overall landscape of the relationship of known and candidate genes. The complete set of mitochondrial genes and FLN are also provided. Thus the FLN and the disease candidates implemented in MitProNet would facilitate the researches in mitochondria and diseases related to this important organelle.

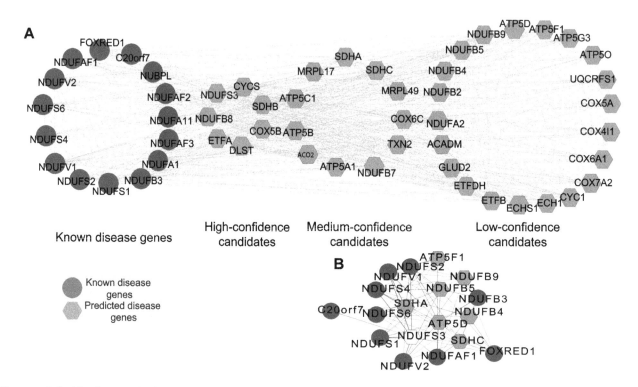

Figure 6. Prioritization results for mitochondrial complex I deficiency. (A) A hypothetical FLN of mitochondrial complex I deficiency. The FLN is comprising of known disease genes (highlighted in red) annotated in OMIM and predicted disease genes (highlighted in greed). The candidates are classified into three levels (high-confidence, middle-confidence and low-confidence) according to their ranking scores. (B) The functional linkage sub-network among the candidate NDUFS3 that has a top score on ranking algorithm for mitochondrial complex I deficiency.

Materials and Methods

An inventory of mammalian mitochondrial proteins

To reduce redundancy, the proteins were transformed into corresponding genes identified unique by Entrez GeneID.

Gold standard sets. The GSP dataset was comprised of human mitochondrial proteins that were curated from the MitoP2 database [88]. To avoid contamination, we only used proteins with supports of sublocalization experiments, and excluded those characterized solely by large-scale proteomic studies. The GSN, on the other hand, was selected from proteins located in other cellular compartments according to Gene Ontology (GO) annotations. For those proteins with multiple subcellular locations, we excluded those with subcellular location in mitochondrial compo-

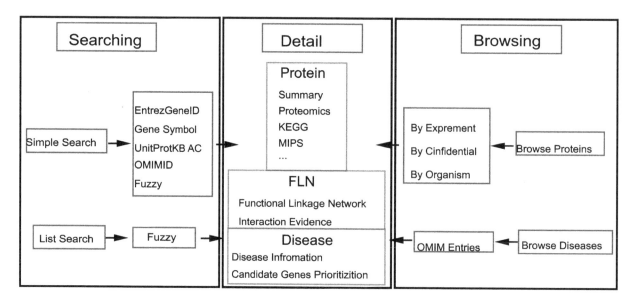

Figure 7. System architecture and main contents of MitProNet. MitProNet is composed of three sections including mitochondrial protein part lists, annotations of mitochondrial protein and disease information.

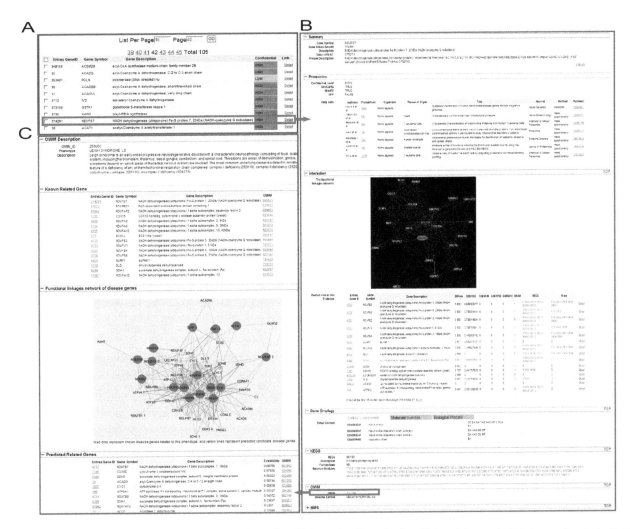

Figure 8. Web pages in MitProNet. (A) A list page of mitochondrial proteins. The mitochondrial proteins can be listed according to proteomic datasets, confidence levels and organisms, respectively. (B) The outcome page for the query protein NDUFS7, an annotated disease gene for Leigh syndrome. The page provides a brief summary of the query protein, subcellular localization evidences and a FLN among the query protein. Moreover, the query protein is annotated according to the information collected from their original sources including GO, KEGG, MIPS and OMIM. (C) The prioritization results for Leigh syndrome. The result page includes a brief description for this phenotype, disease genes and a FLN among these genes. The disease genes are listed dividedly as the known genes and the candidates that are ordered by these ranking scores.

nents or locations from the GSN. As a result, the GSP dataset contained 553 proteins, while the GSN dataset consisted of 9950 non-mitochondrial proteins.

Cross validation and evaluation of machine-learning algorithms. When training the classifiers, the 23 proteomic experiments datasets were considered as 'features'. And for each feature, we assigned a score 1 to each human gene product if the product exists in the dataset, or 0 otherwise. We used the 10-fold cross validation to evaluate prediction performance of these machine-learning classifiers [89]. For each machine-learning classifier, at first, both the GSP and GSN were randomly partitioned into ten equal-sized folds. After that, the machine-learning classifier was trained on nine folds and the remaining one fold was used as a test set to identify the number of positives and negatives. This was repeated ten times with a different fold used for testing each time.

Calculating sensitivity and false discovery rate. Sensitivity is defined as TP/(TP+FN), where TP is the number of true positives and FN is the number of false negatives, respectively, estimated from gold-standard sets. The false discovery

rate (FDR) is the proportion of all predictions that are false; FDR = FP/(FP+TP), where FP represent the number of false positives [29].

Construction of mitochondrial FLN through data integration

To carry out the construction of FLN, each dataset should be transformed into protein pairs with functional linkage. The preprocessing is described in supplementary methods (Method S1) in detail.

Gold standard sets. In this study, we downloaded KEGG pathway [60] and MIPS complex [90] about mitochondrion. The GSP were defined as mitochondrial protein pairs sharing the same KEGG pathway or existing in the same MIPS complex, while the GSN were defined as mitochondrial protein pairs both annotated by KEGG pathway or MIPS complex terms but that do not share any term.

Naïve Bayes for mammalian mitochondrial FLN construction. According the Bayesian theorem, the prior odds

(O_{prior}) of finding a gene pair with functional linkage could be calculated as:

$$O_{\text{prior}} = \frac{P_{pos}}{P_{neg}} \qquad (1)$$

where P_{pos} is the probability that a gene pair functionally relates within all the possible gene pairs while the P_{neg} stands for the probability that a gene pair isn't functionally related. When considering the given n evidences (E) that stands for the functional features, the posterior odds ($O_{posterior}$) of a functional linkage gene pair could be computed as:

$$O_{posterior} = \frac{P(positive|E_1,...,E_n)}{P(negtive|E_1,...,E_n)} = O_{prior} \times LR(E_1,...E_n) \quad (2)$$

where $LR(E_1,...,E_n)$ is the likelihood ratio of the n evidences(E). From Equation 1 and Equation 2, the LR could be calculated as:

$$LR(E_1,...,E_n) = \frac{P(E_1,...,E_n|pos)}{P(E_1,...,E_n|neg)} \qquad (3)$$

If we assume that the evidences are conditionally independent, the composite LR can be calculated simply as following:

$$LR(E_1,...,E_n) = \prod_{i=1}^{n} \frac{P(E_i|pos)}{P(E_i|neg)} \qquad (4)$$

And Equation 4 can also be written as the following:

$$LR(E_1,...,E_n) = \prod_{i=1}^{n} LR(E_i) \qquad (5)$$

Cross validation and cutoff selection. We employed the five-fold cross-validation against the golden standard datasets to evaluate the overall prediction performance under different LR cutoffs. First, both the GSP and GSN datasets were randomly partitioned into five equal-sized folds. After that, the naïve Bayesian classifier was trained on four folds and the remaining one fold was used as a test set to identify the number of positives and negatives. This was repeated five times with a different fold used for testing each time. We used the ratio of true positive to false positive (TP/FP) and the sensitivity to measure the prediction accuracy.

Ranking the mitochondrial disease gene
Average Adjacency Ranking. Given a particular mitochondrial disease, firstly, m genes were extracted randomly from known disease-related mitochondrial genes as seed gene set, and the rest of the genes were treated as unknown ones. Then for every other gene, we compute the adjacency to the m seeds. This process was repeated one hundred times with random samplings of the seed set. Lastly, we calculated the average adjacency with a given disease for each gene:

$$W_i = \frac{1}{n_i} \sum_{K=1}^{n_i} \sum_{j=1}^{m} W_{ij} \qquad (6)$$

where W_i represents the weight of each gene associate with a given

disease and j represents the seed genes, and W_{ij} is the functional linkage weights connecting gene i and seed gene j. n_i is the number of times gene i was not a member in the seed set and k is the iteration number.

PageRank with Priors. PRP mimics a random jump procedure in network, which start with known disease-related genes and randomly jump to candidate genes. When the system jump to a candidate gene, system can continue jumping to other candidate genes or jump back to known disease-related genes and then restart the procedure. After enough jumping, PRP scores each candidate gene based on the probability that system jump to the gene. The iterative stationary probability is:

$$\pi(v)^{(i+1)} = (1-\beta)\left\{ \sum_{u=1}^{d_{in}(v)} p(v|u)\pi^{(i)}(u) \right\} + \beta p_v \qquad (7)$$

where p_v represents the "prior bias" which means the probability to start with a particular genes. $p_v = 1/|R|$ if v in root node set R (known disease-related gene set); $p_v = 0$ otherwise. β is empirically defined on [0, 1], represents a "back probability" which means the probability to jump back to the root node in each step. $d_{in}(v)$ is the in-degree of v. $p(v|u)$ is the probability of arriving node v from u.

K-step Markov. KSM also mimics a random jump procedure that start with disease-related genes and ends after fixed K steps. It computes the relative probability that the system will spend time at any particularnode given that it starts in a set of roots R and ends after K Steps [91]. K keeps a balance between distributions of candidate genes 'biased' toward known disease-related genes. With a larger K, system gets a more steady distribution of candidate genes [75]. The to compute the K-Step Markov importance is:

$$I(t|R) = \left[AP_R + A^2 P_R...A^k P_R \right] \qquad (8)$$

Where A is the transition probability matrix of network, p_R is an vector of initial probabilities for the root set R (known disease genes set), k is the probability transition steps and $I(t|R)$ is the t-th entry in this sum vector.

Heat Kernel Diffusion Ranking. The Heat Kernel Diffusion Ranking approach ranks the candidate genes by diffusing the signal of 'seeds' to the candidate genes through the network based on the weighted edges [92]. The network can be represented as a weighted, simple graph G, where genes are nodes and weighted linkages are weighted edges. Given a graph G, let A be the Adjacency matrix where $a_{ij} = w_{ij}$ and then D can be defined as $D = diag(a_i) = \sum_{j=1}^{n} a_{ij}$. The transition probability matrix W of a random walk on G is defined as $W = D^{-1} A$. Consider $L = I-W$. Given a parameter α, establishing the diffusion rate, and a preference vector p_0, expressing the initial relevance score of each node, the ranking p_α is given by

$$P_\alpha = P_0 (I + \frac{-\alpha}{N} L)^N \qquad (9)$$

where N is the number of iterations.

Evaluation scheme
Leave-one-out cross-validation was conducted to evaluate performance of four ranking algorithms based on four networks. Then, based on the sensitivity and 1-specificity, ROC curves were drawn. In order to find out optimal performance of HKDR, PRW

and SKM, a set of different parameters were empirically selected: HKDR with n = 2, 3, 4, 5, 6, 7; PRW with β = 0.01, 0.05, 0.1, 0.2, 0.3, 0.4,0.5,0.6, 0.7,0.8, 0.9,0.95; SKM with K = 2, 3, 4, 5, 6, 7, 8, 9.

Supporting Information

Figure S1   Measurement of the contributions of diverse datasets for constructing the FLN. (A) GO semantic similarity. (B) Four microarray experiment datasets GSE1133, GSE4330, GSE6210, GSE4726. (C) Protein expression profiles. (D) Protein-protein interaction (PPI), Rosetta Stone (RS), domain-domain interaction (DDI), diseases involvements (DI), genetic interaction(GI). (E) Operons. (F) Phylogenetic profiles. (G) Phenotypic semantic similarity. (H) Shared domains.
(EPS)

Table S1   Ten-fold cross-validation results of machine-learning classifiers in Weka.
(DOC)

Table S2   List of high-confidence mammalian mitochondrial proteins.
(XLS)

Table S3   Coverage of datasets on gold standard set.
(DOC)

Table S4   Pearson correlation coefficients between high coverage datasets on the gold standard set.
(DOC)

Table S5   Optimal parameters and corresponding performances of four algorithms on four networks.
(DOC)

Table S6   Organisms used for phylogenetic profiles.
(XLS)

Method S1   Data source and processing methods of the 11 genomic features to generate FLN.
(DOC)

Author Contributions

Conceived and designed the experiments: XL. Performed the experiments: JW JY. Analyzed the data: JW JY SM YH XH YT XC XL CB. Wrote the paper: JW JY XL. Collected datasets: JW JY SM YH. Constructed the web server and implemented database: JW. Read and approved the final manuscript: JW JY SM YH XH YT XC XL CB.

References

1. Chan DC (2006) Mitochondria: dynamic organelles in disease, aging, and development. Cell 125: 1241–1252.
2. Facecchia K, Fochesato LA, Ray SD, Stohs SJ, Pandey S (2011) Oxidative toxicity in neurodegenerative diseases: role of mitochondrial dysfunction and therapeutic strategies. J Toxicol 2011: 683–728.
3. Shenouda SM, Widlansky ME, Chen K, Xu G, Holbrook M, et al. (2011) Altered mitochondrial dynamics contributes to endothelial dysfunction in diabetes mellitus. Circulation 124: 444–453.
4. Traish AM, Abdallah B, Yu G (2011) Androgen deficiency and mitochondrial dysfunction: implications for fatigue, muscle dysfunction, insulin resistance, diabetes, and cardiovascular disease. Hormone Molecular Biology and Clinical Investigation 8: 431–444.
5. Johri A, Beal MF (2012) Mitochondrial dysfunction in neurodegenerative diseases. Journal of Pharmacology and Experimental Therapeutics 342: 619–630.
6. Salminen A, Ojala J, Kaarniranta K, Kauppinen A (2012) Mitochondrial dysfunction and oxidative stress activate inflammasomes: impact on the aging process and age-related diseases. Cell Mol Life Sci 69: 2999–3013.
7. Distler AM, Kerner J, Hoppel CL (2008) Proteomics of mitochondrial inner and outer membranes. Proteomics 8: 4066–4082.
8. Smith AC, Robinson AJ (2009) MitoMiner, an integrated database for the storage and analysis of mitochondrial proteomics data. Mol Cell Proteomics 8: 1324–1337.
9. Gregersen N, Hansen J, Palmfeldt J (2012) Mitochondrial proteomics–a tool for the study of metabolic disorders. J Inherit Metab Dis 35: 715–726.
10. Elstner M, Andreoli C, Klopstock T, Meitinger T, Prokisch H (2009) The mitochondrial proteome database: MitoP2. Methods Enzymol 457: 3–20.
11. Cotter D, Guda P, Fahy E, Subramaniam S (2004) MitoProteome: mitochondrial protein sequence database and annotation system. Nucleic Acids Res 32: D463–467.
12. Catalano D, Licciulli F, Turi A, Grillo G, Saccone C, et al. (2006) MitoRes: a resource of nuclear-encoded mitochondrial genes and their products in Metazoa. BMC Bioinformatics 7: 36.
13. Basu S, Bremer E, Zhou C, Bogenhagen DF (2006) MiGenes: a searchable interspecies database of mitochondrial proteins curated using gene ontology annotation. Bioinformatics 22: 485–492.
14. Pagliarini DJ, Calvo SE, Chang B, Sheth SA, Vafai SB, et al. (2008) A mitochondrial protein compendium elucidates complex I disease biology. Cell 134: 112–123.
15. Johnson DT, Harris RA, French S, Blair PV, You J, et al. (2007) Tissue heterogeneity of the mammalian mitochondrial proteome. Am J Physiol Cell Physiol 292: C689–697.
16. Ashburner M, Ball CA, Blake JA, Botstein D, Butler H, et al. (2000) Gene ontology: tool for the unification of biology. The Gene Ontology Consortium. Nat Genet 25: 25–29.
17. Oliver S (2000) Guilt-by-association goes global. Nature 403: 601–603.
18. Shen-Orr SS, Milo R, Mangan S, Alon U (2002) Network motifs in the transcriptional regulation network of Escherichia coli. Nat Genet 31: 64–68.
19. Stelzl U, Worm U, Lalowski M, Haenig C, Brembeck FH, et al. (2005) A human protein-protein interaction network: a resource for annotating the proteome. Cell 122: 957–968.
20. Zhang J, Xiang Y, Ding L, Keen-Circle K, Borlawsky TB, et al. (2010) Using gene co-expression network analysis to predict biomarkers for chronic lymphocytic leukemia. BMC Bioinformatics 11 Suppl 9: S5.
21. Snel B, Lehmann G, Bork P, Huynen MA (2000) STRING: a web-server to retrieve and display the repeatedly occurring neighbourhood of a gene. Nucleic Acids Res 28: 3442–3444.
22. Joshi-Tope G, Gillespie M, Vastrik I, D'Eustachio P, Schmidt E, et al. (2005) Reactome: a knowledgebase of biological pathways. Nucleic acids research 33: D428–D432.
23. Stark C, Breitkreutz BJ, Reguly T, Boucher L, Breitkreutz A, et al. (2006) BioGRID: a general repository for interaction datasets. Nucleic Acids Res 34: D535–539.
24. Franke L, van Bakel H, Fokkens L, de Jong ED, Egmont-Petersen M, et al. (2006) Reconstruction of a functional human gene network, with an application for prioritizing positional candidate genes. Am J Hum Genet 78: 1011–1025.
25. Linghu B, Snitkin ES, Hu Z, Xia Y, Delisi C (2009) Genome-wide prioritization of disease genes and identification of disease-disease associations from an integrated human functional linkage network. Genome Biol 10: R91.
26. Reja R, Venkatakrishnan AJ, Lee J, Kim BC, Ryu JW, et al. (2009) MitoInteractome: mitochondrial protein interactome database, and its application in 'aging network' analysis. BMC Genomics 10 Suppl 3: S20.
27. Gu Z, Li J, Gao S, Gong M, Wang J, et al. (2011) InterMitoBase: an annotated database and analysis platform of protein-protein interactions for human mitochondria. BMC Genomics 12: 335.
28. Meisinger C, Sickmann A, Pfanner N (2008) The mitochondrial proteom: From inventory to function. Cell 134: 22–24.
29. Calvo S, Jain M, Xie X, Sheth SA, Chang B, et al. (2006) Systematic identification of human mitochondrial disease genes through integrative genomics. Nat Genet 38: 576–582.
30. Taylor SW, Fahy E, Zhang B, Glenn GM, Warnock DE, et al. (2003) Characterization of the human heart mitochondrial proteome. Nat Biotechnol 21: 281–286.
31. Rezaul K, Wu L, Mayya V, Hwang SI, Han D (2005) A systematic characterization of mitochondrial proteome from human T leukemia cells. Mol Cell Proteomics 4: 169–181.
32. Xie J, Techritz S, Haebel S, Horn A, Neitzel H, et al. (2005) A two-dimensional electrophoretic map of human mitochondrial proteins from immortalized lymphoblastoid cell lines: a prerequisite to study mitochondrial disorders in patients. Proteomics 5: 2981–2999.
33. Ozawa T, Sako Y, Sato M, Kitamura T, Umezawa Y (2003) A genetic approach to identifying mitochondrial proteins. Nat Biotechnol 21: 287–293.
34. Mootha VK, Bunkenborg J, Olsen JV, Hjerrild M, Wisniewski JR, et al. (2003) Integrated analysis of protein composition, tissue diversity, and gene regulation in mouse mitochondria. Cell 115: 629–640.
35. Jin J, Davis J, Zhu D, Kashima DT, Leroueil M, et al. (2007) Identification of novel proteins affected by rotenone in mitochondria of dopaminergic cells. BMC Neurosci 8: 67.
36. Kislinger T, Cox B, Kannan A, Chung C, Hu P, et al. (2006) Global survey of organ and organelle protein expression in mouse: combined proteomic and transcriptomic profiling. Cell 125: 173–186.

37. Da Cruz S, Xenarios I, Langridge J, Vilbois F, Parone PA, et al. (2003) Proteomic analysis of the mouse liver mitochondrial inner membrane. J Biol Chem 278: 41566–41571.

38. Forner F, Foster LJ, Campanaro S, Valle G, Mann M (2006) Quantitative proteomic comparison of rat mitochondria from muscle, heart, and liver. Mol Cell Proteomics 5: 608–619.

39. Reifschneider NH, Goto S, Nakamoto H, Takahashi R, Sugawa M, et al. (2006) Defining the mitochondrial proteomes from five rat organs in a physiologically significant context using 2D blue-native/SDS-PAGE. J Proteome Res 5: 1117–1132.

40. Palmfeldt J, Vang S, Stenbroen V, Pedersen CB, Christensen JH, et al. (2009) Mitochondrial proteomics on human fibroblasts for identification of metabolic imbalance and cellular stress. Proteome Sci 7: 20.

41. Lefort N, Yi Z, Bowen B, Glancy B, De Filippis EA, et al. (2009) Proteome profile of functional mitochondria from human skeletal muscle using one-dimensional gel electrophoresis and HPLC-ESI-MS/MS. J Proteomics 72: 1046–1060.

42. Bousette N, Kislinger T, Fong V, Isserlin R, Hewel JA, et al. (2009) Large-scale characterization and analysis of the murine cardiac proteome. J Proteome Res 8: 1887–1901.

43. Fang X, Wang W, Yang L, Chandrasekaran K, Kristian T, et al. (2008) Application of capillary isotachophoresis-based multidimensional separations coupled with electrospray ionization-tandem mass spectrometry for characterization of mouse brain mitochondrial proteome. Electrophoresis 29: 2215–2223.

44. Zhang J, Li X, Mueller M, Wang Y, Zong C, et al. (2008) Systematic characterization of the murine mitochondrial proteome using functionally validated cardiac mitochondria. Proteomics 8: 1564–1575.

45. Deng WJ, Nie S, Dai J, Wu JR, Zeng R (2010) Proteome, phosphoproteome, and hydroxyproteome of liver mitochondria in diabetic rats at early pathogenic stages. Mol Cell Proteomics 9: 100–116.

46. Wu L, Hwang SI, Rezaul K, Lu LJ, Mayya V, et al. (2007) Global survey of human T leukemic cells by integrating proteomics and transcriptomics profiling. Mol Cell Proteomics 6: 1343–1353.

47. Catherman AD, Durbin KR, Ahlf DR, Early BP, Fellers RT, et al. (2013) Large-scale Top-down Proteomics of the Human Proteome: Membrane Proteins, Mitochondria, and Senescence. Molecular & Cellular Proteomics 12: 3465–3473.

48. Hansen J, Palmfeldt J, Vang S, Corydon TJ, Gregersen N, et al. (2011) Quantitative proteomics reveals cellular targets of celastrol. PLoS One 6: e26634.

49. Chappell NP, Teng PN, Hood BL, Wang G, Darcy KM, et al. (2012) Mitochondrial proteomic analysis of cisplatin resistance in ovarian cancer. J Proteome Res 11: 4605–4614.

50. Chen XL, Cui ZY, Wei SS, Hou JJ, Xie ZS, et al. (2013) Chronic high glucose induced INS-1 beta cell mitochondrial dysfunction: A comparative mitochondrial proteome with SILAC. Proteomics 13: 3030–3039.

51. Hall M, Frank E, Holmes G, Pfahringer B, Reutemann P, et al. (2009) The WEKA data mining software: an update. ACM SIGKDD Explorations Newsletter 11: 10–18.

52. Freund Y, Schapire RE. Experiments with a new boosting algorithm; 1996. 148–156.

53. Guda C, Fahy E, Subramaniam S (2004) MITOPRED: a genome-scale method for prediction of nucleus-encoded mitochondrial proteins. Bioinformatics 20: 1785–1794.

54. Keshava Prasad TS, Goel R, Kandasamy K, Keerthikumar S, Kumar S, et al. (2009) Human Protein Reference Database–2009 update. Nucleic Acids Res 37: D767–772.

55. Brown KR, Jurisica I (2005) Online predicted human interaction database. Bioinformatics 21: 2076–2082.

56. Stein A, Ceol A, Aloy P (2011) 3did: identification and classification of domain-based interactions of known three-dimensional structure. Nucleic Acids Res 39: D718–723.

57. Hunter S, Jones P, Mitchell A, Apweiler R, Attwood TK, et al. (2012) InterPro in 2011: new developments in the family and domain prediction database. Nucleic Acids Res 40: D306–312.

58. Bowers PM, Pellegrini M, Thompson MJ, Fierro J, Yeates TO, et al. (2004) Prolinks: a database of protein functional linkages derived from coevolution. Genome Biol 5: R35.

59. Weiller GF (1998) Phylogenetic profiles: a graphical method for detecting genetic recombinations in homologous sequences. Mol Biol Evol 15: 326–335.

60. Ogata H, Goto S, Sato K, Fujibuchi W, Bono H, et al. (1999) KEGG: Kyoto encyclopedia of genes and genomes. Nucleic acids research 27: 29–34.

61. Cherry JM, Ball C, Weng S, Juvik G, Schmidt R, et al. (1997) Genetic and physical maps of Saccharomyces cerevisiae. Nature 387: 67–73.

62. Smith CL, Goldsmith CA, Eppig JT (2005) The Mammalian Phenotype Ontology as a tool for annotating, analyzing and comparing phenotypic information. Genome Biol 6: R7.

63. Su AI, Wiltshire T, Batalov S, Lapp H, Ching KA, et al. (2004) A gene atlas of the mouse and human protein-encoding transcriptomes. Proc Natl Acad Sci U S A 101: 6062–6067.

64. Edgar R, Domrachev M, Lash AE (2002) Gene Expression Omnibus: NCBI gene expression and hybridization array data repository. Nucleic Acids Res 30: 207–210.

65. Siddiqui AS, Khattra J, Delaney AD, Zhao Y, Astell C, et al. (2005) A mouse atlas of gene expression: large-scale digital gene-expression profiles from precisely defined developing C57BL/6J mouse tissues and cells. Proc Natl Acad Sci U S A 102: 18485–18490.

66. Vianna CR, Huntgeburth M, Coppari R, Choi CS, Lin J, et al. (2006) Hypomorphic mutation of PGC-1beta causes mitochondrial dysfunction and liver insulin resistance. Cell Metab 4: 453–464.

67. Hamosh A, Scott AF, Amberger JS, Bocchini CA, McKusick VA (2005) Online Mendelian Inheritance in Man (OMIM), a knowledgebase of human genes and genetic disorders. Nucleic Acids Res 33: D514–517.

68. Mao F, Dam P, Chou J, Olman V, Xu Y (2009) DOOR: a database for prokaryotic operons. Nucleic Acids Res 37: D459–463.

69. Linghu B, Franzosa EA, Xia Y (2013) Construction of functional linkage gene networks by data integration. Methods Mol Biol 939: 215–232.

70. von Mering C, Krause R, Snel B, Cornell M, Oliver SG, et al. (2002) Comparative assessment of large-scale data sets of protein-protein interactions. Nature 417: 399–403.

71. Bordych C, Eisenhut M, Pick TR, Kuelahoglu C, Weber AP (2013) Co-expression analysis as tool for the discovery of transport proteins in photorespiration. Plant Biol (Stuttg) 15: 686–693.

72. Andersson SG, Zomorodipour A, Andersson JO, Sicheritz-Ponten T, Alsmark UC, et al. (1998) The genome sequence of Rickettsia prowazekii and the origin of mitochondria. Nature 396: 133–140.

73. Scott MS, Barton GJ (2007) Probabilistic prediction and ranking of human protein-protein interactions. BMC Bioinformatics 8: 239.

74. Guan Y, Myers CL, Lu R, Lemischka IR, Bult CJ, et al. (2008) A genomewide functional network for the laboratory mouse. PLoS Comput Biol 4: e1000165.

75. Chen J, Aronow BJ, Jegga AG (2009) Disease candidate gene identification and prioritization using protein interaction networks. BMC Bioinformatics 10: 73.

76. Goncalves JP, Francisco AP, Moreau Y, Madeira SC (2012) Interactogeneous: disease gene prioritization using heterogeneous networks and full topology scores. PLoS One 7: e49634.

77. Kirby DM, Crawford M, Cleary MA, Dahl HH, Dennett X, et al. (1999) Respiratory chain complex I deficiency: an underdiagnosed energy generation disorder. Neurology 52: 1255–1264.

78. Carroll J, Fearnley IM, Shannon RJ, Hirst J, Walker JE (2003) Analysis of the subunit composition of complex I from bovine heart mitochondria. Mol Cell Proteomics 2: 117–126.

79. Murray J, Zhang B, Taylor SW, Oglesbee D, Fahy E, et al. (2003) The subunit composition of the human NADH dehydrogenase obtained by rapid one-step immunopurification. J Biol Chem 278: 13619–13622.

80. Loeffen JL, Triepels RH, van den Heuvel LP, Schuelke M, Buskens CA, et al. (1998) cDNA of eight nuclear encoded subunits of NADH:ubiquinone oxidoreductase: human complex I cDNA characterization completed. Biochem Biophys Res Commun 253: 415–422.

81. Haack TB, Haberberger B, Frisch EM, Wieland T, Iuso A, et al. (2012) Molecular diagnosis in mitochondrial complex I deficiency using exome sequencing. J Med Genet 49: 277–283.

82. Lomax MI, Hsieh CL, Darras BT, Francke U (1991) Structure of the human cytochrome c oxidase subunit Vb gene and chromosomal mapping of the coding gene and of seven pseudogenes. Genomics 10: 1–9.

83. Dwyer TM, Mortl S, Kemter K, Bacher A, Fauq A, et al. (1999) The intraflavin hydrogen bond in human electron transfer flavoprotein modulates redox potentials and may participate in electron transfer. Biochemistry 38: 9735–9745.

84. Horvath R (2012) Update on clinical aspects and treatment of selected vitamin-responsive disorders II (riboflavin and CoQ 10). J Inherit Metab Dis 35: 679–687.

85. Wolfe LA, He M, Vockley J, Payne N, Rhead W, et al. (2010) Novel ETF dehydrogenase mutations in a patient with mild glutaric aciduria type II and complex II-III deficiency in liver and muscle. J Inherit Metab Dis 33 Suppl 3: S481–487.

86. Benit P, Slama A, Cartault F, Giurgea I, Chretien D, et al. (2004) Mutant NDUFS3 subunit of mitochondrial complex I causes Leigh syndrome. J Med Genet 41: 14–17.

87. Triepels R, Smeitink J, Loeffen J, Smeets R, Trijbels F, et al. (2000) Characterization of the human complex I NDUFB7 and 17.2-kDa cDNAs and mutational analysis of 19 genes of the HP fraction in complex I-deficient-patients. Hum Genet 106: 385–391.

88. Prokisch H, Andreoli C, Ahting U, Heiss K, Ruepp A, et al. (2006) MitoP2: the mitochondrial proteome database–now including mouse data. Nucleic Acids Res 34: D705–711.

89. Kohavi R. A study of cross-validation and bootstrap for accuracy estimation and model selection; 1995. 1137–1145.

90. Mewes HW, Frishman D, Guldener U, Mannhaupt G, Mayer K, et al. (2002) MIPS: a database for genomes and protein sequences. Nucleic Acids Res 30: 31–34.

91. White S, Smyth P. Algorithms for estimating relative importance in networks; 2003. ACM. 266–275.

92. Nitsch D, Goncalves JP, Ojeda F, de Moor B, Moreau Y (2010) Candidate gene prioritization by network analysis of differential expression using machine learning approaches. BMC Bioinformatics 11: 460.

4

Relating Diseases by Integrating Gene Associations and Information Flow through Protein Interaction Network

Mehdi Bagheri Hamaneh, Yi-Kuo Yu*

National Center for Biotechnology Information, National Library of Medicine, National Institutes of Health, Bethesda, MD, United States of America

Abstract

Identifying similar diseases could potentially provide deeper understanding of their underlying causes, and may even hint at possible treatments. For this purpose, it is necessary to have a similarity measure that reflects the underpinning molecular interactions and biological pathways. We have thus devised a network-based measure that can partially fulfill this goal. Our method assigns weights to all proteins (and consequently their encoding genes) by using information flow from a disease to the protein interaction network and back. Similarity between two diseases is then defined as the cosine of the angle between their corresponding weight vectors. The proposed method also provides a way to suggest disease-pathway associations by using the weights assigned to the genes to perform enrichment analysis for each disease. By calculating pairwise similarities between 2534 diseases, we show that our disease similarity measure is strongly correlated with the probability of finding the diseases in the same disease family and, more importantly, sharing biological pathways. We have also compared our results to those of MimMiner, a text-mining method that assigns pairwise similarity scores to diseases. We find the results of the two methods to be complementary. It is also shown that clustering diseases based on their similarities and performing enrichment analysis for the cluster centers significantly increases the term association rate, suggesting that the cluster centers are better representatives for biological pathways than the diseases themselves. This lends support to the view that our similarity measure is a good indicator of relatedness of biological processes involved in causing the diseases. Although not needed for understanding this paper, the raw results are available for download for further study at ftp://ftp.ncbi.nlm.nih.gov/pub/qmbpmn/DiseaseRelations/.

Editor: Ozlem Keskin, Koç University, Turkey

Funding: This work was supported by the Intramural Research Program of the National Library of Medicine at the National Institutes of Health. The funder had no role in study design, data collection and analysis, decision to publish, or preparation of the manuscript.

Competing Interests: The authors have declared that no competing interests exist.

* Email: yyu@ncbi.nlm.nih.gov

Introduction

Discovering disease-disease similarities could be helpful in better understanding the underlying causes of diseases and may even be useful for therapeutic purposes, as similar diseases might have similar drug targets. Disease similarities, of course, can be investigated at different levels and from different perspectives. Phenotype similarity is perhaps the most obvious way to classify diseases. This is usually the approach taken in many disease databases including Medical Subject Headings (MESH) [1] and Disease Ontology (DO) [2]. Although this method of classification is very useful, other metrics of similarity could significantly improve our understanding of the biological processes involved in similar diseases. For diseases with genetic causes, disease-disease associations could also be based on whether or not two diseases are associated with the same genes. This would extend the concept of similarity, because different phenotypes could be related to the same set of genes. However, there are similar diseases that do not share gene associations. A similarity metric that could suggest deeper relationships between diseases is therefore desirable.

Network-based similarity measures have gained popularity over the last few years. For example, Goh et al. [3] introduced a human disease network by treating the diseases as nodes and by linking the diseases if they had at least one shared gene association. They showed that their network was clustered according to disease classes, although they did not define a quantitative metric to find the distance between diseases in a given pair. Using a similar approach, Lee et al. [4] constructed a metabolic network, where nodes (diseases) were connected if mutated enzymes associated with them catalyzed adjacent metabolic reactions. They found that connected diseases had higher comorbidity than those without any link between them. Zhang et al. [5] constructed an extended human disease network by adding new gene associations (and so new disease links) inferred based on protein-protein interaction data. Hidalgo et al. [6] created a phenotypic disease network with phenotypes as nodes. The phenotypes were then linked if they had significant comorbidity. They used two different (but related) comorbidity measures based on the disease history data of a large population of patients. On the other hand, Linghu et al. [7] used a network in which the nodes represent genes. They integrated different functional associations, including protein-protein interactions, using a Bayes classifier whose output was then used to weight the links between the genes based on their overall functional associations. For 110 diseases, the disease genes were then prioritized according to their associations with previously known disease genes. They also calculated a measure of similarity between any two diseases based on the mutual predictability of known gene associations of one disease from the known genes related to the other disease. In another study, Suthram *et al.* [8] used mRNA expression and protein-protein interaction networks to find quantitative similarities between 54 human diseases. Mehren *et al.* [9] developed a gene-disease association database by integrating several sources and classified diseases using graph clustering algorithms. They found common functional modules for related diseases, a concept that has been reported in most network-based studies of human diseases [10]. In a recent study, Zitnik *et al.* [11] used a data mining approach to discover disease-disease associations. They introduced relation matrices describing the associations between different types of objects (genes and diseases, for example) and minimized an objective function to factorize these matrices to ones with lower dimensions, consequently clustering the diseases. Zitnik and co-workers used several types of data as constraints in their objective function including protein-protein interactions, although they concluded that these interactions were not as essential as other data in their analysis. Gulbahce *et al.* [12] created a viral disease network and introduced a local impact hypothesis stating that in this network genes associated with virally implicated diseases are located near viral targets. MimMiner, introduced by van Driel *et al.* [13], is another method to relate diseases. Unlike previous approaches, MimMiner uses text mining to assign pairwise similarity scores to more than 5000 diseases.

Although many disease-disease similarity models have been proposed, a method that uses the entire protein interaction network (not just the nearest neighbors) to define pairwise similarity is not yet in use. In this paper a simple similarity measure (called correlation) is defined between any two diseases that have gene associations. In our model a disease-protein network is created by combining disease-gene association and protein-protein interaction databases. In this network the diseases are boundary nodes; i.e. they are not connected to each other, but they are linked to the proteins (products of genes) that are associated with them. The proteins are connected based on their curated binary interactions, and the information flow in the network is modeled by a random walk starting from and ending at each disease [14,15]. Each protein can then be assigned a weight; i.e. the expected number of visits to it. In other words, corresponding to each disease, there is a set of weights associated with the proteins (genes) in the network. From the perspective of using random walk to rank the nodes in the network, our approach is somewhat similar to that of Li and Patra [16]. On the other hand, from the viewpoint of outputting pairwise disease similarities, our method is very similar to that of MimMiner. The method of Li and Patra [16] uses a phenotype similarity network, created using MimMiner similarity scores, in addition to the gene-phenotype and protein interaction networks. Furthermore, their method was developed primarily for gene-disease association prediction. In comparison, the method presented here makes no assumptions about disease-disease similarities. We define the similarity or correlation between any two diseases based on their corresponding gene weights.

We have used our method to calculate correlations between all disease pairs present in the network. We show that higher correlations imply higher probabilities for the diseases to be from the same family of diseases and also higher likelihood of sharing biological pathways. We have compared the results of our method with those of MimMiner since both methods output pairwise disease similarities. It is shown that the results of the two methods complement each other.

We have also compared our method with those of Li and Patra [16] as well as Goh *et al.* [3] in terms of finding "hidden" disease-disease associations. Combining our method with enrichment analysis, we suggest possible disease-pathway associations and find biological pathways that might be shared between different diseases. Finally, we show that clustering diseases based on their correlations increases the number of hits found by the enrichment analysis.

Methods

Disease and gene-disease association databases

Curated disease and disease-gene association data were retrieved (in August 2013) from the Comparative Toxicogenomics Database (CTD) [17], North Carolina State University, Raleigh, NC and Mount Desert Island Biological Laboratory, Salisbury Cove, Maine (URL: http://ctdbase.org/). The CTD disease database merges the hierarchical MESH (Medical Subject Headings) [1] and the flat OMIM (Online Mendelian Inheritance in Man) [18] databases, where OMIM diseases are either merged to the most appropriate MESH terms or are added as children of MESH diseases [19]. The gene-disease associations reported in the CTD database are either based on direct evidences or are inferred. To reduce the uncertainty in the gene-disease associations, we ignored the inferred associations in this study. Also, only the most specific human diseases (the ones with no children) were included in the network.

Protein-protein interaction database

To uncover how gene groups associated with different diseases are related to one another in the context of protein-protein interactions, a protein-protein interaction database is needed. We used ppiTrim [20] to create such a database. By processing iRefindex [21], which incorporates entries from all major protein interaction databases, ppiTrim can produce a protein-protein interaction database in a consistent way and without redundancies [20]. All required input files for ppiTrim were downloaded on June 6 2013, and the program was run on the same day to produce the protein-protein interaction network used in this paper.

Disease-protein network

The disease-protein network was created by combining the CTD gene-disease association database and the protein-protein interaction network produced by ppiTrim. Naturally, only diseases associated with proteins in the ppiTrim-produced database were included in the network. An undirected graph, consisting of 16973 nodes (2548 diseases and 14425 proteins) and 214337 edges, was created by connecting the included diseases to their associated proteins, each of which is a node in the binary interaction network produced by ppiTrim. It was found that for fourteen diseases the associated proteins were disconnected from the rest of the network. These diseases were excluded in the subsequent analysis of the results (leaving 2534 diseases) because the network cannot provide more information about them.

Information flow and disease-disease correlations

Modeling information flow by a random walk with damping, ITMProbe [14,15] is useful in studying information flow in protein networks. Under this method, the random walk starts from one or more *source* nodes and either dissipates or ends at *sink* nodes. Source and sink nodes are also called boundary nodes, while other nodes that are neither sources nor sinks are called *transient* nodes. The ITMProbe program outputs the *expected* number of visits to each transient and sink node by random walkers originated from every source node. In this study ITMProbe was applied to the disease network described in the previous section with all diseases specified as both sources and sinks, all proteins specified as transient nodes, and with a damping factor of 0.85 (for a discussion on the effects of changing the damping factor and also the rational behind using the value 0.85 please see [15]). If we consider the flow of information starting from and ending at a given disease, we can assign a weight (proportional to the expected number of visits) to each protein (transient) node. In other words, for each disease j, there is a corresponding vector of weights $\mathbf{w}_j$ whose dimension equals the number of proteins in the network. Without loss of generality, we always normalize $\mathbf{w}_j$ to have unit length $|\mathbf{w}_j| = 1$ $\forall j$.

The correlation between two diseases j and j' is defined by

$$C_{j,j'} \equiv \mathbf{w}_j \cdot \mathbf{w}_{j'} = \cos(\mathbf{w}_j, \mathbf{w}_{j'}) . \tag{1}$$

The last equality results from $|\mathbf{w}_j| = |\mathbf{w}_{j'}| = 1$. For two disconnected diseases this quantity would vanish, whereas for two diseases with the same connections to the network (diseases associated with the same set of proteins) the correlation would be unity. Disconnected diseases were not included in the analyses and so disease correlations would be positive.

For later convenience, let us also define the average correlation C_j between disease j and the rest of the diseases

$$C_j \equiv \frac{1}{n_d - 1} \sum_{j' \neq j} C_{j,j'} , \tag{2}$$

where n_d is the number of diseases under consideration. We may also define the average pairwise disease correlation $\langle C \rangle$ by

$$\langle C \rangle \equiv \frac{1}{n_d} \sum_{j=1}^{n_d} C_j \tag{3}$$

In this study, we often sort pairwise correlations and bin them. Within such a bin, the average of variable X is generally denoted by X_{bav}, where the subscript bav stands for bin-averaged, with

$$X_{\text{bav}} \equiv \frac{\sum_{j,j'} X_{j,j'} t(C_{j,j'})}{\sum_{j,j'} t(C_{j,j'})} \tag{4}$$

where $t(C)$ is an indicator function taking the value 1 if C is inside the bin of interest and 0 otherwise. When there is only one bin, $t(C_{j,j'}) = 1$ for all $C_{j,j'}$ and $C_{\text{bav}} = \langle C \rangle$ as expected.

Enrichment analysis

In the previous section, the construction of the weight vector associated with a given disease via ITMProbe was described. Evidently, when two diseases have similar weight vectors, they are related from the perspective of protein interaction network. To provide a biological explanation of each weight vector obtained, one would need an enrichment analysis: i.e. one should find the biological terms that best describe the weight vector from an annotated term database such as GO [22] or KEGG [23]. In this study, the enrichment analysis was done using Fisher's exact test, because this is among the most used approaches.

Although there are different implementations of Fisher's exact test, we chose to use Saddlesum [24] because it is an in-house tool and it has already integrated the two term databases (GO and KEGG) of interest for the analyses. For each weight vector, the cutoff (minimum weight included in the analysis) for Fisher's exact

test was chosen to be 0.001 times the mean of the ten largest weights. This variable cutoff was chosen to, in the case of more uniform vectors, prevent including too many genes in the analysis that could result in false term associations. Terms with E-values less than 0.01 were considered significant. For each disease, we used this approach to find the corresponding GO (only terms in the "biological processes" family) and KEGG terms, providing yet another way to compare different diseases.

Clustering

To elucidate the relationships among diseases, a two-stage clustering procedure was employed. At the first stage, the diseases, each associated with a weight vector, were clustered probabilistically based on their vectors' correlations with cluster centers' vectors. An enrichment was then done for each of the final cluster centers. After the first stage, however, many cluster centers were found to be associated with similar, or even identical, sets of GO/KEGG terms. In terms of biological inferences, it was therefore necessary to do the second stage of clustering to group cluster centers based on the similarity of biological terms associated with them.

Although many algorithms are already available for clustering vectors, they usually result in non-overlapping clusters. Many diseases, however, belong to more than one family. Therefore, it is important to have a clustering method that produces overlapping disease classes. Thus, we introduced a probabilistic algorithm in which each disease was iteratively assigned a probability for belonging to a particular cluster. Each cluster was characterized by its center, a vector containing a set of weights for all proteins in the network. For a disease, the probability of being in a cluster was defined to be proportional to the cosine similarity between the weight vectors associated with the disease and the cluster center. The center vectors were initially chosen to coincide with the diseases' weight vectors. In other words, at first the number of clusters was the same as that of the diseases. The initial probabilities were computed using these initial centers. In each iteration the new positions (vectors) of the cluster centers were calculated using

$$\mathbf{v}_i = \sum_j P_{j \to i} \mathbf{w}_j / |\sum_j P_{j \to i} \mathbf{w}_j|, \tag{5}$$

where $P_{j \to i}$ denotes the probability of disease j to be in cluster i, $\mathbf{w}_j$ is the vector associated to the disease j provided by ITMProbe, and the vector $\mathbf{v}_i$ representing cluster i is always normalized to have unit length. The probabilities were then recomputed and used in the next iteration. Since, in this approach, the cluster centers are basically the average of the disease weights, after each iteration they would move closer to each other. For this reason before each iteration the cosine similarities between cluster centers were calculated and the centers that were closer than a cutoff were combined. Specifically, if two centers i_1 and i_2, containing k_1 and k_2 diseases respectively, had the highest cosine similarity and if their similarity was more than the cutoff, the two centers were merged to arrive at a new center vector $\mathbf{v} \propto \left[k_1 \mathbf{v}_{i_1} + k_2 \mathbf{v}_{i_2} \right] / (k_1 + k_2)$ that carries with it $k_1 + k_2$ diseases. This procedure was continued until there was no pair of centers with a similarity bigger than the cutoff, which was set to $1 - \text{std}(C_{j,j'})$ where $\text{std}(C_{j,j'}) = 0.02$ is the standard deviation of the diseases' pairwise correlations (see the Result section). The rational behind this choice is that our model cannot distinguish between two vectors if they are closer than this cutoff. It is worth noting that this process of elimination was performed even before the first iteration, because there were a number of diseases (initial cluster centers) whose correlations was more than the cutoff.

If this iterative method were continued long enough all cluster centers would be combined and there would be only one cluster. The goal of clustering was to group only highly correlated diseases, and so the iterations had to be stopped at an appropriate point. Two observations help us to find this point. First, it is desirable to express the diseases in terms of the lowest number of parameters possible, meaning a lower number of clusters. Second, a disease should be associated with fewest number of clusters possible. This is especially true for diseases that have low correlations with all other diseases (see the Results section). Such a disease should mainly belong to only one cluster (consisting only of that disease). Combining these competing observations, we stopped the iterations when the following quantity was minimized:

$$R = \frac{n_c}{\langle \frac{Q}{C} \rangle} \qquad (6)$$

where $\langle \rangle$ denotes averaging over all diseases, n_c is the number of clusters, $Q_j = \frac{1}{n_c}\sum_{i=1}^{n_c} P_{j\to i}^2$, and C_j (see Eq. (2)) is the average correlation between disease j and all other diseases. For disease j, $1/(n_c Q_j)$ is the participation ratio [14] and Q_j varies between $1/n_c$ (when the disease belongs to only one cluster) and $1/n_c^2$ (when disease is associated with all clusters with the same probability). Therefore, a large Q means the disease is mainly associated with only a few clusters. In the denominator of eq. (6), the contribution of diseases that are far from all others is larger due to the presence of C_j. This would make it unlikely for these diseases to be in more than one cluster when the iteration is stopped. The quantity R went through a minimum after 10 steps (See Fig. S1 in the first section of Supporting Information S1) at which point the iteration was stopped and the final probabilities were calculated. This procedure resulted in 1707 clusters.

Enrichment analysis was then performed for the cluster centers with the same parameters as before, except that the maximum number of genes involved was limited to $\langle n_g \rangle + \mathrm{std}(n_g) \simeq 123$ where $\langle n_g \rangle$ is the average number of genes included in the enrichment analysis per disease and std denotes standard deviation. This limit was imposed because averaging weight vectors results in a vector of more uniformly distributed weights and so the number of included genes could reach much higher values than before. The E-values corresponding to the terms found by the enrichment analysis could be used to create a new vector $\mathbf{e}$ for each disease or cluster center with term association. For this purpose the union of all significant terms associated with all centers was determined. For cluster i the jth component of $\mathbf{e}_i$ was then defined as $-\log(E_{ij})$ if $E_{ij} < 0.01$ and zero otherwise. Here E_{ij} is the E-value corresponding to the jth term when querying SaddleSum using $\mathbf{v}_i$. The $\{\mathbf{e}_i\}$ vectors were then used for the second stage clustering.

In the second stage, the clusters obtained in the first stage were separated into two groups depending on whether or not they had been associated with biological terms. The first (with term associations) and second group had $N_1 = 1301$ and $N_2 = 406$ members respectively. For the first group, the set of vectors $\{\mathbf{e}_i\}$ were then defined and were clustered using a distance-based hierarchical approach. The distance between $\mathbf{e}_i$ and $\mathbf{e}_j$ was defined as $d_{ij} = 1 - \cos(\mathbf{e}_i, \mathbf{e}_j)$, and the cutoff was chosen to be $\langle d \rangle + \mathrm{std}(d)$. To avoid over-clustering, we used a more stringent similarity measure here than the one used before for comparing the terms assigned to different diseases. For disease j the probability of being in the kth new cluster, containing m_k clusters

from the first group, was defined as $P_{j\to k} = \sum_{i=1}^{m_k} P_{j\to i}$ where $P_{j\to i}$ is the probability of disease j belonging to the ith cluster in the first group after the first stage of clustering.

We also used Cfinder [25,26], a program for clustering nodes of a graph, to create overlapping clusters of diseases. The advantage of Cfinde over similar algorithms is that, like the clustering method explained here, it produces overlapping clusters. To use Cfinder, we first created a disease network by connecting each pair of diseases with an edge weighted by their correlation. Using this method, we obtained similar results to those from our clustering algorithm described above. For a detailed description of the procedure and the results please see Supporting Information S1.

Evaluating the accuracy of p-values

The accuracy of p-values reported (by Saddlesum) for the original (not averaged) weight vectors has already been evaluated by Stojmirovic and Yu [24]. The cluster centers, however, are weighted averages of all weight vectors. To investigate whether or not averaging affects the accuracy of the p-values (and consequently the E-values) reported by the enrichment analysis, we took a similar approach to that of Stojmirovic and Yu [24] to calculate the "empirical" p-values and to compare them to the reported ones. Briefly, the gene list was shuffled 672 times and, for each gene list, enrichment analysis was performed for all cluster centers and the reported p-values were recorded. This number (672) was chosen to have approximately 10^{10} weight-term matches, i.e. $n_m = n_c n_t n_l \simeq 10^{10}$, where $n_c = 1707$ is the number of clusters, $n_t = 8719$ is the total number of GO/KEGG terms, and $n_l = 672$ is the number of randomized gene lists. The empirical p-value corresponding to the cutoff value p was then defined as $p_e = \frac{n_p}{n_m}$, where n_p is the total number of reported p-values (for all cluster centers and all gene lists) that are smaller than or equal to p. The results, given in Fig. S2 and section 2 of Supporting Information S1, showed that the reported p-values were indeed accurate.

Results

Statistics of disease-disease correlations

Based on the proteins that the two diseases were connected to, disease pairs were classified into three categories. If both diseases in a pair were connected to the same set of proteins, the pair was assigned to category (1). Members of this category, by definition, had the largest correlation possible (unity) and were equivalent in our study. The number of "independent" (not equivalent to any other) diseases was 1962. If, on the other hand, the two diseases shared some (but not all) associated proteins, the pair was classified in category (2). Category (3) consisted of the rest of the disease pairs (the ones with no shared connections to the protein network). Category (1), (2), and (3) had 978, 5243, and 3203090 pairs respectively.

Using eq. (1), the 3209311 (total number of pairs from all three categories) pairwise correlations $\{C_{j,j'}\}$ were calculated. The median, mean and standard deviation of $\{C_{j,j'}\}$ were 2.8×10^{-8}, 5.4×10^{-4}, and 0.02 respectively. These statistics indicate that although the correlations were generally very small, there was a large number of outliers (disease pairs with high correlations compared to the mean). Obviously, disease pairs in category (1) had the largest correlation possible ($C = 1$). The members of the second category had overall larger correlations (with a mean value of 0.13, a standard deviation of 0.23, and a median of 0.03). However, some pairs in this class had low correlations (with a minimum of 1.5×10^{-5}). In fact 179 disease pairs in the third category (pairs with no common gene associations) had higher correlations than the median correlation of pairs in category (2), indicating that having some shared network connections does not necessarily translate to high correlations, although 95.9% of disease pairs in the second category had larger than average correlations. The maximum correlation of the third category, containing 99.8% of the disease pairs, was 0.237.

Interestingly, the average correlation between a disease and all others varies dramatically. In fact the lowest average correlation (1.4×10^{-8}) was more than five orders of magnitude smaller than the largest, suggesting that some diseases have relatively low correlations with all other diseases. For example, 286 diseases have correlations less than $\langle C \rangle$ with all other diseases.

Interpretations of correlations

To investigate what high correlation between two diseases could imply, a number of quantities were calculated. First, we calculated, for disease pairs with correlations in a certain interval, the probability of being siblings (having the same parents in the CTD/ MESH database). To achieve this, the disease pairs were sorted according their correlations and were divided into bins. The probability I_{bav} (see eq. (4)) of finding a sibling pair in each bin was then defined as the ratio of the number of sibling pairs to the total number of disease pairs in that bin. The number of disease pairs in each bin was 1000.

Second, for each disease pair, we calculated the similarity S between the associated enriched terms, which was defined as the ratio of the number of significant terms shared between the two diseases to the total number of identified significant terms. In other words, for two diseases associated with exactly the same GO and KEGG terms S would be unity, whereas for diseases with no common terms it would vanish. If one or both diseases in a pair did not have any term associated with them, S was not defined. To find the distribution of disease pairs with $S > 0$, the pairs were partitioned into bins as described in the previous paragraph. In each bin, the average probability for two diseases in a pair to hit the same GO/KEGG terms was then defined as S_{bav} (see eq. (4)).

As expected, the enrichment analysis did not find any significant biological terms associated with some of the diseases, implying that, for some disease pairs, calculation of S was not possible. In this study Saddlesum was able to find significant GO/KEGG terms for about 60% of the diseases (1530 out of 2534). For diseases with significant term hits, the average numbers of identified GO and KEGG terms were 34.7 and 5.2, and the standard deviations were 52.2 and 8.2. The large spread of the number of hits was due to the fact that some diseases, with a large number (>10) of connections to the network, had hundreds of GO and tens of KEGG terms associated with them. Overall 3182/ 203 unique GO/KEGG terms were hit by the enrichment analysis.

Interestingly, there was a significant difference between the percentage of pairs with undefined S when disease pairs with low and high correlations are considered. For example, S was undefined for 23% of disease pairs with correlations greater than $\langle C \rangle$, as opposed to 64% for pairs with correlations smaller than $\langle C \rangle$. This can be understood through the fact that the percentage (p) of diseases that had been successfully assigned one or more GO/KEGG terms by Saddlesum was smaller for diseases with very low average correlations. This behavior is shown in Fig. 1. After sorting $\{C_j\}$ into ascending order and placing them in bins each containing 100 diseases, we computed the average C_j in a bin and, in the same bin, the number of diseases N that had one or more GO/KEGG term hits. In Fig. 1, $p = N\%$ is plotted versus the average C_j per bin and the aforementioned behavior is clearly displayed.

Since S was undefined for a large number of disease pairs, the pairs were divided into two sets: with defined (first set) and undefined (second set) term similarities. For the first set (with defined S), Fig. 2 (A) illustrates the behavior of I_{bav-} (in green), I_{bav+} (in blue), and S_{bav} (in red), where I_{bav+} (I_{bav-}) is I_{bav} for pairs with $S > 0$ ($S = 0$). The figure clearly shows, when $C_{bav} > 10^{-6}$, a rise in the probability of a disease pair to have common biological associations as correlation increases. The figure also indicates, when $C_{bav} > 10^{-5}$, that disease pairs with higher correlations are more likely to be siblings if they have $S > 0$. However, the siblings without shared terms have almost a flat (correlation-independent) distribution, although the percentage of such pairs is very small (about 0.5%). One possible explanation for these results is that the increase in the percentage of siblings in highly-correlated diseases is in fact due to an increase in the percentage of the pairs with $S > 0$. In other words, in high correlation regime, most of the siblings are a subset of disease pairs with shared GO/KEGG terms. Figure 2 (B) shows how I_{bav} varies with correlation for the second set of disease pairs (the ones with undefined term similarities).

Using only the terms from the *manually* curated disease-pathway associations [23] in the KEGG DISEASE database (downloaded on December, 12 2013) in place of the terms retrieved from enrichment analyses, term similarities (S) and the probabilities S_{bav}, I_{bav}, I_{bav+}, and I_{bav-} were recalculated. In this KEGG database, one or more OMIM diseases are associated with one or more KEGG pathways. The CTD disease database [17] was used to find the equivalent MESH diseases that were used in our study. The results are shown in Figs. 2 (C) and (D). Overall, the trends observed in these figures are very similar to those shown in 2 (A) and (B): i.e. increase in S_{bav}/I_{bav+} with correlation (for $C_{bav} > 10^{-6}$) and correlation-independence in I_{bav-} for pairs with defined S, and a uniform and then increasing I_{bav} for pairs with undefined term similarities.

We have also computed the degree of overlap between the enrichment analysis and KEGG DISEASE database. For each disease j, out of the total number of annotated KEGG pathway associations $K(j)$ we calculated the number of associations $KE(j)$ that were also reported significant by enrichment analysis. The ratio $s_j \equiv KE(j)/K(j)$, for disease j, measures the agreement between the curated pathway assignment and the enrichment analysis. There were 490 diseases that had been annotated in the KEGG DISEASE database and also had term hits using the enrichment analysis. The average value of s_j for these 490 diseases was found to be 0.48, indicating on average 48% of the annotated terms for each disease in the KEGG DISEASE database were also found using our enrichment analysis.

Comparison with MimMiner and human disease network

MimMiner [13] uses a text mining approach to calculate a pairwise disease similarity score that, like correlation defined here, ranges from 0 to 1. However, since these two measures have very different distributions and are defined based on different concepts, they cannot be directly compared. For this reason we adopted the procedure described in the third section and Fig. S3 of Supporting Information S1 to find equivalent cutoff correlations and MimMiner scores, 5.7×10^{-6} and 0.35 respectively, and to compare the two methods.

In the absence of a true gold standard, we used disease-pathway associations reported by KEEG DISEASE databse [23] to compare the retrieval agreement with KEGG DISEASE from MimMiner and that from our model. This was done by comparing the effectiveness of the two methods to identify disease pairs that

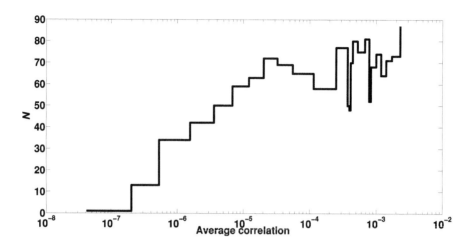

Figure 1. The relation between the results of enrichment analysis and the average correlation C_j. The percentage of diseases for which GO/KEGG terms were identified by Saddlesum as a function of average correlation C_j. To facilitate the calculation, we sorted all C_js in ascending order and placed them into bins each containing 100 diseases. The percentage is then measured by the number N of diseases with GO/KEGG term hit(s) per bin. For very low average correlations N is significantly lower.

are associated with the same biological pathways as annotated in KEGG DISEASE. To make a fair comparison, only disease pairs with defined term similarities S (see the "Interpretations of correlations" section) and with available MimMiner scores were included in this analysis (336610 pairs, about 10% of all pairs). By ranking the disease pairs based on either their correlations or their MimMiner scores, two lists of pairs were created. For each list, the weighted number of disease pairs with common associated

biological terms that were among the first (highest ranking) m pairs was calculated as $M(m) = \sum_{i=1}^{m} S(i)$, where $S(i)$ is the term similarity between diseases of pair i using KEGG DISEASE as the standard. The function $M(m)$ provides a measure for comparing the two methods: a faster rise in $M(m)$ would mean a larger number of pairs with high term similarity have been ranked higher than the others, indicating a better agreement with KEGG DISEASE.

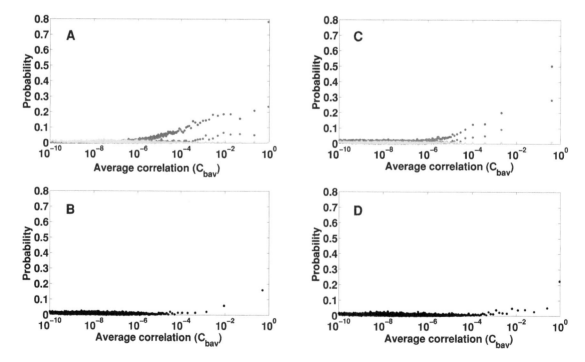

Figure 2. The probabilities of having common term associations or being siblings. (A) The probabilities of finding a pair of diseases with (1) common GO/KEGG terms (red), (2) the same parents and common associations (blue), and (3) the same parents without shared biological terms (green) are shown. Here only pairs with a defined term similarity are considered. (B) For pairs with undefined S (pairs with at least one member not associated with any biological terms), the distribution of siblings is plotted as a function of correlation. (C) and (D) show similar quantities to (A) and (B) respectively, when the biological term associations are directly retrieved from the KEGG DISEASE database.

The results (MimMiner in red, our method in blue) are shown in the inset of Fig. 3 (A). The green curve shows the weighted number $M(m)$ of disease pairs identified (ranked higher than m) by our method, but missed (ranked lower) by MimMiner. Similar trends are observed for both methods, but a better performance (faster rise in $M(m)$) for MimMiner is indicated. This finding is expected, because MimMiner is based on mining the literature, which is also the source of the *manually* curated data in the KEGG DISEASE database. However, an important observation is that the two methods do not find the same pairs, especially in terms of less apparent relationships. To see this feature, we first excluded the disease pairs that were obvious candidates for being related, i.e. sibling diseases and pairs with common gene associations (3847 pairs were excluded leaving 332763). We then recomputed the blue and the green curves, shown in Fig. 3 (A). The closeness between these two curves indicates that for non-apparent relationships, the disease pairs identified by our method are largely missed by MimMiner. In Fig. 3 (A), about 87% of pairs ranked higher than $m = 2500$ (equivalent to a correlation of 2.2×10^{-5} and a MimMiner score of 0.41) by the method presented here were missed by MimMiner.

Given the fact that our method and MimMiner effectively find different pairs, one may wish to look at the quality of retrieval using a different measure other than $M(m)$. In Supporting Information S1, we have described how to find the cutoff cosine similarity and MimMiner score, above which there exists an apparent positive correlation between the similarity/score and the pair relatedness. Denote the number of disease pairs with similarity/score above the cutoff by $\mathcal{N}_c/\mathcal{N}_s$. We defined the normalized rank r as rank divided by $\mathcal{N}_c$ or by $\mathcal{N}_s$ depending on whether cosine similarity or MimMiner score was in use. For a given cutoff term similarity S, we first found among $\mathcal{N}_c$ or $\mathcal{N}_s$ disease pairs with term similarity larger than S, and then computed their average normalized rank $\langle r \rangle$ when all pairs were ranked either by correlation (cosine similarity) or by MimMiner score. Evidently, for large cutoff S, a larger $1/\langle r \rangle$ indicates a better retrieval fidelity. Using this measure, results shown in Fig. 3 (B), our methods seems to provide higher retrieval fidelity. One should bear in mind, however, comparisons of such should always be taken with a grain of salt due to the difficulty in constructing a gold standard and totally impartial datasets.

It is also worth noting that for a very large subset (90%) of disease pairs investigated in this study, KEGG term similarities or MimMiner scores were not defined. However, many of these disease pairs, or the ones with $S = 0$ or with a MimMiner score of zero, may in fact be related. For example, 5090 (97%) of disease pairs in the second category and 838 (86%) of the members of the first category of pairs, which have common gene associations and are more likely to be related, were in this subset. Even diseases with no gene associations or the ones that have been classified in totally different families could share biological pathways or phenotypic similarity. On the one hand, many pairs with documented relationships may not have yet been annotated by KEGG DISEASE or scored by MimMiner, or may have been reported as being not related. Table 1 lists ten example pairs, with correlations much larger than $\langle C \rangle \approx 5.4 \times 10^{-4}$, from all three categories of disease pairs. On the other hand, it is likely that some disease-disease relationships have not yet been discovered. One should keep in mind that the members of the majority of pairs with significant correlations in our study have no obvious relationships (do not share genes and are not siblings), and also their possible relationships have not yet been experimentally verified. From a practical point of view, however, these pairs are more interesting, because they suggest unknown and non-trivial relationships that, if verified, could add to our knowledge about the causes and possibly cures of certain diseases.

From the perspective of finding "related" disease pairs with zero MimMiner score, Li and Patra [16] found 18 non-apparent related pairs while combining a phenotype similarity network (created using MimMiner similarity scores), a gene-phenotype network and a protein interaction network. For 15 out of the 18 pairs, support information for relatedness was provided by Li and Patra [16]. The relatedness evidence for 12 of the 15 pairs is founded on that the member diseases are classified in the same disease class. Using our method, we have found 330 disease pairs with zero MimMiner scores, each of which has its member diseases classified under the same disease family according to MESH. We have also found 43 disease pairs with zero MimMiner scores, each of which has its member diseases share at least one biological pathway according to KEGG DISEASE database.

The proposed method was also compared to the Human Disease Network (HDN), introduced in the pioneering work of Goh *et al.* [3]. In this method the disease network is created by linking diseases that have common gene associations. The method proposed here, however, links the diseases based on their correlations, i.e. the diseases are linked if they have significant correlations (larger than the cutoff 5.7×10^{-6}, as obtained in Supporting Information S1). Interestingly, in our study the

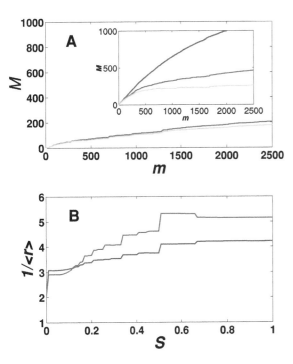

Figure 3. Comparison with MimMiner. (A) The inset figure shows the number (M) of weighted disease pairs with shared KEGG pathways that were ranked higher than m by MimMiner (in red) and or by our method (in blue). Also shown in the inset (in green) is the weighted number of pairs with common term associations missed (ranked lower) by MimMiner, but identified (ranked higher) by our model. In the main panel, the same quantities corresponding to the proposed method are plotted after exclusion of obvious candidates for being related. The closeness between the blue and green curves indicates that the non-apparent candidates found by our method are largely missed by MimMiner. Displayed in panel (B) is the inverse of average normalized rank versus the term similarity cutoff. At large similarity cutoff, the higher the average normalized rank (the smaller $\langle r \rangle$ and thus the larger $1/\langle r \rangle$) the better the agreement between the quality scores (cosine similarity or the MimMiner score) and the KEGG annotation.

Table 1. Examples of relationships between diseases that have undefined (or zero) term similarity and undefined (or zero) MimMiner score, and are from different disease families.

First disease ID	Second disease ID	First disease name	Second disease name	C	N_s	Relationship
MESH:C567070	MESH:C536289	Atypical mycobacteriosis, familial, X-Linked 1	Immunodeficiency without anhidrotic ectodermal dysplasia	1.0	1	Both diseases have been associated with nuclear factor kappa B signaling [31,32].
MESH:C536198	MESH:C536113	Ehlers-Danlos syndrome type 6	Nevo syndrome	1.0	1	These diseases have been suggested to be identical [33].
MESH:C537494	MESH:C566453	Stickler syndrome, type 3	Deafness, autosomal recessive 53	1.0	1	Hearing loss is one of the symptoms of Stickler syndrome, type 3 [34].
MESH:C535407	MESH:D053609	Gamma aminobutyric acid transaminase deficiency	Lethargy	0.9961	1	Lethargy has been reported in pateints with Gamma aminobutyric acid transaminase deficiency [35].
MESH:D016301	MESH:C562440	Alveolar bone loss	Hypophosphatasia, childhood	0.9584	1	These are both tooth/bone diseases.
MESH:C564629	MESH:C538150	Deafness, autosomal recessive 31	Syndactyly Cenani-Lenz type	0.1040	0	Hearing loss has been associated with Cenani-Lenz type of syndactyly [36].
MESH:C536156	MESH:C536601	Keratomalacia	Amaurosis congenita of Leber, type 2	0.0835	0	These are both eye diseases.
MESH:C563906	MESH:C563425	Cardiomyopathy, dilated, 1o	Diabetes mellitus, permanent neonatal	0.0197	0	ATP-sensitive potassium channels have been reported to be involved in both diseases [37,38].
MESH:C564334	MESH:D008527	Acrocapitofemoral dysplasia	Medulloblastoma	0.0168	0	These disease have been associated with Hedgehog signaling pathway [39,40].
MESH:C565334	OMIM:188890	Epilepsy, nocturnal frontal lobe, type 3	Tobacco addiction, susceptibility to	0.0155	0	Both diseases have been associated with mutations in nicotinic acetylcholine receptors [41].

C and N_s denote correlation and the number of common gene associations respectively.

minimum correlation between diseases with shared gene associations was 1.5×10^{-5}. In other words, the links of a disease network created by the method of Goh *et al.* would be a small subset of those of our disease network. To find out if the additional disease-disease relations suggested by the proposed method are supported by the available experimental data, once again KEGG DISEASE database was used. We considered only diseases annotated by KEGG (1272 out of 2534 included diseases) and created three disease networks by linking the diseases using three different connectivity measures, i.e. having shared gene associations, high correlation, and having common pathway associations as annotated by KEGG DISEASE. The total number of links between the diseases in the three networks were 527, 14202, and 45577 respectively. The number of coinciding links between the KEGG-based and the correlation-based networks was 2988, as opposed to 389 when comparing KEGG and HDN networks. In other words, 2599 pairwise disease relations predicted by our method and missed by HDN are supported by the KEGG DISEASE database. Both methods however failed to predict the relationships between a large number of diseases that, according to KEGG DISEASE, have shared biological pathways. On the other hand there are many diseases that have high correlations, but are not reported by KEGG as having common pathway associations. As discussed before, these are not necessarily false positives. The KEGG database does not yet contain many literature-supported relationships (see Table 1 for some examples), but more importantly, there might be many disease relations that have not yet been discovered. An important aspect of the proposed method is that it suggests disease relationships that should be experimentally verified.

Effect of clustering

Based on the hypothesis that highly correlated diseases are more likely to have common pathways, one can use correlation-based clustering to increase the number of hits when searching for biological terms associated with the diseases. Assuming that all diseases in a given cluster share some pathways/processes, one can increase the chance of finding these pathways/processes by weighted averaging of the weight vectors assigned by ITMProbe to the diseases in the cluster. The rational behind this method is that each vector may be contaminated with "noise" and that the "signal" could be amplified by averaging. To accommodate the scenario that a disease might belong to several families, we used a probabilistic clustering method (see Methods) that allowed overlapping clusters and assigned a probability to each disease for being in a particular cluster.

Our iterative approach resulted in 1707 clusters. Enrichment analysis was run for all cluster centers obtained in this stage and found significant hits for 1301 clusters with an average of 70.9/7.5 GO/KEGG terms per cluster, which was higher than the average number of terms found for the diseases. The probabilities of belonging to different clusters were calculated for each disease and were used to determine the percentage of diseases with term hits, defined by

$$\frac{1}{n_d} \sum_{ij} P_{j \to i}\, t(i), \qquad (7)$$

with $t(i)$ being an indicator function taking value 1 when cluster i has a term hit and 0 otherwise. Interestingly, the number of such diseases showed an increase from 60% (when enrichment was directly performed for the diseases) to 85%. For the diseases that had term hits using both methods (direct and through clustering)

the term similarity $\mathcal{Y}$, was calculated using

$$\mathcal{Y} = \frac{1}{n_{d \to T}} \sum_{ij} P_{j \to i}\, \frac{|T_i^c \cap T_j|}{|T_j|}\,, \qquad (8)$$

with $n_{d \to T}$ being the number of diseases that have significant term hits, T_j being the set of terms associated with the jth disease, T_i^c being the set of those assigned to the cluster i, and $|T|$ denoting the number of members in the set T. We found $\mathcal{Y} = 0.41$. This seems to indicate that more than 50% of the terms associated with the diseases were dropped upon merging to clusters and some information might have been lost in the process. What is really important, however, is whether terms of small number of annotated genes are preserved, as these terms are most specific and usually most informative. Upon examining the distribution of minimum GO/KEGG term size (number of annotated genes for that term) when running SaddleSum using diseases directly and using cluster centers, we find that the most informative terms are largely kept in the process. The distribution of the minimum term size is shown in Fig. 4.

To illustrate how clustering through weight vectors may increase the likelihood of associating a disease with terms, we examine the late-onset Parkinson's disease (OMIM:168600). This disease was not associated with any terms when enrichment analysis was directly performed for the disease. After clustering, however, the top four clusters, ranked by their probabilities of including the Parkinson's disease, were associated with the Parkinson's disease pathway. Specifically, the term hits (with

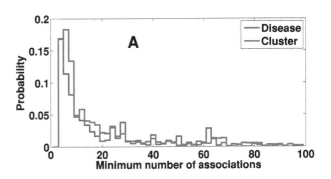

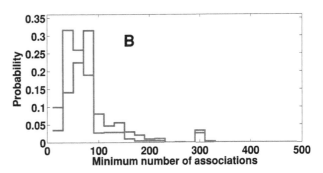

Figure 4. The effect of clustering on the minimum term size. The minimum term size distribution of (A) GO and (B) KEGG terms reported by SaddleSum enrichment analyses when using disease weight vectors directly (red curves) and when using cluster center vectors (blue curves). Not only the most informative (smallest size) terms are preserved during clustering, the clustering procedure seems to shift the minimum term size distribution towards the small end, indicating the likelihood of providing even more specific terms when weight vectors are grouped under the proposed clustering procedure.

Table 2. Terms associated with the cluster with the highest probability to include the Parkinson's disease.

Term ID	Name	E-value
GO:0007268	synaptic transmission	4.22e-12
GO:0019226	transmission of nerve impulse	4.58e-12
GO:0035637	multicellular organismal signaling	2.00e-11
GO:0007267	cell-cell signaling	1.13e-10
GO:0050877	neurological system process	4.54e-10
GO:0001963	synaptic transmission, dopaminergic	5.34e-08
GO:0007270	neuron-neuron synaptic transmission	4.47e-07
GO:0044708	single-organism behavior	8.73e-07
GO:0003008	system process	1.16e-06
GO:0030534	adult behavior	1.69e-06
GO:0001505	regulation of neurotransmitter levels	3.81e-06
GO:0006805	xenobiotic metabolic process	4.11e-06
GO:0071466	cellular response to xenobiotic stimulus	4.59e-06
GO:0009410	response to xenobiotic stimulus	4.59e-06
GO:0044281	small molecule metabolic process	1.62e-05
GO:0007610	behavior	5.39e-05
GO:1901615	organic hydroxy compound metabolic proce	6.38e-05
GO:0023052	signaling	6.72e-05
GO:0044700	single organism signaling	6.72e-05
GO:0065008	regulation of biological quality	7.75e-05
KEGG:hsa04080	Neuroactive ligand-receptor interaction	2.61e-19
KEGG:hsa05010	Alzheimer's disease	2.73e-06
KEGG:hsa05012	Parkinson's disease	8.69e-06

E-values smaller than 1e-4) for the cluster with the highest probability (13%) are listed in Table 2, which include Parkinson's disease, Alzheimer's disease and other neurological processes.

Retinitis Pigmentosa is an eye disease, with many different types, which is characterized by progressive retinal degeneration. As a second example, we examined the cluster and term associations for type 7 of this disease (MESH:C564284), which had no term hit before clustering. The disease was in multiple clusters (with relatively high probabilities $\sim 10\%$) that were associated with the phototransduction pathway. Given in Table 3 are the terms associated with the cluster with the highest probability (10%), which are related to phototransduction, detection of light and response to light. The phototransduction pathway, along with the Retinal metabolism (KEGG:hsa00830) and Spliceosome (hsa03040) pathways, has been indeed annotated to be related to this disease by the KEGG DISEASE database. Figure 5 visualizes the clusters that contain Parkinson's disease and Retinitis Pigmentosa 7.

As a third example, the associations of the Knobloch syndrome (MESH:C537209) were also investigated. This is another eye disease that is characterized by different abnormalities, including cataracts, dislocated lenses, vitreoretinal degeneration, and retinal detachment [27]. Unlike the other two examples, this disease is primarily a member of one cluster with a probability of 62%, and the second highest probability was much smaller (3%). According to the KEGG DISEASE database, the pathways involved in this disease are focal adhesion (KEGG:hsa04510), ECM-receptor interaction (KEGG:hsa04512), and cell adhesion molecules (KEGG:hsa04514). Focal adhesion, and ECM-receptor interac-

tion were in fact among the KEGG terms that were found to be associated with the cluster. Due to the relatively large number of associated terms with this cluster, the whole list is not given.

The first two examples indicate that there was a high degree of overlap between the GO/KEGG terms associated with different clusters. In other words, several clusters have phototransduction or Parkinson's disease pathways associated with them. For this reason a second round of clustering was performed, which was based on term similarity rather than disease correlations through weight vectors, and the probabilities of belonging to each new cluster was calculated. The second stage clustering reduced the number of clusters with term associations to 217, which substantially reduced the term overlap. For example, the highest new membership probability (P_{max}) for Parkinson's disease became 31%, and the new cluster, as expected, was associated with the Parkinson's disease pathway. Similarly, after the second stage clustering, Retinitis Pigmentosa type 7 was primarily in one cluster with the probability of 44% (which is associated with phototransduction) and all other probabilities were less than 15%. On other hand, P_{max} for the Knobloch syndrome only increased modestly to 73%. The large reduction of the number of clusters is consistent with the view that many diseases share common modules or biological pathways.

Discussion

Disease networks can provide valuable information when investigating if and how two given diseases are related. In this paper a simple network-based measure, referred to as correlation, is introduced to explore possible relations between any two genetic

Table 3. Terms associated with the cluster with the highest probability to include the Retinitis Pigmentosa type 7.

Term ID	Name	E-value
GO:0007603	phototransduction, visible light	5.64e-09
GO:0009584	detection of visible light	9.95e-09
GO:0007602	phototransduction	1.69e-08
GO:0009583	detection of light stimulus	2.51e-08
GO:0009582	detection of abiotic stimulus	1.12e-07
GO:0009581	detection of external stimulus	2.98e-07
GO:0051606	detection of stimulus	3.66e-06
GO:0022400	regulation of rhodopsin mediated signali	1.07e-05
GO:0016056	rhodopsin mediated signaling pathway	1.31e-05
GO:0009416	response to light stimulus	1.47e-05
GO:0009314	response to radiation	1.45e-04
GO:0071482	cellular response to light stimulus	9.23e-04
GO:0008277	regulation of G-protein coupled receptor	5.23e-03
GO:0071478	cellular response to radiation	5.23e-03
KEGG:hsa04744	Phototransduction	8.51e-04

diseases. The correlation between two diseases is defined (eq. (1)) by the inner product of their corresponding weight vectors. The weight vector associated with a disease is based on the flow of information from that disease back to itself in a disease-protein network created by integrating the available disease-gene associations and protein-protein interactions. The results obtained are therefore reflective of the data available. Although specific cases might be sensitive to the data employed, the general trend obtained should remain robust.

Our results suggest that diseases with higher correlations are more likely to be phenotypically related (be children of the same parent), and to share biological pathways as determined by enrichment analysis. Specifically, our result shows that most siblings with high correlations share at least some GO/KEGG terms. In fact, siblings with large mutual correlations are mostly a subset of diseases that have been assigned common biological terms. We also find that when enrichment analysis does not return a shared pathway (when S_{bav} is very small), only less than 1% of disease pairs are siblings. However, correlation between diseases seems to be more an indicator of similarity of the involved biological processes than an absolute measure of phenotypic overlap. This is evidenced by a small but steady presence of sibling disease pairs with extremely low correlations. This is consistent with the view that high correlation indicates shared pathways and thus likely shared pathophenotypes, while pathophenotypic similarities might not require high correlations.

Different genes/processes may cause the same or closely related phenotypes when they effectively influence the same pathway. The clustering procedure used in this paper is aimed to find this type of event and group them together. When diseases are caused by different genes that are parts of the same pathway, the likelihood for their weight vectors to resemble one another is apparently higher than when they do not share the same pathway. Suitably averaging those weight vectors (eq. (5)) leads to a cluster center that may better represent the pathway. This procedure also has the effect of reducing the "noise" and enhancing the "signal" of the weight vectors, which is evidenced by the increase, from 60% to 85%, of the percentage of diseases having significant term hits upon first stage clustering.

On the other hand, since our data sources only include the disease-gene associations and the protein interactions, the regulatory effects were not included explicitly. This presents a limitation as well as points the direction for future improvement. For example, from a gene-centric point of view, being a part of one or multiple pathways, a gene could result in multiple diseases either through different states (overactive vs. underactive) in a single pathway or through influencing multiple pathways. Largely absent from the disease-protein network currently used, these subtle effects do exist. Table 4 shows a set of five diseases with shared connections to the network through low density lipoprtein receptor-related protein 5 (*LRP5*, OMIM:603506). As far as the disease-protein network is concerned these diseases are equivalent and have perfect correlations. That is, these five diseases are naturally grouped together under our method even though their annotations do not suggest such a grouping. One of the diseases (MESH:C566619) is a member of the family of eye diseases characterized by incomplete development of the retinal vasculature, and the others are musculoskeletal diseases associated with high (MESH:C536527, MESH:C536748, and MESH:C536056) and low (MESH:C536063) bone densities (MESH:C536056 and MESH:C536748 are sibling diseases). Interestingly, osteoporosis-pseudoglioma syndrome is a disease that is characterized by both low bone density and eye abnormalities. The top (with lowest E-value) GO/KEGG term assigned to these diseases by enrichment analysis was "Wnt signaling pathway", i.e. GO:0016055/KEGG:hsa04310 (obviously, all diseases in this set had the same terms associated with them, because they share the same connection to the network). This pathway, through mutations in *LRP5*, has been indeed reported to be involved in development of diseases related to both bone density and also some eye abnormalities [28–30]. In these studies MESH:C536527, MESH:C536748, and MESH:C536056 have been associated with an increase in Wnt signaling, whereas underactive Wnt signaling has been reported to cause MESH:C536063 and MESH:C566619.

Although MimMiner [13] uses a totally different approach to measure disease-disease similarity, like our method, it provides pairwise scores for a large number of diseases. Therefore, it is

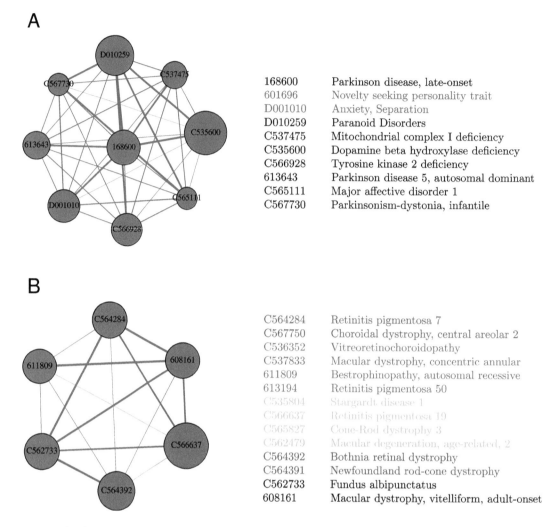

168600	Parkinson disease, late-onset
601696	Novelty seeking personality trait
D001010	Anxiety, Separation
D010259	Paranoid Disorders
C537475	Mitochondrial complex I deficiency
C535600	Dopamine beta hydroxylase deficiency
C566928	Tyrosine kinase 2 deficiency
613643	Parkinson disease 5, autosomal dominant
C565111	Major affective disorder 1
C567730	Parkinsonism-dystonia, infantile

C564284	Retinitis pigmentosa 7
C567750	Choroidal dystrophy, central areolar 2
C536352	Vitreoretinochoroidopathy
C537833	Macular dystrophy, concentric annular
611809	Bestrophinopathy, autosomal recessive
613194	Retinitis pigmentosa 50
C535804	Stargardt disease 1
C566637	Retinitis pigmentosa 19
C565827	Cone-Rod dystrophy 3
C562479	Macular degeneration, age-related, 2
C564392	Bothnia retinal dystrophy
C564391	Newfoundland rod-cone dystrophy
C562733	Fundus albipunctatus
608161	Macular dystrophy, vitelliform, adult-onset

Figure 5. Two example clusters. The clusters that include Parkinson's disease (OMIM:168600) and Retinitis pigmentosa 7 (MESH:C564284) are shown in panels (A) and (B) respectively. In each case, only diseases with membership probabilities larger than 5% are shown. The size of each node (circle) is proportional to the probability of membership of that node in the cluster. For a disease pair, the thickness of the line linking the diseases is proportional to $1 + \log_{10} \frac{C}{C_m}$, where C is the correlation between the two diseases and C_m is the minimum correlation between all diseases shown in each cluster. The names and IDs of the members of each cluster are also given. Diseases whose names are written in the same color (other than black) have exactly the same gene associations and so are equivalent in our study. Equivalent diseases are represented by one node in the figure. For example, the node identified by C566637 in panel (B) represents the four diseases whose names are in green, i.e. C535804, C566637, C565827, and C562479.

perhaps the most suitable method to be compared with the approach presented in this paper. A comparison between the performances of the two methods in identifying disease pairs with shared associated KEGG pathways indicated that the results of the two approaches are largely complementary. In other words, each method can provide valuable information about relationships between diseases that cannot be obtained from the other. It should be noted, however, that the KEGG DISEASE database, which

Table 4. An example of a set of diseases that are associated with the same gene, but some have different phenotypes.

Disease ID	Disease annotation	Disease family
MESH:C566619	Exudative Vitreoretinopathy 4	Eye diseases
MESH:C536527	Van Buchem disease type 2	Musculoskeletal diseases
MESH:C536063	Osteoporosis-pseudoglioma syndrome	Musculoskeletal diseases
MESH:C536748	Worth syndrome	Musculoskeletal disease
MESH:C536056	Osteopetrosis autosomal dominant type 1	Musculoskeletal diseases

was assumed as a gold standard for making such comparison, is underdeveloped and manually curated. A more complete database that does not bias towards either text-mining or gene-disease associations is needed for a sound comparison among methods outputting pairwise disease similarities.

Although using protein interaction data in conjunction with finding disease relations is not new, utilizing the information flow to find for each disease its corresponding ITM (information transduction modules) in the context of protein-protein interaction is novel. There are a number of directions that we can potentially look into but did not do so because of the lack of a comprehensive gold standard to assess them. For example, the clustering procedure proposed can be turned into a tool to classify diseases based on the underlying protein interactions. Also, it would be interesting to examine clusters without any term hits but containing multiple diseases. This might help in finding the common cause among seemingly unrelated diseases. In addition, it can be valuable to examine clusters with significant term hits but whose member diseases do not yet have annotated cause. The term hits in this case may shed some light in searching for the underlying cause of the disease. Even though we did not pursue further analyses along those directions, we have, however, compiled the clustering results and make them available for download. If properly used, these compiled results form a database for finding candidates of not-yet-solved problems in disease cause and mutual relations.

Another interesting finding of the study was the higher rate of failure of the enrichment analysis to find significant GO/KEGG terms associated with diseases that had very low average correlations with the others. This is perhaps due to the incompleteness of the network, i.e. missing protein-protein interactions or gene-disease associations. Such missing nodes would prevent both ITMProbe from finding correlated diseases and Saddlesum from assigning biological terms. Improvement in the databases used in this study to create the disease network could change the results for diseases with missing connections. However, such improvements seem to be less likely to significantly change the relations that are already embedded in the network. For this reason high disease correlations seem to be more informative. In other words, a high correlation between two diseases is suggestive of a relationship between the two, but a low correlation may just reflect that there is not enough information in the network. Even for very highly correlated ($C > 0.1$) diseases, our approach still could not find common pathways for all disease pairs. This could still be due to incompleteness of the network or because of the fact that our method uses a rather simple measure to investigate possible disease relations.

In summary, we have proposed to use network-based correlations between diseases as a measure of diseases similarity. Higher correlations could be interpreted as a higher probability for the disease pairs to have common biological pathways/processes. Despite its simplicity and limitations, the simple approach employed seems to be able to, in most cases, distinguish between disease pairs with and without shared GO/KEGG biological terms as well as properly group diseases sharing similar biological processes/pathways.

Supporting Information

Supporting Information S1 All supporting information are given in this file, including a description of how cutoffs were calculated for MimMiner score and correlation, the results of the evaluation of the accuracy of the p-values, and also the results of clustering using Cfinder. **Figure S1**, Finding the optimum number of clusters. Figure shows (A) the number of clusters, and (B) R, as a function of number of iterations. R is minimized after 10 iterations. **Figure S2**, Empirical p-values vs p-value cutoffs. The empirical values were calculated by shuffling the gene list 672 times. **Figure S3**, The probability of finding shared KEGG pathways is plotted (in red) as a function of average MimMiner score (a) or average correlation (b). The blue line shows the fitted piecewise function. The separation points are considered the cutoffs above which the scores or correlations are significant. (PDF)

Author Contributions

Conceived and designed the experiments: YKY. Performed the experiments: MH. Analyzed the data: MH YKY. Wrote the paper: MH YKY.

References

1. Coletti MH, Bleich HL (2001) Medical subject headings used to search the biomedical literature. J Am Med Inform Assoc 8: 317–323.
2. Schriml LM, Arze C, Nadendla S, Chang YW, Mazaitis M, et al. (2012) Disease Ontology: a backbone for disease semantic integration. Nucleic Acids Res 40: D940–946.
3. Goh KI, Cusick ME, Valle D, Childs B, Vidal M, et al. (2007) The human disease network. Proc Natl Acad Sci USA 104: 8685–8690.
4. Lee DS, Park J, Kay KA, Christakis NA, Oltvai ZN, et al. (2008) The implications of human metabolic network topology for disease comorbidity. Proc Natl Acad Sci USA 105: 9880–9885.
5. Zhang X, Zhang R, Jiang Y, Sun P, Tang G, et al. (2011) The expanded human disease network combining protein-protein interaction information. Eur J Hum Genet 19: 783–788.
6. Hidalgo CA, Blumm N, Barabasi AL, Christakis NA (2009) A dynamic network approach for the study of human phenotypes. PLoS Comput Biol 5: e1000353.
7. Linghu B, Snitkin ES, Hu Z, Xia Y, Delisi C (2009) Genome-wide prioritization of disease genes and identification of disease-disease associations from an integrated human functional linkage network. Genome Biol 10: R91.
8. Suthram S, Dudley JT, Chiang AP, Chen R, Hastie TJ, et al. (2010) Network-based elucidation of human disease similarities reveals common functional modules enriched for pluripotent drug targets. PLoS Comput Biol 6: e1000662.
9. Bauer-Mehren A, Bundschus M, Rautschka M, Mayer MA, Sanz F, et al. (2011) Gene-disease network analysis reveals functional modules in mendelian, complex and environmental diseases. PLoS ONE 6: e20284.
10. Barabasi AL, Gulbahce N, Loscalzo J (2011) Network medicine: a network-based approach to human disease. Nat Rev Genet 12: 56–68.
11. Zitnik M, Janjic V, Larminie C, Zupan B, Przulj N (2013) Discovering disease-disease associations by fusing systems-level molecular data. Sci Rep 3: 3202.
12. Gulbahce N, Yan H, Dricot A, Padi M, Byrdsong D, et al. (2012) Viral perturbations of host networks reflect disease etiology. PLoS Comput Biol 8: e1002531.
13. van Driel MA, Bruggeman J, Vriend G, Brunner HG, Leunissen JA (2006) A text-mining analysis of the human phenome. Eur J Hum Genet 14: 535–542.
14. Stojmirovic A, Yu YK (2007) Information flow in interaction networks. J Comput Biol 14: 1115–1143.
15. Stojmirovic A, Yu YK (2012) Information flow in interaction networks II: channels, path lengths, and potentials. J Comput Biol 19: 379–403.
16. Li Y, Patra JC (2010) Genome-wide inferring gene-phenotype relationship by walking on the heterogeneous network. Bioinformatics 26: 1219–1224.
17. Davis AP, Murphy CG, Johnson R, Lay JM, Lennon-Hopkins K, et al. (2013) The Comparative Toxicogenomics Database: update 2013. Nucleic Acids Res 41: D1104–1114.
18. Amberger J, Bocchini C, Hamosh A (2011) A new face and new challenges for Online Mendelian Inheritance in Man (OMIM). Hum Mutat 32: 564–567.
19. Davis AP, Wiegers TC, Rosenstein MC, Mattingly CJ (2012) MEDIC: a practical disease vocabulary used at the Comparative Toxicogenomics Database. Database (Oxford) 2012: bar065.
20. Stojmirovic A, Yu YK (2011) ppiTrim: constructing non-redundant and up-to-date interactomes. Database (Oxford) 2011: bar036.
21. Razick S, Magklaras G, Donaldson IM (2008) iRefIndex: a consolidated protein interaction database with provenance. BMC Bioinformatics 9: 405.
22. Ashburner M, Ball CA, Blake JA, Botstein D, Butler H, et al. (2000) Gene ontology: tool for the unification of biology. The Gene Ontology Consortium. Nat Genet 25: 25–29.
23. Kanehisa M, Goto S (2000) KEGG: kyoto encyclopedia of genes and genomes. Nucleic Acids Res 28: 27–30.

24. Stojmirovic A, Yu YK (2010) Robust and accurate data enrichment statistics via distribution function of sum of weights. Bioinformatics 26: 2752–2759.

25. Palla G, Derenyi I, Farkas I, Vicsek T (2005) Uncovering the overlapping community structure of complex networks in nature and society. Nature 435: 814–818.

26. Farkas I, Abel D, Palla G, Vicsek T (2007) Weighted network modules. New J Phys 9: 180–197.

27. Aldahmesh MA, Khan AO, Mohamed JY, Alkuraya H, Ahmed H, et al. (2011) Identification of ADAMTS18 as a gene mutated in Knobloch syndrome. J Med Genet 48: 597–601.

28. Boyden LM, Mao J, Belsky J, Mitzner L, Farhi A, et al. (2002) High bone density due to a mutation in LDL-receptor-related protein 5. N Engl J Med 346: 1513–1521.

29. Gong Y, Slee RB, Fukai N, Rawadi G, Roman-Roman S, et al. (2001) LDL receptor-related protein 5 (LRP5) affects bone accrual and eye development. Cell 107: 513–523.

30. Toomes C, Bottomley HM, Jackson RM, Towns KV, Scott S, et al. (2004) Mutations in LRP5 or FZD4 underlie the common familial exudative vitreoretinopathy locus on chromosome 11q. Am J Hum Genet 74: 721–730.

31. Filipe-Santos O, Bustamante J, Haverkamp MH, Vinolo E, Ku CL, et al. (2006) X-linked susceptibility to mycobacteria is caused by mutations in NEMO impairing CD40-dependent IL-12 production. J Exp Med 203: 1745–1759.

32. Orange JS, Levy O, Brodeur SR, Krzewski K, Roy RM, et al. (2004) Human nuclear factor kappa B essential modulator mutation can result in immunodeficiency without ectodermal dysplasia. J Allergy Clin Immunol 114: 650–656.

33. Voermans NC, Bonnemann CG, Lammens M, van Engelen BG, Hamel BC (2009) Myopathy and polyneuropathy in an adolescent with the kyphoscoliotic type of Ehlers-Danlos syndrome. Am J Med Genet A 149A: 2311–2316.

34. Sirko-Osadsa DA, Murray MA, Scott JA, Lavery MA, Warman ML, et al. (1998) Stickler syndrome without eye involvement is caused by mutations in COL11A2, the gene encoding the alpha2(XI) chain of type XI collagen. J Pediatr 132: 368–371.

35. Medina-Kauwe LK, Tobin AJ, De Meirleir L, Jaeken J, Jakobs C, et al. (1999) 4-Aminobutyrate aminotransferase (GABA-transaminase) deficiency. J Inherit Metab Dis 22: 414–427.

36. Seven M, Yüksel A, Özkılıç A, Elçioğlu N (2000) A variant of Cenani-Lenz type syndactyly. Genet Couns 11: 41–47.

37. Babenko AP, Polak M, Cave H, Busiah K, Czernichow P, et al. (2006) Activating mutations in the ABCC8 gene in neonatal diabetes mellitus. N Engl J Med 355: 456–466.

38. Bienengraeber M, Olson TM, Selivanov VA, Kathmann EC, O'Cochlain F, et al. (2004) ABCC9 mutations identified in human dilated cardiomyopathy disrupt catalytic KATP channel gating. Nat Genet 36: 382–387.

39. Hellemans J, Coucke PJ, Giedion A, De Paepe A, Kramer P, et al. (2003) Homozygous mutations in IHH cause acrocapitofemoral dysplasia, an autosomal recessive disorder with cone-shaped epiphyses in hands and hips. Am J Hum Genet 72: 1040–1046.

40. Mullor JL, Sanchez P, Ruiz i Altaba A (2002) Pathways and consequences: Hedgehog signaling in human disease. Trends Cell Biol 12: 562–569.

41. Miwa JM, Freedman R, Lester HA (2011) Neural systems governed by nicotinic acetylcholine receptors: emerging hypotheses. Neuron 70: 20–33.

5

Prediction of Interactions between Viral and Host Proteins Using Supervised Machine Learning Methods

Ranjan Kumar Barman[1], Sudipto Saha[3]*, Santasabuj Das[1,2]*

1 Biomedical Informatics Centre, National Institute of Cholera and Enteric Diseases, Kolkata, West Bengal, India, **2** Division of Clinical Medicine, National Institute of Cholera and Enteric Diseases, Kolkata, West Bengal, India, **3** Bioinformatics Centre, Bose Institute, Kolkata, West Bengal, India

Abstract

Background: Viral-host protein-protein interaction plays a vital role in pathogenesis, since it defines viral infection of the host and regulation of the host proteins. Identification of key viral-host protein-protein interactions (PPIs) has great implication for therapeutics.

Methods: In this study, a systematic attempt has been made to predict viral-host PPIs by integrating different features, including domain-domain association, network topology and sequence information using viral-host PPIs from VirusMINT. The three well-known supervised machine learning methods, such as SVM, Naïve Bayes and Random Forest, which are commonly used in the prediction of PPIs, were employed to evaluate the performance measure based on five-fold cross validation techniques.

Results: Out of 44 descriptors, best features were found to be domain-domain association and methionine, serine and valine amino acid composition of viral proteins. In this study, SVM-based method achieved better sensitivity of 67% over Naïve Bayes (37.49%) and Random Forest (55.66%). However the specificity of Naïve Bayes was the highest (99.52%) as compared with SVM (74%) and Random Forest (89.08%). Overall, the SVM and Random Forest achieved accuracy of 71% and 72.41%, respectively. The proposed SVM-based method was evaluated on blind dataset and attained a sensitivity of 64%, specificity of 83%, and accuracy of 74%. In addition, unknown potential targets of hepatitis B virus-human and hepatitis E virus-human PPIs have been predicted through proposed SVM model and validated by gene ontology enrichment analysis. Our proposed model shows that, hepatitis B virus "C protein" binds to membrane docking protein, while "X protein" and "P protein" interacts with cell-killing and metabolic process proteins, respectively.

Conclusion: The proposed method can predict large scale interspecies viral-human PPIs. The nature and function of unknown viral proteins (HBV and HEV), interacting partners of host protein were identified using optimised SVM model.

Editor: Dinesh Gupta, International Centre for Genetic Engineering and Biotechnology (ICGEB), India

Funding: This project was supported by the Indian Council of Medical Research [extramural project (IRIS ID: 2013-1551G)]. SS thanks the Department of Biotechnology for a Ramalingaswami fellowship (BT/RLF/Re-entry/11/2011). The funders had no role in study design, data collection and analysis, decision to publish, or preparation of the manuscript.

Competing Interests: The authors have declared that no competing interests exist.

* Email: ssaha4@jcbose.ac.in (SS); dasss@icmr.org.in (SD)

Introduction

Viral pathogens affect their eukaryotic host partly by interacting with the proteins of the host cells [1]. Virus-host PPIs are crucial for better understanding of the mechanisms and pathogenesis of infectious diseases [2]. Several computational methods have been proposed to predict protein-protein interactions, but most are designed for intra-species PPIs and only a few for inter-species PPIs. Widely used machine-learning methods for PPIs are SVM, Naïve Bayes and Random forest [3,4,5]. Shen et al. used protein sequence information to predict human PPIs by employing SVM with a kernel function and a conjoint triad method, in which the best model predicted with an average accuracy of 83.90% [6]. Guo et al. predicted yeast PPIs with an accuracy of 88.09% using auto covariance (AC) and support vector machines (SVM) [7]. In contrast, Wu et al. predicted yeast PPIs by mining the knowledge of functional associations from the GO-based annotations [8]. Jansen et al. has developed a Bayesian networks approach to predict PPIs in yeast [4], while Lin et al. shows that Random Forest (RF) model may be more effective than Bayesian networks for predicting PPIs [5]. In addition, a number of computational methods are also available in order to predict PPIs based on domain information [9–11]. However, relatively few methods have so far been proposed to predict interspecies (specifically host-pathogen) PPIs [3,12–16]. For example, Cui et al. used relative frequency of amino acid triplets of protein sequence to predict the interactions between two types of viruses (hepatitis C virus and human papillomaviruses) and human proteins [3]. Proposed SVM methods of Cui et al. had an accuracy yield over 80%. Dyer et al. also proposed a method to predict physical interactions between human and HIV proteins based on a number of features, such as domain profiles, protein sequence k-mers and properties of human proteins in a human PPI network [12]. At a precision value of 70%, their method achieved recall (sensitivity) values of 40%.

In this paper, we have made an attempt to predict viral-host (inter-species) PPIs based on three well-known supervised machine-learning methods, namely SVM, Naïve Bayes and Random Forest using significantly diverse biological information like protein sequence, domain-domain associations, disorder regions, degree and amino-acids composition of viral and human proteins. The viral-host PPIs dataset were obtained from VirusMINT, a viral protein interaction database [17]. We have shown that only four features can able to predict viral-host PPIs with high degree of accuracy, which is comparable to the existing prediction models for viral-host PPIs. Furthermore we have shown that the viral protein amino acids composition (methionine, serine and valine) plays an important role in viral-host PPIs. An attempt was made to predict unknown PPIs between hepatitis B virus (HBV)-human proteins and hepatitis E virus (HEV)-human proteins using our proposed SVM optimal model. Predicted significant protein pairs were grouped using hierarchical clustering analysis (HCA) and validated using gene ontology enrichment analysis. Overall, the proposed support vector machines (SVM)-based machine learning technique was able to predict unknown viral-host protein interaction pairs with reasonable accuracy, which may be subjected to experimental validation.

Materials and Methods

2.1 Datasets

2.1.1 Data preparation. The dataset used were obtained from "VirusMINT: a viral protein interaction database" (ftp://mint.bio.uniroma2.it/pub/virusmint/MITAB/current/2012-10-26-mint-viruses-binary.mitab26.txt) [17]. VirusMINT database emphasises on interaction between human and some of the medically significant viruses: human immunodeficiency virus 1 (HIV-1), simian virus 40 (SV40), hepatitis B virus (HBV), hepatitis C virus (HCV), papilloma virus. Unique and positive 1,146 viral-host PPIs were derived from initial 2,707 interactions, after eliminating 1,224 repetitive interactions (Vprot A-Hprot B and Hprot B-VprotA) and 337 interacting protein pairs not having any "InterPro" domain hit. Out of these, 1,035 interactions were found between viral and human proteins and 111 interactions between viral proteins and proteins of others species including mouse, rat, dog and bovine. Furthermore, non-redundant interaction analysis based on the homologous viral proteins present in training and testing sets showed that 0.77% of the viral-human PPIs were redundant (shown in Table S1). We used cd-hit-2d webserver (http://weizhong-lab.ucsd.edu/cdhit_suite/cgi-bin/index.cgi?cmd=cd-hit-2d) at 85% sequence identity level to find the homologous proteins present in the training and testing sets [18]. Since, large numbers of viral-human PPIs were distinct (99.23%) and there were only few (1,035) viral-human PPIs in the initial set, we considered all 1,035 positive interactions between the viral and human proteins as training and testing datasets in our 5-fold cross-validation study (Table S2).

2.1.2 Negative training and testing dataset. Ben-Hur et al. [19] proposed that in the case of predicting protein-protein interactions, a simple uniform random choice of non-interacting protein pairs yield an unbiased estimate of the true distribution. In absence of experimentally proven non-interacting protein pairs, which are considered as an ideal negative dataset, we choose random 1,035 viral-human protein pairs that were not found in the positive training and testing datasets in our study as the negative dataset. In order to avoid prediction bias, we generated a negative dataset with the same number of viral-human PPIs (positive:negative = 1:1) as the positive dataset (Table S3).

2.1.3 Blind dataset. 111 positive interactions between viral and non-human species proteins, which were not used in 5-fold cross-validation, were considered as a blind dataset to avoid overfitting problem in building our optimal model for predictions (Table S4). Non-redundant interaction analysis, based on the homologous proteins present in the training and blind sets showed that 8.11% interactions between the viral and non-human species proteins were redundant (shown in Table S1). Therefore we removed 9 redundant interactions between the viral and non-human species proteins from the blind dataset. Like the negative training and test dataset, 102 negative viral and non-human species protein pairs were also generated (Table S5).

2.1.4 Independent dataset. In order to predict unknown viral and human PPIs, we focused on some of the medically significant viruses, such as hepatitis B and hepatitis E. Instead of taking all the proteins of hepatitis B, we concentrated on the proteins of hepatitis B virus genotype C that is prevalent in the eastern India [20]. Thus, reviewed 4 hepatitis B virus proteins (genotype C) with InterPro domain hits were obtained from Swiss-Prot [21]. Begum et al. observed that hepatitis E virus genotype 4 'e' is prevalent in the Northern India [22], while Caron et al. found that genotype 1 of hepatitis E virus is most prevalent in the Asian countries [23]. Hence, reviewed 3 hepatitis E virus proteins (genotype 4 'e') and 3 hepatitis E virus proteins (genotype 1) with InterPro domain hits were retrieved from Swiss-Prot. Reviewed 17,615 human proteins with InterPro domain hits were also retrieved from Swiss-Prot.

2.2 Machine Learning Techniques (MLT)

We focused on three well-known supervised machine learning methods, such as SVM, Naïve Bayes and Random Forest that were used for predicting PPIs [3,4,5].

Table 1. List of best 4 features selected based on categorical regression method.

Features	Beta	Bootstrap (1000) Estimate of Std. Error	df	F	Sig. (P_Value)
Average domain-domain association score	0.511	0.016	1.000	982.607	0.000
Virus Methionine	0.070	0.021	1.000	10.911	0.001
Virus Serine	0.106	0.021	1.000	25.838	0.000
Virus Valine	0.094	0.023	1.000	16.829	0.000

Table 2. Comparison of performance between selected best 4 features vs all 44 features.

Method	All Features			Selected Features		
	Accuracy (%)	Area under ROC curve	F1 Score (%)	Accuracy (%)	Area under ROC curve	F1 Score (%)
Naïve Bayes	67.48	0.66	56.72	68.50	0.71	54.35
SVM	**68.00**	**0.72**	**65.04**	**71.00**	**0.73**	**69.41**
Random Forest	71.69	0.77	67.13	72.41	0.76	66.39

2.2.1 SVM. Support Vector Machines (SVM)-based method is defined over a vector space where the problem is to find a decision surface that maximizes the margin between data points in the two classes. We have used SVM^{light} tool provided by T. Joachims [24], which allows users to select various parameters and various kernel functions like linear, polynomial, radial basis function (RBF), sigmoid to find optimal parameters for each task.

2.2.2 Naïve Bayes. Naïve Bayes model computes subsequent probabilities for a given hypothesis (present/absence) assuming that the features that describe data instances are conditionally independent. Its performance is comparable to other supervised learning methods. We used Waikato Environment for Knowledge Analysis (WEKA) machine learning tool box to perform Naïve Bayes classification [25]. Since WEKA does not allow us to select different parameters set for Naïve Bayes classification, we used the default parameters set.

2.2.3 Random Forest. Random Forest (RF) classifier "grows" several Decision Trees (DTs) simultaneously where each node uses a random subset of the features. Each tree in the Random Forest classifies the new object, and "votes" for that class. The forest elects the classification based on majority vote (over all the trees in the forest). We obtained Random Forest (RF) classifier from the WEKA machine learning tool box. Optimal parameters were used for evaluation of the method.

2.3 Feature Vectors

We focused on forty-four features of protein pairs to produce feature vectors (Table S6). First, occurrence frequency of viral-host domain-domain association was used since domain-domain association plays an important role in protein-protein interactions [11]. Second, common domains observed in virus and host proteins were chosen and represented as binary format [0,1] (absence and presence of common domain observed in virus and host proteins in a particular protein pair represented by 0 and 1, respectively). Third, maximum degree of viral or human protein for a given viral-human protein pair was selected. Degrees of human proteins were collected from APID2NET (a Cytoscape plugin) and viral protein degrees were collected from viral-host PPIs. APID2NET provided us all possible PPIs from BIND, BioGrid, DIP, HPRD, IntAct and MINT databases [26]. Fourth, average percentages of disorder regions of protein pairs were selected, because intrinsically disordered proteins were found to be implicated in numerous cellular pro-cesses including signal transduction, transcriptional regulation and PPIs. We used "ESpritz: accurate and fast prediction

Table 3. SVM based performance on testing dataset (5-fold cross-validation) using parameters t = 2 (RBF kernel), and g = 1, c = 0.1, j = 2.

Threshold	Sensitivity (%)	Specificity (%)	Accuracy (%)	PPV (%)	MCC
0.8	37	91	64	76	0.32
0.7	46	89	67	78	0.39
0.6	52	85	68	76	0.40
0.5	59	80	69	75	0.42
0.4	**67**	**74**	**71**	**72**	**0.44**
0.3	69	70	70	69	0.42
0.2	73	65	69	68	0.41
0.1	76	59	68	65	0.38
0	80	51	66	62	0.35
-0.1	81	46	64	60	0.32
-0.2	83	40	62	58	0.28
-0.3	85	36	60	57	0.26
-0.4	87	29	58	55	0.23
-0.5	89	25	57	54	0.20
-0.6	89	20	54	53	0.15
-0.7	91	16	53	52	0.12
-0.8	91	11	51	51	0.08

Table 4. Comparison of performance measures among Naïve Bayes, SVM and Random Forest methods on testing dataset using 5-fold cross-validation technique in our study.

Methods	Sensitivity (%)	Specificity (%)	Accuracy (%)	PPV (%)	MCC	Area under ROC curve	F1 Score (%)
Naïve Bayes	37.49	99.52	68.50	98.80	0.47	0.71	54.35
SVMlight	67.00	74.00	71.00	72.00	0.44	0.73	69.41
Random Forest	55.66	89.08	72.41	82.26	0.48	0.76	66.39

of protein disorder" to gather percentage of disorder regions of proteins [27]. Finally, amino acid compositions of viral and host proteins were selected as a fifth to twenty-fourth and twenty-fifth to forty-fourth features of our proposed feature vectors. Since, Roy et al. proposed that amino acid composition (AAC) monomers feature is crucial for predicting PPIs [28].

2.4 Infer domain-domain associations

We inferred viral and host domain-domain associations from interacting protein pairs. Our goal was to find the frequency of a certain viral and host domain-domain association present in protein pairs. We collected all protein related "InterPro" domains from Protein Knowledgebase, UniProtKB (http://www.uniprot. org/) [21]. After retrieving the "InterPro" domain information, we computed viral domain-human domain association matrix (rows and columns represented host and viral domain names, respectively), using similar approach proposed by Sprinzak et al. [29]. The range of domain-domain association varies between 30 and 1, where 0 represents no association. We tried with two domain-domain association scores, such as Maximum Domain-Domain Association Score (MDDAS) and Average Domain-Domain Association Score (ADDAS). The MDDAS and ADDAS were calculated using following equations:

$$MDDAS \ of \ protein \ pair \ (P_i, P_j)$$
$$= Maximum(N_{mn}) \ \dots\dots\dots(i)$$

$Where \ N_{mn} \ is \ the \ total \ no. \ of \ protein$
$pairs \ that \ contain \ domain \ pair \ (d_m, d_n).$

$$ADDAS \ of \ protein \ pair \ (P_i, P_j)$$
$$= Maximum(N_{mn})/T_{ij} \dots\dots\dots(ii)$$

$Where \ T_{ij} \ is \ the \ all \ possible \ domain \ pairs \ of \ protein \ pair \ (P_i, P_j).$

2.5 Amino acid composition

Amino acid composition is the percentage of each amino acid present in a protein. Percentage of all twenty natural amino acids was calculated using the following equations:

$$Percentage \ of \ a \min o \ acid \ i$$
$$= \frac{total \ no. \ of \ a \min o \ acids(i)}{total \ no. \ of \ a \min o \ acids \ in \ proteins} \times 100\%$$

$[Where \ i \ can \ be \ any \ natural \ amino \ acid]$

2.6 Feature selection

The feature selection was performed by regression for categorical data method with beta coefficient >0.00 and p-value<0.05 for selection of best features using SPSS statistical analysis software, version 20 (SPSS, Chicago, IL, USA). The beta

Table 5. Comparison of proposed method with other viral-host PPIs prediction methods.

Performance Mesaure	Dyer et al. Dataset*		Performance Mesaure	Cui et al. Dataset*		
	Dyer et al. [12]	Proposed SVM Model		Shen et al. [6]	Proposed SVM Model	Cui et al. [3]
Sensitivity (%)	40.00	87.05	Accuracy (%)	78.00	80.00	82.00

*Partial dataset.

coefficient value is a measure of how strongly each "predictor variable" influences the "criterion variable". The higher beta coefficient value implies greater impact of the "predictor variable" on the "criterion variable".

2.7 5-fold cross-validation

We used 5-fold cross-validation to estimate performance of all methods. In 5-fold cross-validation, the dataset has been partitioned into 5 equally (or nearly equally) sized segments or folds. Consequently, 5 times of training and testing were performed such that each time a different fold of the data is held-out for testing while the remaining four folds are used for training. The overall performance of a method was calculated using average performance over five folds.

2.8 Performance measures

2.8.1 Threshold Dependent. Sensitivity (also referred to as recall), specificity, accuracy, PPV (Positive Prediction Value, also referred to as precision), Matthew's correlation coefficient (MCC) and F1 score were computed on 5-fold cross validation step. All the performance measures were based on a balanced dataset of 1:1 positive vs. negative examples. Sensitivity, specificity, accuracy, PPV, MCC and F1 score were calculated by the following equations:

$$Sensitivity = \frac{TP}{TP+FN} \times 100\%; \; Specificity = \frac{TN}{TN+FP} \times 100\%$$

$$Accuracy = \frac{TP+TN}{TP+FP+TN+FN} \times 100\%;$$

$$PPV = \frac{TP}{TP+FP} \times 100\%$$

$$MCC = \frac{TP \times TN - FP \times FN}{\sqrt{((TP+FP) \times (TP+FN) \times (TN+FP) \times (TN+FN))}}$$

$$F1 = 2 \times \frac{Sensitivity \times PPV}{Sensitivity + PPV} \times 100\%$$

Where, *True Positive (TP)* : *interacting protein pairs correctly identified as a PPIs.*
False Positive (FP) : *non−interacting protein pairs incorrectly identified as a PPIs.*
True Negative (TN) : *non−interacting protein pairs correctly identified as a non−interactingprotein pairs.*

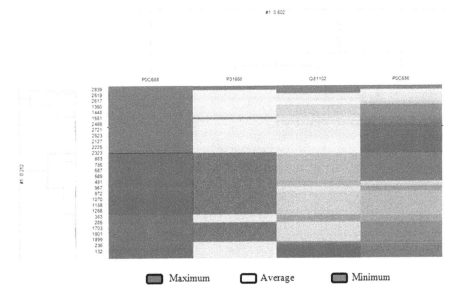

Figure 1. Hierarchical clustering of highly predicted SVM score of HBV-human protein pairs. Hierarchical clustering analysis was done using TIBCO Spotfire software with complete linkage clustering method, cosine correlation distance measure, average value ordering weight, scale between 0 and 1 normalization and empty value replace by 0 for both (row and column) dendrogram. The high, average and low SVM predicted scores are marked in red, white and blue, respectively.

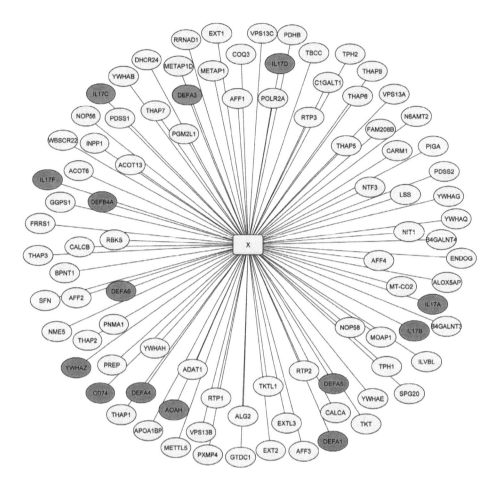

Figure 2. A network of HBX-human protein interactions predicted by our proposed method. The network visualized by Cytoscape 3.0.2 [35]. The HBX protein is represented by cyan node. The significant gene ontology enriched human proteins are representing by salmon node, whereas other human proteins are representing by slate grey node.

Table 6. The Gene Ontology Biological Process enrichment analysis on interacting human protein partners of HBV proteins using DAVID server.

Hepatitis B virus protein	GO term~Biological Process	Human protein
C	GO:0022406~membrane docking	SCFD1, SCFD2, VPS45, STXBP1, STXBP2, STXBP3
	GO:0006835~dicarboxylic acid transport	SLC1A4, SLC1A5, SLC1A2, SLC1A3,SLC1A1
	GO:0006865~amino acid transport	SLC1A4, CPT1B, SLC1A5, SLC1A2,CPT2, SLC1A3, XK, SLC1A1
X	GO:0001906~cell killing	DEFA6, DEFA5, DEFA4, DEFA3, DEFA1
	GO:0009620~response to fungus	DEFA6, DEFA5, DEFA4, DEFA3, DEFA1
	GO:0006952~defense response	YWHAZ, DEFB4A, CD74, IL17C,IL17D, IL17A, IL17B, DEFA6, AOAH, DEFA5, DEFA4, IL17F, DEFA3, DEFA1
P	GO:0051186~cofactor metabolic process	NAMPT, ACO2, HMGCR, ACO1,IREB2, GIF, PNP, SOD2, SDHA,GSS, MTHFS, PGLS, PANK2, PANK3, FXN, PANK1, NARFL, CTNS, NAPRT1, FH
	GO:0006732~coenzyme metabolic process	NAMPT, ACO2, HMGCR, ACO1, PNP, SOD2, SDHA, GSS, MTHFS, PGLS, PANK2, PANK3, PANK1, CTNS, NAPRT1, FH

Significant biological process annotation terms were filtered by FDR (false discovery rate) <0.05.

False Negative (FN) : interacting protein pairs incorrectly identified as a non − interactingprotein pairs.

2.8.2 Threshold independent. From Receiver Operating Characteristic (ROC) plot, area under ROC curve was computed on 5-fold cross validation step.

2.9 Hierarchical clustering analysis (HCA)

Hierarchical clustering analysis was done using TIBCO Spotfire software [30]. The input matrix was viral-host SVM prediction scores obtained from the best optimized model. Following parameters were used for HCA: complete linkage clustering method, cosine correlation distance measure, average value ordering weight, scale between 0 and 1 normalization and empty value replace by 0 for both (row and column) dendrogram.

2.10 GO Enrichment analysis

The Database for Annotation, Visualization and Integrated Discovery (DAVID) web server was used to identify significantly enriched gene ontology (GO) annotation terms in predicted interacting human protein partners of hepatitis B and E viruses [31]. We consider only GO biological process annotation terms of level greater than 2 with significant false discovery rate (FDR) value<0.05.

Results and Discussion

3.1 Selection of optimal features

We started with 44 features of a specific viral-host protein pair and tried with different subsets of features in order to achieve maximum accuracy with nearly equal sensitivity and specificity of our proposed method (Table S6, S7). Interestingly, we observed that four features with beta coefficient>0.00 and p-value<0.05 showed reasonably decent accuracy of 71%, sensitivity of 67% and specificity of 74% in proposed SVM method (shown in Table 1). As shown in Table 2, selected four features achieved slightly higher accuracy than all the forty-four features used together. Although disordered regions play a significant role in protein-protein interactions, it was not selected as the best feature based on our feature selection with regression (beta coefficient and p-value). Methionine residue interacts with aromatic residues and plays a specific role in stabilization of protein structure and may be associated with number of mutation and age related diseases [32]. Serine residues are crucial for serine/threonine protein phosphatises and control many cell functions [33], while valine residue was shown to play a vital role in modulating syncytium formation during infection [34].

3.2 Performance of SVM, NB and RF using 5-fold cross validation

In order to achieve optimal sensitivity, specificity and accuracy, we tried different kernels and parameters using SVM. The linear and polynomial kernel function showed high specificity, but low sensitivity, whereas the sigmoid kernel function exhibited poor sensitivity (Table S8). In contrast, the radial basis function (RBF) showed reasonable sensitivity of 67%, specificity of 74% and accuracy of 71% as shown in Table 3. We tried with different parameters in WEKA for Random Forest (shown in Table S9). SVM had nearly equal sensitivity (67%) and specificity (74%), whereas Naïve Bayes and Random Forest showed lower sensitivity (37.49% for NB, 55.66% for RF), but higher specificity (99.52% for NB, 89.08% for RF) (Table 4). As shown in Table 4, Random Forest perform better in terms of accuracy, MCC and area under ROC curve, whereas SVM perform better in terms of sensitivity

and F1 score. We are more concerned about the recall and precision, since they are directly proportional to the true positives. As shown in Table 4, recall score of SVM (67%) is better than RF (55.66%), while precision score of RF (82.26%) is better than SVM (72%). Therefore we computed the F1 score. F1 score of SVM and RF shows that, SVM (69.41) performs slightly better than RF (66.39). Therefore, we used the best SVM model for further study.

3.3 Assessment on blind dataset using SVM based method

In order to avoid bias in the performance of our proposed model, we tested it on blind dataset, not used in training or testing. Consequently, 204 protein pair between viral proteins and non-human species (mouse, rat, dog, bovine etc.) was considered as a blind dataset. We used the same parameters and cut-off (threshold) for each approach. As shown in Table 3, threshold value of 0.4 generated reasonable accuracy on the test dataset using 5-fold cross-validation technique in our study. At this threshold value, sensitivity of 64%, specificity of 83%, and accuracy of 74% was achieved on the blind dataset.

3.4 Comparison with other predictions methods for virus-host PPIs

Dyer et al. developed a method to predict HIV-human PPIs using SVM classifier with linear kernel on different combinations of protein features, including domain profiles, protein sequence k-mers and properties of human proteins in a human PPI network [12]. They predicted PPIs with a precision of 70% and a recall (also referred to as sensitivity) value greater of 40% using a combination of protein sequence four-mers, protein domains and PPI network information with 1:25 ratio of positive example (PE) to negative example. We obtained only 332 positive interactions instead of 1028 interactions reported by Dyer et al. between human and HIV proteins [12]. As shown in Table 5, our proposed method achieved the sensitivity of 87% whereas Dyer et al. achieved 40%.

Cui et al. worked on a similar problem of HCV-human and HPV-human PPIs using an SVM model with RBF kernel and relative frequency of amino acid triplets of a protein sequence. They have used 11 HCV (lead to 695 interactions) and 9 HPV proteins (lead to 252 interactions) [3]. From the available datasets in the supplementary tables, we can extract 1 HCV protein (leads to 10 positive and 9 negative interactions) and 1 HPV protein (leads to 9 positive and 7 negative interactions) from Swiss-Prot. Our proposed method achieved accuracy of 80% on this sparsely available dataset (shown in Table 5), whereas Shen et al. and Cui et al. achieved accuracy of 78% and 82%, respectively [3,6].

3.5 Prediction of unknown HBV-human and HEV-human PPIs

Hepatitis B virus and hepatitis E virus proteins were used in order to predict unknown viral-human PPIs. All possible combinations of hepatitis B virus and human protein pairs (17615 * 4) were predicted by our proposed model (Table S10). The predicted SVM score greater than 0.58 of hepatitis B virus-human protein pairs (n = 8411) was used for HCA. As shown in Figure 1, P0C688 (gene name P) was far apart from the other three hepatitis B viral proteins, out of which P31868 (gene name S) and Q81102 (gene name C) are closely associated. Similarly, all combinations of hepatitis E virus and human protein pairs (17615 * 6) were predicted by the proposed model (Table S11). The highly predicted SVM score of hepatitis E virus and human protein pairs were used for HCA. In HEV-host interaction pairs (n = 20,375) clustering analysis, Q9IVZ7 (gene name ORF3 and genotype 4) was far apart from the other five hepatitis E viral proteins, out of which Q9IVZ8 (gene name ORF2, genotype 4), Q9IVZ9 (gene name ORF1, genotype 4), P33424 (gene name ORF1, genotype 1) and P33426 (gene name ORF2, genotype 1) were closely associated (Figure S1).

Finally, the human proteins present in high confidence (red area) of hierarchical clustering analysis (Shown in Figure 1 and Figure S1) were used for further gene ontology enrichment analysis. The analysis on interacting human protein partners of hepatitis B virus (Shown in Figure 2 and Figure S2, S3) showed probable functions of viral "X protein" (UniProtKBId: P0C686), "C protein" (UniProtKBId: Q81102) and "P protein" (UniProtKBId: P0C688) (shown in Table 6 and full data on Table S12-S14). As shown in Table 6, HBV "C proteins" probably plays a significant role in membrane docking, while "X protein" and "P protein" function in cell killing and modulating metabolic processes of host proteins, respectively.

Similar study as hepatitis B virus proteins was also done with hepatitis E virus proteins (Shown in Figure S4-S9), where ORF1 (genotype 1) is probably involved in many biological processes including regulation of cytoskeleton organization, nitrogen compound biosynthetic process and translation (Shown in Table S15 and full data on Table S16-S21).

Conclusion

Here, we proposed three supervised machine learning-based techniques for predicting viral-host (across species) PPIs by incorporating potential biological information of protein pairs including domain-domain associations score, degree, percentage of disorder regions and amino acid compositions. Initially, we started with 44 features and predicted four best features, which were domain-domain association and methionine, serine and valine amino acid composition of viral proteins using categorical regression model (beta coefficient$>$0.00 and p-values$<$0.05). There are biological interpretations of these residues of viral proteins for their importance in viral-host PPIs. For example, methionine, serine and valine may be involved in stabilization of the protein structure, serine/threonine protein phosphatases and modulating syncytium formation during infection, respectively. It was observed that Random Forest perform better in terms of accuracy, MCC and area under ROC curve, while the proposed SVM method performs better in terms of sensitivity and F1 score. Performance of the proposed SVM method was evaluated on the blind dataset of 204 viral-host protein pairs (102 positive and 102 negative viral-host protein pairs), which achieved a sensitivity of 64%, specificity of 83%, and accuracy of 74%. In addition, unknown HBV-human and HEV-human PPIs were predicted using optimised SVM model and were grouped by HCA and further validated by GO enrichment analysis. Hepatitis B virus interacting human proteins show distinct GO biological process terms; for example, "X-protein" probably interferes with cell defence mechanism, whereas "P-protein" binds to metabolic pathways. The predicted viral-human PPIs give us hint about the possible role of viral proteins in the pathogenesis process.

Supporting Information

Figure S1 Hierarchical clustering of highly predicted SVM score of HEV-human protein pairs. Hierarchical clustering analysis was done using TIBCO Spotfire software with complete linkage clustering method, cosine correlation distance measure, average value ordering weight, scale between 0 and 1 normalization and empty value replace by 0 for both (row and column) dendrogram. The high, average and low SVM predicted scores are marked in red, white and blue, respectively. (PDF)

Figure S2 A network of HBC-human protein interactions predicted by our proposed method. The network visualized by Cytoscape 3.0.2 [35]. The HBC protein is representing by cyan node. The significant gene ontology enriched human proteins are representing by salmon node, whereas other human proteins are representing by slate grey node. (PDF)

Figure S3 A network of HBP-human protein interactions predicted by our proposed method. The network visualized by Cytoscape 3.0.2 [35]. The HBP protein is representing by cyan node. The significant gene ontology enriched human proteins are representing by salmon node, whereas other human proteins are representing by slate grey node. (PDF)

Figure S4 A network of HEORF1 (Genotype 1)-human protein interactions predicted by our proposed method. The network visualized by Cytoscape 3.0.2 [35]. The HEORF1 (Genotype 1) protein is representing by cyan node. The significant gene ontology enriched human proteins are representing by salmon node whereas other human proteins are representing by slate grey node. (PDF)

Figure S5 A network of HEORF2 (Genotype 1)-human protein interactions predicted by our proposed method. The network visualized by Cytoscape 3.0.2 [35]. The HEORF2 (Genotype 1) protein is representing by cyan node. The significant gene ontology enriched human proteins are representing by salmon node whereas other human proteins are representing by slate grey node. (PDF)

Figure S6 A network of HEORF3 (Genotype 1)-human protein interactions predicted by our proposed method. The network visualized by Cytoscape 3.0.2 [35]. The HEORF3 (Genotype 1) protein is representing by cyan node. The significant gene ontology enriched human proteins are representing by salmon node whereas other human proteins are representing by slate grey node. (PDF)

Figure S7 A network of HEORF1 (Genotype 4)-human protein interactions predicted by our proposed method. The network visualized by Cytoscape 3.0.2 [35]. The HEORF1 (Genotype 4) protein is representing by cyan node. The significant gene ontology enriched human proteins are representing by salmon node whereas other human proteins are representing by slate grey node. (PDF)

Figure S8 A network of HEORF2 (Genotype 4)-human protein interactions predicted by our proposed method. The network visualized by Cytoscape 3.0.2 [35]. The HEORF2 (Genotype 4) protein is representing by cyan node. The significant gene ontology enriched human proteins are representing by salmon node whereas other human proteins are representing by slate grey node. (PDF)

Figure S9 A network of HEORF2 (Genotype 4)-human protein interactions predicted by our proposed method. The network visualized by Cytoscape 3.0.2 [35]. The HEORF2

(Genotype 4) protein is representing by cyan node. The significant gene ontology enriched human proteins are representing by salmon node whereas other human proteins are representing by slate grey node.
(PDF)

Table S1 Statistic of homologous protein present in the training and testing sets as well as from the blind datasets.
(XLSX)

Table S2 Positive interactions dataset used in this study to build optimal model for prediction. The positive interactions dataset used in the study were obtained from VirusMINT.
(XLSX)

Table S3 Negative interactions dataset used in this study to build optimal model for prediction. The negative interactions dataset used in the study were chosen using random protein pairs which are not found in interacting protein pairs.
(XLSX)

Table S4 Positive interactions dataset used in this study as a positive blind dataset. The positive blind dataset used in the study were obtained from VirusMINT.
(XLSX)

Table S5 Negative interactions dataset used in this study as a negative blind dataset. The negative blind dataset used in the study were chosen using random protein pairs which are not found in interacting protein pairs (positive blind dataset).
(XLSX)

Table S6 All 44 input features.
(XLSX)

Table S7 SVM performance measures based on different subsets of features. Optimal parameters were used for respective subset of features.
(XLSX)

Table S8 Several SVM kernel-wise performance measures (sensitivity and specificity) on different models. Optimal parameters and threshold were used for respective kernel. In Model 1, 1st, 2nd, 3rd and 4th folds were used for training and 5th fold was kept for testing. In Model 2, 1st, 2nd, 3rd and 5th folds were used for training and 4th fold was left out for testing. In Model 3, 1st, 2nd, 4th, 5th folds were used for training and 3rd fold for testing. In Model 4, 1st, 3rd, 4th, 5th folds were used for training and 2nd fold was used for testing. In Model 5, 2nd, 3rd, 4th, 5th folds were used for training and 1st fold was kept aside for testing.
(XLSX)

Table S9 Different parameters used in Random Forest using WEKA.
(XLSX)

Table S10 Predicted scores of HBV-human protein-protein association by proposed optimal model.
(XLSX)

Table S11 HEV-human protein-protein association predicted scores by proposed optimal model.
(XLSX)

Table S12 GO enrichment analysis on interacting human protein partners of HBV X proteins using DAVID server. Significant biological process terms were chosen by $P_{Value} < 0.05$.

(XLSX)

Table S13 GO enrichment analysis on interacting human protein partners of HBV C proteins using DAVID server. Significant biological process terms were chosen by $P_{Value} < 0.05$.
(XLSX)

Table S14 GO enrichment analysis on interacting human protein partners of HBV P proteins using DAVID server. Significant biological process terms were chosen by $P_{Value} < 0.05$.
(XLSX)

Table S15 GO enrichment analysis on interacting human protein partners of HEV proteins using DAVID server. Significant biological process annotation terms were filter by FDR (false discovery rate) < 0.05.
(XLSX)

Table S16 GO enrichment analysis on interacting human protein partners of HEV ORF1 (Genotype 1) proteins using DAVID server. Significant biological process terms were chosen by $P_{Value} < 0.05$.
(XLSX)

Table S17 GO enrichment analysis on interacting human protein partners of HEV ORF2 (Genotype 1) proteins using DAVID server. Significant biological process terms were chosen by $P_{Value} < 0.05$.
(XLSX)

Table S18 GO enrichment analysis on interacting human protein partners of HEV ORF3 (Genotype 1) proteins using DAVID server. Significant biological process terms were chosen by $P_{Value} < 0.05$.
(XLSX)

Table S19 GO enrichment analysis on interacting human protein partners of HEV ORF1 (Genotype 4) proteins using DAVID server. Significant biological process terms were chosen by $P_{Value} < 0.05$.
(XLSX)

Table S20 GO enrichment analysis on interacting human protein partners of HEV ORF2 (Genotype 4) proteins using DAVID server. Significant biological process terms were chosen by $P_{Value} < 0.05$.
(XLSX)

Table S21 GO enrichment analysis on interacting human protein partners of HEV ORF3 (Genotype 4) proteins using DAVID server. Significant biological process terms were chosen by $P_{Value} < 0.05$.
(XLSX)

Acknowledgments

This project was supported by Indian Council of Medical Research [extramural project (IRIS ID: 2013-1551G)]. SS thanks Department of Biotechnology for Ramalingaswami fellowship (BT/RLF/Re-entry/11/2011).

Author Contributions

Conceived and designed the experiments: SD SS. Performed the experiments: RKB SS. Analyzed the data: RKB SS SD. Contributed to the writing of the manuscript: RKB SS SD.

References

1. Arnold R, Boonen K, Sun MG, Kim PM (2012) Computational analysis of interactomes: Current and future perspectives for bioinformatics approaches to model the host–pathogen interaction space. Methods 57: 508–518.

2. Zhou H, Jin J, Wong L (2013) Progress in computational studies of host-pathogen interactions. Journal of Bioinformatics and Computational Biology 11(2): 1230001 (26 pages).

3. Cui G, Fang C, Han K (2012) Prediction of protein-protein interactions between viruses and human by an SVM model. BMC Bioinformatics 13 Suppl 7:S5.

4. Jansen R, Yu H, Greenbaum D, Kluger Y, Krogan NJ, et al. (2003) A Bayesian networks approach for predicting protein-protein interactions from genomic data. Science 302(5644): 449–53.

5. Lin N, Wu B, Jansen R, Gerstein M, Zhao H (2004) Information assessment on predicting protein-protein interactions. BMC Bioinformatics 5: 154.

6. Shen J, Zhang J, Luo X, Zhu W, Yu K, et al. (2007) Predicting protein-protein interactions based only on sequences information. Proc Natl Acad Sci 104(11): 4337–41.

7. Guo Y, Yu L, Wen Z, Li M (2008) Using support vector machine combined with auto covariance to predict protein-protein interactions from protein sequences. Nucleic Acids Res 36(9): 3025–30.

8. Wu X, Zhu L, Guo J, Zhang DY, Lin K (2006) Prediction of yeast protein-protein interaction network: insights from the Gene Ontology and annotations. Nucleic Acids Res 34(7): 2137–50.

9. Binny Priya S, Saha S, Anishetty R, Anishetty S (2013) A matrix based algorithm for Protein-Protein Interaction prediction using Domain-Domain Associations. Journal of Theoretical Biology 326: 36–42.

10. Hayashida M, Kamada M, Song J, Akutsu T (2011) Conditional random field approach to prediction of protein–protein interactions using domain information. BMC Systems Biology 5 (Suppl 1):S8.

11. Memišević V, Wallqvist A, Reifman J (2013) Reconstituting protein interaction networks using parameter-dependent domain-domain interactions. BMC Bioinformatics 14: 154.

12. Dyer MD, Murali TM, Sobral BW (2011) Supervised learning and prediction of physical interactions between human and HIV proteins. Infect Genet Evol 11(5): 917–23.

13. Davis FP, Barkan DT, Eswar N, McKerrow JH, Sali A (2007) Host pathogen protein interactions predicted by comparative modeling. Protein Sci 16(12): 2585–96.

14. Tastan O, Qi Y, Carbonell JG, Klein-Seetharaman J (2009) Prediction of interactions between HIV-1 and human proteins by information integration. Pac Symp Biocomput, 516–27.

15. Qi Y, Tastan O, Carbonell JG, Klein-Seetharaman J, Weston J (2010) Semi-supervised multi-task learning for predicting interactions between HIV-1 and human proteins. Bioinformatics 26(18):i645–52.

16. Doolittle JM, Gomez SM (2011) Mapping protein interactions between Dengue virus and its human and insect hosts. PLoS Negl Trop Dis 5(2):e954.

17. Chatr-aryamontri A, Ceol A, Peluso D, Nardozza A, Panni S, et al. (2009) VirusMINT: a viral protein interaction database. Nucleic Acids Res 37 (Database issue): D669–73.

18. Li W, Godzik A (2006) Cd-hit: a fast program for clustering and comparing large sets of protein or nucleotide sequences. Bioinformatics 22(13): 1658–9.

19. Ben-Hur A, Noble WS (2006) Choosing negative examples for the prediction of protein-protein interactions. BMC Bioinformatics 7 Suppl 1:S2.

20. Datta S (2008) An overview of molecular epidemiology of hepatitis B virus (HBV) in India. Virology Journal 5: 156.

21. UniProt Consortium (2013) Update on activities at the universal protein resource (UniProt) in 2013. Nucleic Acids Res 41:D43–D47.

22. Begum N, Polipalli SK, Husain SA, Kar P (2010) Molecular analysis of swine hepatitis E virus from north India. Indian J Med Res 132: 504–508.

23. Caron M, Enouf V, Than SC, Dellamonica L, Buisson Y, et al. (2006) Identification of genotype 1 hepatitis E virus in samples from swine in Cambodia. Journal of Clinical Microbiology 44(9): 3440–2.

24. Joachims T (2002) Learning to Classify Text Using Support Vector Machines. Dissertation, Kluwer.

25. Hall M, Frank E, Holmes G, Pfahringer B, Reutemann P, et al. (2009) The WEKA Data Mining Software: An Update. SIGKDD Explorations 11: Issue 1.

26. Hernandez-Toro J, Prieto C, De Las Rivas J (2007) APID2NET: unified interactome graphic analyser. Bioinformatics 23(18): 2495–7.

27. Walsh I, Martin AJ, Di Domenico T, Tosatto SC (2012) ESpritz: accurate and fast prediction of protein disorder. Bioinformatics 28(4): 503–9.

28. Roy S, Martinez D, Platero H, Lane T, Werner-Washburne M (2009) Exploiting Amino Acid Composition for Predicting Protein-Protein Interactions. PLoS One 4(11):e7813.

29. Sprinzak E, Margalit H (2001) Correlated sequence-signatures as markers of protein-protein interaction. J Mol Biol 311: 681–692.

30. TIBCO Spotfire 5.5, Available: http://spotfire.tibco.com/. Accessed 2014.

31. Huang da W, Sherman BT, Lempicki RA (2009) Systematic and integrative analysis of large gene lists using DAVID bioinformatics resources. Nature Protocols 4(1): 44–57.

32. Valley CC, Cembran A, Perlmutter JD, Lewis AK, Labello NP, et al. (2012) The Methionine-aromatic Motif Plays a Unique Role in Stabilizing Protein Structure. Journal of Biological Chemistry 287: 34979–34991.

33. Depaoli-Roach AA, Park IK, Cerovsky V, Csortos C, Durbin SD, et al. (1994) Serine/threonine protein phosphatases in the control of cell function. Advances in Enzyme Regulation 34: 199–224.

34. Wilson DW, Davis-Poynter N, Minson AC (1994) Mutations in the cytoplasmic tail of herpes simplex virus glycoprotein H suppress cell fusion by a syncytial strain. Journal of Virology 68(11): 6985–93.

35. Smoot ME, Ono K, Ruscheinski J, Wang PL, Ideker T (2011) Cytoscape 2.8: new features for data integration and network visualization. Bioinformatics 27(3): 431–2.

6

A Network-Based Classification Model for Deriving Novel Drug-Disease Associations and Assessing Their Molecular Actions

Min Oh[1], Jaegyoon Ahn[2], Youngmi Yoon[1]*

1 Department of Computer Engineering, Gachon University, Seongnam, Korea, **2** Department of Integrative Biology and Physiology, University of California Los Angeles, Los Angeles, California, United States of America

Abstract

The growing number and variety of genetic network datasets increases the feasibility of understanding how drugs and diseases are associated at the molecular level. Properly selected features of the network representations of existing drug-disease associations can be used to infer novel indications of existing drugs. To find new drug-disease associations, we generated an integrative genetic network using combinations of interactions, including protein-protein interactions and gene regulatory network datasets. Within this network, network adjacencies of drug-drug and disease-disease were quantified using a scored path between target sets of them. Furthermore, the common topological module of drugs or diseases was extracted, and thereby the distance between topological drug-module and disease (or disease-module and drug) was quantified. These quantified scores were used as features for the prediction of novel drug-disease associations. Our classifiers using Random Forest, Multilayer Perceptron and C4.5 showed a high specificity and sensitivity (AUC score of 0.855, 0.828 and 0.797 respectively) in predicting novel drug indications, and displayed a better performance than other methods with limited drug and disease properties. Our predictions and current clinical trials overlap significantly across the different phases of drug development. We also identified and visualized the topological modules of predicted drug indications for certain types of cancers, and for Alzheimer's disease. Within the network, those modules show potential pathways that illustrate the mechanisms of new drug indications, including propranolol as a potential anticancer agent and telmisartan as treatment for Alzheimer's disease.

Editor: Alexey Porollo, Cincinnati Childrens Hospital Medical Center, United States of America

Funding: This research was supported by the Basic Science Research Program, through the National Research Foundation of Korea (NRF), funded by the Ministry of Education, Science and Technology (NRF-2010-0008639). The funders had no role in study design, data collection and analysis, decision to publish, or preparation of the manuscript.

Competing Interests: The authors have declared that no competing interests exist.

* Email: ymyoon@gachon.ac.kr

Introduction

Drugs cure diseases by targeting the proteins related to the phenotypes arising from the disease. However, drug development does not precisely follow the "one gene, one drug, one disease" paradigm, which has been challenged in many cases. The concept of polypharmacology was proposed for drugs acting on multiple targets rather than one target [1,2]. The polypharmacological concept can lead to drug repositioning, which involves finding new indications for existing drugs or side effects due to the molecular mechanisms that may underlie a chemical–disease connection [3,4].

In order to decipher how drugs exert their effect on diseases, it is important to understand how a drug acts on targets related to a disease phenotype, how a gene module causes an abnormal phenotype, and how, in consequence, the targets and causative genes interact with each other. Furthermore, it is of great importance to investigate how drugs exert their activities directly or indirectly via such gene modules, how patho-phenotypes are influenced by the abnormality of gene modules, and how drugs and disease phenotypes are associated on the basis of gene modules [5]. With this understanding, identifying and analyzing

how a drug and a disease are actually associated at the molecular level plays a crucial role in the prediction of new drug indications.

Currently, computational methods to predict potential drug-disease interactions can be divided into the drug-centric approach, the disease-centric approach, and the drug-disease mutual approach.

With the drug-centric approach, opportunities are sought to repurpose drugs using accumulated chemical or pharmaceutical knowledge. Keiser et al. applied an integrated chemical similarity approach to drug repositioning using structural similarities among drug compounds and knowledge of established compound-target relationships [6]. However, many physiological effects cannot be predicted by chemical properties alone because drugs undergo complex, largely uncharacterized metabolic transformations as they are metabolized and physiologically distributed [7].

The disease-centric approach mainly utilizes the characteristics of diseases from the perspective of disease management, symptomatology, or pathology [7]. This approach builds a diseasome, or group of diseases, by incorporating established knowledge about diseases, or it finds and uses the common characteristics of diseases associated with an existing drug. Hu and Agarwal established a disease-similarity network using gene expression profiles and incorporated into this network a body of knowledge about drugs [8]. Suthram et al. constructed a disease network and discovered functional modules common to diseases that are enriched for pluripotent drug target genes [9]. This disease-only-based approach relies heavily on data denoting the characteristics of diseases, and it can be affected by the quality of the data. Therefore, outcomes could be restricted according to the means used to measure gene expression profiles or phenotypic profiles, which represent the characteristics of diseases.

The drug-disease mutual approach is a combination of the two approaches described above. It can infer new therapeutic relationships between drugs and diseases by directly matching the biomolecular or chemical properties of drugs, or processed data pertaining to these properties, with the property data or processed data of diseases. Alternatively, it can infer relationships indirectly using related or higher-level data or representations of drugs and diseases. Utilizing knowledge of both drugs and diseases can be a complementary and successful strategy; in particular, this approach can overcome missing knowledge with regard to the pharmacology of a drug, such as unknown or additional targets [7].

Among drug-disease mutual approaches, one study that directly matched the properties of drugs and diseases constructed a signature of a drug and a signature of a disease using gene expression microarrays. This approach identified new therapeutic potentials of the drug by matching the two signatures [10]. Another attempt introduced the concept of a co-module, which is a representation of a drug-gene-disease relationship [11]. A network-based gene-closeness profile was defined to relate the drug to the disease, and new drug-disease associations were identified.

As another drug-disease mutual approach, Gottlieb et al. indirectly utilized the properties of drugs and diseases [12]. Based on the observation that similar drugs are indicated for similar diseases, they constructed drug-drug and disease-disease similarity measures and exploited these measures to construct classification features, with the subsequent learning of a classification rule. For reproducible implementation, it is limited to gather all the required properties of drugs and diseases.

With the increasing number and variety of high-throughput datasets, functional genetic networks are becoming more accurate and complete. These networks make it possible to understand how drugs and diseases are associated at the molecular level. If the network features of drug-disease associations can be properly selected, these can be used to infer novel indications or side effects of existing drugs with increased accuracy, providing more concrete evidence.

Here, we propose scoring methods to quantify drug-disease relationship. Network adjacencies of drug-drug, disease-disease were quantified. Furthermore distance between topological module of drug-drug and disease, and distance between topological module of disease-disease and drug were quantified. These quantified scores were used as features for the prediction of novel drug-disease associations. Our method obtains an AUC of 0.845 when predicting drug-disease associations and shows better performance compared to other methods. We confirmed that our prediction method covers a number of current clinical trials (over 34%). Also, we extracted the topological modules of novel predictions, which involve propranolol for certain types of cancer and telmisartan for Alzheimer's disease. The module of propranolol shows significant enrichment in cancer pathways and putative inhibition mechanism of cancer growth and proliferation. In addition, the module of telmisartan indicates its therapeutic action related with inhibition of the defective signaling that usually occurred in Alzheimer's disease. Our approach provides promising drug-disease relationships for drug repositioning and reveals potential mechanism of them.

Methods

The proposed method consists of two processes, as shown in Figure 1. In the first stage, the degree of the drug-disease association is scored by means of adjacency-based inference and module-distance-based inference. Detailed descriptions of the adjacency-based inference and module-distance-based inference methods are given in section "Two methods for scoring drug-disease associations". In the second stage, the scores from the first stage are regarded as features characterizing the drug-disease relationship; a classifier is subsequently built using these features by means of learning. With this classifier, predictions are made regarding whether an unknown drug-disease pair has an association. Finally, new drug-disease associations are discovered. The details of this stage are given in section "Characterizing a drug-disease relationship via features".

Datasets

We used an integrative genetic network that combines three types of protein and gene networks. They comprised 152,388 protein-protein interactions from the Online Predicted Human Interaction Database (version 2.0) [13], 13,106 gene regulation data consisting of 13,046 activation and 1,085 inhibition interactions, and 16,302 inferred protein-protein interactions from known protein complexes in the human pathways of the Pathway Interaction Database [14]. We assumed that each protein in a protein complex also has an interaction that is equal to the interactions of the protein complex. For instance, we can get four binary interactions when a protein complex with four proteins has an interaction with a gene [44]. In total, we used 177,672 interactions for the integrative genetic network, with 15,804 unique proteins. To integrate the networks, we mapped the proteins and genes with UniProt ID.

We obtained drugs and their targets from DrugBank [15]. We selected drugs FDA approved as candidate drugs having multiple targets (≥ 2) on the integrative genetic network, resulting in 832 drugs and 4,889 drug-target relationships. Diseases and their susceptible genes were sourced from OMIM (Online Mendelian Inheritance in Man) [16]. We secured 239 diseases having multiple susceptible genes (≥ 2) in the integrative network and 4,013 disease-gene relationships. We note that, since drug targets and disease susceptibility genes are not completely revealed and the networks including PPI and the gene regulatory network are incomplete, we used drugs and diseases with multiple target genes and susceptibility genes.

From the Comparative Toxicogenomics Database (CTD), we obtained known drug-disease associations (Sep. 2013, downloaded) [17]. The CTD consists of two types of chemical (drug)-disease relations: curated and inferred. We used only curated relations. We mapped chemicals to their DrugBank identifiers and diseases to their OMIM identifiers. Through the mapping processing, we obtained 5,201 drug-disease associations. We selected drug-disease instances with identifiers belonging to the list of drugs and diseases in our set. Finally, 1,295 known associations consisting of 377 drugs and 80 diseases were used in this study.

Two methods for scoring drug-disease associations

1) Adjacency-Based Inference. We devised a scoring method for drug-disease relationships based on known drug-disease associations and adjacencies. In this method, three

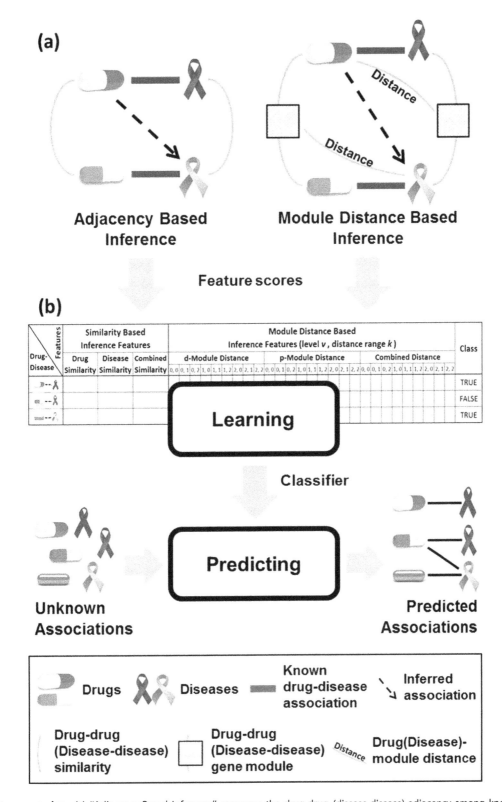

Figure 1. System overview. (a) "Adjacency-Based Inference" measures the drug-drug (disease-disease) adjacency among known drug-disease associations, and infers new drug-disease association. "Module-Distance-Based Inference" derives drug-drug (disease-disease) gene module among known drug-disease associations, measures the distance between the gene module and disease (drug), and infers new drug-disease association. (b) Drug-disease relationship represented by score becomes features. Various machine learning based classifiers are built with those features, and predict unknown drug-disease relationship.

approaches are used: drug-adjacency-based inference, disease-adjacency-based inference, and combined adjacency inference.

The basic idea of drug-adjacency-based inference stems from the hypothesis that if there is a known association between a drug and a disease, another similar drug would also have an association with the disease. Figure 2 (a) describes this concept. When d' is the drug and p is the disease in a known association, d, which is adjacent to d', can be inferred to have an association with p. It can be said that the inferred association is stronger if the adjacency score is higher; therefore, the drug-drug adjacency score of d and d' is used as the measure of the inferred association.

The disease-adjacency-based inference approach infers a drug-disease association based on disease-disease adjacency. It is described in Figure 2 (b).

The combined-adjacency inference approach combines the scores from drug-adjacency-based inference and disease-adjacency-based inference. Although the ways in which drug-target proteins and disease-genes work during the biological process differ, the same scoring method is applied. This method prevents one of the scores from being given a greater weight and considers both compounds' mechanisms of action and disease molecular pathologies.

These three approaches are heavily affected by the degree of adjacencies of the drug-drug or disease-disease relations. A higher adjacency score implies a tighter drug-disease association. Measuring the adjacency score is crucial. We used target proteins for drugs and disease genes for diseases. In the integrative genetic network, the closeness of the association between each target protein set of two drugs is measured, and this measure is used as the drug-drug adjacency score. Also, the closeness of disease gene sets in the integrative network is used for the disease-disease adjacency score. Figures 2 (d) and (e) display examples of drug-drug (disease-disease) adjacency.

Figure 2 (f) displays the process of scoring the drug-drug adjacency in Figure 2 (d). In the integrative genetic network, when finding a shortest path for each pair of proteins, one protein from drug d and another protein from drug d' are shown. There are three types of paths: R0, R1, and R2. In R0, the target protein of d and the target protein of d' are identical. In R1, the target protein of d and the target protein of d' are directly connected. In R2, the

target protein of d and target protein of d' are indirectly connected by means of another protein between them. If the length of a path is shorter, the score is higher. The scores of all pairs are summed and then scaled according to the number of targets. The scaled score becomes the drug-drug adjacency score, as follows:

$$Adj(d,\, d') = \frac{\displaystyle\sum_{t_i \in T(d)}\ \sum_{t_j \in T(d')} S_{t_i\, t_j}}{scaling\ factor}, \qquad (1)$$

Here, t_i is a target protein of drug d in its target set $T(d)$, and t_j is a target protein of drug d'; $scaling\ factor$ essentially denotes the number of proteins in $T(d)$ multiplied by the number of proteins in $T(d')$, which allows $Adj(d,\, d')$ to vary from 0 to 1. Additionally, $S_{t_i\, t_j}$ is the shortest path between t_i and t_j, and it is one of three types of paths, i.e., R0, R1, and R2. We set the score of R0 to the reciprocal number of median degree of a network, which is one sixth. We then set R1 and R2 to the square of R0 and the cube of R0, respectively. Among the drug-drug adjacency scores for multiple drugs, the maximum value becomes the final score for the association between d and p, as follows:

$$A(d,\, p) = \max_{1 \le k \le n}\left(Adj\left(d, d'_k\right)\right)\ \ d \ne d'_k, \qquad (2)$$

In this equation, n is the number of drugs that have a known association with disease p. We computed disease-disease adjacency scores in a similar manner using disease genes in the network. The scores for the association between d and p from the drug-drug and disease-disease adjacency scores are combined into a single score by computing their weighted geometric mean, as follows:

$$C(d,\, p) = \sqrt{A_D(d,\, p)\ \times\ A_P(d,\, p)}. \qquad (3)$$

Each $A_D(d,\, p)$ and $A_P(d,\, p)$ value indicates each maximum drug-drug and disease-disease adjacency score for the association between d and p.

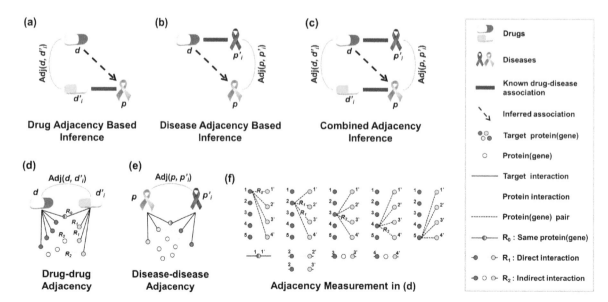

Figure 2. Adjacency-Based Inference.

2) Module-Distance-Based Inference.   We selected a topologically related gene set that is called a topological module. The topological module that is shared by two drugs is extracted for a particular disease. One drug is from a known drug-disease association and the other is among the candidate drugs for the disease. This topological module common to two drugs is called d-module, and is used for repurposing a new drug-disease association. In the same manner, a topological module that is common to two diseases is called p-module, and is extracted to discover new drug-disease associations. We applied three approaches: d-module distance-based inference, p-module distance-based inference, and a combined module-distance inference method.

From the two drugs $\langle d, d' \rangle$, the common topological drug-drug gene module (d-module) is derived and the distance between the d-module and the disease is measured, as described in Figure 3 (a). This module distance indicates how closely common features of the drugs actually relate to the disease. In our assumption, the higher the distance value, the more likely it is that the pathway of the drug in the known drug-disease association is shared with the other drug's expected pathway.

Inferring drug-disease association by means of the drug-drug topological module is named as d-module distance-based inference. As shown in Figure 3 (d), every interaction in each target protein set from d and d' is displayed in the integrative genetic network, and the d-module is extracted according to the level parameter v. The distance between the d-module at each level and the disease is measured according to the distance parameter k. In Figure 3 (e), the module distance between the d-module in level 2 and the disease is measured. Given drugs d and d' and disease p, the d-module distance is computed as follows:

$$Mdis(d, d') = \frac{\sum\limits_{t_i \in Mod_v(d, d')} \sum\limits_{g_j \in T(p)} S^k_{t_i g_j}}{scaling\ factor}, \qquad (4)$$

Here, t_i is a protein of d-module $Mod_v(d, d')$ and g_j is a gene of disease p; *scaling factor* essentially denotes the number of proteins in $Mod_v(d, d')$ multiplied by the number of genes in $T(p)$, which allows $Mdis(d, d')$ to vary from 0 to 1. Also, $S^k_{t_i g_j}$ is the score of the path between t_i and g_j; k is a fixed length of the path, which is used to calculate $S^k_{t_i g_j}$. The value of $S^k_{t_i g_j}$ changes according to k, as follows:

$$S^k_{t_i g_j} = \begin{cases} R0 & \text{if } k=0 \\ R1 & \text{if } k=1 \\ R2 & \text{if } k=2 \end{cases} . \qquad (5)$$

In the case of $k = 0$, only the intersections between t_i and g_j receive a score. When $k = 1$, only the paths whose length is one receive a score. When $k = 2$, only the paths whose length is two are scored. We set the score of $R0$ to the inverse number of median degree of a network, which is one sixth. We then set $R1$ and $R2$ to the square of $R0$ and the cube of $R0$, respectively. Among the module distances for multiple drugs, the maximum values become the final score for the association between d and p, as follows:

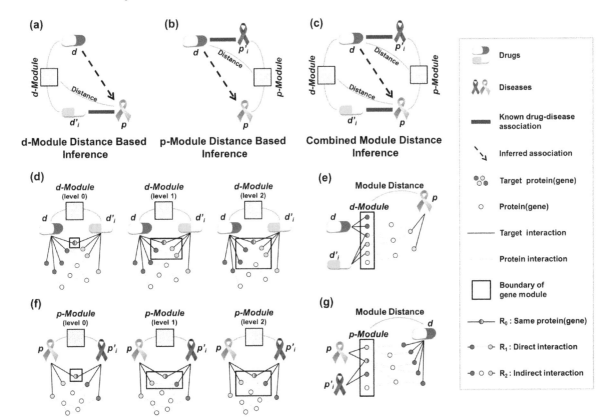

Figure 3. Module-Distance-Based Inference.

$$M(d, p) = \max_{1 \leq i \leq n} \left(Mdis(d, d'_i) \right) \quad d \neq d'_i, \tag{6}$$

In this equation, n denotes the number of drugs with a known association with disease p. When the d-module of d and d' is closely related to the disease genes, we can expect that the two drugs show a similar biological function.

In the p-module distance-based inference method, the common topological disease-disease gene module (p-module) from two diseases is extracted, as shown in Figures 3 (b), (f), and (g). The p-module represents the pathological molecules shared by the two diseases. The shorter the module distance between the p-module and the drug, the more safely it can be assumed that the drug also works for the other disease. The method of calculating the distance follows the d-module-based inference method.

The combined module-distance inference method is a combination of the previously described d-module distance-based inference and the p-module distance-based inference methods. It is expressed as follows:

$$C(d, p) = \sqrt{M_D(d, p) \times M_P(d, p)}, \tag{7}$$

Here, $M_D(d, p)$ and $M_P(d, p)$ indicate the maximum of the d-module distance-based inference and p-module distance-based inference methods, respectively. It makes the two methods complementary while considering not only the molecular activity shared by the drugs but also the causative genes shared by the diseases (Figure 3 (c)).

Characterizing a drug-disease relationship via features

The scores from adjacency-based inference and module-distance-based inference are converted into features, and a classifier predicting a new drug-disease association is thereby learned. The drug-adjacency-based inference score, disease-adjacency-based inference score, and combined adjacency inference score from the adjacency-based inference method become the first three features for each drug-disease relationship.

The d-module distance-based inference scores, p-module distance-based inference scores, and combined module-distance inference scores from the module-distance-based inference are used as 27 features. In the process of calculating the module, we set the level parameter v to range from 0 to 2, which denotes the scope of the module. When calculating the shortest path between proteins in a module and the target proteins (disease genes) of a drug (disease), we set the distance parameter k to range from 0 to 2, which determines the shortest paths (R0~R2 in Figure 3). Accordingly, 27 features are generated from the module-distance-based inference method. Each drug-disease pair has 30 features overall, including 3 features from the adjacency-based inference. We note that calculating all 30 features for 2,590 positive and negative associations takes about five minutes on Intel Core i7 CPU (3.50 GHz).

The training set used for 10-fold cross-validation includes 1,295 known drug-disease associations as a positive set and randomly generated drug-disease pairs as a negative set. The negative set is randomly generated from drugs and diseases in the positive set, taking the same size as the positive set. We note that a random negative set might give optimistic results, but it is challenging to create an exact negative set. To obtain a precise AUC score, 10-fold cross-validation is independently conducted 10 times. Each cross-validation is conducted with different random negative sets, and generates an AUC score. We then average the resulting AUC scores. Table S1 shows one sample of an actual training data set. A classifier is learned with these features. Ten-fold cross-validation is done for a performance evaluation.

Results

We selected 1,295 known drug-disease associations from the CTD and their elements, along with 377 drugs and 80 diseases. The integrative genetic network used here consists of a gene regulation database and inferred and experimental protein interaction databases. For each drug-disease pair, specific feature scores are calculated using adjacency-based inference and module-distance-based inference on top of the integrative genetic network, and a classifier is learned with the feature scores. This classifier predicts unknown drug-disease associations.

Performance evaluation

For performance testing, independent 10-fold cross-validations were conducted ten times. The training set in each 10-fold cross-validation consisted of a positive set of true drug-disease associations and a negative set of randomly generated drug-disease pairs. Each training set was arbitrarily separated into 10 parts (trained on nine of them and tested on the remaining one), and the process was repeated ten times for cross-validation. This procedure was applied to all of the ten different training sets, and a random negative set was respectively generated for each of them. The resulting AUC scores of the ten 10-fold cross-validations were averaged. We used C4.5, Multilayer Perceptron and Random Forest, as implemented in Weka v3.6 [18]. The 10-fold cross-validation results with the 10 training data sets are shown in Figure 4 and Table S2. The highest AUC (area under the ROC) is 0.855.

In our study, two methods were implemented for scoring the features. Figure 5 shows the AUC when each method was used alone and when both methods were used. The performance when both methods were used exceeds that when only one method was used. In Table S3, we evaluated the contribution of all features, when both methods were used. We also used the integrative genetic network, which consists of gene and protein interaction databases. Figure 6 shows the AUC of three individual networks and the integrative network. Using PPI only results in a higher AUC compared to the use of gene regulation data alone or the use of the inferred PPI alone. Additionally, using integrative networks shows a slightly better AUC for C4.5 compared to the use of PPI alone.

Comparison with other methods

We compared our method with two previous methods, PREDICT [12] and CMap [19]. PREDICT observes similar drugs that are indicated for similar diseases based on multiple drug-drug and disease-disease similarity measurements. PREDICT uses a gold standard set of drug-disease associations as known associations. We obtained our classification result using this gold standard set and compared their classification performances. Out of 1,933 associations, 247 known associations, composed of 179 drugs and 80 diseases that have multiple targets and multiple susceptibility genes, were used to make the training set. The AUC score of our method shows slight better performance (AUC = 0.917) than that of PREDICT (AUC = 0.900).

CMap searches for drug response gene-expression profiles that relate to the disease signature and predicts drug-disease associations. We downloaded 21 disease gene signatures out of 80 diseases in our study from ArrayExpress [42]. Given that CMap is restricted to include only signatures having up-regulated or down-regulated genes and to include signatures not exceeding 1,000

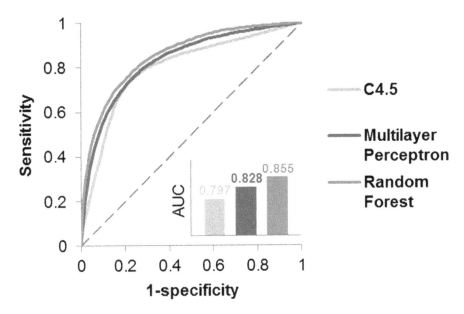

Figure 4. Ten-fold cross-validation.

genes, we were able to secure only five disease signatures. The number of drugs used in both our method and CMap is 201. There are 31 known drug-disease associations between 201 drugs and five diseases. These 31 known associations and their elements, 30 drugs and four diseases, were used for a comparison. Even with this limited number of known drug-disease associations, the AUC in our method is 0.991, while the AUC of CMap is 0.360. In the Table S7, we show AUC reports of CMap based on the different proportion of input genes.

New predictions

Among 30,160 drug-disease pairs (377 drugs and 80 diseases), we predicted 6,143 novel drug-disease associations using the classifier (Table S4). We compared our predictions and clinical trials for validation, and 7,854 unique drug-disease associations were obtained from a registry of publicly and privately conducted worldwide clinical studies (http://clinicaltrials.gov/). The MeSh term was used to map the conditions of the clinical trials to an MIM number and to map interventions to DrugBank entries. Out of 7,854 associations, 942 associations involve drugs and diseases that are present in our data set. The coverage rate of the 6,143 predicted associations with respect to 942 clinical trial associations is 36.2 percent as shown in Table 1 (Fisher's exact $P = 4.02E-30$). In order to validate that the prediction is not trivial due to

structurally same drugs, we computed chemical similarity between known drugs and predicted drugs for same disease. We calculated Tanimoto score between drugs based on their fingerprints downloaded from DrugBank. Among all the drug pairs, only 0.58% of them show chemical similarity (Tanimoto coefficient > 0.7) as displayed in Figure S1.

We also examined extended drug-disease pairs that consist of 832 FDA-approved drugs extracted from DrugBank and 239 diseases taken from OMIM. Among 198,848 drug-disease pairs, 26,909 associations were predicted using our classifier (Table S5). Out of 7,854 clinical trials, 1,100 associations involve drugs and diseases that are present in the extended data set. The coverage rate of the predicted associations with respect to 1,100 clinical trial associations is 36.8 percent (Fisher's exact $P = 3.71E-84$). These two coverage rates are relatively high compared to that reported by Gottlieb et al., who demonstrated coverage of 27 percent [12].

Discussion

Path types used in the genetic network

To measure the network traits between genes derived from drug and disease, we defined three types of paths between two genes: R0, R1 and R2. First, R0 denotes that the two genes are identical. Second, R1 indicates that the two genes are linked by direct

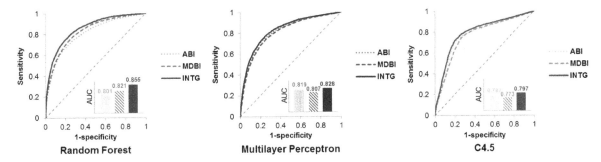

Figure 5. Performance evaluation of each method. Results from the adjacency-based inference (ABI) method, the module-distance-based inference (MDBI) method, and the integrated method of ABI and MDBI (INTG) are compared.

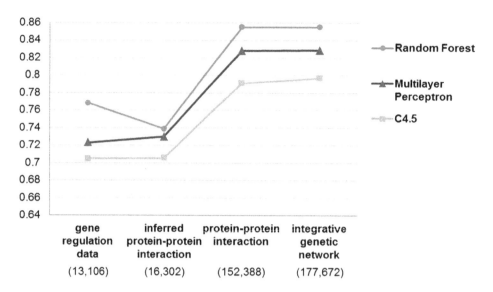

Figure 6. AUC comparison of three individual networks and the integrated network.

interaction. Third, R2 means that indirect interaction through another gene connects two genes (see method section). We conduct the independent experiments using only single type of gene pair and comparing them with original method which uses all types of gene pairs. Table S6 shows that it is better to consider all the paths together than consider only single type of path in improving classification performance. Additionally, in most cases, the classifier using only R2 shows higher performance than the classifier using only R0 or R1. Table S3 indicates the feature contribution of these paths by changing k value, which means alteration of path type. The features of which k = 2, meaning only R2 being used, generally show higher rank than other features where k = 1 or k = 0, and the features of which k = 1 usually have higher information gain than other features where k = 0.

Beta-adrenergic antagonist as a potential cancer treatment

First, we focused on a potential therapeutic agent as a cancer treatment. Out of 26,909 new predictions, 5,809 cancer-specific associations were selected according to the disease category from earlier work [20]. Nearly 5 percent of the drugs in these associations target beta-adrenergic receptors. Several recent epidemiological studies have shown that the use of beta-blockers reduces the progression and secondary formation of cancer and improves the potential for relapse-free survival in patients with cancer [21–24]. Also, Al-wadei et al. indicated that the inhibition of beta-adrenergic signaling can lead to potential anti-cancer drug development [25]. Our 162 cancer-related predicted associations include 21 drugs out of the 24 approved beta-adrenergic blockers from DrugBank. Generally, beta-adrenergic blockers are known as antihypertensive agents, and their blood pressure regulation pathways are well known. However, a few studies recently explained the mechanism of pathways related to the initiation and progression of cancer.

To elucidate the role of beta-blockers as a cancer treatment, we displayed a predicted gene network extending from propranolol to cancers (Figure 7). The gene regulatory network and disease-gene are used to construct the network. Propranolol is an antihypertensive agent that is also a beta-adrenergic antagonist. Several cancers are predicted to have an association with propranolol in our study. Propranolol inhibits beta-adrenergic signaling, as shown

in Figure 7. It is known that the stimulated beta-adrenergic receptor activates $G\alpha_s$ guanine nucleotide-binding protein, resulting in the activation of adenylyl cyclase and the subsequent formation of cyclic adenosine $3',5'$-monophosphate (cAMP) [26]. In addition, the cAMP activation of protein kinase A (PKA) phosphorylates cyclic AMP-responsive element-binding protein (CREB) and transactivates epidermal growth factor receptor (EGFR) [27]. Also, cAMP activates serine/threonine-protein kinase B-raf (BRAF) and mitogen-activated protein kinase (MAPK), leading to the PKA-dependent activation of downstream kinases such as Src kinase (Src) and focal adhesion kinase (FAK) [28]. These kinases and proteins are proto-oncogenes or play a role in the network as an activator of proto-oncogenes, as shown in Figure 7. The gene set in the network was significantly enriched in the cancer pathways such as *prostate cancer* (P = 1.24E-28, FDR = 5.64E-27), *pathways in cancer* (P = 1.11E-27, FDR = 2.53E-26), and *glioma* (P = 2.4E-26, FDR = 4.37E-25), using KEGG biological pathways.

Figure 7 shows that several genes, including HRAS, RAC1, AKT1, and PIK3CA, which usually undergo somatic mutations in specific cancers, play an important role in the pathway. In Figure 7, the cAMP-dependent kinase regulatory chain family and PKA activate GPCRs to cause the activation of HRAS. Also, RAC1, AKT1, and PIK3CA are stimulated by various genes, including FAK, Src, G-beta, the gamma family, and the gene group in the middle of the network. When PIK3CA and AKT1 are downstream effectors of RAS, both HRAS and RAC1 are RAS superfamily of small GTPases that cause cancer growth, invasion, and metastasis [29]. In summary, our network explains the major mechanism of the beta-adrenergic signaling pathway that is related to the RAS superfamily and its downstream genes that cause specific cancers.

Elucidating the mechanism of telmisartan with regard to Alzheimer's disease

Many therapeutic agents for Alzheimer's disease aim to achieve symptomatic benefits; however, currently no disease-modifying therapies are approved. Drug repositioning for Alzheimer's disease is being considered as an efficient strategy, with several classes of repositioning drugs presented in recent studies. Among the potential repositioning drugs, peroxisome proliferator-activated

Table 1. Statistics of the Predictions in the Clinical Trials.

Phases	377 drugs and 80 diseases				832 drugs and 239 diseases			
	Associations in clinical trials	Associations in prediction	Coverage	p-value	Associations in clinical trials	Associations in prediction	Coverage	p-value
Total	942	341	36.2%	4.02E-30	1100	405	36.8%	3.71E-84
Phase 1	243	79	32.5%	3.13E-24	287	93	32.4%	2.11E-07
Phase 1/2	124	37	29.8%	4.61E-50	150	41	27.3%	5.17E-28
Phase 2	466	159	34.1%	0.006636	532	180	33.8%	0.006982
Pahse 2/3	60	15	25.0%	2.23E-71	66	17	25.8%	1.01E-46
Phase 3	249	57	22.9%	5.40E-36	297	73	24.6%	3.94E-13
Phase 4	214	55	25.7%	3.25E-37	244	75	30.7%	1.53E-12
Unlisted	251	77	30.7%	3.46E-25	296	95	32.1%	5.51E-07

Total numbers are unique associations, excluding redundancy between phases.

receptor-γ (PPARγ) agonists and angiotensin receptor blockers have been spotlighted for Alzheimer's disease [30].

Telmisartan is a therapeutic agent that is prescribed for hypertension. It is known as a unique angiotensin II receptor blocker with PPARγ agonistic properties, and it was predicted to be associated with Alzheimer's disease in our study. Amyloid-β (Aβ) deposition is a key pathological hallmark of Alzheimer's disease, and it was reduced by the use of low doses of telmisartan in an Alzheimer's disease mouse model in vivo [31,43]. In addition, the epidemiological feasibility of telmisartan was recently identified in a small Alzheimer's patient cohort [32].

The potential pathways of the PPARγ agonist in Alzheimer's disease are fairly well known. PPARγ agonists regulate multiple processes, including Aβ homeostasis through the suppression of BACE1 expression, energy metabolism, insulin sensitivity, dyslipidemia, and microglial inflammatory responses [33,34]. Meanwhile, animal studies suggest that angiotensin receptor blockers decrease Aβ oligomerization [35]. Hajjr et al. provided the first autopsy evidence that angiotensin receptor blockers are associated with reduced amyloid accumulation and Alzheimer's disease-related pathological change [36]. However, the mechanisms of angiotensin receptor blocker have not been annotated and thus need to be clarified.

We depicted the potential pathway starting from telmisartan to Alzheimer's disease through a gene regulatory network. In Figure 8, angiotensin II receptor (AGTR1) activates MTG1 to translocate group A genes, which activate the transcription of alpha-2-macroglobulin (A2M). On the other hand, group B genes activated by MTG1 inhibit the transcription of A2M. In other words, the signaling of the angiotensin receptor interrupts the transcription of A2M, which mediates the clearance and degradation of Aβ. The angiotensin II receptor blocker telmisartan may work such that the transcription of A2M is not interrupted. In addition, PPARγ, denoted as PPARG, inhibits the transcription of IL2, which plays a role similar to that of MTG1 in that it partially activates genes in groups A and B.

A GO enrichment analysis for group B shows that the insulin receptor signaling pathway (P = 1.43e-06; FDR = 3.48e-05), the cellular response to an insulin stimulus (P = 3.17e-06; FDR = 6.55e-05) and the response to an insulin stimulus (P = 5.88e-06; FDR = 8.58e-05) are ranked as significantly enriched biological processes. Insulin signaling has a direct role in the development of neurodegenerative diseases [37], and insulin administration improves memory [38]. However, defective insulin signaling is a characteristic feature of the AD brain, and oligomeric amyloid-β induces insulin resistance in the brain [39,40]. In this regard, Figure 8 shows a potential mechanism of telmisartan, i.e., showing how it blocks the defective insulin signaling cascade, resulting in the inhibition of A2M transcription.

The Aβ peptide mediates synapse loss through cAMP-response element binding protein (CREB) signaling, and the altered CREB signaling plays a crucial role in cognitive dysfunction [41]. In Figure 8, group B, IL2, and their downstream components activate CREB1, resulting in the activation of NOS3 transcription. NOS3 was reported to show various polymorphisms and a significant association with Alzheimer's disease. Therefore, another potential mechanism of telmisartan may be explained by the inhibition of altered CREB signaling.

Thus, we suggest that the potential mechanism of the angiotensin receptor blocker and the PPARγ agonist is related to defective insulin signaling and altered CREB signaling. Therefore, telmisartan is expected to be a robust candidate drug for Alzheimer's disease.

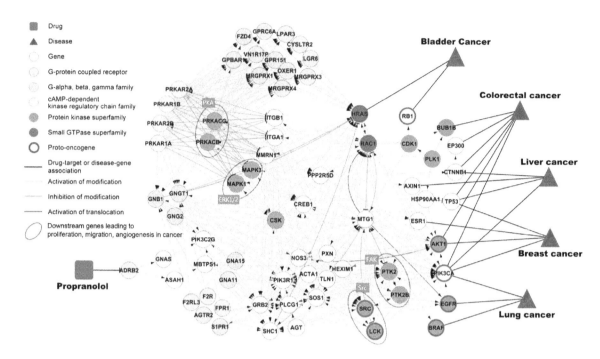

Figure 7. Gene regulatory network between ADRB2 and cancer-specific genes.

Conclusion

We proposed a method of drug repositioning based on feature extraction from integrative genetic networks and known drug-disease associations. Our method showed high classification accuracy in terms of large-scale prediction of drug indications, and it obtained novel predictions that were validated by their overlap with clinical trials. Furthermore, we discussed interesting examples of novel drug-disease associations at the molecular level. Based on shared pathways of existing drug-disease associations, we can infer more specifically how newly predicted drugs work on the disease. However, there is a limitation in inferring whether a particular drug indication is adverse or effective. For future work,

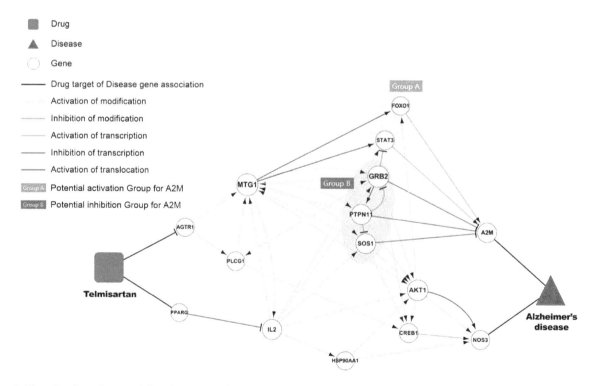

Figure 8. Visualization of potential pathway associated with targets of telmisartan and genes related to Alzheimer's disease.

we plan to develop our method for predicting the side effects of existing drugs.

Supporting Information

Figure S1   The distribution of Tanimoto scores.
(TIF)

Table S1   One sample of actual training data set.
(XLSX)

Table S2   The result of ten 10-fold cross-validation runs.
(XLSX)

Table S3   Feature contribution.
(XLSX)

Table S4   New predictions for 377 drugs and 80 diseases.
(XLSX)

Table S5   New predictions for 832 drugs and 239 diseases.
(XLSX)

Table S6   Effect of path types toward the prediction accuracy.
(XLSX)

Table S7   The AUC reports according to proportion of up- and down-regulated genes.
(DOCX)

Author Contributions

Conceived and designed the experiments: MO JA YY. Performed the experiments: MO. Analyzed the data: MO JA YY. Contributed reagents/materials/analysis tools: MO. Wrote the paper: MO YY.

References

1. Hopkins AL (2008) Network pharmacology: the next paradigm in drug discovery. Nature chemical biology 4(11): 682–690.
2. Cheng F, Liu C, Jiang J, Lu W, Li W, et al. (2012) Prediction of drug-target interactions and drug repositioning via network-based inference. PLoS computational biology 8(5): e1002503.
3. Chong CR, Sullivan DJ (2007) New uses for old drugs. Nature 448: 645–646.
4. Campillos M, Kuhn M, Gavin AC, Jensen LJ, Bork P (2008) Drug target identification using side-effect similarity. Science 321(5886): 263–266.
5. Schadt EE, Friend SH, Shaywitz DA (2009) A network view of disease and compound screening. Nature reviews Drug discovery 8(4): 286–295.
6. Keiser MJ, Setola V, Irwin JJ, Laggner C, Abbas AI, et al. (2009) Predicting new molecular targets for known drugs. Nature 462(7270): 175–181.
7. Dudley JT, Deshpande T, Butte AJ (2011) Exploiting drug–disease relationships for computational drug repositioning. Briefings in bioinformatics 12(4): 303–311.
8. Hu G, Agarwal P (2009) Human disease-drug network based on genomic expression profiles. PLoS One 4(8): e6536.
9. Suthram S, Dudley JT, Chiang AP, Chen R, Hastie TJ, et al. (2010) Network-based elucidation of human disease similarities reveals common functional modules enriched for pluripotent drug targets. PLoS computational biology 6(2): e1000662.
10. Sirota M, Dudley JT, Kim J, Chiang AP, Morgan AA, et al. (2011) Discovery and preclinical validation of drug indications using compendia of public gene expression data. Science translational medicine 3(96): 96ra77–96ra77.
11. Zhao S, Li S (2012) A co-module approach for elucidating drug–disease associations and revealing their molecular basis. Bioinformatics 28(7): 955–961.
12. Gottlieb A, Stein GY, Ruppin E, Sharan R (2011) PREDICT: a method for inferring novel drug indications with application to personalized medicine. Molecular systems biology 7(1): 496.
13. Brown KR, Jurisica I (2007) Unequal evolutionary conservation of human protein interactions in interologous networks. Genome biology 8(5): R95.
14. Schaefer CF, Anthony K, Krupa S, Buchoff J, Day M, et al. (2009) PID: the pathway interaction database. Nucleic acids research (suppl 1): D674–D679.
15. Knox C, Law V, Jewison T, Liu P, Ly S, et al. (2011) DrugBank 3.0: a comprehensive resource for 'omics' research on drugs. Nucleic acids research (suppl 1): D1035–D1041.
16. McKusick VA (2007) Mendelian Inheritance in Man and its online version, OMIM. American journal of human genetics 80(4): 588.
17. Davis AP, Murphy CG, Saraceni-Richards CA, Rosenstein MC, Wiegers TC, et al. (2009) Comparative Toxicogenomics Database: a knowledgebase and discovery tool for chemical–gene–disease networks. Nucleic acids research (suppl 1): D786–D792.
18. Hall M, Frank E, Holmes G, Pfahringer B, Reutemann P, et al. (2009) The WEKA data mining software: an update. ACM SIGKDD explorations newsletter 11(1): 10–18.
19. Lamb J, Crawford ED, Peck D, Modell JW, Blat IC, et al. (2006) The Connectivity Map: using gene-expression signatures to connect small molecules, genes, and disease. science 313(5795): 1929–1935.
20. Goh KI, Cusick ME, Valle D, Childs B, Vidal M, et al. (2007) The human disease network. Proceedings of the National Academy of Sciences 104(21): 8685–8690.
21. Wang HM, Liao ZX, Komaki R, Welsh JW, O'Reilly MS, et al. (2013) Improved survival outcomes with the incidental use of beta-blockers among patients with non-small-cell lung cancer treated with definitive radiation therapy. Annals of oncology 24(5): 1312–1319.
22. Barron TI, Connolly RM, Sharp L, Bennett K, Visvanathan K (2011) Beta blockers and breast cancer mortality: a population-based study. Journal of clinical oncology 29(19): 2635–2644.
23. Melhem-Bertrandt A, Chavez-MacGregor M, Lei X, Brown EN, Lee RT, et al. (2011) Beta-blocker use is associated with improved relapse-free survival in patients with triple-negative breast cancer. Journal of clinical oncology 29(19): 2645–2652.
24. Powe DG, Voss MJ, Zänker KS, Habashy HO, Green AR, et al. (2010) Beta-blocker drug therapy reduces secondary cancer formation in breast cancer and improves cancer specific survival. Oncotarget 1(7): 628.
25. Al-Wadei HA, Ullah MF, Al-Wadei MH (2012) Intercepting neoplastic progression in lung malignancies via the beta adrenergic (β-AR) pathway: Implications for anti-cancer drug targets. Pharmacological Research 66(1): 33–40.
26. Dohlman HG, Thorner J, Caron MG, Lefkowitz RJ (1991) Model systems for the study of seven-transmembrane-segment receptors. Annual review of biochemistry 60(1): 653–688.
27. Schuller HM (2009) Is cancer triggered by altered signalling of nicotinic acetylcholine receptors? Nature Reviews Cancer 9(3): 195–205.
28. Cole SW, Sood AK (2012) Molecular pathways: beta-adrenergic signaling in cancer. Clinical cancer research 18(5): 1201–1206.
29. Vigil D, Cherfils J, Rossman KL, Der CJ (2010) Ras superfamily GEFs and GAPs: validated and tractable targets for cancer therapy. Nature Reviews Cancer 10(12): 842–857.
30. Corbett A, Pickett J, Burns A, Corcoran J, Dunnett SB, et al. (2012) Drug repositioning for Alzheimer's disease. Nature Reviews Drug Discovery 11(11): 833–846.
31. Tsukuda K, Mogi M, Iwanami J, Min LJ, Sakata A, et al. (2009) Cognitive deficit in Amyloid-β–injected mice was improved by pretreatment with a low dose of telmisartan partly because of peroxisome proliferator-activated receptor-γ activation. Hypertension 54(4): 782–787.
32. Kume K, Hanyu H, Sakurai H, Takada Y, Onuma T, et al. (2012) Effects of telmisartan on cognition and regional cerebral blood flow in hypertensive patients with Alzheimer's disease. Geriatrics & gerontology international 12(2): 207–214.
33. Landreth G, Jiang Q, Mandrekar S, Heneka M (2008) PPARγ agonists as therapeutics for the treatment of Alzheimer's disease. Neurotherapeutics 5(3): 481–489.
34. Kummer MP, Heneka MT (2008) PPARs in Alzheimer's disease. PPAR research 2008: 8.
35. Wang J, Ho L, Chen L, Zhao Z, Zhao W, et al. (2007) Valsartan lowers brain β-amyloid protein levels and improves spatial learning in a mouse model of Alzheimer disease. Journal of Clinical Investigation 117(11): 3393–3402.
36. Hajjar I, Brown L, Mack WJ, Chui H (2012) Impact of angiotensin receptor blockers on Alzheimer disease neuropathology in a large brain autopsy series. Archives of neurology 69(12): 1632–1638.
37. Plum L, Schubert M, Brüning JC (2005) The role of insulin receptor signaling in the brain. Trends in Endocrinology & Metabolism 16(2): 59–65.
38. Craft S, Asthana S, Cook DG, Baker LD, Cherrier M, et al. (2003) Insulin dose-response effects on memory and plasma amyloid precursor protein in Alzheimer's disease: interactions with apolipoprotein E genotype. Psychoneuro-endocrinology 28(6): 809–822.
39. Talbot K, Wang HY, Kazi H, Han LY, Bakshi KP, et al. (2012) Demonstrated brain insulin resistance in Alzheimer's disease patients is associated with IGF-1 resistance, IRS-1 dysregulation, and cognitive decline. The Journal of clinical investigation 122(4): 1316.
40. Bomfim TR, Forny-Germano L, Sathler LB, Brito-Moreira J, Houzel JC, et al. (2012) An anti-diabetes agent protects the mouse brain from defective insulin signaling caused by Alzheimer's disease–associated Aβ oligomers. The Journal of clinical investigation 122(4): 1339.

41. Saura CA, Valero J (2011) The role of CREB signaling in Alzheimer's disease and other cognitive disorders. Reviews in the neurosciences 22(2): 153–169.

42. Parkinson H, Kapushesky M, Kolesnikov N, Rustici G, Shojatalab M, et al. (2009) ArrayExpress update–from an archive of functional genomics experiments to the atlas of gene expression. Nucleic acids research (suppl 1): D868–D872.

43. Shindo T, Takasaki K, Uchida K, Onimura R, Kubota K, et al. (2011) Ameliorative effects of telmisartan on the inflammatory response and impaired spatial memory in a rat model of Alzheimer's disease incorporating additional cerebrovascular disease factors. Biological & pharmaceutical bulletin 35(12): 2141–2147.

44. Ahn J, Yoon Y, Park C, Shin E, Park S (2011) Integrative gene network construction for predicting a set of complementary prostate cancer genes. Bioinformatics 27(13): 1846–1853.

Identification of Potential Serum Proteomic Biomarkers for Clear Cell Renal Cell Carcinoma

Juan Yang[1]꜒, Jin Yang[2]꜒, Yan Gao[2], Lingyu Zhao[1], Liying Liu[1], Yannan Qin[1], Xiaofei Wang[1], Tusheng Song[1], Chen Huang[1]*

1 Key Laboratory of Environment and Genes Related to Diseases of the Education Ministry, Department of Genetics and Molecular Biology, Medical School of Xi'an Jiaotong University, Xi'an, China, **2** Department of Medical Oncology, First Affiliated Hospital of Medical School of Xi'an Jiaotong University, Xi'an, China

Abstract

Objective: To investigate discriminating protein patterns and serum biomarkers between clear cell renal cell carcinoma (ccRCC) patients and healthy controls, as well as between paired pre- and post-operative ccRCC patients.

Methods: We used magnetic bead-based separation followed by matrix-assisted laser desorption ionization (MALDI) time-of-flight (TOF) mass spectrometry (MS) to identify patients with ccRCC. A total of 162 serum samples were analyzed in this study, among which there were 58 serum samples from ccRCC patients, 40 from additional paired pre- and post-operative ccRCC patients (n = 20), and 64 from healthy volunteers as healthy controls. ClinProTools software identified several distinct markers between ccRCC patients and healthy controls, as well as between pre- and post-operative patients.

Results: Patients with ccRCC could be identified with a mean sensitivity of 88.38% and a mean specificity of 91.67%. Of 67 m/z peaks that differed among the ccRCC, healthy controls, pre- and post-operative ccRCC patients, 24 were significantly different (P<0.05). Three candidate peaks, which were upregulated in ccRCC group and showed a tendency to return to healthy control values after surgery, were identified as peptide regions of RNA-binding protein 6 (RBP6), tubulin beta chain (TUBB), and zinc finger protein 3 (ZFP3) with the m/z values of 1466.98, 1618.22, and 5905.23, respectively.

Conclusion: MB-MALDI-TOF-MS method could generate serum peptidome profiles of ccRCC, and provide a new approach to identify potential biomarkers for diagnosis as well as prognosis of this malignancy.

Editor: Jörg D. Hoheisel, Deutsches Krebsforschungszentrum, Germany

Funding: This work was supported by The National Natural Science Foundation of China (grant number: 81200845) and the Key Science and Technology Program of Shaanxi province (grant number: 2012 K13-02-01). The funders had no role in study design, data collection and analysis, decision to publish, or preparation of the manuscript.

Competing Interests: The authors have declared that no competing interests exist.

* Email: hchen@mail.xjtu.edu.cn

꜒ These authors contributed equally to this work.

Introduction

Clear cell renal cell carcinoma (ccRCC) is a renal cortical tumor typically characterized by malignant epithelial cells with a clear cytoplasm and a compact-alveolar (nested) or acinar growth pattern interspersed with intricate, arborizing vasculature. ccRCC represents over 80% of renal cell carcinomas (RCCs) [1], which are the most common form of kidney cancer, accounting for 3% of all cancer diagnoses and more than 100,000 deaths worldwide each year [2]. The most effective treatment for ccRCC is currently surgical resection, partial nephrectomy is considered for tumors smaller than 4 cm in diameter (stage pT1a) and radical nephrectomy for tumors larger than 4 cm [3]. However, ccRCC is associated with a fast rate of extrarenal growth, metastasis (most commonly to the lung, liver, bone or brain) and mortality [1,4]. The survival rate of ccRCC patients decreases with increasing disease stage [5,6]. Therefore, the early detection of ccRCC would significantly improve patient diagnosis and outcome.

The identification of biomarkers for the early detection of cancer could lead to the development of efficient treatments, reduce suffering, and lower mortality rates [7]. However, there are currently no biomarkers for the reliable screening of patients with ccRCC [8–10]. Human serum contains a complex array of peptides, and some of these could serve as biomarkers because their presence/absence or relative abundance may be correlated with certain diseases and could thus be useful for prognosis or diagnosis [11–13]. The identification of differentially expressed peptides and proteins by mass spectrometry (MS) combined with software-generated models capable of discriminating between the spectra of patients with ccRCC and healthy controls could lead to the identification of potential new biomarkers for ccRCC. Here, we report on the use of magnetic bead-based purification approaches coupled with MALDI-TOF MS for the comparative analysis of sera from patients with ccRCC and healthy controls, as well as ccRCC patients who underwent surgical resection. And

Table 1. Clinico-pathological characteristics of 78 clear cell renal cell cancers.

Characters	Number of cases	% of cases
Age Median years (range)	58.53 (33–74)	
Gender		
Male	58	74
Female	20	26
Furhman grade		
G1	6	7.5
G2	54	69
G3	12	15
G4	0	0
unknown	6	7.5
TNM stage		
Stage I	48	62
Stage II	30	38
Stage III	0	0
Stage IV	0	0
T stage		
T1	48	62
T2	30	38
T3	0	0
T4	0	0
Lymph node metastasis		
N0	78	100
N1	0	0
Distant metastasis		
M0	78	100
M1	0	0

potential serum biomarkers for detection of ccRCC were then identified by LC-ESI-MS/MS.

Materials and Methods

Patients and sample preparation

The study protocol was approved by the Ethics Committee and the Human Research Review Committee of Xi'an Jiaotong University, and each subject has been provided signed informed consent before the work. All samples were collected from the First Affiliated Hospital of Xi'an Jiaotong University between January 1st, 2010 and December 12th, 2011. Clinical data were retrospectively collected from medical record reviews and electronic records, and tumor histology were obtained from pathology. Patients with a known history of other tumors and those with obvious inflammatory diseases were excluded. For clinical variables, age at diagnosis, sex, and tumor stage and Fuhrman grade were considered. Tumor stage was defined according to the seventh edition of the American Joint Committee on Cancer (AJCC) cancer staging manual.

The 64 control serum samples were obtained from healthy donors recruited for this study including 32 men and 32 women with an average age of 51.7 years (range, 31–78 years). Serum samples of ccRCC groups were obtained from 58 ccRCC patients before surgical operation, including 44 men and 14 women with an average age of 54.6 years (range, 33–74 years). Besides, 40

serum samples were obtained from 20 additional paired pre- and post-treatment ccRCC patients, with their post-operative serum samples collected three days after surgery. Clinico-pathological characteristics of all patients were shown in Table 1.

All blood samples were drawn from non-fasting subjects in a sitting position. The samples were collected in 10 cc serum separator tubes and kept at 4°C for 1 h, then centrifuged at 3000 g for 20 min at 4°C. The serum samples were distributed into 500 μL aliquots and stored at −80°C until use.

MS analysis: WCX fractionation and MALDI-TOF MS analysis

Samples were separated by magnetic bead-based weak cation-exchange chromatography (MB–WCX) using ClinProt purification reagent sets from Bruker Daltonics. MB-WCX purifications were performed using the Bruker Magnetic Separator according to the manufacturer's protocol. The details of this experiment were reported previously [13]. To prepare the MALDI target, 1 μL of a mixture containing 10 μL of 0.3 g/L α-cyano-4-hydroxy cinnamic acid (HCCA) in 2:1 ethanol/acetone (volume/volume) and 1 μL of the eluted peptide fraction were spotted onto the MALDI AnchorChip (Bruker Daltonics, Germany) sample target platform (384 spots). To evaluate the reproducibility of the assay, all serum samples were spotted in triplicate. Air-dried targets were measured immediately using a calibrated Autoflex III MALDI-TOF MS

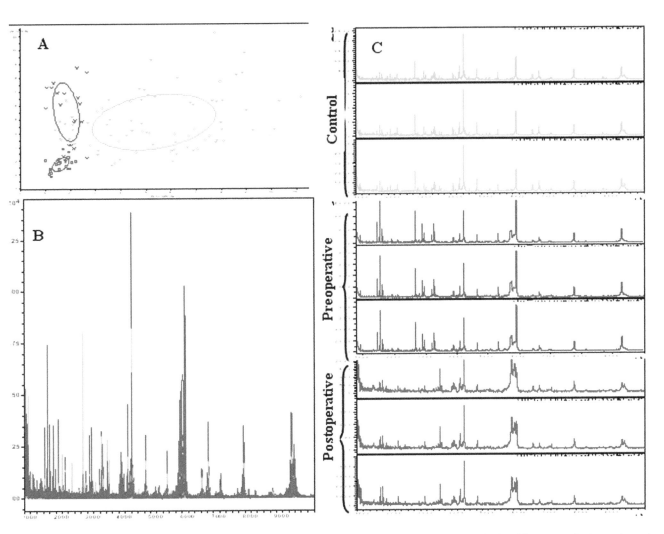

Figure 1. Comparative profiling of serum peptides from preoperative ccRCC patients (red), healthy controls (green), and postoperative ccRCC patients (blue). A. Bivariate plot of preoperative ccRCC patients (red), healthy controls (green), and postoperative ccRCC patients (blue) in the component analysis. **B.** Overall sum of the spectra in the mass range from 1000 to 10,000 Da obtained from the three groups described above. **C.** Representative mass spectra of one ccRCC patient (including preoperative (red) and postoperative (blue) serum samples) and a healthy control (green) (three spectra per sample) in the mass range from 1000 to 10,000 Da showing low variability between replicates of each sample.

(Bruker) with FlexControl software (version 3.0, Bruker) and optimized measuring protocols. Matrix suppression up to 1000 Da, with a mass range of 1000–10,000 Da was set as the default. Instrument calibration parameters were determined using standard peptide and protein mixtures. All measurements were performed in a blinded manner, including patient and control sera in one mixed approach.

Data analyses were performed using the programs Flex analysis v3.0 and ClinProTools v2.2 (Bruker Daltonics, Germany). ClinProTools v2.2 uses a standard data preparation workflow including spectra pretreatment, peak picking, and peak calculation operations, and was applied for the recognition of peptide patterns in this study. For statistical analyses, a k-nearest neighbor genetic algorithm, as implemented in the software suite, was used to identify statistically significant differences in protein peaks among the groups analyzed.

For statistical analyses, three different algorithms of mathematical models were used: Genetic Algorithm (GA), Supervised Neural Network (SNN) and Quick Classifier (QC). The GA algorithm is derived from evolutionary survival in which the best peak clusters are combined into a new feature and the poor clusters are discarded. This process is iteratively repeated until the optimal peak combination is found. The SNN algorithm maximizes the distance of multiple local peak clusters specific to each group. Clusters that provide greater separation are prioritized over those with low separation. Finally, the QC algorithm generates an average spectrum for each group with weighted p-values for each peak. Based on the peak weights, spectra are categorized into either group along with a likeliness value.

Peptide identification by LC-ESI-MS/MS

After completing the statistical analysis, differentially expressed peptides were identified. The peptide sequences of potential m/z peaks were first determined by LC-ESI-MS/MS, and the identified sequence data were then subjected to a Mascot database search to identify the corresponding full-length protein matches.

Table 2. Mean levels of differentially expressed proteins among ccRCC patients (n = 58), healthy controls (n = 64), pre- and post-operatives ccRCC samples (n = 20).

Mass	PTTA	ccRCC n = 58	Control n = 64	Preoperative n = 20	Postoperative n = 20
1866.63 ↓	<0.000001	3.13±1.25	5.72±2.63	1.98±0.56	1.73±0.36
3317.00 ↓	<0.000001	2.29±1.36	4.95±1.94	3.18±1.46	2.14±0.44
6433.36 ↓	<0.000001	2.48±1.27	4.30±2.34	2.90±1.12	1.73±0.43
1780.10 ↓	<0.000001	2.17±0.72	3.17±1.05	1.92±1.07	1.97±0.78
1061.34 ↓	<0.000001	2.69±1.03	6.19±3.28	2.05±0.92	2.27±0.86
5752.25 ↑	1.87E-05	2.38±1.37	1.28±0.34	2.93±2.7	5.92±3.91
5724.76 ↑	5.87E-05	1.75±0.82	0.77±0.19	2.22±3.23	5.67±2.27
5844.36 ↑	7.36E-05	2.69±1.27	1.70±0.53	2.83±1.83	7.24±3.53
5739.69 ↑	8.06E-05	2.07±0.98	0.94±0.22	2.49±3.31	6.50±5.06
5818.83 ↑	9.31E-05	2.67±1.14	1.33±0.3	2.45±1.64	6.21±3.54
5905.23 ↑	**0.00027**	**19.54±4.21**	**13.26±3.64**	**21.10±5.29**	**12.32±3.66**
1466.98 ↑	**0.00041**	**4.72±1.23**	**2.72±0.78**	**6.49±3.75**	**2.56±0.89**
4091.86 ↑	0.00041	9.79±3.57	6.03±2.61	9.23±3.49	5.47±1.94
1714.57 ↑	0.00043	7.71±3.26	1.78±0.7	5.87±1.56	7.44±3.78
1618.22 ↑	**0.0019**	**6.37±2.08**	**4.32±2.77**	**7.10±2.87**	**5.11±2.01**
1981.69 ↓	0.00403	1.88±0.50	5.43±2.02	1.62±0.44	1.69±0.41
1077.14 ↓	0.00895	2.19±1.23	3.37±2.33	2.09±1.38	2.84±0.76
1547.00 ↑	0.0981	9.17±4.82	5.91±3.63	10.15±3.15	5.25±2.7

Statistical Analysis

Statistical analyses were conducted using GraphPad Prism v5.0 (GraphPad Software). Correlations between the mean expressions of all detected m/z peaks with Clinicopathological characteristics of ccRCC patients were evaluated using the Spearman rank-order correlation coefficient. All data were expressed as the mean ± SD. The results were considered statistically significant if $P<0.05$. We conducted a power analysis using G*Power [14] to determine the minimum sample size needed to detect a 1.5 fold differences in mean expression level between ccRCC patients and healthy controls, as well as between pre- and post-operative patients. 16 samples of each population were required to achieve a power of 0.85 at $\alpha = 0.05$ (two-tailed test).

Results

The serum peptidome profiles of 58 ccRCC patients, 64 healthy controls, and 20 paired pre-and post-oprerative ccRCC patients were analyzed in the current study. We evaluated changes at the peptidome level in the serum samples of ccRCC patients compared with those of healthy controls, and analyzed differences among preoperative ccRCC patients, postoperative ccRCC patients, and healthy controls. Analysis of spectra (screened from different groups) using ClinProTools version 2.2 led to the identification of distinct proteomic patterns between ccRCC and healthy controls, as well as between pre- and post-operative samples, and the identification of three potential serum proteomic biomarkers for ccRCC.

MALDI spectrum generation and assay reproducibility

Prefractionation of serum samples by MB-WCX and MALDI-TOF MS identified up to 67 peaks, of which 24 showed significantly different m/z peaks among ccRCC patients, healthy controls, pre- and post-operative ccRCC samples, with P<0.05

according to the Wilcoxon rank sum test (Table 2). In the component analysis, a bivariate plot of preoperative ccRCC patients (red), healthy controls (green), and postoperative ccRCC patients (blue) showed few overlapping regions among the three groups (Fig. 1A). Overall, the three groups showed protein profiles ranging from 1 to 10 kDa (Fig. 1B). Within this mass range, a significant number of differentially expressed proteins or peptides could be detected.

Evaluation of the reproducibility and stability of the mass spectra in triplicate samples showed closely reproducible peaks (Fig. 1C). In addition, mass spectra differed among healthy controls (green), preoperative ccRCC patients (red) and postoperative serum samples (blue).

Quality Control

To evaluate the reproducibility and stability of the mass spectra, all samples were analyzed in triplicate. The intra-assay variation of each MALDI ProteinChip assay was determined by MALDI profiling of 10 aliquots of each serum sample spotted randomly onto 10 of the 384 wells of the ProteinChip arrays along with all analytical samples. The mean value of the coefficient of variance (CV) for all 67 WCX peaks was 17.27%, with maximum and minimum values of 21.13% and 9.7%, respectively.

Correlation of m/z peaks expression with Clinicopathological characteristics of ccRCC

Correlation between the mean expression of all detected m/z peaks and Clinicopathological characteristics of ccRCC patients were analyzed separately. While the mean expression levels of m/z peaks were found to be significantly correlated with gender difference for ccRCC ($r_s = 0.972$, p = 0.000), and it was also correlated with tumor stage ($r_s = 0.954$, p = 0.000), as well as correlate with Furhman grade (G1 vs G2: $r_s = 0.911$, p<0.000; G1

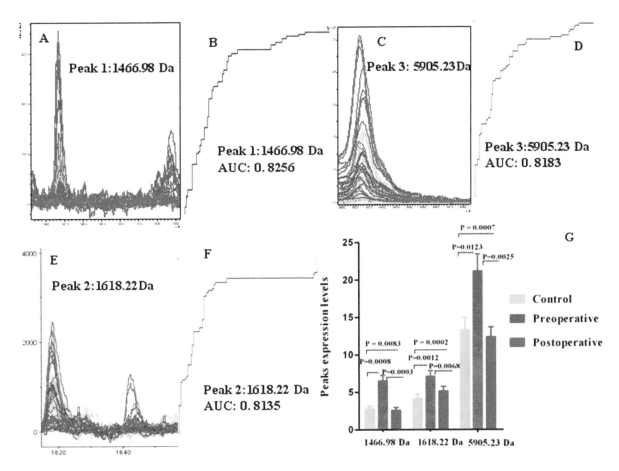

Figure 2. Representative spectra of three selected serum protein peaks of preoperative ccRCC patients (red), healthy controls (green), and postoperative ccRCC patients (blue). A, C, E. Comparison of the spectra of peaks 1466.98, 5905.23 and 1618.22 in the three groups described above. **B, D and F.** Receiver operating characteristic (ROC) curves for the three selected peaks are shown together with their area under the curve (AUC) values. **G.** Average expression levels of three selected peaks in preoperative ccRCC patients (red), healthy controls (green), and postoperative ccRCC patients (blue) and their P values. Values are expressed as the mean ± SD.

vs G3: $r_s = 0.729$, p<0.000; and G2 vs G3: $r_s = 0.748$, p<0.000, respectively).

Peak selection and model testing

ClinProTools analysis identified up to 67 peaks, of which 18 showed significantly different m/z peaks among different groups. Of these, seven were downregulated and 11 were upregulated in preoperative ccRCC patients compared with healthy controls. Seven peptide peaks (m/z: 1061.34, 5905.23, 1466.98, 4091.86, 1618.22, 1077.14, 1547.00) showed similar values to those of healthy controls when compared with the postoperative group (Table 2), of which two (m/z: 1061.34 and 1077.14) were downregulated and five were upregulated in the ccRCC patient group.

Utilizing spectral data from ccRCC and healthy controls, three different classification models for the two groups were generated using GA, SNN and QC algorithms. These models discriminate both groups with high sensitivity and specificity. Based on the GA algorithm model, ccRCC patients could be discriminated from healthy controls with 92.71% sensitivity and 98.31% specificity. The sensitivity and specificity of the SNN model was 89.88% and 92.84% respectively, and these values were 82.55% and 83.82% for the QC model. The mean values of sensitivity and specificity of the three models were 88.38% and 91.67%, respectively.

ClinProt data and the GA algorithm model showed that the preoperative ccRCC patient group could be discriminated from healthy controls and postoperative ccRCC patients with sensitivity of 91.2% and specificity of 93.6%. The five peaks used in GA model included two downregulated peaks (m/z values: 1866.63 and 1061.34) and three upregulated peaks (m/z values: 1466.98, 1618.22, and 5905.23) in the preoperative ccRCC group. The three upregulated peaks showed a tendency to return to healthy control values after surgery. Comparison of the spectra of these three peaks among preoperative ccRCC patients (red), healthy controls (green) and postoperative ccRCC patients (blue) and their receiver operating characteristic (ROC) curves were shown in Fig. 2. The area under the curve (AUC) values of these three peaks were 0.8256 (m/z: 1466.98), 0.8135 (m/z: 1618.22), and 0.8183 for m/z peaks 5905.23. (Fig. 2.)

Identification of ccRCC serum biomarkers

A higher concentration of peptides with m/z values of 1466.98, 1618.22, and 5905.23 were evident in the spectra of the preoperative ccRCC group as compared with that of healthy controls (P<0.001), and they all showed a tendency to return to healthy control values when compared with the postoperative ccRCC group (Fig. 2). LC-ESI-MS/MS and the Mascot database (Figs. 3, 4 and 5) were used for MS/MS fragmentation of these three peptides, which resulted in the identification of the sequences

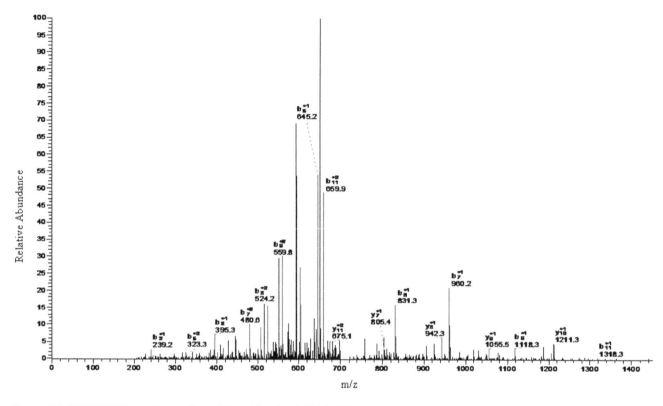

Figure 3. LC-ESI-MS/MS spectrum of peptides with m/z of 1466.98 Da.

KEDIDTSSKGGCVQ, AILVDLEPGTMDSVR, and IHTGE-NPYECSECGKAFRYSSALVRHQRIHTGEKPLNGIGMSKS-SLRVTTELN. The three peptides were found to be regions of RNA-binding protein 6 (RBP6), tubulin beta chain (TUBB), and zinc finger protein 3 (ZFP3) (Table 3).

Discussion

The detection and diagnosis of RCC in an early stage, which remain a challenge for oncologists, could significantly reduce mortality from this cancer. In approximately 30% of RCC cases, metastatic disease is present at the time of initial diagnosis [13]. Although significant advances have been made in the treatment of RCC, few biomarkers for detection of this disease have been identified. Most current biomarker studies have relied on DNA markers [15–17], gene expression [18,19] and microRNAs [20]. Nevertheless, there are no biomarkers for routine clinical use in ccRCC. Proteomic research is currently used worldwide in the search for biomarkers in different types of cancer [21,22]. In ccRCC, Most reported proteomic studies on ccRCC or RCC were mainly from tissue specimens [23–28]. Noriyuki H *et al* (2013) identified nuclear N-myc downstream-regulated gene 1 as a prognostic tissue biomarker candidate in renal cell carcinoma [23]. Jones EE *et al* (2014) revealed a panel of 108 proteins that had potential disease-specific expression patterns based on MALDI imaging MS profiling of proteins in the 2–20 kDa range, from 20 matched ccRCC and distal nontumor tissues [24]. With a complex array of peptides, human serum could be value of diagnostic or prognostic markers identification [29,30]. MALDI-TOF MS is being widely applied in the analysis of serum samples for the diagnosis of human diseases and the identification of potential biomarkers [22]. Magnetic bead-based fractionation followed by MALDI-TOF MS, combined with advanced bioinformatics

(ClinProTools software) can identify biomarkers and improve the reproducibility of mass spectra, making it a valuable tool for clinical proteomic studies [21,31]. Moreover, repeated sampling enables a better understanding of time-dependent changes that occur in response to specific stimuli, as well as information on disease progression and response to treatment [32–34].

In the current study, we used MB-WCX fractionation followed by MALDI-TOF MS techniques combined with ClinProTools software to analyze the serum proteomic profiles of patients with ccRCC, and generated numerous discriminating m/z peaks that could accurately distinguish cancer patients from healthy individuals, and preoperative from postoperative ccRCC patients. The ClinProTools provided predictive models for ccRCC versus healthy controls, and also between pre- and post-operative ccRCC. Moreover, the cross validation and recognition capacities of these models were 88.38% of sensitivity and 91.67% of specificity, 91.2% of sensitivity and 93.6% of specificity, respectively. We identified 18 potential biomarkers for distinguishing ccRCC patients from healthy controls. Some discriminating m/z peaks were upregulated in ccRCC patients (e.g., 1466.98, 1618.22, and 5905.23), while others were downregulated (e.g., 1061.34 and 1077.14). And three candidate peaks (1466.98, 1618.22, and 5905.23) were identified that were upregulated in the preoperative ccRCC group and had a tendency to return to healthy control values 3 days after surgery. These three peaks therefore were not only potential biomarkers for detection of ccRCC, but could also be potential markers of the response to treatment, although further studies with longer follow-up times and large cohort validation are necessary to verify these results. These three potential ccRCC serum biomarkers were identified as peptide regions of RBP6, TUBB, and ZFP3.

TUBB (beta-tubulins) is encoded by a multigene family and heterodimerize to form microtubules. Ma *et al* (2012) compared

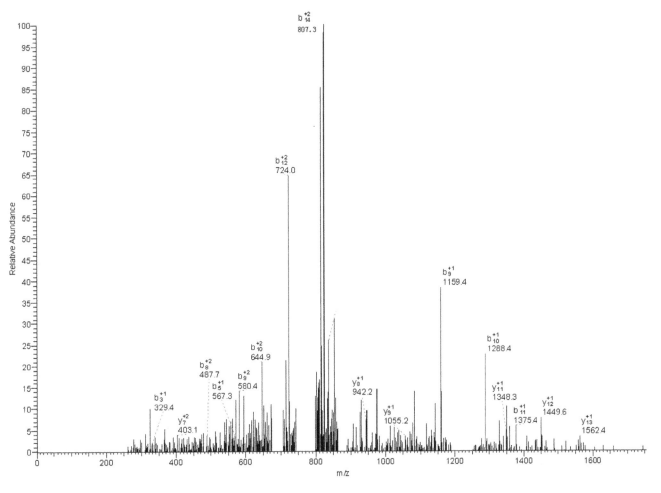

Figure 4. LC-ESI-MS/MS spectrum of peptides with m/z of 1618.22 Da.

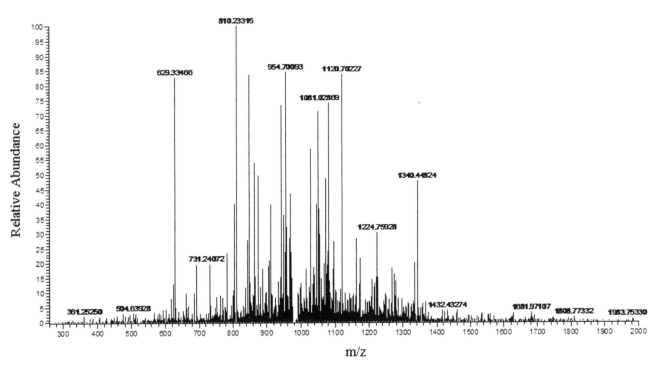

Figure 5. LC-ESI-MS/MS spectrum of peptides with m/z of 5905.23 Da.

Table 3. Sequence identification of the three ccRCC serum protein biomarkers.

Mass m/z	Peptide sequence	International protein Index	Identity
1466.98 Da	KEDIDTSSKGGCVQ	IPI:IPI00297723.2	RBP6 (RNA-binding protein 6)
1618.22 Da	AILVDLEPGTMDSVR	IPI:IPI00011654.2	TUBB (Tubulin beta chain)
5905.23 Da	IHTGENPYECSECGKAFRYSSALVRHQRIHTGEKPLNGIGMSKSSLRVTTELN	IPI:IPI00642617.3	ZFP3 (Zinc finger protein 3)

the average expression ratios of ten target reference genes in the degraded RCC samples across the time course, and the result indicated that TUBB was upregulated in malignant ccRCC tissues compared with nonmalignant renal tissues from 16 ccRCC patients [35]. This is in accordance with our finding that peptides of TUBB (m/z: 1618) were identified as being upregulated in ccRCC patients. TUBB was also reported to be associated with other cancers, and could decrease expression in uveal melanomas that subsequently metastasized compared with those that did not [36]. In cases of ovarian carcinoma, high expression levels of class III beta-tubulin appeared to be associated with earlier recurrence [37]. Several widely used anticancer drugs base their activity on β-tubulin binding, microtubule dynamics alteration, and cell division blockage [38,39]. In addition, Zinc Finger Proteins (ZFPs) contain a conserved structural motif that mediates binding to protein, DNA and RNA [40]. Despite their abundance and important roles in human development and disease, few ZFPs have been characterized in detail [41]. RNA-binding proteins (RBPs) play a critical role in the post-transcriptional regulation of RNAs and are involved in many biological processes from embryogenesis to the regulation of cytokines in the immune system [42].

In conclusion, the results of the present study showed that MB-WCX fractionation followed by MALDI-TOF MS combined with ClinProTools software shows high sensitivity and specificity for the screening of serum proteins and could be a potential tool for the identification of patients with ccRCC. To the best of our knowledge, this is the first study to report the identification of three peptide regions (m/z values of 1466.98, 1618.22, and 5905.23) corresponding to RBP6, TUBB, and ZFP3, respectively.

The method proposed in the present study could generate serum peptidome profiles of ccRCC and provide a new approach to identify potential biomarkers for diagnosis as well as prognosis of this malignancy. Our analyses are mainly limited by a limited number of cases, especially between pre- and post-operative comparsions, as well as lack of validation cohort. We will enlarge the patient cohort, and include other types of RCC as well as other renal diseases in further studies. The study is further limited by the lack of validation of the identified biomarkers as well as their functional data, for there were no corresponding peptide antibodies available for this study, although validated biomarkers are recognized as a priority in the management of patients with ccRCC [13]. The present method could also be of value for the identification of prognostic or predictive markers of response to treatment. Future studies will be aimed at developing antibodies against the three candidate markers identified and verifying their efficacy using multiple samples.

Acknowledgments

The authors would like to thank three anonymous reviewers for their suggestions and comments on the earlier version of this manuscript.

Author Contributions

Conceived and designed the experiments: Juan Yang CH. Performed the experiments: Juan Yang LL TS. Analyzed the data: Juan Yang LZ XW YQ. Contributed reagents/materials/analysis tools: Jin Yang YG. Wrote the paper: Juan Yang.

References

1. Gupta K, Miller JD, Li JZ, Russell MW, Charbonneau C (2008) Epidemiologic and socioeconomic burden of metastatic renal cell carcinoma (mRCC): a literature review. Cancer Treat Rev 34: 193–205.

2. Motzer RJ, Hutson TE, Tomczak P, Michaelson MD, Bukowski RM, et al. (2009) Overall survival and updated results for sunitinib compared with interferon alfa in patients with metastatic renal cell carcinoma. J Clin Oncol 27: 3584–3590.

3. Lam JS, Belldegrun AS, Pantuck AJ (2006) Long-term outcomes of the surgical management of renal cell carcinoma. World J Urol 24: 255–266.

4. Zigeuner R, Hutterer G, Chromecki T, Imamovic A, Kampel-Kettner K, et al. (2010) External validation of the Mayo Clinic stage, size, grade, and necrosis (SSIGN) score for clear-cell renal cell carcinoma in a single European centre applying routine pathology. Eur Urol 57: 102–109.

5. Cheng L, Zhang S, MacLennan GT, Lopez-Beltran A, Montironi R (2009) Molecular and cytogenetic insights into the pathogenesis, classification, differential diagnosis, and prognosis of renal epithelial neoplasms. Hum Pathol 40: 10–29.

6. Haleem JI, Timothy JW, Timothy DV (2011) Cancer biomarker discovery: Opportunities and pitfalls in analytical methods, Electrophoresis 32: 967–975.

7. Ljungberg B (2007) Prognostic markers in renal cell carcinoma. Curr Opin Urol 17: 303–308.

8. Seliger B, Dressler SP, Lichtenfels R, Kellner R (2007) Candidate biomarkers in renal cell carcinoma. Proteomics 7: 4601–4612.

9. Craven RA, Vasudev NS, Banks RE (2013) Proteomics and the search for biomarkers for renal cancer. Clin Biochem 46: 456–465.

10. Banville N, Burgess JK, Jaffar J, Tjin G, Richeldi L, et al. (2014) A Quantitative proteomic Approach to Identify Significantly Altered Protein Networks in the Serum of Patients with Lymphangioleiomyomatosis (LAM). Plos One 9: e 105365.

11. Liu J, Li Y, Wei L, Yang X, Xie Z, et al. (2014) Screening and identification of potential biomarkers and establishment of the diagnostic serum proteomic model for the Traditional Chinese Medicine Syndromes of tuberculosis. J Ethnopharmacol 155: 1322–1331.

12. Liu AN, Sun P, Liu JN, Yu CY, Qu HJ, et al. (2014) Analysis of the differences of serum of serum protein mass spectrometry in patients with triple negative breast cancer and non-triple negative breast cancer. Tumour Biol. E-pub ahead of print.

13. Weiss RH, Kim K (2012) Metabolomics in the study of kidney diseases. Nature Reviews. Nephrology 8: 22–33.

14. Faul F, Erdfelder E, Lang AG, Buchner A (2007) G*Power 3: a flexible statistical power analysis program for the social, behavioral, and biomedical sciences. Behav Res Methods 39: 175–191.

15. Cremona M, Espina V, Caccia D, Veneroni S, Colecchia M, et al. (2014) Stratification of clear cell renal cell carcinoma by signaling pathway analysis. Expert Rev Proteomics 11: 237–249.

16. Varela I, Tarpey P, Raine K, Huang D, Ong CK, et al. (2011) Exome sequencing identifies frequent mutation of the SWI/SNF complex gene PBRM1 in renal carcinoma. Nature 469: 539–542.

17. Dalgliesh GL, Furge K, Greenman C, Chen L, Bignell G, et al. (2010) Systematic sequencing of renal carcinoma reveals inactivation of histone modifying genes. Nature 463: 360–363.

18. Sanjmyatav J, Steiner T, Wunderlich H, Diegmann J, Gajda M, et al. (2011) A specific gene expression signature characterizes metastatic potential in clear cell renal cell carcinoma. J Urol 186: 289–294.

19. Brannon AR, Reddy A, Seiler M, Arreola A, Pruthi R, et al. (2010) Molecular stratification of clear cell renal cell carcinoma by consensus clustering reveals distinct subtypes and survival patterns. Genes Cancer 1: 152–163.
20. Yi Z, Fu Y, Zhao S, Zhang X, Ma C (2010) Differential expression of miRNA patterns in renal cell carcinoma and nontumorous tissues. J Cancer Res Clin Oncol 136: 855–862.
21. Yang J, Song YC, Dang CX, Song TS, Liu ZG, et al. (2012) Serum peptidome profiling in patients with gastric cancer. Clin Exp Med 12: 79–87.
22. Cho WC (2011) Proteomics and translational medicine: molecular biomarkers for cancer diagnosis, prognosis and prediction of therapy outcome. Expert Rev Proteomics 8: 1–4.
23. Hosoya N, Sakumotoa M, Nakamura Y, Narisawa T, Bilim T, et al. (2013) Proteomics identified nuclear N-myc downstream-regulated gene 1 as a prognostic tissue biomarker candidate in renal cell carcinoma. Biochim Biophys 1834: 2630–2639.
24. Jones EE, Powers TW, Neely BA, Cazares LH, Troyer DA, et al. (2014) MALDI imaging mass spectrometry profiling of proteins and lipids in clear cell renal cell carcinoma. 14: 924–935.
25. Raimondo F, Salemi C, Chinello C, Fumagalli D, Morosi L, et al. (2012) Proteomic analysis in clear cell renal cell carcinoma: identification of differentially expressed protein by 2-D DIGE. Mol Biosyst 8: 1040–1051.
26. White NM, Masui O, Desouza LV, Krakovska O, Metias S, et al. (2014) Quantitative proteomic analysis reveals potential diagnostic markers and pathways involved in pathogenesis of renal cell carcinoma. Oncotarget 5: 506–518.
27. Atrih A, Mudaliar MAV, Zakikhani P, Lamont DJ, Huang JTJ, et al. (2014) Quantitative proteomics in resected renal cancer tissue for biomarker discovery and profiling. Brit J Cancer 110: 1622–1633.
28. Steurer S, Seddiqi AS, Singer JM, Bahar AS, Eichelberg C, et al. (2014) MALDI imaging on tissue microarrays identifies molecular features associated with renal cell cancer phenotype. Anticancer Res 34: 2255–2261.
29. Mustafa A, Gupta S, Hudes GR, Egleston BL, Uzzo RG, et al. (2011) Serum amino acid levels as a biomarker for renal cell carcinoma. J Urol 186: 1206–1212.
30. Qi YJ, Ward DG, Pang C, Wang QM, Wei W, et al. (2014) Proteomic profiling of N-linked glycoproteins identifies ConA-binding procathepsin D as a novel serum biomarker for hepatocellular carcinoma.Proteomics 14: 186–195.
31. Schaub NP, Jones KJ, Nyalwidhe JO, Cazares LH, Karbassi ID, et al. (2009) Serum Proteomic Biomarker Discovery Reflective of Stage and Obesity in Breast Cancer Patients. J Am Coll Surg 22: 1–9.
32. Saude E, Sykes B (2007) Urine stability for metabolomic studies: effects of preparation and storage. Metabolomics 3: 19–27.
33. Nicholson JK, Lindon JC (2008) Systems biology: Metabonomics. Nature 455: 1054–1056.
34. Davis VW, Bathe OF, Slupsky CM, Sawyer MB, et al. (2011) Metabolomics and surgical oncology: Potential role for small molecule biomarkers. J Surg Oncol 103: 451–459.
35. Ma Y, Dai HL, Kong XM, Wang LM (2012) Impact of thawing on reference expression stability in renal cell carcinoma samples. Diagn Mol Pathol 21: 157–163.
36. Linge A, Kennedy S, O'Flynn D, Beatty S, Moriarty P, et al. (2012) Differential expression of fourteen proteins between uveal melanoma from patients who subsequently developed distant metastases versus those who did not. Biochem Molec 53: 4634–4643.
37. Ohishi Y, Oda Y, Basaki Y, Kobayashi H, Wake N, et al. (2007) Expression of beta-tubulin isotypes in human primary ovarian carcinoma. Gynecol Oncol 105: 586–592.
38. Berrieman HK, Lind MJ, Cawkwell L (2004) Do beta-tubulin mutations have a role in resistance to chemotherapy? Lancet Oncol 5: 158–164.
39. Bhattacharya R, Cabral F (2004) A ubiquitous beta-tubulin disrupts microtubule assembly and inhibits cell proliferation. Mol Biol Cell 15: 3123–3131.
40. Khalfallah O, Ravassard P, Lagache CS, Fligny C, Serre A, et al (2009) Zinc finger protein 191 (ZNF191/Zfp191) is necessary to maintain neural cells as cycling progenitors. Stem cells 27: 1643–1653
41. Al-Kandari W, Jambunathan S, Navalgund V, Koneni R, Freer M, et al. (2007) ZXDC, a novel zinc finger protein that binds CIITA and activates MHC gene transcription. Mol Immunol 44: 311–321.
42. Colegrove-Otero LJ, Minshall N, Standart N (2005) RNA-binding proteins in early development. Crit Rev Biochem Mol Biol 40: 21–73.

Structural and Functional Characterization of Cleavage and Inactivation of Human Serine Protease Inhibitors by the Bacterial SPATE Protease EspPα from Enterohemorrhagic *E. coli*

André Weiss, Hanna Joerss, Jens Brockmeyer*

Institute of Food Chemistry, Westfälische Wilhelms-Universität Münster, Münster, Germany

Abstract

EspPα and EspI are serine protease autotransporters found in enterohemorrhagic *Escherichia coli*. They both belong to the SPATE autotransporter family and are believed to contribute to pathogenicity via proteolytic cleavage and inactivation of different key host proteins during infection. Here, we describe the specific cleavage and functional inactivation of serine protease inhibitors (serpins) by EspPα and compare this activity with the related SPATE EspI. Serpins are structurally related proteins that regulate vital protease cascades, such as blood coagulation and inflammatory host response. For the rapid determination of serpin cleavage sites, we applied direct MALDI-TOF-MS or ESI-FTMS analysis of coincubations of serpins and SPATE proteases and confirmed observed cleavage positions using in-gel-digest of SDS-PAGE-separated degradation products. Activities of both serpin and SPATE protease were assessed in a newly developed photometrical assay using chromogenic peptide substrates. EspPα cleaved the serpins α1-protease inhibitor (α1-PI), α1-antichymotrypsin, angiotensinogen, and α2-antiplasmin. Serpin cleavage led to loss of inhibitory function as demonstrated for α1-PI while EspPα activity was not affected. Notably, EspPα showed pronounced specificity and cleaved procoagulatory serpins such as α2-antiplasmin while the anticoagulatory antithrombin III was not affected. Together with recently published research, this underlines the interference of EspPα with hemostasis or inflammatory responses during infection, while the observed interaction of EspI with serpins is likely to be not physiologically relevant. EspPα-mediated serpin cleavage occurred always in flexible loops, indicating that this structural motif might be required for substrate recognition.

Editor: Stefan Bereswill, Charité-University Medicine Berlin, Germany

Funding: The study was supported by the Deutsche Forschungsgemeinschaft (DFG) (http://dfg.de), grant BR 4258/1-1. The funder had no role in study design, data collection and analysis, decision to publish, or preparation of the manuscript.

Competing Interests: The authors have declared that no competing interests exist.

* Email: jbrockm@uni-muenster.de

Introduction

Enterohemorrhagic *Escherichia coli* (EHEC) cause severe diseases in humans worldwide. Shiga toxins are regarded as their main virulence factor. However, EHEC possess various further virulence factors that mediate adherence or interfere with host defense [1,2]. One of these additional virulence factors is the plasmid-encoded extracellular serine protease EspP which belongs to the serine protease autotransporter of *Enterobacteriaceae* (SPATE) family [3]. Five subtypes of EspP have been described (EspPα-EspPε [4,5], from which the translocation-competent and proteolytically active subtype EspPα (Uniprot Accession Number: Q7BSW5) is associated with highly virulent strains and isolates from patients with severe disease [4,6]. EspPα exhibits serine protease activity. In addition to porcine pepsin A and EHEC-Hemolysin [3,7], EspPα cleaves the human plasma proteins apolipoprotein A-I, the complement factors C3 and C5, and coagulation factor V [3,8,9]. EspPα-mediated cleavage of complement factors has been demonstrated to significantly reduce complement activation [9]. In addition, the degradation of factor

V has been suggested to interfere with blood coagulation possibly leading to prolonged bleeding during EHEC infection [3].

The *E. coli* secreted protease, island-encoded (EspI) is a further member of the SPATE family and is secreted by Shiga toxin-producing *E. coli* (STEC) [8]. Notably, EspI has been found in less pathogenic *E. coli* serotypes [8,10,11]. The physiological function of EspI is yet unknown and to date only two substrates have been identified, namely porcine pepsin A and human apolipoprotein A-I [8].

Serine protease inhibitors (serpins) are structurally closely related proteins which modulate different important protease cascades by irreversible inactivation of serine proteases. They are involved in inflammatory host defense, complement activation, and blood coagulation [12,13]. Serpins share an exposed reactive center loop (RCL) that serves as a pseudosubstrate for the target protease. Cleavage of the reactive serpin bond initiates a conformational rearrangement of the serpin structure that leads to distortion and inactivation of the target protease by formation of an irreversible covalent serpin-protease complex [14]. α1-protease

Inhibitor (α1-PI, Uniprot Accession Number: P01009) is the archetypal member of the serpin family and the most abundant serpin in human plasma. Its main physiological target is neutrophil elastase [15]. α1-antichymotrypsin, (α1-AC, Uniprot Accession Number: P01011) which is closely related to α1-PI, [16,17] mainly inhibits cathepsin G and mast cell chymases [15,18]. α2-antiplasmin (α2-AP, Uniprot Accession Number: P08697) is the main physiological inhibitor of plasmin and thus influences fibrinolysis following blood coagulation [19,20]. Antithrombin III (ATIII, Uniprot Accession Number: P01008) inhibits thrombin, FIXa, and FXa - proteases of the blood coagulation pathway - which is considerably faster in the presence of its cofactor heparin [21–24]. Angiotensinogen (AGT, Uniprot Accession Number: P01019) is a non-inhibitory serpin that does not target proteases [25]. Via proteolytic processing by renin, AGT releases the vasopressor peptide angiotensin I which is further converted to angiotensin II [26,27]. An overview of serpin functions and nomenclature is given in Table 1.

Serpins are therefore highly relevant concerning their regulatory function as pseudosubstrates that inactivate serine proteases by formation of serpin-enzyme-complexes. In addition, cleavage of serpins without formation of an inhibitory complex has been described in literature for different metalloproteases. The human matrix metalloproteinase-3, e.g., cleaves α1-AC, α2-AP, and plasminogen activator inhibitor-1 [28,29] while human matrix metalloproteinase-9 cleaves α1-PI [30]. The bacterial 56-kDa proteinase from *Serratia marcescens* also cleaves α1-PI, α2-AP, ATIII, and C1 esterase inhibitor (C1-INH) [31,32]. C1-INH is also specifically cleaved by StcE, a metalloprotease found in highly pathogenic EHEC [33]. Surprisingly, interference of StcE with C1-INH also results in enhanced inhibition of complement-mediated lysis irrespective of cleavage of this serpin [34,35]. Interference with serpin function in the human host during bacterial infection is therefore a further pathogenicity mechanism.

Notably, we describe here the specific cleavage of various serpins from human plasma by the bacterial serine protease EspPα and compare this activity with the related SPATE EspI. Presented data further support the hypothesis that EspPα mediates virulence by interaction with key regulatory proteins of host defense and blood coagulation. In addition, we developed a photometrical assay for the analysis of serpin activity and applied matrix assisted laser desorption ionization-time of flight-mass spectrometry (MALDI-TOF-MS) and electrospray ionisation-fourier transform mass spectrometry (ESI-FTMS) for the direct elucidation of proteolytic cleavage sites.

Materials and Methods

Pseudonymized residual sample material from voluntary blood donations from the Transfusion medicine of the University Clinics Münster was used. Blood donors approved prior to donation that residual sample material can be used for scientific studies. The Ethics Committee of the Medical Faculty of the University of Münster was informed and approved the study design.

Proteins

EspPα was purified from clone HB101 (WB4–5k) containing *espP* from *E. coli* O157:H7 strain EDL933 [3]. The inactive EspP mutant S263A served as a negative control [36] and EspI was purified in the same way from clone DH5α/pZH4 containing *espI* from *E. coli* O91:H$^-$ strain 4797/97 [4,8]. Protein precipitation from culture supernatants was performed as described previously [4]. Briefly, protein pellets were dissolved in 20 mM Tris buffer containing 50 mM NaCl (pH 6.5). Proteins were purified using HiPrep 16/10 DEAE FF, HiTrap Benzamidine FF (HS), and HiPrep 16/60 Sephacryl S-200 HR columns (GE Healthcare) according to the manufacturers instructions. Protein preparations were diluted to 1 μg/μL with phosphate buffered saline (PBS, 100 mM NaCl, 4.5 mM KCl, 7.0 mM Na$_2$HPO$_4$, 3.0 mM KH$_2$PO$_4$, pH 7.4).

Purified serpins were purchased from Merck Millipore and dissolved according to the manufacturers instructions in the following buffers: α1-PI, 30 mM Na$_3$PO$_4$, 300 mM NaCl, pH 6.5, α1-AC, 20 mM Tris, 250 mM NaCl, 4.5 mM KCl, 7.0 mM Na$_2$HPO$_4$, 3.0 mM KH$_2$PO$_4$, pH 7,4, α2-AP, 20 mM Bis-Tris, 200 mM NaCl, pH 6.4, ATIII, 100 mM NaCl, 4.5 mM KCl, 7.0 mM Na$_2$HPO$_4$, 3.0 mM KH$_2$PO$_4$, pH 7.4, AGT, 50 mM Na$_3$PO$_4$, 150 mM NaCl, pH 7.0.

Plasma fractionation

Plasma samples (fresh frozen plasma, FFP) were stabilized with 17–23% (v/v) citrate-phosphate-dextrose (CPD) and were derived from whole blood donations using standard separation procedures for blood banks.

Plasma was diluted with 20 mM Na$_3$PO$_4$ buffer (pH 7.0) and depleted using HiTrap Protein A FF and HiTrap Blue HP (GE

Table 1. Serpins used in this study.

Serpin	Systematic name	Main Target proteases	Function	Reference
α1-Protease Inhibitor	SERPINA1	neutrophil elastase	Protection of tissue during inflammation, deficiency results in emphysema	[15,56–58]
α1-Antichymotrypsin	SERPINA3	Cathepsin G, mast cell chymases	Deficiency may result in emphysema, possible contribution to Alzheimer	[15,18,59–61]
Angiotensinogen	SERPINA8	-	Non-inhibitory, renin substrate, release of angiotensin I	[25,62]
α2-Antiplasmin	SERPINF2	plasmin	Regulation of fibrinolysis	[19,45]
Antithrombin III	SERPINC1	thrombin, FIXa, FXa	Most important inhibitor of the coagulation pathway	[21–24]

Given are the systematic serpin name, target proteases, and general function.

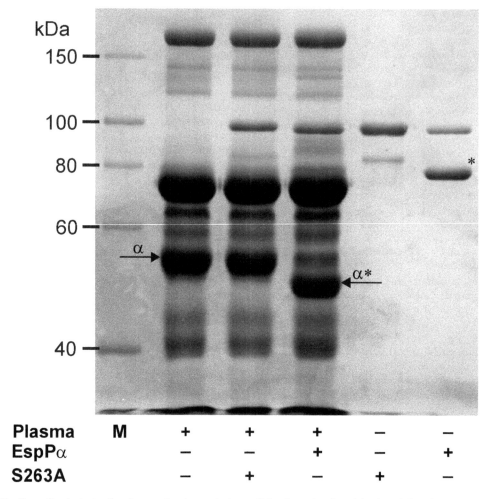

Figure 1. Identification of substrates in plasma. Fractionated plasma (25 μg) was incubated (15 h, 37°C) with EspPα or S263A (1.5 μg) and separated via SDS-PAGE using a glycine buffer. M, molecular weight marker, *, EspPα autodegradation product, α, α1-PI, α*, α1-PI degradation product.

Healthcare) according to the manufacturers instructions. The depleted plasma was further fractionated using HiPrep 16/10 DEAE FF via gradient elution ranging from 100% buffer A (20 mM Tris, 50 mM NaCl, pH 8.0) to 70% buffer B (20 mM Tris, 500 mM NaCl, pH 8.0). The protein fraction eluting from 15–40% buffer B was used for further experiments.

Cleavage of Substrates

To determine cleavage of substrates by EspPα or EspI, fractionated plasma (25 μg) or serpins (5 μg or 10 μg) were incubated (15 h, 37°C) with 1.5 μg of purified protease in 30 μL PBS buffer. ATIII was incubated in the same way after addition of 25 μg/mL (4.8 units/mL) unfractionated heparin (Merck). Proteins were either separated via sodium dodecyl sulfate-polyacrylamide gel electrophoresis (SDS-PAGE), digested in-gel and analyzed using matrix assisted laser desorption ionization-time of flight-mass spectrometry (MALDI-TOF-MS) or subjected directly to MS analysis.

SDS-PAGE

After denaturation, proteins were separated on a 7.5% SDS-PAGE gel using a glycine (19.2 mM) containing buffer [37] or on a 13.3% SDS-PAGE gel using a tricine (100 mM) containing buffer [38] and stained with Coomassie Blue.

In-gel-digestion

In-gel-digestion was performed as described before [39]. Briefly, gel pieces were cut out, proteins were reduced using dithiothreitol (10 mM), alkylated with iodoacetamide (55 mM) and digested (15 h, 37°C) with trypsin (13 ng/μL, Promega). Peptides were extracted and desalted using ZipTip C_{18} Pipette Tips (Merck Millipore) according to the manufacturers instructions. Peptides were eluted with 40% acetonitrile (MeCN)/1% formic acid (FA) and 70% MeCN/1% FA (5 μL each) and eluates were combined.

Mass spectrometric analysis

In-gel-digests or incubation mixtures (0.5 μL) were mixed with 0.5 μL α-cyano-4-hydroxycinnamic acid (Sigma-Aldrich, 10 μg/μL in 50% MeCN/1% trifluoroacetic acid) and 0.5 μL of the mixture were spotted on a MALDI target (MTP 384 target plate ground steel, Bruker). Samples were analyzed using a Bruker autoflex speed in positive mode.

To determine the accurate masses of the largest α2-AP fragment, the incubation mixture was desalted using ZipTip C_{18} Pipette Tips as described before and measured using a Thermo

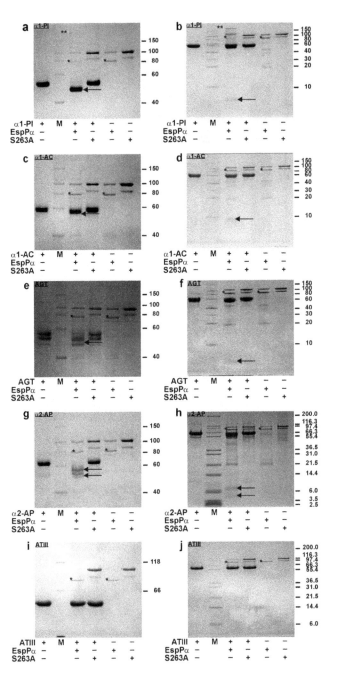

Figure 2. Cleavage of various serpins by EspPα. Serpins (5 μg) were incubated (15 h, 37°C) with EspPα or S263A (1.5 μg). Degradation products were separated via SDS-PAGE using a glycine buffer (a, c, e, g, i) or tricine buffer system (b, d, f, h, j). Proteolytic serpin fragments formed by EspPα are indicated by an arrow. a, b α1-PI is degraded to a large and small fragment (~45 kDa and ~4 kDa, respectively), c, d cleavage of α1-AC in two fragments, e, f the AGT band with the highest molecular weight is cleaved in two fragments, g, h large and small fragments (~55–57 kDa and ~4–7 kDa) formed by α2-AP cleavage. i, j ATIII is not cleaved by EspPα. Incubation of α1-PI with EspPα leads to a weak formation of an inhibitory enzyme-serpin complex as marked by **. M, molecular weight marker, *, autodegradation product of EspPα.

LTQ Orbitrap XL in positive static nanospray mode (sheath gas flow rate 15 arb.u., aux gas flow rate 10 arb.u., sweep gas flow rate 5 arb.u.).

Determination of EspPα and α1-PI activity

Potential functional consequences of the interaction between α1-PI and EspPα were analyzed by measuring the activities of both proteins after coincubation. To investigate effects of α1-PI on EspPα protease activity, both proteins were incubated together (15 h, 37°C) at equimolar concentrations. Preincubated (15 h, 37°C) EspPα or α1-PI were used as controls. The remaining EspPα protease activity was then determined by the incubation (15 h, 37°C) of an aliquot containing 1 μg EspPα (either preincubated alone or with α1-PI) with 2 mM of the chromogenic peptide substrate Suc-Ala-Ala-Pro-Leu-pNA (Bachem) in 100 μL PBS (pH 7.4) and 5% dimethyl sulfoxide (DMSO). Active EspPα releases para-nitroaniline (pNA) from the peptide which is detected at 405 nm using a FLUOstar Optima plate reader (BMG Labtech). PBS was used as a buffer control.

The effect of EspPα-mediated cleavage on α1-PI serpin activity was determined by coincubation (15 h, 37°C) of α1-PI and EspPα in a molar ratio of 4:1. Again, incubations (15 h, 37°C) of EspPα or α1-PI alone were used as controls. To assess remaining serpin activity of α1-PI, the coincubation mixture and controls were incubated (5 h, 37°C) with trypsin (Promega) at a molar ratio of α1-PI and trypsin of 4:1. Active α1-PI inhibits trypsin protease activity. The remaining serpin activity was therefore assessed indirectly by determination of reduced trypsin activity using aliquots of coincubation mixtures and controls containing 0.25 μg trypsin and incubation (2 h, 37°C) with 2 mM of the chromogenic peptide Bz-Arg-pNA (Bachem) in 100 μL PBS (pH 7.4) containing 5% DMSO. Active trypsin releases pNA and absorbance was measured at 405 nm using a FLUOstar Optima plate reader. PBS was used as a buffer control.

Results and Discussion

Purification of EspPα and S263A

EspPα and the inactive EspPα mutant S263A were purified from culture supernatants using ammonium sulfate precipitation and liquid chromatography. Purity was verified via SDS-PAGE (Fig. 1, lane 5 and 6). EspPα shows a band at ~104 kDa representing the intact EspPα and a band at ~80 kDa which was identified by MALDI-TOF-MS as autoproteolysis product. S263A samples showed a pronounced protein band at ~104 kDa and a weaker band at ~85 kDa which was identified as a truncated form of S263A. The autoproteolyis product of EspPα remains active even after long term incubation (Figure S1). Proteolytic activity of purified EspPα and the inactive S263A were assessed using a chromogenic oligopeptide substrate. As expected, all EspPα samples were proteolytically active while S263A showed no proteolytic activity (Fig. S1).

Identification of EspPα substrates in plasma

To identify physiological relevant substrates of EspPα, fractionated plasma was incubated either with EspPα or the EspPα negative control S263A (Fig. 1, lane 3 and 4). Incubation with EspPα resulted in loss of a pronounced 50 kDa band in plasma and the occurrence of a degradation product with a molecular weight of ~45 kDa in SDS-PAGE. The according protein band was digested in-gel and subjected to MALDI-TOF-MS analysis and unambiguously identified as α1-PI (Aldente score 235.7, sequence coverage 69% to α1-PI (UniProtKB: P01009)).

EspPα cleaves various serpins

To determine if further serpins are cleaved by EspPα, different serpins were incubated with EspPα or S263A and cleavage was monitored by SDS-PAGE. EspPα degrades α1-PI, α1-AC, and the

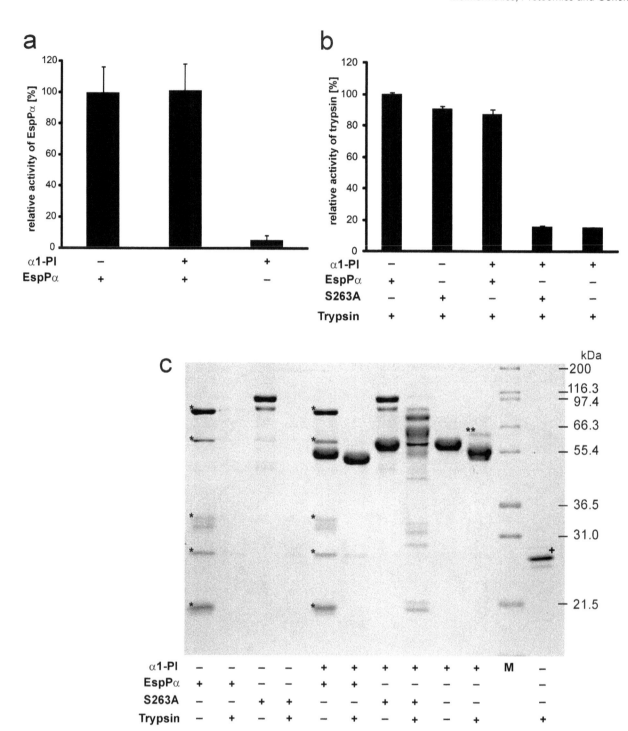

Figure 3. Activity of EspPα and α1-PI after coincubation. a, Determination of EspPα activity. EspPα and α1-PI were preincubated (15 h, 37°C) at equimolar concentrations and remaining activity of EspPα was analyzed by incubation of an aliquot of the mixture with the chromogenic substrate Suc-Ala-Ala-Pro-Leu-pNA. Activity was measured via released *para*-nitroaniline and normalized to EspPα. n = 9 for EspPα and EspPα+α1-PI or n = 6 for α1-PI, respectively. b, α1-PI activity (measured as inhibitory potential on trypsin) after incubation with EspPα. α1-PI and EspPα or S263A were preincubated at a molar ratio of serpin:enzyme = 4:1. Remaining inhibitory activity of α1-PI on trypsin was analyzed by incubation at a molar ratio of α1-PI:trypsin = 4:1. Trypsin activity was measured via release of *para*-nitroaniline from the chromogenic substrate Bz-Arg-pNA. c, SDS-PAGE analysis of conincubations. α1-PI, EspPα, S263A, and trypsin were incubated as in b) and mixtures were separated via SDS-PAGE (12% SDS-PAGE gel, glycine buffer). M, molecular weight marker, *, EspPα autodegradation product, **, inhibitory complex of α1-PI and trypsin, +, trypsin was directly subjected to SDS-PAGE without incubation.

non-inhibitory serpin AGT into a large (>40 kDa) and a small (< 10 kDa) fragment (Fig. 2 a–f), while incubation of α2-AP leads to several degradation products (Fig. 2 g, h). None of the incubations led to pronounced formation of an inhibitory serpin-enzyme complex. Interestingly, the anticoagulatory serpin ATIII was not degraded by EspPα (Fig. 2i, j).

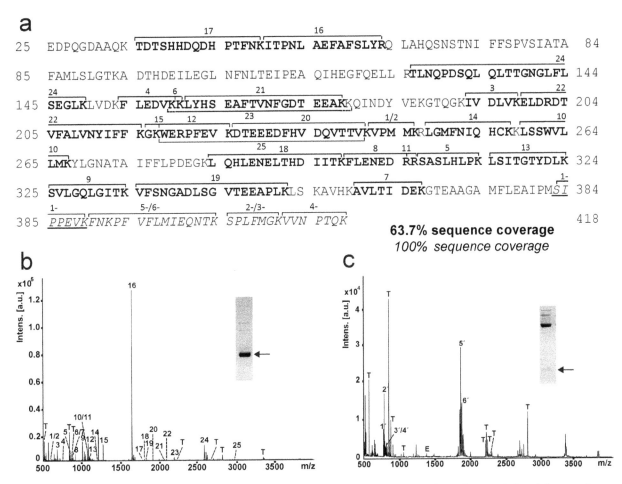

a

```
                     17                 16
  25  EDPQGDAAQK TDTSHHDQDH PTFNKITPNL AEFAFSLYRQ LAHQSNSTNI FFSPVSIATA  84
                                                                    24
  85  FAMLSLGTKA DTHDEILEGL NFNLTEIPEA QIHEGFQELL RTLNQPDSQL QLTTGNGLFL 144
       24         4    6           21                   3        22
 145  SEGLKLVDKF LEDVKKLYHS EAFTVNFGDT EEAKQINDY VEKGTQGKIV DLVKELDRDT 204
       22         15   12       23        20        1/2      14        10
 205  VFALVNYIFF KGKWERPFEV KDTEEEDFHV DQVTTVKVPM MKRLGMFNIQ HCKKLSSWVL 264
       10                       25  18            8    11    5           13
 265  LMKYLGNATA IFFLPDEGKL QHLENELTHD IITKFLENED RRSASLHLPK LSITGTYDLK 324
        9              19                      7                1-
 325  SVLGQLGITK VFSNGADLSG VTEEAPLKLS KAVHKAVLTI DEKGTEAAGA MFLEAIPMSI 384
       1-          5-/6-            2-/3-     4-
 385  PPEVKFNKPF VFLMIEQNTK SPLFMGKVVN PTQK                            418
```

63.7% sequence coverage
100% sequence coverage

b

c

Figure 4. Peptide mapping of EspPα cleavage products of α1-PI. α1-PI fragments were subjected to tryptic in-gel-digest and generated peptides were analyzed via MALDI-TOF-MS. a, Sequence coverage of α1-PI fragments. Peptides of the large fragment are given in bold and numbered 1–25. Peptides of the small fragment are given in italics and numbered 1'–6'. Note the newly formed N-terminus of the small fragment (SIPPEVK, underlined). b, MALDI-TOF-MS spectrum of the large fragment of α1-PI. Inset: SDS-PAGE gel, glycine buffer. Fragment used for peptide mapping is marked by arrow. c, MALDI-TOF-MS spectrum of the small fragment of α1-PI. Inset: SDS-PAGE gel, tricine buffer. Fragments used for peptide mapping are marked by arrow. α1-PI peptides are numbered according to a, T, trypsin autoproteolysis products, E, EspPα autoproteolysis products.

Activity of α1-PI and EspPα after incubation

We next determined the functional consequences of the coincubation of serpin and SPATE protease by use of the *bona fide* serpin α1-PI and EspPα. The remaining EspPα-activity following incubation with α1-PI was assessed in a photometrical assay using the chromogenic EspPα substrate Suc-Ala-Ala-Pro-Leu-pNA. Incubation with α1-PI had no influence on the proteolytic activity of EspPα (Fig. 3a), demonstrating that α1-PI does not target EspPα.

Table 2. Serpin cleavage sites determined by MALDI-TOF-MS.

Serpin	m/z determined	Theoretical mass	Deviation (ppm)	Sequence	Position
α1-PI	4133.333	4133.234	+24	[380]IPM-SIP[385]	Reactive bond
α1-AC	4623.419	4623.495	−16	[381]TLL-SAL[386]	Reactive bond
AGT	4299.351	4299.293	+14	[444]QQL-NKP[449]	Reactive center loop
α2-AP	2181.123	2181.097	+12	[45]SPL-TLL[50]	N-terminal extension
α2-AP	3489.789	3489.788	<1	[458]QSL-KGF[463]	C-terminal extension
α2-AP	3602.870	3602.872	−1	[457]LQS-LKG[462]	C-terminal extension
α2-AP	5308.3 (average)	5307.9 (average)	+75	[442]REL-KEQ[447]	C-terminal extension

Given are masses determined by MALDI-TOF-MS directly after incubation of serpin with EspPα, theoretical masses, mass deviation, according sequence, and position inside the serpin sequence. Numeration is according to the serpin precursor.

Table 3. α2-AP cleavage site determined by ESI-FTMS.

m/z determined	m/z theoretical	Charge state (z)	Deviation (ppm)	Sequence	Position
884.9507	884.9512	6	−1	^{442}REL-KEQ^{447}	C-terminal extension
758.6725	758.6736	7	−1	^{442}REL-KEQ^{447}	C-terminal extension
663.9624	663.9653	8	−4	^{442}REL-KEQ^{447}	C-terminal extension

Given are masses of the large α2-AP fragment as determined by nanospray ESI-FTMS.

The remaining inhibitory potential of α1-PI following incubation with EspPα was analyzed using trypsin as a serpin target. Although neutrophil elastase is the physiological target for α1-PI, trypsin also forms an irreversible inhibitory complex with the serpin and can therefore be used as an indicator for α1-PI activity [40]. Active α1-PI inhibits the proteolytic activity of trypsin and consequently loss of α1-PI serpin activity results in high proteolytic activity in the assay. Trypsin activity was determined by photometrical detection of the cleavage of the trypsin substrate Bz-Arg-pNA.

Incubation of trypsin with α1-PI or α1-PI preincubated with S263A resulted in nearly complete loss of trypsin activity (Fig. 3b), demonstrating that the employed α1-PI shows high serpin activity and that the inactive EspPα mutant S263A does not affect α1-PI. In contrast, α1-PI preincubated with EspPα did not reduce trypsin activity in the following assay (Fig. 3b). This demonstrates that EspPα-mediated α1-PI cleavage leads to loss of the inhibitory serpin activity. Corresponding results were obtained using SDS-PAGE (Fig. 3c). Incubation of α1-PI with trypsin leads to the formation of a serpin-enzyme-complex (Fig. 3c, lane 10). After incubation with EspPα, α1-PI is not able to form this complex with trypsin. Instead, the large α1-PI fragment is further degraded by trypsin (Fig. 3c, lane 6). EspPα as well as S263A were completely degraded when incubated with trypsin, demonstrating that neither EspPα nor S263A directly interfere with trypsin activity (Fig. 3c, lanes 2 and 4). In addition, α1-PI does not interact with S263A (no serpin enzyme complex) (Fig. 3c, lane 7) but is cleaved by EspPα (Fig. 3c, lane 5). The addition of trypsin to the mixture of α1-PI and S263A led to incomplete degradation and occurrence of several degradation bands in SDS-PAGE. This is due to the fact that degradation of S263A by trypsin and the inhibition of trypsin by α1-PI occur in parallel resulting in only incomplete S263A degradation (Fig. 3c, lane 8).

EspPα cleaves inside the reactive center loop

The loss of activity of α1-PI but not EspPα is based on cleavage of α1-PI without formation of an inhibitory serpin-enzyme-complex. To further understand how EspPα-mediated cleavage affects the inhibitory function, we determined the cleavage sites in α1-PI and the other serpins included in this study. To this end, large and small fragments of cleaved serpins were separated using SDS-PAGE, in-gel-digested and subjected to MALDI-TOF-MS analysis. Figure 4 shows the peptide mapping of EspPα cleavage products of α1-PI. The large α1-PI fragment consists of the N-terminal part of the serpin (Fig. 4a and b), while the C-terminal part from residue 383 to 418 forms the small fragment (Fig. 4 a, and c). EspPα cleavage occurs at the active site of the serpin between ^{382}Met and ^{383}Ser as demonstrated by the occurrence of the non-tryptic peptide 1'(SIPPEVK) and the complete sequence coverage for the small fragment (Fig. 4c). Sequence coverage of degradation products of the other serpins are given in Figure S2.

Direct MALDI-TOF-MS analysis of small fragments

Not all cleavage sites can be identified via in-gel-digest. Tryptic peptides might be too small when cleavage occurs close to lysine or arginine residues or when several cleavage sites are in close proximity to each other. As all small fragments formed by EspPα-cleavage show a molecular weight below 10 kDa, we applied direct MALDI-TOF-MS analysis to determine the exact mass of the small serpin fragments to elucidate and confirm cleavage sites (Fig. 5). For the small α1-PI fragment we observed a signal for the proton adduct of the α1-PI sequence ^{383}Ser-^{418}Lys (m/z 4133.333) confirming the cleavage site determined via in-gel-digest. In addition, signals representing the Na$^+$ adduct and the oxidized Na$^+$ adduct of the according α1-PI fragment sequence were observed (Fig. 5a). α1-AC shows a similar spectrum with a pronounced signal at m/z 4623.419 demonstrating cleavage C-terminal of ^{383}Leu at the reactive bond (Fig. 6b), which is in good accordance with data from in-gel-digest (Figure S2). For AGT, we already observed three bands in SDS-PAGE (intact AGT and two non-proteolytic fragments) when incubated without protease (Fig. 2e). Accordingly, signals of two small AGT fragments were observed in MALDI-TOF-MS (Fig. 5c, right lane). Incubation with EspPα led to degradation of intact AGT and occurrence of the corresponding small fragment in MALDI-TOF-MS (Fig. 2e and Fig. 5c, left lane). For α2-AP, proteolytic cleavage into several fragments is observed in SDS-PAGE (see Fig. 2g and 2h and Fig. 5d) after incubation with EspPα. Four distinct signals are seen in the MS spectrum indicating 4 cleavage sites. As the resolution for the signal at m/z 5308.3 is too low to determine the monoisotopic mass, we measured this sample in addition via nanospray-ESI-FTMS. Table 2 summarizes EspPα cleavage sites and their positions within the respective serpin. Measurement of α2-AP after incubation with EspPα via nanospray ESI-FTMS is described in Table 3.

α1-PI and α1-AC are cleaved at their reactive bonds (position of reactive sites are described in [41,42]), leading to loss of serpin function. In both molecules the reactive bonds are exposed in the RCL and serve as pseudosubstrates for the targeted proteases. In case of EspPα, the serpins are not able to form a stable inhibitor-enzyme-complex and therefore release the intact EspPα after cleavage. Although AGT as non-inhibitory serpin does not contain a reactive bond, it is structurally closely related to the other serpins and is also cleaved in the RCL, indicating that a reactive bond is not necessary for EspPα-mediated serpin degradation. This is further underlined for α2-AP, which is cleaved at four positions outside the RCL (for RCL position see [43]). Cleavage sites are located at the N- and C-terminal extensions 25 aa downstream the N-terminus and 46, 31, and 30 aa upstream the C-terminus (see Table 2). Intriguingly, both the N- and C-terminal extensions are vital for the functional relevant binding of α2-AP to other proteins [19,44,45].

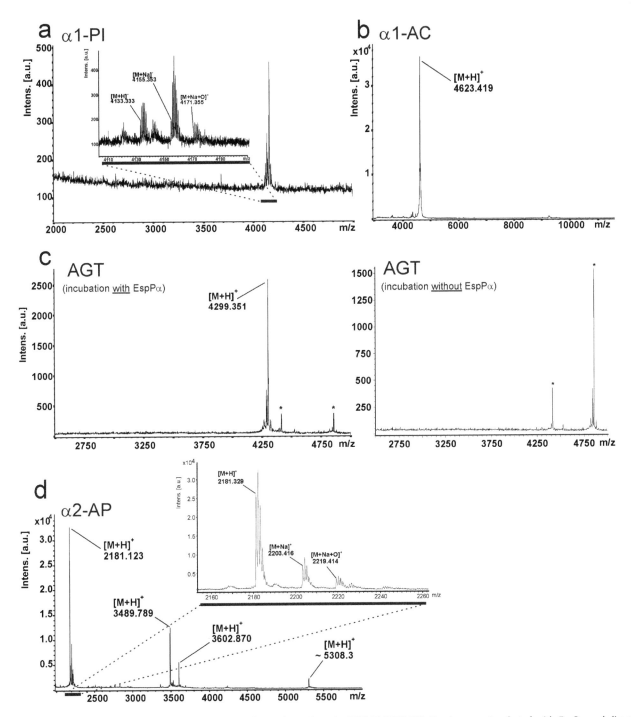

Figure 5. Direct analysis of the small cleavage product of serpins via MALDI-TOF-MS. Serpins were incubated with EspPα and directly analyzed via MALDI-TOF-MS. a, MALDI-TOF-MS spectrum of α1-PI fragment. Inset: Detailed view of the signal representing the small α1-PI fragment. b, MALDI-TOF-MS spectrum of α1-AC. c, MALDI-TOF-MS spectrum of AGT. Left lane: Spectrum after incubation with EspPα, right lane: Spectrum after incubation of AGT without EspPα. *, signals represent non-proteolytic fragments also found after incubation of AGT without EspPα. d, MALDI-TOF-MS spectrum of α2-AP. Inset: Detailed view of the m/z window 2160–2260 representing signals (M H), (M Na), (M Na+O) of the cleavage site in the N-terminal extension of α2-AP are exemplarily shown. (M H), proton adduct of small serpin fragment, (M Na), Na adduct of small serpin fragment, (M Na+O), Na adduct oxidized at one methionine residue.

Cleavage of serpins by EspI

Purified EspI samples showed a protein band at ~110 kDa (intact EspI) as well as two EspI autoproteolysis products at ~50 and 45 kDa, respectively. Similar to EspPα, autoproteolysis products remain active. Serpins were incubated with purified EspI in the same way as described for EspPα. Incubation of α1-PI and α1-AC with EspI led to degradation of these serpins. Notably, EspI also forms a pronounced inhibitory complex with both protease inhibitors resulting in only incomplete serpin degradation (Fig. 6 a-d). In contrast to EspPα, EspI does not cleave α2-AP and

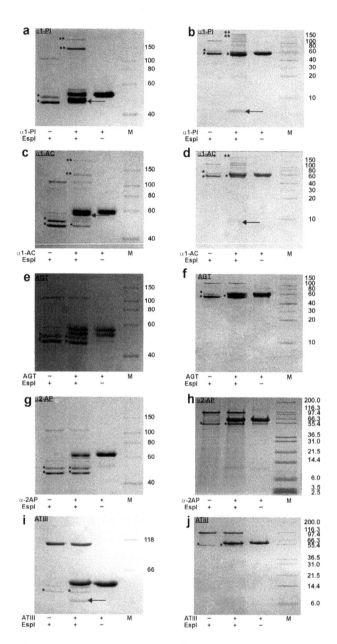

Figure 6. Cleavage of serpins by EspI. Serpins (5 μg) were incubated (15 h, 37°C) with EspI (1.5 μg). Degradation products were separated via SDS-PAGE using a glycine buffer (a, c, e, g, i) or a tricine buffer (b, d, f, h, j). a, b α1-PI is cleaved into two fragments (~45 kDa and ~4 kDa), c, d α1-AC is cleaved into two fragments, e, f AGT is not cleaved by EspI, g, h α2-AP is not cleaved by EspI, i, j ATIII is cleaved only with very low efficiency. Note the formation of inhibitory serpin-enzyme-complexes after incubation with α1-PI and α1-AC. M, molecular weight marker, *, autodegradation product of EspI, **, inhibitory serpin-EspI-complex. Serpin fragments are indicated by an arrow.

AGT (Fig. 6e–h). Cleavage of ATIII occurred only with very low efficiency (Fig. 6i) and might not be relevant under physiological conditions.

To determine the cleavage sites of α1-PI and α1-AC, we subjected incubation mixtures of serpins and EspI to direct MALDI-TOF-MS analysis. Serpin cleavage occurred at the reactive bond leading to signals at m/z 4155.400 (α1-PI, 20 ppm deviation according to calculated m/z) and 4623.509

(α1-AC, 19 ppm deviation according to calculated m/z), respectively (data not shown).

Conclusions

EspPα is an EHEC virulence factor that belongs to the SPATE family. As suggested for SPATEs in general, EspPα most likely mediates its virulence via cleavage and inactivation of host proteins. Here, we present a method for the rapid determination of EspPα-mediated cleavage sites in various human plasma serpins via MALDI-TOF-MS as well as a photometrical assay to analyze serpin functionality after proteolytic cleavage. Concerning the functional consequences, degradation of α2-AP might lead to bleeding disorders. This serpin is the primary physiological inhibitor of plasmin and deficiency has been shown to result in uncontrolled fibrinolysis and severe hemorrhagic complication [44,45]. α2-AP harbors a 42 aa N-terminal and a 55 aa C-terminal extension [19,46]. While the N-terminal extension is cross-linked to fibrin, the very C-terminal [491]Lys residue mediates binding to plasmin [47]. EspPα cleaves between [47]Leu and [48]Thr releasing part of the N-terminal extension and at three different sites inside the C-terminal extension leading to release of a polypeptide containing [491]Lys. Together, this most likely leads to loss of function of α2-AP. The role of α1-PI in thrombosis is not well understood. However, α1-PI is able to inhibit activated protein C. In pediatric ischemic stroke patients elevated levels of α1-PI have been found and were discussed to contribute to this thrombotic disease in children [48,49]. ATIII is the main anticoagulatory serpin. Although it is able to interfere with virtually all proteolytic coagulation factors, its main targets are thrombin, FIXa, and FXa. Intriguingly, it is the only serpin in this study that is not cleaved by EspPα. Despite the structural similarity of serpins, EspPα specifically cleaves only selected serpins. More specific, procoagulatory serpins such α2-AP and α1-PI are efficiently degraded while the anticoagulatory ATIII is not affected at all. Together with data demonstrating that EspPα cleaves coagulation factor V [3], this underlines the hypothesis that interference with blood coagulation (and possibly also inflammatory host responses) [50] might be one of the major functions of EspPα which might contribute to formation of hemorrhages observed during EHEC infection.

Having a closer look at EspPα cleavage sites, it is notable that more than 70% (5 of 7) of cleavage sites identified in this study occur after Leu. This is in good accordance to already reported EspPα cleavage sites [3,9,7,51], indicating that substrate cleavage is most favorable C-terminal to Leu. In α2-AP, cleavage also occurs after [459]Ser. This residue, however, is positioned next to [460]Leu after which EspPα cleaves, too. The second non-Leu cleavage site is C-terminal to [382]Met in α1-PI. The [382]Met-[383]Ser bond, however, is the reactive bond exposed in the RCL and required to react with target proteases. Similarly, α1-AC is cleaved at the reactive bond that consists of a Leu-Ser motif which is also located in the exposed RCL. Cleavage of the non-inhibitory AGT shows that a reactive bond is not strictly required for substrate recognition by EspPα but cleavage also occurs inside the corresponding reactive center loop. In contrast, α2-AP is not cleaved in the RCL but inside the N- and C-terminal extensions which are vital for α2-AP functionality. Though the crystal structure of α2-AP has only been solved for a N-terminally truncated murine form, it seems that the C-terminal extension consists of a flexible loop because it could not be modeled into electron density maps [52]. Perhaps, this structural flexibility seen in the reactive center loops and in the C-terminal extension of α2-AP is required for substrate recognition by EspPα. Figure 7 shows crystal structures of the serpins that are cleaved by EspPα [52–55].

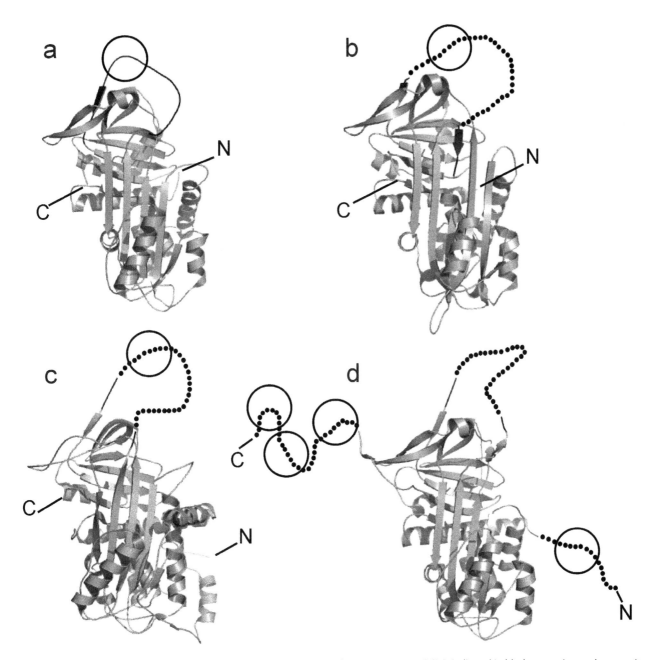

Figure 7. Crystal structures of serpins cleaved by EspPα. Serpins are shown as cartoons. RCL is indicated in black, approximate cleavage sites are encircled. Non-resolved parts of the crystal structures are indicated by dots (c, RCL of AGT, d, RCL of α2-AP and the N- and C-terminal extension of α2-AP). a, human α1-PI, b, cleaved human α1-AC, the RCL is indicated by dots, c, human angiotensinogen, d, murine truncated α2-AP$_{\Delta 43}$, the N-terminal extension of native α2-AP is indicated by dots.

EspI shows significant differences in substrate specificity compared to EspPα. α1-PI and α1-AC are also cleaved at their reactive bonds which should lead to loss of function of these serpins. However, serpin cleavage and release of the protease is not complete for EspI, most probably due to the pronounced formation of an inhibitory serpin-enzyme-complex of EspI with α1-PI and α1-AC. In contrast, EspPα completely degrades both serpins and forms only small amounts of the inhibitory complex only with α1-PI which does not significantly reduce EspPα activity. In addition, AGT and α2-AP, which are degraded by EspPα at positions other than the reactive bond, are not degraded by EspI. Concerning the functional differences of both SPATE proteases,

EspPα is able to cleave serpins specifically within accessible loop structures and is notably not inhibited by the analyzed serpins, while EspI is only able to interact with the reactive bond of α1-PI and α1-AC. The latter interactions show equilibria between EspI inhibition and serpin degradation. Taking into account the high amounts of serpins such as α1-PI in plasma, EspI activity might be strongly reduced in this milieu in vivo, while serpin degradation and inactivation might be a relevant function of EspPα also during infection.

In summary, we established a rapid method to determine cleavage sites of small proteolytic fragments via MALDI-TOF-MS. Functional implications have been investigated in a newly developed photometrical assay using chromogenic peptide substrates. EspPα degrades and thereby inactivates different plasma serpins which, in case of α2-AP, might lead to bleeding disorders or in case of α1-PI and α1-AC might interfere with the acute phase reaction during inflammatory host response. Cleavage occurs in flexible regions most favorable C-terminal to Leu. Comparison of EspPα and EspI indicate different functions of this SPATE also in vivo.

Supporting Information

Figure S1   Activity of EspPα and S263A. a, Determination of EspPα and S263A activity directly after purification. EspPα or S263A was incubated (15 h, 37°C) with the chromogenic substrate Suc-Ala-Ala-Pro-Leu-pNA. Activity was measured via released *para*-nitroaniline and normalized to EspPα. PBS was used as control. n = 2, b, Determination of EspPα activity after preincubation. Purified EspPα was preincubated for 15 h at 37°C resulting in the formation of autoproteolysis products (see Fig. 3c, lane1). To assess remaining proteolytic activity of autoproteolysis products the preincubated sample was incubated with the

chromogenic substrate Suc-Ala-Ala-Pro-Leu-pNA (15 h, 37°C). Again, activity was measured via released *para*-nitroaniline and normalized to EspPα. PBS was used as control. n = 2.
(TIFF)

Figure S2   Peptide mapping of EspPα cleavage products of the serpins. Serpin fragments were subjected to in-gel-digest and analyzed via MALDI-TOF-MS. Peptides of the large fragment are given in bold. Peptides of the small fragments are given in italics, a, sequence coverage of α1-AC fragments, b, sequence coverage of AGT, c, sequence coverage of α2-AP. Note that in the small fragments of AGT and α2-AP no serpin peptides were found.
(TIF)

Author Contributions

Conceived and designed the experiments: AW HJ JB. Performed the experiments: AW HJ. Analyzed the data: AW HJ JB. Contributed reagents/materials/analysis tools: AW JB. Contributed to the writing of the manuscript: AW JB.

References

1. Law D (2000) Virulence factors of Escherichia coli O157 and other Shiga toxin-producing E. coli. Journal of Applied Microbiology 88(5): 729–745.
2. Karmali MA (2004) Infection by Shiga toxin-producing Escherichia coli: an overview. Molecular Biotechnology 26(2): 117–122.
3. Brunder W, Schmidt H, Karch H (1997) EspP, a novel extracellular serine protease of enterohaemorrhagic Escherichia coli O157:H7 cleaves human coagulation factor V. Molecular Microbiology 24(4): 767–778.
4. Brockmeyer J, Bielaszewska M, Fruth A, Bonn ML, Mellmann A, et al. (2007) Subtypes of the plasmid-encoded serine protease EspP in Shiga toxin-producing Escherichia coli: distribution, secretion, and proteolytic activity. Applied and Environmental Microbiology 73(20): 6351–6359.
5. Bielaszewska M, Stoewe F, Fruth A, Zhang W, Prager R, et al. (2009) Shiga toxin, cytolethal distending toxin, and hemolysin repertoires in clinical Escherichia coli O91 isolates. Journal of Clinical Microbiology 47(7): 2061–2066.
6. Khan AB, Naim A, Orth D, Grif K, Mohsin M, et al. (2009) Serine protease espP subtype alpha, but not beta or gamma, of Shiga toxin-producing Escherichia coli is associated with highly pathogenic serogroups. Int Journal of Medical Microbiology 299(4): 247–254.
7. Brockmeyer J, Aldick T, Soltwisch J, Zhang W, Tarr PI, et al. (2011) Enterohaemorrhagic Escherichia coli haemolysin is cleaved and inactivated by serine protease EspPalpha. Environmental Microbiology 13(5): 1327–1341.
8. Schmidt H, Zhang WL, Hemmrich U, Jelacic S, Brunder W, et al. (2001) Identification and characterization of a novel genomic island integrated at selC in locus of enterocyte effacement-negative, Shiga toxin-producing Escherichia coli. Infection and Immunity 69(11): 6863–6873.
9. Orth D, Ehrlenbach S, Brockmeyer J, Khan AB, Huber G, et al. (2010) EspP, a serine protease of enterohemorrhagic Escherichia coli, impairs complement activation by cleaving complement factors C3/C3b and C5. Infection and Immunity 78(10): 4294–4301.
10. dos Santos LF, Irino K, Vaz TM, Guth BE (2010) Set of virulence genes and genetic relatedness of O113: H21 Escherichia coli strains isolated from the animal reservoir and human infections in Brazil. Journal of Medical Microbiology 59(Pt 6): 634–640.
11. Toszeghy M, Phillips N, Reeves H, Wu G, Teale C, et al. (2012) Molecular and phenotypic characterisation of Extended Spectrum beta-lactamase CTX-M Escherichia coli from farm animals in Great Britain. Research in Veterinary Science 93(3): 1142–1150.
12. Gettins PG (2002) Serpin structure, mechanism, and function. Chemical Reviews 102(12): 4751–4804.
13. Rau JC, Beaulieu LM, Huntington JA, Church FC (2007) Serpins in thrombosis, hemostasis and fibrinolysis. Journal of Thrombosis and Haemostasis 5 Suppl 1: 102–115.
14. Huntington JA, Read RJ, Carrell RW (2000) Structure of a serpin-protease complex shows inhibition by deformation. Nature 407(6806): 923–926.
15. Beatty K, Bieth J, Travis J (1980) Kinetics of association of serine proteinases with native and oxidized alpha-1-proteinase inhibitor and alpha-1-antichymotrypsin. The Journal of Biological Chemistry 255(9): 3931–3934.
16. Marshall CJ (1993) Evolutionary relationships among the serpins. Philosophical Transactions of the Royal Society of London. Series B, Biological Sciences 342(1300): 101–119.
17. Chandra T, Stackhouse R, Kidd VJ, Robson KJ, Woo SL (1983) Sequence homology between human alpha 1-antichymotrypsin, alpha 1-antitrypsin, and antithrombin III. Biochemistry 22(22): 5055–5061.
18. Schechter NM, Sprows JL, Schoenberger OL, Lazarus GS, Cooperman BS, et al. (1989) Reaction of human skin chymotrypsin-like proteinase chymase with plasma proteinase inhibitors. Journal of Biological Chemistry 264(35): 21308–21315.
19. Coughlin PB (2005) Antiplasmin: the forgotten serpin? The FEBS Journal 272(19): 4852–4857.
20. Favier R, Aoki N, de Moerloose P (2001) Congenital alpha(2)-plasmin inhibitor deficiencies: a review. British Journal of Haematology 114(1): 4–10.
21. Rogers SJ, Pratt CW, Whinna HC, Church FC (1992) Role of thrombin exosites in inhibition by heparin cofactor II. Journal of Biological Chemistry 267(6): 3613–3617.
22. Mauray S, de Raucourt E, Talbot JC, Dachary-Prigent J, Jozefowicz M, et al. (1998) Mechanism of factor IXa inhibition by antithrombin in the presence of unfractionated and low molecular weight heparins and fucoidan. Biochimica et Biophysica Acta 1387(1–2): 184–194.
23. Izaguirre G, Zhang W, Swanson R, Bedsted T, Olson ST (2003) Localization of an antithrombin exosite that promotes rapid inhibition of factors Xa and IXa dependent on heparin activation of the serpin. Journal of Biological Chemistry 278(51): 51433–51440.
24. Olson ST, Swanson R, Raub-Segall E, Bedsted T, Sadri M, et al. (2004) Accelerating ability of synthetic oligosaccharides on antithrombin inhibition of proteinases of the clotting and fibrinolytic systems. Comparison with heparin and low-molecular-weight heparin. Thrombosis and Haemostasis 92: 929–39.
25. Stein PE, Tewkesbury DA, Carrell RW (1989) Ovalbumin and angiotensinogen lack serpin S-R conformational change. Biochemical Journal 262(1): 103–107.
26. Arakawa K, Minohara A, Yamada J, Nakamura M (1968) Enzymatic degradation and electrophoresis of human angiotensin I. Biochimica et Biophysica Acta 168(1): 106–112.
27. Lentz KE, Skeggs LT, Jr., Woods KR, Kahn JR, Shumway NP (1956) The amino acid composition of hypertensin II and its biochemical relationship to hypertensin I. The Journal of Experimental Medicine 104(2): 183–191.
28. Mast AE, Enghild JJ, Nagase H, Suzuki K, Pizzo SV, et al. (1991) Kinetics and physiologic relevance of the inactivation of alpha 1-proteinase inhibitor, alpha 1-antichymotrypsin, and antithrombin III by matrix metalloproteinases-1 (tissue collagenase), -2 (72-kDa gelatinase/type IV collagenase), and -3 (stromelysin). The Journal of biological chemistry 266: 15810–15816.
29. Lijnen HR, Arza B, Van Hoef B, Collen D, Declerck PJ (2000) Inactivation of plasminogen activator inhibitor-1 by specific proteolysis with stromelysin-1 (MMP-3). The Journal of biological chemistry 275: 37645–37650.
30. Lijnen HR, Van Hoef B, Collen D (2001) Inactivation of the serpin alpha(2)-antiplasmin by stromelysin-1. Biochimica et biophysica acta 1547: 206–213.
31. Liu Z, Zhou X, Shapiro SD, Shipley JM, Twining SS, et al. (2000) The serpin alpha1-proteinase inhibitor is a critical substrate for gelatinase B/MMP-9 in vivo. Cell 102: 647–655.
32. Virca GD, Lyerly D, Kreger A, Travis J (1982) Inactivation of human plasma alpha 1-proteinase inhibitor by a metalloproteinase from Serratia marcescens. Biochimica et biophysica acta 704: 267–271.
33. Molla A, Akaike T, Maeda H (1989) Inactivation of various proteinase inhibitors and the complement system in human plasma by the 56-kilodalton proteinase from Serratia marcescens. Infection and immunity 57: 1868–1871.
34. Lathem WW, Grys TE, Witowski SE, Torres AG, Kaper JB, et al. (2002) StcE, a metalloprotease secreted by Escherichia coli O157:H7, specifically cleaves C1 esterase inhibitor. Molecular microbiology 45: 277–288.

35. Lathem WW, Bergsbaken T, Welch RA (2004) Potentiation of C1 esterase inhibitor by StcE, a metalloprotease secreted by Escherichia coli O157:H7. The Journal of experimental medicine 199: 1077–1087.

36. Brockmeyer J, Spelten S, Kuczius T, Bielaszewska M, Karch H (2009) Structure and Function Relationship of the Autotransport and Proteolytic Activity of EspP from Shiga Toxin-Producing *Escherichia coli*. PLoS ONE 4(7): e6100.

37. Laemmli UK (1970) Cleavage of structural proteins during the assembly of the head of bacteriophage T4. Nature 227(5259): 680–685.

38. Schagger H, von Jagow G (1987) Tricine-sodium dodecyl sulfate-polyacrylamide gel electrophoresis for the separation of proteins in the range from 1 to 100 kDa. Analytical Biochemistry 166(2): 368–379.

39. Shevchenko A, Tomas H, Havlis J, Olsen JV, Mann M (2006) In-gel digestion for mass spectrometric characterization of proteins and proteomes. Nature Protocols 1(6): 2856–2860.

40. Thelwell C, Marszal E, Rigsby P, Longstaff C (2011) An international collaborative study to establish the WHO 1st international standard for alpha-1-antitrypsin. Vox Sanguinis 101(1): 83–89.

41. Johnson D, Travis J (1978) Structural evidence for methionine at the reactive site of human alpha-1-proteinase inhibitor. Journal of Biological Chemistry 253(20): 7142–7144.

42. Morii M, Travis J (1983) Amino acid sequence at the reactive site of human alpha 1-antichymotrypsin. Journal of Biological Chemistry 258(21): 12749–12752.

43. Potempa J, Shieh BH, Travis J (1988) Alpha-2-antiplasmin: a serpin with two separate but overlapping reactive sites. Science 241(4866): 699–700.

44. Collen D, Wiman B (1979) Turnover of antiplasmin, the fast-acting plasmin inhibitor of plasma. Blood 53(2): 313–324.

45. Aoki N (1984) Genetic abnormalities of the fibrinolytic system. Seminars in Thrombosis and Hemostasis 10(1): 42–50.

46. Holmes WE, Nelles L, Lijnen HR, Collen D (1987) Primary structure of human alpha 2-antiplasmin, a serine protease inhibitor (serpin). Journal of Biological Chemistry 262(4): 1659–1664.

47. Hortin GL, Gibson BL, Fok KF (1988) Alpha 2-antiplasmin's carboxy-terminal lysine residue is a major site of interaction with plasmin. Biochemical and Biophysical Research Communications 155(2): 591–596.

48. Heeb MJ, Griffin JH (1988) Physiologic inhibition of human activated protein C by alpha 1-antitrypsin. Journal of Biological Chemistry 263(24): 11613–11616.

49. Burghaus B, Langer C, Thedieck S, Nowak-Gottl U (2006) Elevated alpha1-antitrypsin is a risk factor for arterial ischemic stroke in childhood. Acta Haematologica 115(3–4): 186–191.

50. Weiss A, Brockmeyer J,(2012) Prevalence, biogenesis, and functionality of the serine protease autotransporter EspP. Toxins 5(1): 25–48.

51. Dutta PR, Cappello R, Navarro-Garcia F, Nataro JP (2002) Functional Comparison of Serine Protease Autotransporters of Enterobacteriaceae. Infection and Immunity 70(12): 7105–7113.

52. Law RH, Sofian T, Kan WT, Horvath AJ, Hitchen CR, et al. (2008) X-ray crystal structure of the fibrinolysis inhibitor alpha2-antiplasmin. Blood 111(4): 2049–2052.

53. Elliott PR, Pei XY, Dafforn TR, Lomas DA (2000) Topography of a 2.0 A structure of alpha1-antitrypsin reveals targets for rational drug design to prevent conformational disease. Protein Science 9(7): 1274–1281.

54. Pearce MC, Powers GA, Feil SC, Hansen G, Parker MW, et al. (2010) Identification and characterization of a misfolded monomeric serpin formed at physiological temperature. Journal of Molecular Biology 403(3): 459–467.

55. Zhou A, Carrell RW, Murphy MP, Wei Z, Yan Y, et al. (2010) A redox switch in angiotensinogen modulates angiotensin release. Nature 468(7320): 108–111.

56. Lomas DA, Evans DL, Finch JT, Carrell RW (1992) The mechanism of Z alpha 1-antitrypsin accumulation in the liver. Nature 357(6379): 605–607.

57. Lomas DA, Finch JT, Seyama K, Nukiwa T, Carrell RW (1993) Alpha 1-antitrypsin Siiyama (Ser53->Phe). Further evidence for intracellular loop-sheet polymerization. Journal of Biological Chemistry 268(21): 15333–15335.

58. Lomas DA, Elliott PR, Sidhar SK, Foreman RC, Finch JT, et al. (1995) alpha 1-Antitrypsin Mmalton (Phe52-deleted) forms loop-sheet polymers in vivo. Evidence for the C sheet mechanism of polymerization. Journal of Biological Chemistry 270(28): 16864–16870.

59. Abraham CR, Selkoe DJ, Potter H (1988) Immunochemical identification of the serine protease inhibitor alpha 1-antichymotrypsin in the brain amyloid deposits of Alzheimer's disease. Cell 52(4): 487–501.

60. Kamboh MI, Sanghera DK, Ferrell RE, DeKosky ST (1995) APOE*4-associated Alzheimer's disease risk is modified by alpha 1-antichymotrypsin polymorphism. Nature Genetics 10(4): 486–488.

61. Nielsen HM, Minthon L, Londos E, Blennow K, Miranda E, et al. (2007) Plasma and CSF serpins in Alzheimer disease and dementia with Lewy bodies. Neurology 69(16): 1569–1579.

62. Poulsen K, Haber E, Burton J (1976) On the specificity of human renin. Studies with peptide inhibitors. Biochimica et Biophysica Acta 452(2): 533–537.

Systems Level Analysis and Identification of Pathways and Networks Associated with Liver Fibrosis

Mohamed Diwan M. AbdulHameed[1], Gregory J. Tawa[1], Kamal Kumar[1], Danielle L. Ippolito[2], John A. Lewis[2], Jonathan D. Stallings[2], Anders Wallqvist[1]*

1 Department of Defense Biotechnology High Performance Computing Software Applications Institute, Telemedicine and Advanced Technology Research Center, U.S. Army Medical Research and Materiel Command, Fort Detrick, Maryland, United States of America, 2 U.S. Army Center for Environmental Health Research, Fort Detrick, MD, United States of America

Abstract

Toxic liver injury causes necrosis and fibrosis, which may lead to cirrhosis and liver failure. Despite recent progress in understanding the mechanism of liver fibrosis, our knowledge of the molecular-level details of this disease is still incomplete. The elucidation of networks and pathways associated with liver fibrosis can provide insight into the underlying molecular mechanisms of the disease, as well as identify potential diagnostic or prognostic biomarkers. Towards this end, we analyzed rat gene expression data from a range of chemical exposures that produced observable periportal liver fibrosis as documented in DrugMatrix, a publicly available toxicogenomics database. We identified genes relevant to liver fibrosis using standard differential expression and co-expression analyses, and then used these genes in pathway enrichment and protein-protein interaction (PPI) network analyses. We identified a PPI network module associated with liver fibrosis that includes known liver fibrosis-relevant genes, such as tissue inhibitor of metalloproteinase-1, galectin-3, connective tissue growth factor, and lipocalin-2. We also identified several new genes, such as perilipin-3, legumain, and myocilin, which were associated with liver fibrosis. We further analyzed the expression pattern of the genes in the PPI network module across a wide range of 640 chemical exposure conditions in DrugMatrix and identified early indications of liver fibrosis for carbon tetrachloride and lipopolysaccharide exposures. Although it is well known that carbon tetrachloride and lipopolysaccharide can cause liver fibrosis, our network analysis was able to link these compounds to potential fibrotic damage before histopathological changes associated with liver fibrosis appeared. These results demonstrated that our approach is capable of identifying early-stage indicators of liver fibrosis and underscore its potential to aid in predictive toxicity, biomarker identification, and to generally identify disease-relevant pathways.

Editor: Christina Chan, Michigan State University, United States of America

Funding: The authors were supported by the Military Operational Medicine Research Program and the U.S. Army's Network Science Initiative, U.S. Army Medical Research and Materiel Command (USAMRMC, http://mrmc.amedd.army.mil), Ft. Detrick, MD. This research was supported in part by an appointment to the Postgraduate Research Participation Program at the U.S. Army Center for Environmental Health Research (USACEHR, http://usacehr.amedd.army.mil) administered by the Oak Ridge Institute for Science and Education through an interagency agreement between the U.S. Department of Energy and USAMRMC. The funders had no role in study design, data collection and analysis, decision to publish, or preparation of the manuscript.

Competing Interests: The authors have declared that no competing interests exist.

* Email: awallqvist@bhsai.org

Introduction

Exposure to toxic chemicals can lead to liver injury through a variety of mechanisms, such as oxidative stress, the immune response, activation of apoptotic pathways, and necrosis [1]. Liver fibrosis is a common pathologic feature observed in a wide spectrum of liver injuries [2,3] and is marked by inflammation and excessive accumulation of extracellular matrix (ECM) components [4]. Liver fibrosis results in scar formation and, if unresolved, leads to cirrhosis, portal hypertension, and liver failure [4]. Liver fibrosis typically starts with apoptosis or necrosis of hepatocytes, which causes reactive oxygen species generation. This process leads to the release of inflammatory mediators and ultimately results in activation of hepatic stellate cells [3], the main ECM-producing cells in the liver. This activation of hepatic stellate cells is the key pathogenic mechanism of liver fibrosis [3–6]. Activated hepatic

stellate cells lead to further inflammation and ECM generation, which results in the replacement of liver parenchymal cells with ECM [5]. Despite recent progress, our understanding of the molecular mediators of liver fibrosis remains incomplete, and we are still in the process of identifying such mediators [7,8].

Although fibrotic damage is reversible, there are no approved drugs or treatments for liver fibrosis. Key in understanding damage and control of fibrosis is accurate diagnosis or early indicators of damage. The gold standard for diagnosing fibrosis is currently via liver biopsy. This invasive method has many limitations, such as inter- and intra-observer variations and sampling variability [9]. Thus, there is a need to identify sensitive, specific, and non-invasive biomarkers of liver fibrosis. Identification of such biomarkers will improve diagnosis and allow better clinical management of the disease. In the military, this capability would aid in field assessment and potentially enable timely evacuation or guide return-to-duty

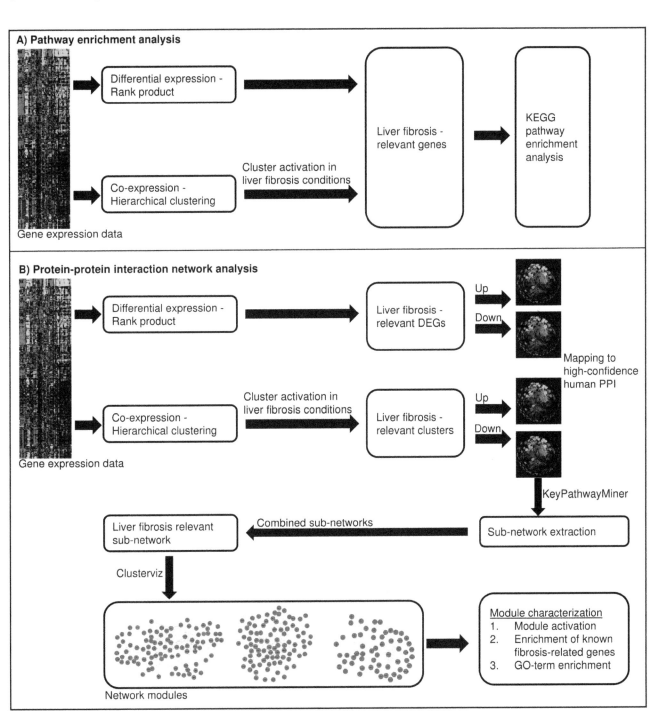

Figure 1. Workflow used in this study to identify pathways and networks associated with liver fibrosis.

decisions. Elucidation of the pathways and networks associated with liver fibrosis will provide insight into the molecular mechanisms of this disease and, importantly, help us to identify mechanism-based biomarkers of liver damage.

Computational systems biology approaches are now routinely used to analyze gene expression data and to gain insight into the molecular mechanisms of many diseases [10–15]. Pathway enrichment analysis provides a biological interpretation of gene lists obtained from microarray data using manually curated pathway databases, such as the Kyoto Encyclopedia of Genes

and Genomes (KEGG) and Reactome [16,17]. The BioSystems database [18,19] provides an integrated resource of pathways from several major pathway databases, including KEGG and Reactome. Huang et al. [20] have summarized the various tools and statistical methods available for pathway enrichment analysis and their utility in elucidating the mechanisms of diseases [21–23]. In literature related to liver fibrosis, the work of Affo et al. [24] utilizes KEGG pathway analysis in identifying the role of focal adhesion and cytokine-cytokine receptor interaction pathways in alcoholic hepatitis. Although widely used, pathway analysis has

Table 1. Chemical exposure conditions that produced periportal liver fibrosis.

Chemical	Dose (mg/kg)	Duration (days)	Histopathology (severity score)
1-Naphthylisothiocyanate	30	7	2
1-Naphthylisothiocyanate	60	7	2
4,4'-Methylenedianiline	81	5	2
N-Nitrosodimethylamine	10	5	2

some limitations [25]. Foremost, pathway analysis relies exclusively on experimentally confirmed, curated data. Only a small fraction of human genes map to curated pathway collections (e.g., KEGG [14]); thus, pathway analysis is inherently biased against the discovery of new molecular mediators. For example, we found that only 5,870 of the ~21,000 human genes mapped to the 229 KEGG pathways [26]. Moreover, the gene networks comprising the pathway maps are not mutually exclusive, and the same gene can occur in many pathways [25]. Integrated analyses of gene expression data with protein-protein interaction (PPI) networks enable us to partly overcome the limitations associated with pathway analysis. The potential of this integrated approach has been shown in identifying biomarkers for breast cancer [27] and in understanding the molecular mechanisms of dilated cardiomyopathy [28], hepatitis C virus infection [29], and cancer and heart disease [30–34]. However, to the best of our knowledge, no such integrated analysis has been reported for liver fibrosis.

Thus, our goal in this study was to identify liver fibrosis-relevant pathways and networks based on an integrated gene expression and PPI network analysis. We analyzed the gene expression data from a range of chemical exposure conditions that produced periportal liver fibrosis in DrugMatrix, a publicly available toxicogenomics database [35]. We carried out differential expression and co-expression analysis using rank product and hierarchical clustering, respectively, to identify genes associated with liver fibrosis [36]. We then examined these genes in two separate analyses. In the first analysis, we identified the KEGG pathways associated with liver fibrosis. In the second analysis, we integrated the gene expression data with the high-confidence human PPI network to obtain liver fibrosis-relevant sub-networks [37]. We identified a PPI network module associated with liver fibrosis that includes known liver fibrosis-relevant genes like *Timp1*, *Lgals3*, *Ctgf*, and *Lcn2*, along with several new genes. We further analyzed the expression pattern of the genes in the PPI network across a wide range of 640 chemical exposure conditions in DrugMatrix and linked compounds such as carbon tetrachloride to potential fibrotic damage before histopathological evidence of liver fibrosis appeared. These results illustrate the potential of our approach to aid in toxicity prediction and biomarker discovery.

Materials and Methods

Figure 1 shows the overall workflows used in this study to identify liver fibrosis-relevant pathways and interaction networks. We used two separate but complementary approaches to map the overall gene transcriptional response to liver fibrosis via *1)* enrichment analysis of knowledge-based pathways association, and *2)* integration of gene expression data with PPI networks to identify interaction networks. These fibrosis-relevant interaction networks can be considered as *de novo* pathways.

Data collection and processing

We used DrugMatrix, a publicly available toxicogenomics database [35,38], for our analyses. DrugMatrix is a large collection of gene expression, hematology, histopathology, and clinical chemistry data obtained from Sprague-Dawley rats after exposure to a range of chemicals, including industrial chemicals, toxicants, and drugs with multiple time intervals, dose ranges, and tissues for each chemical [35]. A chemical exposure condition in DrugMatrix data refers to exposure to a particular chemical at a particular dose and time. We downloaded the DrugMatrix liver gene expression data generated using Affymetrix GeneChip Rat Genome 230 2.0 Arrays from the National Institute of Environmental Health Sciences (NIEHS) server (https://ntp.niehs.nih.gov/drugmatrix/ index.html). Of the 2,218 microarray CEL files in this database, 1,941 were chemical exposures and 277 were controls. We performed background correction, quantile normalization, and summarization using the robust multi-array average method in R/ BioConductor package *affy* [26,39–41]. We then used the BioConductor package *ArrayQualityMetrics* to assess the quality of the microarray data and removed 155 outlier arrays [42]. We reprocessed the remaining arrays using the robust multi-array average method and used this final normalized data for all our analyses. With the BioConductor package *genefilter*, we carried out non-specific filtering of the genes [43]. We removed probe sets without Entrez ID or with low variance across chemical exposures based on inter-quartile range. We additionally filtered probe sets based on "Present" calls using the BioConductor package *affy*, retaining only the probe sets for which at least 25% of the chemical exposures had "Present" calls for all replicates within the chemical exposure condition. After calculating the average intensity between the replicates of a chemical exposure condition, we computed log-ratios for each gene between treatments and their corresponding controls. Our final log-ratio matrix contained 7,826 genes and 640 chemical exposure conditions.

Identifying genes relevant to liver fibrosis

The DrugMatrix database provides histopathology data associated with each chemical exposure condition. There were five chemical exposure conditions that produced liver periportal fibrosis with a histopathology score >1 (**Figure S1**). We carried out a quality check by clustering the replicates of these five chemical exposures, along with their respective controls. Ideally, the chemical exposures and controls would have clustered separately. But all replicates of 5-day exposures to Crotamiton-750 mg/kg clustered with controls rather than with other treatments (**Figure S1**). Hence, we excluded this condition from the liver fibrosis-producing condition set. **Table 1** lists the four chemical exposure conditions that were used in this study as the liver fibrosis-producing condition set. All replicates of the four chemical exposure conditions had a histopathology score of 2. We used the rank product method to identify differentially expressed genes (DEG) [36]. Rank product is a non-parametric, permutation-based method that has been widely used in many studies [36,44,45]. With this method, the fold-change values were converted into rank, and then the significance of the obtained rank, including the false discovery rate (FDR) p-value, was calculated. We used the BioConductor package *RankProd* for this analysis [44]. This method produces separate lists of up-regulated and down-regulated genes. We considered all genes with an FDR <0.05 to be significantly differentially expressed. We carried out the rank product analysis separately for each of the four chemical exposure conditions; the genes that were significantly differentially expressed in at least two of the four chemical exposure conditions that produced periportal liver fibrosis were considered as fibrosis-relevant DEGs.

We carried out hierarchical clustering using the R package *hclust* to identify co-expressed genes [21]. We clustered the genes in the log-ratio matrix using their $\log_2$ ratio values across 640 chemical exposure conditions. We used the Pearson correlation and the average linkage method to perform the clustering, and the R package *dynamicTreeCut* for automated extraction of clusters [46]. The dynamic tree cut algorithm implements an automated iterative process to identify and split sub-clusters from a dendrogram until the minimum cluster size threshold is reached [46]. We used the *cutreeDynamic* function in this package with *minimum cluster size* set to *16*, *method* set to *hybrid*, and *deepsplit* set to *True*.

We calculated cluster activation scores to identify liver fibrosis-relevant clusters in order to identify clusters whose constituent genes show altered gene expression (either up- or down-regulation) in chemical exposure conditions that produce periportal liver fibrosis. To calculate a cluster activation score, we first normalized the log-ratio values of each gene across 640 chemical exposure conditions by converting them into Z-scores. The Z-score of gene i under chemical exposure condition j is given by

$$Z_{i,j} = \frac{X_{i,j} - \mu_i}{\sigma_i}, \tag{1}$$

where $X_{i,j}$ is the log-ratio value for gene i under chemical exposure condition j, μ_i is the average log ratio for gene i across all 640 chemical exposure conditions, and σ_i is the standard deviation of the log ratio for gene i across all 640 chemical exposure conditions. Next, we obtained the cluster activation scores for liver fibrosis by averaging the Z-scores of all genes within a cluster and across all chemical exposure conditions that produced periportal liver fibrosis. The activation score A_c of cluster c in liver fibrosis is given by

$$A_c = \frac{1}{N_c N_f} \sum_{i \in c}^{N_c} \sum_{j}^{N_f} Z_{i,j}, \tag{2}$$

where N_c is the number of genes associated with cluster c, N_f is the number of chemical exposure conditions that produce periportal liver fibrosis, and $Z_{i,j}$ is the Z-score of gene i under chemical exposure conditions that produce periportal liver fibrosis j. We used an absolute cluster activation score cutoff of 2, corresponding to the 95th percentile of the probability density distribution, and selected genes in clusters above this threshold as liver fibrosis-relevant co-expressed genes.

In general, to find disease-relevant genes, either differential expression or co-expression analysis is commonly used. Utilization of both approaches together will allow us to overcome the limitation associated with each approach. Hence, we combined the gene list from both approaches, i.e., differential expression and co-expression, to form the liver fibrosis-relevant gene set.

Pathway enrichment analysis

We used the Database for Annotation, Visualization, and Integrated Discovery (DAVID) tool to perform KEGG pathway enrichment analysis [47]. The pathways below a Benjamini-Hochberg FDR <0.05 were considered to be significantly enriched. We used the liver fibrosis-relevant gene set, i.e., the combined list of genes determined by differential expression and co-expression analysis to be relevant to liver fibrosis, for the pathway enrichment analysis. Then we separately used the up-regulated genes and down-regulated genes in this set and carried out pathway enrichment analysis.

PPI network analysis

Yu et al. [37] used the interaction detection based on shuffling (IDBOS) approach to generate a high-confidence PPI network and showed that the resulting high-confidence PPIs reduce the noise inherent in aggregated PPIs. In their work, they created experimentally derived high-confidence PPI networks for both rats and humans [37]. The rat high-confidence PPI network contained 1,001 nodes, whereas the human high-confidence PPI network consisted of 14,230 nodes. We chose to use the human high-confidence PPI network due to its much larger coverage. The rat probe-set identifiers were mapped to their corresponding human gene identifiers using orthology mapping tools [26,48,49]. This approach followed the work by Zhang et al. [50] for mapping other species' gene expression data to a human PPI network. We utilized the KeyPathwayMiner program in Cytoscape 2.8 to obtain the liver fibrosis-relevant sub-network [51–53]. Key-PathwayMiner attempts to find maximally connected sub-networks for the input query genes with gene expression data using the ant-colony optimization algorithm [51]. We used KeyPathwayMiner with *ant-colony optimization algorithm, node exceptions (K)* set to *100*, and *case exceptions (L)* set to *0*. The *node exception (K)* value provided the allowance for genes that were not present in the input gene set and the *case exception (L)* defined the number of conditions in which the input gene may not be active. We separately ran KeyPathwayMiner using the up-regulated and down-regulated genes in liver fibrosis-relevant DEGs and co-expressed genes and extracted the sub-networks. We then combined the four sub-networks into one final liver fibrosis-relevant sub-network. This sub-network was created by the union of the four sub-networks. We did not use the intersection of the four sub-networks, as it contained only two nodes. Finally, we clustered this liver fibrosis-relevant sub-network using the topology-based network clustering program in Cytoscape, Clusterviz. We used the *EAGLE algorithm* in Clusterviz with default parameters [54]. As implemented, Clusterviz generated 11 network modules.

Module characterization

We mapped the proteins in the PPI network modules to the rat gene expression Z-score matrix. We calculated the activation scores (**Equation 2**) for the 11 network modules under conditions that caused periportal liver fibrosis. The method is the same as described above, except that here we used the genes in the module instead of cluster. We collected 28 known liver fibrosis-relevant genes from literature (*Set 1*) [3,5,55,56]. Of these 28 genes, 26 mapped to the high-confidence human PPI network. We also collected genes that are known to be associated with liver cirrhosis from the Comparative Toxicogenomics Database (CTD) (*Set 2*) [57]. Of the 126 genes with direct evidence of an association with

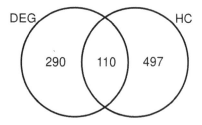

Figure 2. Number of fibrosis-relevant genes from differential and co-expression analysis. Number of genes in the liver fibrosis-relevant differentially expressed gene list and liver fibrosis-relevant co-expressed gene list and the overlap between them.

liver cirrhosis, 95 mapped to the high-confidence human PPI network. We used the Fisher exact test to calculate the enrichment of these genes (*Set 1* and *Set 2*) in each PPI network module. The module genes were also characterized by gene ontology (GO) biological process-term enrichment using the DAVID tool. We used the Revigo tool to visualize the GO enrichment results [58] and analyzed the network modules in terms of activation in liver fibrosis, enrichment with known liver fibrosis-relevant genes, and enrichment of liver fibrosis-relevant GO terms. Based on this analysis, we prioritized one PPI network module (M5) as a liver fibrosis-relevant network module.

We used two statistical significance tests to analyze whether the network module M5 was obtained by random chance. Our null hypothesis was that the observed number of nodes ($M5_{nodes}$) and edges ($M5_{edges}$) in module M5 were obtained by random chance. In the first analysis, we randomly selected 92 proteins from the human PPI network and counted the number of nodes (R_{nodes}) and edges (R_{edges}) of the largest connected component. This process was repeated 1,000 times. We computed the number of times $R_{nodes} \geq M5_{nodes}$, denoted as $N_{randnode}$. Similarly, we

computed the number of times $R_{edges} \geq M5_{edges}$, denoted as $N_{randedge}$. Then we computed the probability of obtaining a similar number of nodes by random chance using $P = N_{randnode}/1,000$, and the probability of obtaining a similar number of edges by random chance using $P = N_{randedge}/1,000$. In the second analysis, we shuffled the human PPI network and then mapped the proteins in the M5 network to this randomized network. We preserved the average node degree during network shuffling. Similar to the first analysis, we extracted the largest connected component, counted the number of nodes and edges, and calculated the probability of obtaining M5 parameters by random chance. We analyzed the overall robustness of the M5 module by comparing the modules generated from a reduced number of samples to those generated from the full dataset. We left out one quarter of the samples from the differential gene expression dataset and analyzed the remaining samples, repeating this procedure four times and leaving out each quarter of the data once. We then compared the overlap of the final module proteins from these four analyses with the module M5 proteins and found an average overlap of 72%. This showed that our method identified roughly the same genes even when samples were missing. We analyzed the expression of genes in module M5 in chemical exposures that produced periportal liver fibrosis across different time periods of exposures. Among the four chemical exposure conditions that produced periportal liver fibrosis, earlier time points were not available for exposure with N-nitroso dimethylamine at 10 mg/kg. Data were available for exposures to 1-naphthyl isothiocyanate at 30 mg/kg and 60 mg/kg at all time points, i.e., 0.25 day, 1 day, and 3 days, and for exposures to 4,4'-methylene dianiline at 1 day and 3 days. We mapped the expression profile of genes in module M5 across different time periods using the average $\log_2$ ratio of the available chemical exposure data at that time point. Finally, the genes in the prioritized network module M5 were used to cluster the 640 chemical exposure conditions in the DrugMatrix database. We used the clustering software *cluster3* for this purpose [59]. We

Table 2. KEGG[a] pathway enrichment for all genes relevant to liver fibrosis.

Pathway	Count	BH[b]
Prion diseases	11	0.003
Leukocyte transendothelial migration	19	0.008
Focal adhesion	26	0.007
Fc gamma R-mediated phagocytosis	16	0.006
Pyruvate metabolism	10	0.010
Viral myocarditis	15	0.011
Antigen processing and presentation	14	0.035
Systemic lupus erythematosus	14	0.037
Chemokine signaling pathway	21	0.035
Complement and coagulation cascades	12	0.033
Regulation of actin cytoskeleton	24	0.031
ECM[c]-receptor interaction	13	0.030
PPAR[d] signaling pathway	12	0.029
Arginine and proline metabolism	10	0.035
Glycerolipid metabolism	9	0.035

[a]Kyoto Encyclopedia of Genes and Genomes
[b]Benjamini-Hochberg false discovery rate
[c]Extracellular matrix
[d]Peroxisome proliferator-activated receptor

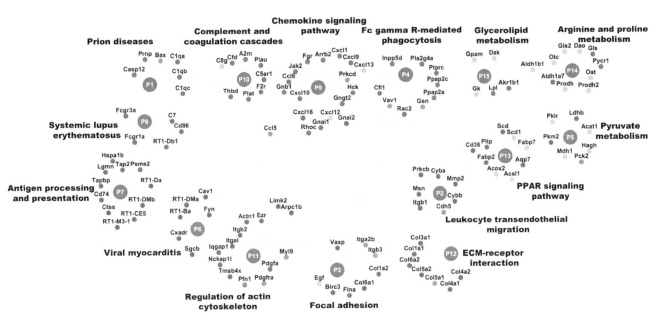

Figure 3. Genes that mapped to the enriched Kyoto Encyclopedia of Genes and Genomes (KEGG) pathways. The average $\log_2$ fold-change ratio across chemical exposures that produced periportal liver fibrosis was used as the gene expression value. Genes with average $\log_2$ fold-change ratios >0.6 are colored in red. Genes with average $\log_2$ fold-change ratios <−0.6 are colored in green. Genes whose average $\log_2$ fold-change ratios are between 0.6 and −0.6 are colored in grey.

evaluated the specificity of network module M5 using the average Z-score across the genes in M5 for each of the 640 chemical exposure conditions.

External validation

In order to further demonstrate the relevance of network module M5 in liver fibrosis, we evaluated the M5 genes in two external datasets (GSE13747 and GSE6929) from the Gene Expression Omnibus (GEO). Both datasets used Affymetrix GeneChip Rat Genome 230 2.0 Arrays. In GSE13747, liver fibrosis was produced by bile duct ligation [60]. Six replicates of liver fibrosis samples and six controls were available. In GSE6929, liver cirrhosis was induced by inhalation of carbon tetrachloride [13]. Four replicates of liver cirrhosis controls and four replicates of sunitinib (SU11248)-treated samples were available. We used the same steps described above for the DrugMatrix database to preprocess these two external datasets and calculated the $\log_2$ ratio between the treatment and controls. Finally, we matched the genes in module M5 and calculated the correlation between the average $\log_2$ ratios in the four chemical exposure conditions that produced periportal liver fibrosis from the DrugMatrix database with the $\log_2$ ratio from external datasets.

Results and Discussion

Identification of liver fibrosis-relevant genes

We used differential gene expression and co-expression analysis to identify liver fibrosis-relevant genes. As outlined in the Materials and Methods section, we analyzed the chemical exposures in the DrugMatrix data that produced periportal liver fibrosis with a histopathological score >1. We used the rank product approach and identified 400 liver fibrosis-relevant DEGs, of which 192 genes were significantly up-regulated, and 208 genes were significantly down-regulated. Here, we used an FDR <0.05 as the cutoff value to select DEGs. This cutoff-based approach knowingly excludes DEGs to minimize false positives and may not capture the

complete picture of the disease or processes being studied [61–63]. For example, many genes that do not meet the cutoff criteria can be involved in the same pathway as the DEGs and provide insights into the altered disease process [63]. Identifying co-expressed genes by means of gene clustering is an alternative approach that does not use cutoffs or thresholds at the individual gene level. Instead, genes are clustered based on their expression profiles across a wide range of exposures in order to identify gene sets that are expected to have similar functions, i.e., participate in related pathways [64]. We used hierarchical clustering to cluster 7,826 genes based on their log-ratio values across 640 chemical exposure conditions, which yielded 210 gene clusters containing an average of 37 genes each. Unlike differential expression analysis, these co-expressed genes were not linked or associated with liver fibrosis or any other particular disease. We used the cluster activation scores defined in **Equation 2** to establish the connection between the gene clusters and liver fibrosis. We found 565 genes in the nine clusters with activation scores >2 and 42 genes in the two clusters with activation scores <−2. The genes in these clusters were used as liver fibrosis-relevant co-expressed genes. Finally, we combined the differentially expressed and co-expressed gene lists to create a liver fibrosis-relevant gene set with 897 genes. There were 110 genes in common between the differentially expressed and co-expressed genes, and we show the overlap between these two sets as a Venn diagram (**Figure 2**). **Table S1** provides the list of 897 genes, along with log-ratio values, in the four chemical exposure conditions that produced periportal liver fibrosis.

Pathway enrichment analysis

Table 2 lists the significantly enriched KEGG pathways derived from the liver fibrosis-relevant gene set. These pathways include leukocyte transendothelial migration, focal adhesion, chemokine signaling, regulation of the actin cytoskeleton pathway, and ECM-receptor interaction, and they mainly represent liver fibrosis-related processes. These processes are consistent with previous reports for liver fibrosis. Injured liver cells and activated

Table 3. Activation of network modules and enrichment of known genes relevant to liver fibrosis.

Module	Activation score in Liver fibrosis	No. of genes in module	No. of fibrosis genes (set 1)[a]	p-value (set 1)[a]	No. of cirrhosis genes (set 2)[b]	p-value (set 2)[b]
M1	1.47	150	1	0.24	1	1
M2	0.95	144	3	0.002	5	0.003
M3	1.67	127	2	0.02	0	1
M4	1.92	110	1	0.18	1	0.52
M5	2.12	92	6	1.28E-08	13	3.25E-14
M6	0.77	81	1	0.14	2	0.10
M7	1.09	58	-	1	1	0.32
M8	1.38	49	-	1	2	0.04
M9	0.36	46	-	1	1	0.27
M10	0.87	40	-	1	1	0.24
M11	1.38	34	-	1	1	0.20

[a]Liver fibrosis-relevant genes collected from literature [3,5,55,56]
[b]Liver cirrhosis-relevant genes collected from the Comparative Toxicogenomics Database (CTD)

hepatic stellate cells release chemokines that recruit leukocytes to the site of injury [65]. An elevated expression of chemokines and chemokine receptors has been reported in both animal models and clinical cases of liver fibrosis [66]. Hepatic stellate cell migration is an essential process in fibrosis [67]. Hepatic stellate cells are also known to express adhesion molecules and α-smooth muscle actin

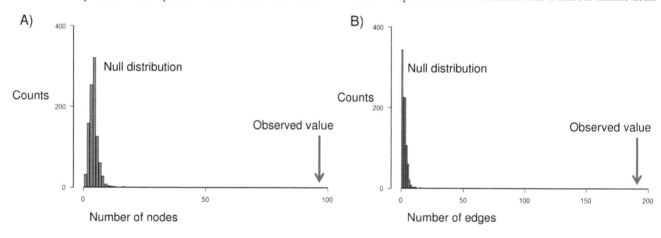

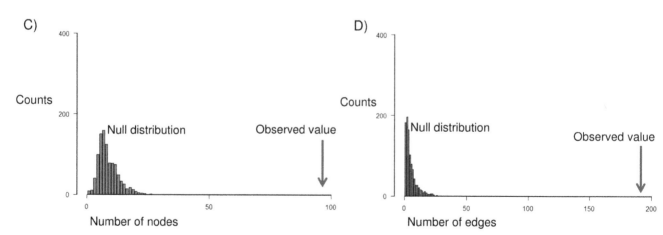

Figure 4. Statistical significance analysis of network module M5. A) The comparison of number of nodes in M5 to that from random sampling analysis. B) The comparison of number of edges in M5 to that from random sampling analysis. C) The comparison of number of nodes in M5 to the number present in shuffled protein-protein interaction (PPI) networks. D) The comparison of number of edges in M5 to the number present in shuffled PPI networks.

(SMA) [4]. Activated hepatic stellate cells are known to secrete prion protein [68,69]. Innate immunity and adaptive immunity play a major role in hepatic stellate cell activation and propagation of liver fibrosis [70]. Anticoagulant drugs and peroxisome proliferator-activated receptor (PPAR)-γ agonists were reported to have an anti-fibrotic effect in experimental liver fibrosis, which is consistent with the enriched pathways, such as complement and coagulation cascades, and the PPAR signaling pathway [70].

We analyzed genes involved in the enriched pathways to ascertain whether we could provide new, testable hypotheses. Genes were mapped to the enriched KEGG pathways (**Figure 3**) and **Table S2** provides the complete list of these genes with pathway information, and their average log_2 ratio in chemical exposures that produced periportal liver fibrosis. The log_2 ratio is commonly used in microarray data analysis, and a value of 0.6 corresponds to a ~1.5 fold-change (up-regulation) in gene expression. *Col1a1*, *Col1a2*, *Col4a1*, *Col4a2*, *Col5a2*, *Itgb1*, *Plat*, *Plau*, *Pdgfa*, *Ezr*, and *Msn* were up-regulated and have an average log_2 ratio >0.6. These genes are known to be altered in liver fibrosis [4,67,71]. We also analyzed the gene list (**Table S2**) for potential new candidates and found genes such as *Lgmn* and *Limk2*, which were up-regulated in all four chemical exposure conditions that produced periportal liver fibrosis. These genes are potential candidates for further exploration. Next, we carried out pathway enrichment analyses using the up- and down-regulated genes separately (**Tables S3** and **S4**). The significantly up-regulated pathways were related to liver fibrosis-relevant processes, whereas the significantly down-regulated pathways were related to

metabolism. The down-regulation of metabolism-related pathways could either be related to external factors, such as altered food intake, or could also be an indication of reduced liver function.

Although pathway enrichment analyses are useful and provide an overview of biological processes associated with our gene list, the method has some well-known limitations. Pathway analysis is based on curated data and is limited to the information present in the underlying knowledge database. Moreover, the pathway enrichment based on an over-representation analysis approach treats pathways as simple gene lists without accounting for network connectivity [72]. As such, pathway analysis has limited utility in identifying new molecular mediators or new pathways. Using the connectivity information or pathway topology may provide us with alternative approaches to capture the most relevant pathways associated with a disease.

PPI network analysis

Based on the premise that proteins that are closely connected to each other in a network are more likely to be involved in similar processes, an integration of gene expression with PPI networks has been used to identify disease-specific networks [73]. Such networks have been proposed as *de novo* pathways and partly remedy the limitations of KEGG pathway analysis [25]. Consequently, we mapped the liver fibrosis-relevant genes to a high-confidence human PPI network. Out of 897 fibrosis-relevant genes, 606 mapped to the human PPI network. We extracted a liver fibrosis-relevant sub-network with 902 nodes (proteins) and 2,527 edges using KeyPathwayMiner. Out of the 606 fibrosis-relevant genes,

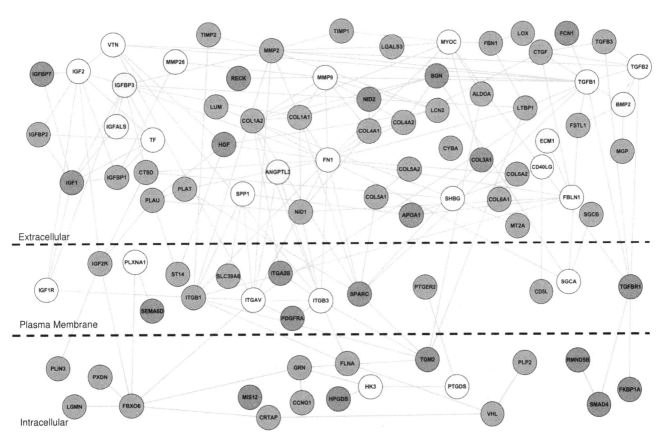

Figure 5. Liver fibrosis-relevant network module M5. Proteins encoded by genes with average log_2 fold-change ratios>0.6 are colored in red. Proteins encoded by genes with average log_2 fold-change ratios <−0.6 are colored in green. Proteins encoded by genes with average log_2 fold-change ratios between 0.6 and −0.6<−0.6 are colored in grey. Proteins without corresponding gene expression data are shown as white circles.

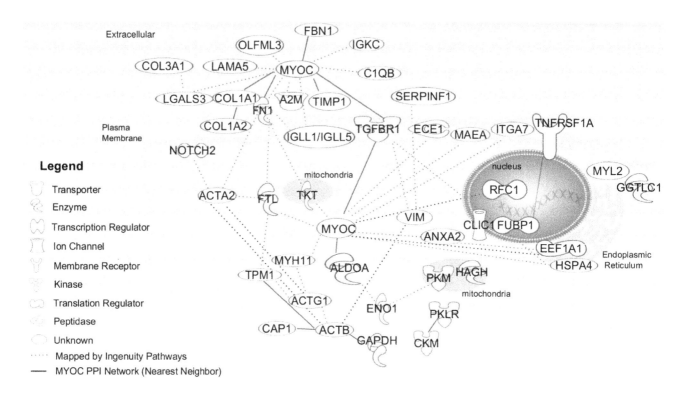

Figure 6. Myocilin interaction network. First neighbors of myocilin (MYOC) in the entire high-confidence human protein-protein interaction (PPI) network.

573 were present in this sub-network. Finally, we clustered the 902 proteins in the liver fibrosis-relevant sub-network into 11 PPI network modules. **Data S1 – S3** provides all the input and Cytoscape session files associated with PPI network analysis. **Table S5** provides the mapping of rat probe IDs to human gene IDs, and **Table S6** provides the protein membership in the PPI network modules, along with their gene expression data.

Module characterization and potential application

We further analyzed the 11 network modules in terms of activation in chemical exposure conditions that produced periportal liver fibrosis, enrichment with known liver fibrosis-relevant genes, and enrichment of GO terms. **Table 3** shows the module activation scores $1A_1c$ calculated using **Equation 2** for chemical exposure conditions that produced periportal liver fibrosis. Module M5 with 92 genes coding for proteins was the highest activated module with an activation score of 2.12. Next, we analyzed the enrichment of known fibrosis-relevant genes collected from the literature (**Table S7**) [3,5,55,56]. **Table 3** shows the p-values for enrichment of known fibrosis-relevant genes in the network modules. Modules M2, M3, and M5 were enriched with liver fibrosis-relevant genes with p-values <0.05. Module M5 had the lowest enrichment p-value with six known fibrosis-associated genes coding for proteins mapped to this module: TIMP1, APOA1, CTGF, LGALS3, TGFB1, and MMP-2. CTD provides curated information on genes associated with a disease [57]. Liver fibrosis was not curated in CTD, but genes associated with liver cirrhosis, the final stage of liver fibrosis, were available. We further analyzed the enrichment of genes associated with liver cirrhosis collected from CTD (**Table S8**) and found module M5 to be the module most enriched with liver cirrhosis-related genes (**Table 3**). We also characterized the 11 modules using the GO biological process-term enrichment analysis. We found that module M5 was

enriched with liver fibrosis-relevant GO terms such as *ECM organization* and *Wound healing*. **Figure S2** shows the enriched GO biological process-terms for module M5, and **Table S9** provides the entire list of enriched GO terms for all modules. Based on activation in liver fibrosis-producing conditions, enrichment of known liver fibrosis-relevant genes, and liver fibrosis-relevant GO terms, we selected module M5 as the top liver fibrosis-relevant network module identified from this analysis.

First, we ascertained whether the network module M5 could be observed by random chance. As outlined in the Materials and Methods section, we carried out two statistical significance tests based on random sampling and permutation of the network. The average node degree of the network was 12.76 and it was preserved during network shuffling. Our null hypothesis was that the number of nodes and edges in M5 could be obtained by random chance. In both tests, M5 was significantly different from random occurrence, and the probability of finding this module by random chance was essentially zero (**Figure 4**).

Second, we plotted the PPI network for the protein products encoded by the genes in module M5 (**Figure 5**). Many of the molecular interactions captured were already reported to be associated with liver fibrosis, validating the computational approach. For example, the network map shows connectivity between the genes coding for matrix metalloproteinases and the up-regulated genes encoding TIMP1, COLs and FBN1 (**Figure 5**) [4]. This network map supports a published disease mechanism where increased expression of the gene coding for TIMP1, the negative regulator of matrix metalloproteinases, leads to an increased accumulation of ECM proteins, e.g., collagens (COLs). The network module M5 included other genes encoding the proteins implicated in the pathogenesis of fibrosis (e.g., LUM, CTGF, LGALS3, LCN2, PDGFR, PLAT, and LOX) [4,74–76]. Genes coding for the integrin receptors (ITG) in this network

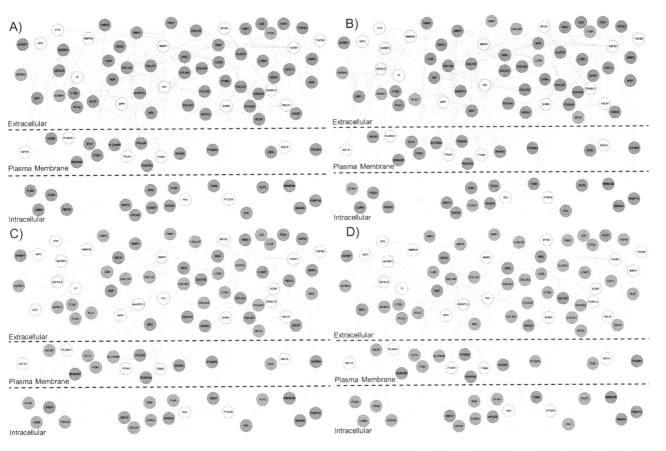

Figure 7. Activation of genes encoding proteins in liver fibrosis-relevant network module M5 at different time points. Genes encoding proteins with average $\log_2$ fold-change ratios >0.6 are colored in red. Genes encoding proteins with average $\log_2$ fold-change ratios <−0.6 are colored in green. Genes encoding proteins with average $\log_2$fold-change ratios between 0.6 and −0.6<−0.6 are colored in grey. A) Activation at 0.25-day exposure. The mapped expression profile is the average $\log_2$ ratio in 1-naphthyl isothiocyanate 30 mg/kg and 60 mg/kg, at 0.25-day exposure. B) Activation at 1 day of exposure. The mapped expression profile is the average $\log_2$ ratio in 1-naphthyl isothiocyanate 30 mg/kg and 60 mg/kg, and 4,4′-Methylenedianiline 81 mg/kg, at 1 day of exposure. C) Activation at 3 days of exposure. The mapped expression profile is the average $\log_2$ ratio in 1-naphthyl isothiocyanate 30 mg/kg and 60 mg/kg, and 4,4′-Methylenedianiline 81 mg/kg, at 3 days of exposure. D) Activation at >3 days of exposure. The mapped expression profile is the average $\log_2$ ratio across chemical exposures that produced liver fibrosis.

interact with ECM proteins that support their known role as mediators of pro-fibrogenic signaling of ECM proteins [77]. In addition to retrieving genes coding for proteins that are already known to be associated with liver fibrosis, this network module also retrieved some potential new candidate protein products. One such candidate is LGMN (average $\log_2$ ratio of 1.24), a cysteine protease that functions in ECM remodeling, but has no known associations with liver fibrosis [78]. Another candidate is the gene encoding PLIN3 (mannose-6-phosphate receptor binding protein), with an average $\log_2$ ratio of 1.29; it interacts with IGF2R in the M5 network. PLIN3 is known to play a role in the pathogenesis of steatosis, and is also reported to play a role in PGE2 production [79,80].

Third, we used the network analysis to identify proteins encoded by genes that do not change in expression, but may form PPI networks. Among the 92 proteins in M5, 30 are K-node exceptions obtained using KeyPathwayMiner, based on their connection to fibrosis-relevant proteins. In protein network analysis, the concept of guilt by association is well known [14]. If a protein is known to be associated with many proteins involved in a biological process, it can be hypothesized to play a role or to be related to the biological process associated with these proteins. For example, the genes encoding osteopontin (SPP1) and vitronectin (VTN) are

known to be associated with liver fibrosis [4,81]. These genes did not reach the fold-change threshold in our preprocessed expression dataset, but the network interactions with other fibrosis-relevant protein products that did change in expression predicted them to be fibrosis-relevant (**Figure 5**). In addition to retrieving known proteins, the network analysis also identified MYOC, a protein with no reported association with liver fibrosis. MYOC is a secreted glycoprotein involved in the pathogenesis of glaucoma [82]. MYOC interacts with many up-regulated genes in the high-confidence human PPI network, including genes that encode TIMP1, LGALS3, FBN1, COL1A2, and COL3A1 (**Figure 6**). Based on its connection with many liver fibrosis-relevant proteins, MYOC is a new testable candidate in fibrosis diagnosis and/or pathogenesis. We also analyzed the expression profile of genes in module M5 in chemical exposures that produced periportal liver fibrosis across different time periods of exposures (**Figure 7**). Only a few genes were activated (i.e., $\log_2$ ratio >0.6) after 0.25 day and 1 day, but at ≥3 days, most of the genes exhibited increased expression (**Figure 7**).

We performed hierarchical clustering of the 640 chemical exposure conditions present in the DrugMatrix database using the expression data for the genes in module M5. We identified a single cluster that included all four chemical exposure conditions that

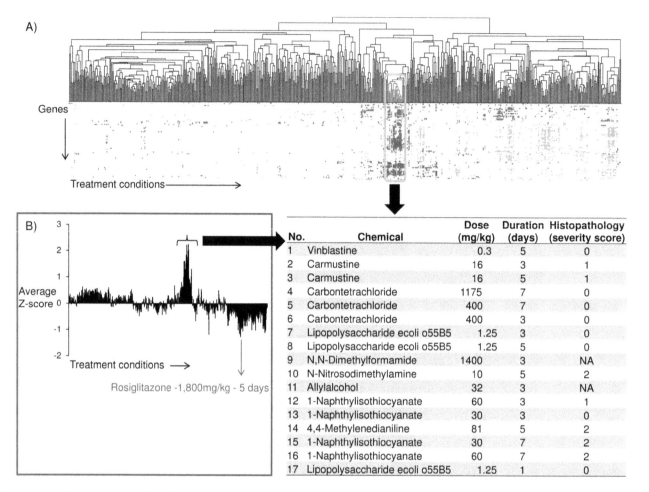

No.	Chemical	Dose (mg/kg)	Duration (days)	Histopathology (severity score)
1	Vinblastine	0.3	5	0
2	Carmustine	16	3	1
3	Carmustine	16	5	1
4	Carbontetrachloride	1175	7	0
5	Carbontetrachloride	400	7	0
6	Carbontetrachloride	400	3	0
7	Lipopolysaccharide ecoli o55B5	1.25	3	0
8	Lipopolysaccharide ecoli o55B5	1.25	5	0
9	N,N-Dimethylformamide	1400	3	NA
10	N-Nitrosodimethylamine	10	5	2
11	Allylalcohol	32	3	NA
12	1-Naphthylisothiocyanate	60	3	1
13	1-Naphthylisothiocyanate	30	3	0
14	4,4-Methylenedianiline	81	5	2
15	1-Naphthylisothiocyanate	30	7	2
16	1-Naphthylisothiocyanate	60	7	2
17	Lipopolysaccharide ecoli o55B5	1.25	1	0

Figure 8. Analysis of genes in liver fibrosis-relevant network module M5. A) Hierarchical clustering of 640 chemical exposures using genes in liver fibrosis-relevant network module M5. The conditions that clustered with four liver fibrosis-producing conditions are highlighted and listed. Genes with Z-scores>2 are colored in red. Genes with Z-scores <−2 are colored in green. Genes with Z-scores between 2 and -2 are colored in yellow. NA in the table represents that histopathological data was not available for that chemical exposure condition. B) Average Z-scores across the genes in module M5 for each of the 640 chemical exposure conditions.

were initially identified with grade 2 periportal liver fibrosis (**Figure 8a**). Most of the 17 conditions in the cluster were associated with compounds that cause fibrosis. Vinblastine, carmustine, and allyl alcohol all had conditions within the drug matrix dataset that caused periportal fibrosis, but they did not meet the histopathology grade 2 threshold used for our initial

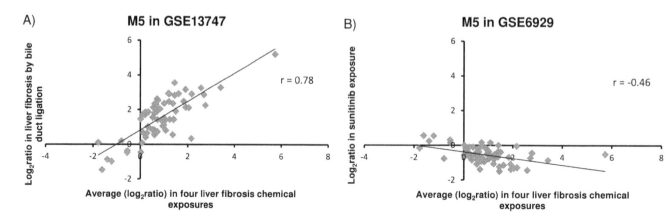

Figure 9. Validation with external datasets. M5 gene expression compared with external datasets. A) GSE13747 represents liver fibrosis produced by bile duct ligation. B) GSE6929 represents sunitinib (SU11248) treatment in liver cirrhosis.

analysis. Carbon tetrachloride and lipopolysaccharide exposures were present in this cluster (**Figure 8a**). Analysis of the literature shows that both carbon tetrachloride and lipopolysaccharide are well known agents that cause liver fibrosis [66]. In the DrugMatrix database, a 28-day exposure study with carbon tetrachloride showed histopathological evidence of liver fibrosis; however, gene expression data were not available for this time point. Furthermore, all but one (crotamiton-750 mg/kg exposure) of the chemicals in the DrugMatrix database that caused periportal fibrosis and had Affymetrix arrays for the liver were identified in this cluster. The data associated with the crotamiton-750 mg/kg exposure most likely represents an outlier in the underlying dataset. Thus, our analysis has the potential to generate gene sets that could predict the early onset of liver fibrosis, consistent with earlier reports that gene expression analysis of liver biopsy samples could complement histopathological data [83].

To evaluate whether network module M5 activation was specific for liver fibrosis, we plotted the average Z-scores of M5 genes in each of the 640 chemical exposure conditions (**Figure 8b**). The M5 genes were mostly activated in liver fibrosis-related chemical exposure conditions and showed very low activation in other chemical exposure conditions (**Figure 8b**). Exposure to rosiglitazone, a PPAR-γ agonist, produced an opposing effect on expression of the genes encoding the protein products in M5 (i.e., down-regulation as measured by $\log_2$ ratio values; **Figure S3**). This result was consistent with the evidence that PPAR-γ agonists prevented liver fibrosis in experimental models [70].

Finally, we used two external gene expression datasets (GEO datasets GSE13747 and GSE6929) to evaluate the relevance of module M5 in liver fibrosis. The overlap of genes between the M5 module and the processed GSE13747 and GSE6929 datasets was 66 and 65, respectively. In the study associated with the GSE13747 dataset, liver fibrosis was induced using bile duct ligation. We observed the predicted positive correlation of gene expression data between the DEGs in this dataset and the genes in module M5 (r = 0.78, **Figure 9a**). In the GSE6929 dataset, sunitinib, a multi-kinase inhibitor, was used to treat experimental liver cirrhosis. As expected, we observed a negative correlation (r = −0.46) between $\log_2$ ratio gene expression data from GSE6929 and M5 (**Figure 9b**). Thus, the external dataset further supports the validity of M5 genes in liver fibrosis.

In this work, we showed the utility of integrating gene expression and PPI network analyses to understand the molecular-level details of liver fibrosis and to identify biomarker candidates. The presented computational approach has general applicability for characterizing adverse health effects. For example, our approach can be used to computationally predict toxicity pathways associated with specific diseases. Recent work has suggested that the body's response to toxic exposures and the resulting disease progression proceed through a finite set of toxicity pathways [84]. As stated earlier, the available knowledge-based pathway databases are limited in coverage and susceptible to annotation biases towards well-studied diseases. The computational approach presented in this work enabled us to partly overcome these limitations and develop de novo pathways linked to a specific disease. In particular, we believe that our approach could be used to further understand the molecular basis of adverse health effects such as acute kidney injury, liver cholestasis, liver steatosis, and myocardial infarction. More recently, the concept of adverse outcome pathways (AOPs) has been proposed as a novel tool in mechanism-based predictive toxicology [85]. The AOPs provide a flow-chart-like mechanistic representation of adverse health effects and consist of a molecular initiating event and a series of intermediate key events that lead to the final adverse outcome [85]. Our computational approach will be applicable to the development of novel AOPs; we believe that network modules, such as the liver fibrosis module (M5) identified in this work, could be used either as a key molecular event or used to quantify the reversibility or point of departure of key events in AOPs.

Conclusion

We have carried out systems-level analyses of gene expression data for periportal liver fibrosis. We found that both pathway and network analyses provided insights into the molecular mechanisms of liver fibrosis. Network analyses allowed us to generate de novo pathways and overcome the limitations of analyses based on KEGG pathways. We identified a liver fibrosis-relevant network module that was enriched with known liver fibrosis-relevant genes predictive of liver fibrosis and validated it with external data. The systems approach used in this study allowed us to generate liver fibrosis-relevant pathways and have the potential to predict mechanism-based biomarker candidates.

Supporting Information

Figure S1 Dendrogram from the clustering of chemical exposure conditions that produced periportal liver fibrosis with histopathological scores >1 and their controls.
(DOCX)

Figure S2 Treemap view of gene ontology (GO) biological process-term enrichment for genes in the network module M5.
(DOCX)

Figure S3 Activation of proteins in liver fibrosis-relevant network module M5 in rosiglitazone-1,800 mg/kg, at 5 days of exposure.
(DOCX)

Table S1 The list of fibrosis-relevant genes from differential expression and co-expression analysis.
(XLSX)

Table S2 The list of genes that mapped to enriched Kyoto Encyclopedia of Genes and Genomes (KEGG) pathways and their gene expression data in chemical exposures that produced liver fibrosis.
(XLSX)

Table S3 The list of enriched up-regulated pathways for the liver fibrosis-relevant gene set.
(XLSX)

Table S4 The list of enriched down-regulated pathways for the liver fibrosis-relevant gene set.
(XLSX)

Table S5 Mapping of rat probe IDs to human gene IDs.
(XLSX)

Table S6 Protein membership in the protein-protein interaction (PPI) network modules and their gene expression data.
(XLSX)

Table S7 List of 28 fibrosis-relevant genes collected from literature.
(XLSX)

Table S8 List of 126 experimental liver cirrhosis-relevant genes collected from the Comparative Toxicogenomics Database (CTD).
(XLSX)

Table S9 List of enriched gene ontology (GO) terms for the protein-protein interaction (PPI) network modules, M1-M11.
(XLSX)

Data S1 Cytoscape session file associated with network module generation. The Cytoscape session file includes 1) high-confidence human PPI network data, 2) the liver fibrosis-relevant parent sub-network, and 3) 11 modules obtained by clustering the liver fibrosis-relevant parent sub-network.
(ZIP)

Data S2 KeyPathwayMiner input gene list.
(XLSX)

Data S3 Cytoscape session file associated with generation of the liver fibrosis-relevant parent sub-network. The Cytoscape session file includes 1) four sub-networks obtained

from KeyPathwayMiner, 2) merged nodes of four sub-networks, 3) high-confidence human PPI network data, and 4) the final liver fibrosis-relevant parent sub-network.
(ZIP)

Acknowledgments

We thank Drs. Xueping Yu and Nela Zavaljevski for valuable discussions and comments on the manuscript. *Disclaimer*: The opinions and assertions contained herein are the private views of the authors and are not to be construed as official or as reflecting the views of the U.S. Army or of the U.S. Department of Defense. Citations of commercial organizations or trade names in this report do not constitute an official Department of the Army endorsement or approval of the products or services of these organizations. This paper has been approved for public release with unlimited distribution.

Author Contributions

Conceived and designed the experiments: MDMA GJT DLI JAL JDS AW. Performed the experiments: MDMA. Analyzed the data: MDMA GJT DLI AW. Contributed reagents/materials/analysis tools: KK. Wrote the paper: MDMA GJT DLI JAL JDS AW.

References

1. Malhi H, Gores GJ (2008) Cellular and molecular mechanisms of liver injury. Gastroenterology 134: 1641–1654.
2. Sebastiani G (2012) Serum biomarkers for the non-invasive diagnosis of liver fibrosis: the importance of being validated. Clin Chem Lab Med 50: 595–597.
3. Brenner DA (2009) Molecular pathogenesis of liver fibrosis. Trans Am Clin Climatol Assoc 120: 361–368.
4. Bataller R, Brenner DA (2005) Liver fibrosis. J Clin Invest 115: 209–218.
5. Baranova A, Lal P, Birerdinc A, Younossi ZM (2011) Non-invasive markers for hepatic fibrosis. BMC Gastroenterology 11: 91.
6. Kisseleva T, Brenner DA (2008) Mechanisms of fibrogenesis. Exp Biol Med (Maywood) 233: 109–122.
7. Mukhopadhyay P, Rajesh M, Cao Z, Horváth B, Park O, et al. (2014) Poly (ADP-ribose) polymerase-1 is a key mediator of liver inflammation and fibrosis. Hepatology 59: 1998–2009.
8. Qiu L, Lin J, Ying M, Chen W, Yang J, et al. (2013) Aldose reductase is involved in the development of murine diet-induced nonalcoholic steatohepatitis. PLOS ONE 8: e73591.
9. Adams LA (2011) Biomarkers of liver fibrosis. J Gastroenterol Hepatol 26: 802–809.
10. Miller JA, Oldham MC, Geschwind DH (2008) A systems level analysis of transcriptional changes in Alzheimer's disease and normal aging. J Neurosci 28: 1410–1420.
11. Afshari CA, Hamadeh HK, Bushel PR (2011) The evolution of bioinformatics in toxicology: advancing toxicogenomics. Toxicol Sci 120 Suppl 1: S225–S237.
12. Shi Z, Derow CK, Zhang B (2010) Co-expression module analysis reveals biological processes, genomic gain, and regulatory mechanisms associated with breast cancer progression. BMC Syst Biol 4: 74.
13. Tugues S, Fernandez-Varo G, Munoz-Luque J, Ros J, Arroyo V, et al. (2007) Antiangiogenic treatment with sunitinib ameliorates inflammatory infiltrate, fibrosis, and portal pressure in cirrhotic rats. Hepatology 46: 1919–1926.
14. Feala JD, Abdulhameed MD, Yu C, Dutta B, Yu X, et al. (2013) Systems biology approaches for discovering biomarkers for traumatic brain injury. J Neurotrauma 30: 1101–1116.
15. Tawa GJ, AbdulHameed MD, Yu X, Kumar K, Ippolito DL, et al. (2014) Characterization of chemically induced liver injuries using gene co-expression modules. PLOS ONE 9: e107230.
16. Kanehisa M, Goto S, Sato Y, Furumichi M, Tanabe M (2012) KEGG for integration and interpretation of large-scale molecular data sets. Nucleic Acids Res 40: D109–D114.
17. Matthews L, Gopinath G, Gillespie M, Caudy M, Croft D, et al. (2009) Reactome knowledgebase of human biological pathways and processes. Nucleic Acids Res 37: D619–D622.
18. Geer LY, Marchler-Bauer A, Geer RC, Han L, He J, et al. (2010) The NCBI BioSystems database. Nucleic Acids Res 38: D492–D496.
19. Pan Y, Cheng T, Wang Y, Bryant SH (2014) Pathway analysis for drug repositioning based on public database mining. J Chem Inf Model 54: 407–418.
20. Huang da W, Sherman BT, Lempicki RA (2009) Bioinformatics enrichment tools: paths toward the comprehensive functional analysis of large gene lists. Nucleic Acids Res 37: 1–13.
21. Kim MJ, Pelloux V, Guyot E, Tordjman J, Bui LC, et al. (2012) Inflammatory pathway genes belong to major targets of persistent organic pollutants in adipose cells. Environ Health Perspect 120: 508–514.
22. Smid M, Wang Y, Zhang Y, Sieuwerts AM, Yu J, et al. (2008) Subtypes of breast cancer show preferential site of relapse. Cancer Res 68: 3108–3114.
23. Huang L, Heinloth AN, Zeng ZB, Paules RS, Bushel PR (2008) Genes related to apoptosis predict necrosis of the liver as a phenotype observed in rats exposed to a compendium of hepatotoxicants. BMC Genomics 9: 288.
24. Affo S, Dominguez M, Lozano JJ, Sancho-Bru P, Rodrigo-Torres D, et al. (2013) Transcriptome analysis identifies TNF superfamily receptors as potential therapeutic targets in alcoholic hepatitis. Gut 62: 452–460.
25. Lehne B, Schlitt T (2012) Breaking free from the chains of pathway annotation: de novo pathway discovery for the analysis of disease processes. Pharmacogenomics 13: 1967–1978.
26. Gentleman RC, Carey VJ, Bates DM, Bolstad B, Dettling M, et al. (2004) Bioconductor: open software development for computational biology and bioinformatics. Genome Biol 5: R80.
27. Chuang HY, Lee E, Liu YT, Lee D, Ideker T (2007) Network-based classification of breast cancer metastasis. Mol Syst Biol 3: 140.
28. Zhu W, Yang L, Du Z (2009) Layered functional network analysis of gene expression in human heart failure. PLOS ONE 4: e6288.
29. Reiss DJ, Avila-Campillo I, Thorsson V, Schwikowski B, Galitski T (2005) Tools enabling the elucidation of molecular pathways active in human disease: application to hepatitis C virus infection. BMC Bioinformatics 6: 154.
30. Camargo A, Azuaje F (2007) Linking gene expression and functional network data in human heart failure. PLOS ONE 2: e1347.
31. Azuaje F, Devaux Y, Wagner DR (2010) Coordinated modular functionality and prognostic potential of a heart failure biomarker-driven interaction network. BMC Syst Biol 4: 60.
32. Azuaje FJ, Dewey FE, Brutsaert DL, Devaux Y, Ashley EA, et al. (2012) Systems-based approaches to cardiovascular biomarker discovery. Circ Cardiovasc Genet 5: 360–367.
33. Xiao Y, Xu C, Xu L, Guan J, Ping Y, et al. (2012) Systematic identification of common functional modules related to heart failure with different etiologies. Gene 499: 332–338.
34. Huan J, Wang L, Xing L, Qin X, Feng L, et al. (2014) Insights into significant pathways and gene interaction networks underlying breast cancer cell line MCF-7 treated with 17beta-estradiol (E2). Gene 533: 346–355.
35. Ganter B, Snyder RD, Halbert DN, Lee MD (2006) Toxicogenomics in drug discovery and development: mechanistic analysis of compound/class-dependent effects using the DrugMatrix database. Pharmacogenomics 7: 1025–1044.
36. Breitling R, Armengaud P, Amtmann A, Herzyk P (2004) Rank products: a simple, yet powerful, new method to detect differentially regulated genes in replicated microarray experiments. FEBS Lett 573: 83–92.
37. Yu X, Wallqvist A, Reifman J (2012) Inferring high-confidence human protein-protein interactions. BMC Bioinformatics 13: 79.
38. DrugMatrix (nd) National Institute of Environmental Health Sciences. Available: https://ntp.niehs.nih.gov/drugmatrix/index.html.
39. Irizarry RA, Bolstad BM, Collin F, Cope LM, Hobbs B, et al. (2003) Summaries of Affymetrix GeneChip probe level data. Nucleic Acids Res 31: e15.
40. Gautier L, Cope L, Bolstad BM, Irizarry RA (2004) Affy – analysis of Affymetrix GeneChip data at the probe level. Bioinformatics 20: 307–315.
41. R Development Core Team (2011) R: A language and environment for statistical computing. Vienna, Austria: R Foundation for Statistical Computing. Available: http://www.R-project.org/.

42. Kauffmann A, Gentleman R, Huber W (2009) ArrayQualityMetrics – a bioconductor package for quality assessment of microarray data. Bioinformatics 25: 415–416.

43. Gentleman R, Carey V, Huber W, Hahne F (2013) genefilter: methods for filtering genes from microarray experiments, R. package version 1.40.0. Seattle, WA: Bioconductor. Available: http://bioc.ism.ac.jp/2.11/bioc/html/genefilter.html.

44. Hong F, Breitling R, McEntee CW, Wittner BS, Nemhauser JL, et al. (2006) RankProd: a bioconductor package for detecting differentially expressed genes in meta-analysis. Bioinformatics 22: 2825–2827.

45. Vinuela A, Snoek LB, Riksen JA, Kammenga JE (2010) Genome-wide gene expression analysis in response to organophosphorus pesticide chlorpyrifos and diazinon in *C. elegans*. PLOS ONE 5: e12145.

46. Langfelder P, Zhang B, Horvath S (2008) Defining clusters from a hierarchical cluster tree: the Dynamic Tree Cut package for R. Bioinformatics 24: 719–720.

47. Huang DW, Sherman BT, Lempicki RA (2009) Systematic and integrative analysis of large gene lists using DAVID bioinformatics resources. Nat Protoc 4: 44–57.

48. Yu C, Desai V, Cheng L, Reifman J (2012) QuartetS-DB: a large-scale orthology database for prokaryotes and eukaryotes inferred by evolutionary evidence. BMC Bioinformatics 13: 143.

49. Eppig JT, Blake JA, Bult CJ, Kadin JA, Richardson JE, et al. (2012) The Mouse Genome Database (MGD): comprehensive resource for genetics and genomics of the laboratory mouse. Nucleic Acids Res 40: D881–D886.

50. Zhang J, Yang Y, Wang Y, Zhang J, Wang Z, et al. (2011) Identification of hub genes related to the recovery phase of irradiation injury by microarray and integrated gene network analysis. PLOS ONE 6: e24680.

51. Alcaraz N, Friedrich T, Kotzing T, Krohmer A, Muller J, et al. (2012) Efficient key pathway mining: combining networks and OMICS data. Integr Biol (Camb) 4: 756–764.

52. Baumbach J, Friedrich T, Kotzing T, Krohmer A, Muller J, et al. (2012) Efficient algorithms for extracting biological key pathways with global constraints. Proceedings of the 14th Annual Conference on Genetic and Evolutionary Computation. Philadelphia, Pennsylvania, USA: Association for Computing Machinery. pp. 169–176.

53. Saito R, Smoot ME, Ono K, Ruscheinski J, Wang PL, et al. (2012) A travel guide to Cytoscape plugins. Nat Methods 9: 1069–1076.

54. Shen H (2009) Detect overlapping and hierarchical community structure in networks. Physica A 388: 1706–1712.

55. Henderson NC, Mackinnon AC, Farnworth SL, Poirier F, Russo FP, et al. (2006) Galectin-3 regulates myofibroblast activation and hepatic fibrosis. Proc Natl Acad Sci U S A 103: 5060–5065.

56. Page S, Birerdinc A, Estep M, Stepanova M, Afendy A, et al. (2013) Knowledge-based identification of soluble biomarkers: hepatic fibrosis in NAFLD as an example. PLOS ONE 8: e56009.

57. Davis AP, Murphy CG, Johnson R, Lay JM, Lennon-Hopkins K, et al. (2013) The Comparative Toxicogenomics Database: update 2013. Nucleic Acids Res 41: D1104–1114.

58. Supek F, Bosnjak M, Skunca N, Smuc T (2011) REVIGO summarizes and visualizes long lists of gene ontology terms. PLOS ONE 6: e21800.

59. de Hoon MJ, Imoto S, Nolan J, Miyano S (2004) Open source clustering software. Bioinformatics 20: 1453–1454.

60. Moreno M, Chaves JF, Sancho-Bru P, Ramalho F, Ramalho LN, et al. (2010) Ghrelin attenuates hepatocellular injury and liver fibrogenesis in rodents and influences fibrosis progression in humans. Hepatology 51: 974–985.

61. Subramanian A, Tamayo P, Mootha VK, Mukherjee S, Ebert BL, et al. (2005) Gene set enrichment analysis: a knowledge-based approach for interpreting genome-wide expression profiles. Proc Natl Acad Sci U S A 102: 15545–15550.

62. Pan KH, Lih CJ, Cohen SN (2005) Effects of threshold choice on biological conclusions reached during analysis of gene expression by DNA microarrays. Proc Natl Acad Sci U S A 102: 8961–8965.

63. Nam D, Kim SY (2008) Gene-set approach for expression pattern analysis. Brief Bioinform 9: 189–197.

64. Lee HK, Hsu AK, Sajdak J, Qin J, Pavlidis P (2004) Coexpression analysis of human genes across many microarray data sets. Genome Res 14: 1085–1094.

65. Friedman SL (2000) Molecular regulation of hepatic fibrosis, an integrated cellular response to tissue injury. J Biol Chem 275: 2247–2250.

66. Liedtke C, Luedde T, Sauerbruch T, Scholten D, Streetz K, et al. (2013) Experimental liver fibrosis research: update on animal models, legal issues and translational aspects. Fibrogenesis Tissue Repair 6: 19.

67. Li L, Wang JY, Yang CQ, Jiang W (2012) Effect of RhoA on transforming growth factor beta1-induced rat hepatic stellate cell migration. Liver Int 32: 1093–1102.

68. Ikeda K, Kawada N, Wang YQ, Kadoya H, Nakatani K, et al. (1998) Expression of cellular prion protein in activated hepatic stellate cells. Am J Pathol 153: 1695–1700.

69. Kitada T, Seki S, Ikeda K, Nakatani K, Sakaguchi H, et al. (2000) Clinicopathological characterization of prion: a novel marker of activated human hepatic stellate cells. J Hepatol 33: 751–757.

70. Wynn TA, Ramalingam TR (2012) Mechanisms of fibrosis: therapeutic translation for fibrotic disease. Nat Med 18: 1028–1040.

71. Okayama T, Kikuchi S, Ochiai T, Ikoma H, Kubota T, et al. (2008) Attenuated response to liver injury in moesin-deficient mice: impaired stellate cell migration and decreased fibrosis. Biochim Biophys Acta 1782: 542–548.

72. Khatri P, Sirota M, Butte AJ (2012) Ten years of pathway analysis: current approaches and outstanding challenges. PLOS Comput Biol 8: e1002375.

73. Ulitsky I, Krishnamurthy A, Karp RM, Shamir R (2010) DEGAS: de novo discovery of dysregulated pathways in human diseases. PLOS ONE 5: e13367.

74. Krishnan A, Li X, Kao WY, Viker K, Butters K, et al. (2012) Lumican, an extracellular matrix proteoglycan, is a novel requisite for hepatic fibrosis. Lab Invest 92: 1712–1725.

75. Schuppan D, Kim YO (2013) Evolving therapies for liver fibrosis. J Clin Invest 123: 1887–1901.

76. Zhang LP, Takahara T, Yata Y, Furui K, Jin B, et al. (1999) Increased expression of plasminogen activator and plasminogen activator inhibitor during liver fibrogenesis of rats: role of stellate cells. J Hepatol 31: 703–711.

77. Mallat A, Lotersztajn S (2013) Cellular mechanisms of tissue fibrosis. 5. Novel insights into liver fibrosis. Am J Physiol Cell Physiol 305: C789–C799.

78. Morita Y, Araki H, Sugimoto T, Takeuchi K, Yamane T, et al. (2007) Legumain/asparaginyl endopeptidase controls extracellular matrix remodeling through the degradation of fibronectin in mouse renal proximal tubular cells. FEBS Lett 581: 1417–1424.

79. Okumura T (2011) Role of lipid droplet proteins in liver steatosis. J Physiol Biochem 67: 629–636.

80. Nose F, Yamaguchi T, Kato R, Aiuchi T, Obama T, et al. (2013) Crucial role of perilipin-3 (TIP47) in formation of lipid droplets and PGE2 production in HL-60-derived neutrophils. PLOS ONE 8: e71542.

81. Koukoulis GK, Shen J, Virtanen I, Gould VE (2001) Vitronectin in the cirrhotic liver: an immunomarker of mature fibrosis. Hum Pathol 32: 1356–1362.

82. Tamm ER (2002) Myocilin and glaucoma: facts and ideas. Prog Retin Eye Res 21: 395–428.

83. Heinloth AN, Boorman GA, Foley JF, Flagler ND, Paules RS (2007) Gene expression analysis offers unique advantages to histopathology in liver biopsy evaluations. Toxicol Pathol 35: 276–283.

84. Hartung T, McBride M (2011) Food for Thought... on mapping the human toxome. ALTEX 28: 83–93.

85. Vinken M (2013) The adverse outcome pathway concept: a pragmatic tool in toxicology. Toxicology 312: 158–165.

10

Computational Approaches for Predicting Biomedical Research Collaborations

Qing Zhang[1¤], Hong Yu[1,2]*

1 Department of Quantitative Health Sciences, University of Massachusetts Medical School, Worcester, Massachusetts, United States of America, 2 VA Central Massachusetts, Leeds, Massachusetts, United States of America

Abstract

Biomedical research is increasingly collaborative, and successful collaborations often produce high impact work. Computational approaches can be developed for automatically predicting biomedical research collaborations. Previous works of collaboration prediction mainly explored the topological structures of research collaboration networks, leaving out rich semantic information from the publications themselves. In this paper, we propose supervised machine learning approaches to predict research collaborations in the biomedical field. We explored both the semantic features extracted from author research interest profile and the author network topological features. We found that the most informative semantic features for author collaborations are related to research interest, including similarity of out-citing citations, similarity of abstracts. Of the four supervised machine learning models (naïve Bayes, naïve Bayes multinomial, SVMs, and logistic regression), the best performing model is logistic regression with an ROC ranging from 0.766 to 0.980 on different datasets. To our knowledge we are the first to study in depth how research interest and productivities can be used for collaboration prediction. Our approach is computationally efficient, scalable and yet simple to implement. The datasets of this study are available at https://github.com/qingzhanggithub/medline-collaboration-datasets.

Editor: Neil R. Smalheiser, University of Illinois-Chicago, United States of America

Funding: The research reported in this publication was supported in part by the National Institutes of Health (NIH) the National Institute of General Medical Sciences under award number 5R01GM095476 and the National Center for Advancing Translational Sciences under award number UL1TR000161. The funders had no role in study design, data collection and analysis, decision to publish, or preparation of the manuscript.

Competing Interests: The authors have declared that no competing interests exist.

* Email: hong.yu@umassmed.edu

¤ Current address: Currently at eBay Inc., San Jose, California, United States of America

Introduction

Millions of researchers contribute to biomedical research, collectively publishing tens of millions of research papers. These research papers interlink researchers into a complex co-authorship network. Biomedical research is a fast-growing interdisciplinary field that frequently requires high degree of collaboration. It has been found that the average number of collaborators in the biomedical field is twice that in physics and more than four times that in mathematics [1]. Such collaborations span basic, translational, and clinical research. Successful collaborations often yielded high impact work [2–5] such as the Gene Ontology [6].

The importance of scientific collaboration has motivated the development of researcher profile platforms, most of which focus on facilitating institutional collaborations. Such platforms, including the Harvard Catalyst Profile [7], SciVal Experts [8], and ProQuest Pivot [9], integrate research and collaboration information—including publication history, co-authorship connections, research topics, and funding information—making it easier to find potential collaborators. In addition, semantic Web resources, including VIVO [10], have been developed to provide a general scheme to describe researcher profiles so that the profiles can be embedded in particular applications. Online communities, including BiomedExperts [11], allow users to upload their personal profiles and help them make new connections. Few systems, however, have the functionality of recommending collaborators automatically. Such services, on the other hand, may be important for researchers, especially junior researchers, whose work depends upon successful collaborations. Automatically recommending collaborators may offer an attractive alternative to traditional ways of finding a collaborator, such as socializing at a scientific conference or being introduced by a mutual colleague.

We formulate research collaboration prediction as a link prediction problem in the context of a co-authorship network. Since joint publication is one of the most effective representations of collaboration, a co-authorship indicates a collaborative relation. Our goal is illustrated in Figure 1, where author s has collaborated with authors a, c, and e and we would like to know the probability that s will collaborate with b, f, and d.

Link prediction has been studied in social networks. Liben-Nowell and Kleinberg [12] used various topological features for link prediction. For example, two researchers are more likely to collaborate with each other when they have common collaborators. Al Hasan, et al. [13] compared different machine learning models and learning features for predicting author collaboration. They found that support vector machines (SVMs) performed the best and shortest distance (i.e., the minimum number of edges that separate two authors) is a top topological feature. In addition, they explored node attributes (e.g., author's productivity and the research similarity between two authors) as additional features and concluded that they are top features for the prediction. Backstrom

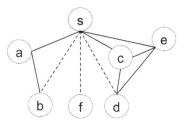

Figure 1. An illustration of automatic research collaboration recommendation. The graph shows a co-authorship network in which the nodes are authors and the links represent co-authorship. The solid lines represent existing co-authorships. Our study is to build a computational model to predict whether author *s* will collaborate with authors *b*, *f*, and *d* based on their existing research and collaborations.

and Leskovec [14] explored a supervised random walk model and found it outperformed decision tree and logistic regression models in predicting new friends in Facebook. The aforementioned work provides important understanding of link predictions problem formulation. However most of the work mainly explored network topological structures. On the other hand, the rich semantic information (research work and collaboration activities), well documented by their publications, have been explored little.

In this paper we report the development and evaluation of Automatic Research Collaboration Recommendation (ARCR) system to predict new author collaborations. ARCR was built on supervised machine learning models. Our contributions are: We explored rich learning features, which were derived from the semantic content of an author's research profile. Our supervised machine learning models and learning features are computationally efficient, making them applicable to the big data challenge of scientific collaboration recommendation. In addition, we evaluated our approaches on various datasets reflecting data sparseness, a common problem in the real world. Finally, we provided in-depth analysis of important features including research interest and mutual collaborators that contribute to biomedical collaborations.

Background

Author collaboration prediction can be considered as a case of modeling evolving networks. Significant amount of theoretical work is based on network structures and their evolution. Early work modeled the network as random graph, where the establishment of a connection follows Poisson distribution [15]. Later the scale-free model was proposed, in which the probability of a new node connecting with a given node is proportional to the degree of the node [16]. This phenomenon, which is also called preferential attachment, has been observed in many evolving networks, including social networks [1,17], World Wide Web [18], and the protein-protein interaction network [19]. Additional findings, including "first-mover-advantage" [20] and "the-fit-get-richer" [21], have enriched further the scale-free model.

"Small world" is another important characteristic in various networks [22,23]. A social network, including a co-authorship network, consists of both structured (close neighbors) and random contacts and one can navigate from one node to another with very few steps. Newman [17] found that only five to six steps are needed to navigate from one randomly chosen scientist to another in a community. In addition, social networks appear assortative, meaning that nodes tend to connect to other nodes with similar characteristics (e.g., the degree [24]).

In computer science, author collaboration prediction is often formulated as a link prediction problem. Early work focuses on topology-based prediction that utilizes network structure, including the connectivity and similarity of neighbor nodes. As stated earlier in this paper, Liben-Nowell and Kleinberg [12] comprehensively evaluated a collection of topological predictors, including number of common co-author and random walk, for the link prediction in co-authorship network of physics field. The work is one of the foundations of many later studies[13,14], including ours.

Much work in link prediction explored supervised machine learning models and different learning features. As stated earlier, Al Hasan et al. [13] explored naïve Bayes and SVMs. They explored topological features (e.g., the number of common co-authors) and simple semantic features (e.g., the overlap of the keywords of two author's publication profiles). Sun et al [25] studied topological features in the heterogeneous networks consisting both co-authorship and citation relations to predict co-authorship in the DBLP data sets. Backstrom and Leskovec [14] applied supervised machine learning to predict the strength of a connection. The predicted weight is subsequently used to guide the random walk. The stationary probability of landing on a particular node is considered as the chance of a connection from the starting node. Wang et al [26] modeled the local topological structure by Markov Random Field to infer the co-occurrence probability of two nodes, and subsequently integrated with other topological and semantic features for link prediction.

Co-authorship networks have been widely studied. For example, Newman [17] compared the co-authorship network in biomedicine with that in physics and observed differences. He made several observations. In the biomedical domain, it is less common that two researchers collaborate when they have a mutual collaborator than in physics. The networks are scale free: the network structure is dominated by many "little" people with few collaborators, instead of a few people with many collaborators. He also observed that two researchers are more likely to collaborate if they have had a strong history of collaborations, either between themselves or with others) [27].

Several studies showed that co-authorship networks in the biomedical domain exhibit different characteristics than network in other domains. Newman [1] showed that biomedical research has the highest degree of collaboration, in comparison with the physics and mathematics domains. Huang, et al. [28] observed that the collaboration pattern and its evolution in the computer science domain are more similar to the mathematics domain than to biology. Ding [29] found that, in the information retrieval field, productive authors tend to collaborate with and cite researchers who have the same research interests.

Factors that lead to successful collaborations have also been studied, including various social and environmental factors: leadership, geographical proximity, and the personalities of the team members. For example, one study concluded that a leader in a research field typically plays an important "broker" role to bridge people from different disciplines [5]. Physical proximity between first and last author was found to be positively related to the impact of collaboration, measured by the citation received [30]. They concluded that close geographical distance is important for the outcome of the collaboration. International collaborations, however, are found to be a positive factor for the impact of a work. As shown in [31], the average number of citations increases with the number of affiliated countries. Certain characteristics of team members, such as openness and flexibility, also contribute to the success of the collaboration [5,32].

Our work is closely related to the work of Al Hasan, et al. [13]. However, unlike their approach which mainly explored topological features, we explored rich semantic features derived from the author's research profile, including publication history similarity, citation similarity, and common co-authors, and we show that these semantic features significantly improve the research collaboration predictions.

Materials and Methods

We formulate research collaboration prediction as a classification task and therefore explore supervised learning approaches. In the following we first describe the supervised machine learning models we used and then the feature set.

Supervised Machine Learning Models

We explored four supervised machine learning models: naïve Bayes, naïve Bayes multinomial, Support Vector Machines (SVMs), and logistic regression, which are all commonly used for classification tasks. A naïve Bayes classifier is a probabilistic classifier based on Bayes' theorem with the naïve assumption that the features are independent from each other, given the instance label [33]. The naïve Bayes multinomial model assumes the conditional probability of the feature, given a class, follows a multinomial distribution [34]. SVMs are based on the concept of maximum margin decision planes that define generalizable decision boundaries for classification and regression. An SVM constructs a hyperplane to maximize the margin between the data points and the hyperplane, often after mapping the data points to a higher-dimensional space in which they are linearly separable or close to it [35]. We explore an SVM model with the widely-used linear kernel for its efficiency. Logistic regression estimates discrete or continuous value parameters to predict discrete category values. The probabilities that describe the possible class of a single instance are trained as a function of explanatory variables, using a logistic function[33]. These four classifiers are not only the well-studied models in a variety of classification tasks [36], but also widely available in open source software communities. In addition we used K-nearest neighbor model (KNN) as it particularly learns non-linear decision boundaries and is easy to interpret [36,37]. We use data mining software Weka [38] to build and evaluate naïve Bayes, naïve Bayes multinomial and logistic regression models, LIBSVM [39] for SVM, and the python machine learning package Scikit [40] for KNN.

Features

There are many reasons why two researchers collaborate, e.g., geographically close proximity (e.g., within or outside institutes) [30], proximity in the network (e.g., two researchers who have colleagues in common are more likely to collaborate) [12], and proximity in research (two researchers with the same goal in research may collaborate). Topological proximity has long been studied and considered as a factor of establishing connection in social networks. Semantic features, on the other hand, integrate specific domain knowledge of the nodes and have not yet been fully explored, which are the major contributions of this paper. In the following we will first describe the topological features, and then the semantic features.

Co-authorship Network Connectivity. Newman [1] observed that two scientists with a common collaborator are more likely to co-author a paper than two scientists who have no common collaborator. We therefore explored this feature called *numCommonCoauthor* of authors x and y, which is defined as

$$numCommonCoauthor(x,y) = |\Gamma(x) \cap \Gamma(y)|$$

where $\Gamma(.)$ is the set of coauthors, and the feature value is number of co-authors two researchers have in common.

coAuthorJaccard and *Adamic* are two extensions of the common coauthor feature, both of which have been studied by Liben-Nowell et al [12]. *coauthorJaccard* is the number of co-authors two researchers have in common normalized by the total number of their unique co-authors. *Adamic* was first introduced by Adamic et al in [41] to measure the similarity of two web pages. The idea is that two web pages are more similar if they have common web pages that link both. Web pages that are exclusive to the two web pages are weighted more than those that also link to other web pages. *Adamic* was explored for link prediction [12,13], although its contribution to link prediction remain inconsistent among different studies. While Liben-Nowell and Kleinberg [12] found that *Adamic* was one of the most valuable features, Hasan et al [13] did not report any performance improvement. Here we adopted *Adamic* to measure the similarity of two researchers by their common neighbors. For authors x and y,

$$Adamic(x,y) = \sum_{z \in \Gamma(x) \cap \Gamma(y)} \frac{1}{\log |\Gamma(z)|}$$

where z is the common neighbor (co-author) of x and y. The larger the value, the more similar x and y are. The higher number of the common co-authors is, the higher the *Adamic* value is. Each common co-author is also weighted by their exclusiveness to authors x and y. The less inclusive a common co-author is, the higher its *Adamic* value. Assuming that researcher z is the only common neighbor of x and y and that z has no other connections other than x and y (or x and y are the only co-authors of z), the *Adamic* value of x and y is $1/\log(2)$. On other hand if z has 3 connections in addition to x and y, the value becomes $1/\log(5)$. Therefore x and y are more similar in the former case.

A feature commonly used for describing the small-world characteristics in a network is clustering coefficient [27], which we designated the feature as *sumClusteringCoef*. It is the sum of each researcher's clustering coefficient, a measure of the probability that a researcher's collaborators have collaborations among themselves. The higher the clustering coefficient the closer the nodes in the network are connected.

We also included the feature *sumCoauthor*, which is the sum of each researcher's average number of unique co-authors per year. *SumCoauthor* represents how active a researcher is in collaboration with others, which is defined as

$$sumCoauthor(x,y) = avgCoauthor(x) + avgCoauthor(y)$$

where $avgCoauthor(.)$ is the average number of unique co-author per year.

Research Profile Similarity. It was reported that the keyword overlap from two author's publication history was more effective than topological features [13]. We therefore explored research profile similarity as additional features. To do so, we first built a research profile for every author. Specifically the research profile of an author comprises of three components of all his/her publications: abstracts, the assigned Medical Subject Headings (MeSH) terms, and the citations. We speculate that these components represent the author's research interests: abstract is the summary of an article by the author(s); MeSH terms represent main topics of the article; and out-citing citations (other articles cited by the article) show the relevant background information of the article while in-citing citations (other articles that cite the article in question) represent the recognition of the work by peers.

We used the classical vector space model (TF*IDF weighted) to build the research profile. Assuming that the publication collection of author s by a certain year is D, the TF-IDF for term t in D is calculated by

$$tfidf(t,D) = idf(t,D) * \sum_{d \in D} tf(t,d)$$

where $idf(t, D)$ is the inverse document frequency of term t which is calculated from the entire MEDLINE database and $tf(t,d)$ is the term frequency of term t in collection D. Using the aforementioned formula, we built three vector space models to represent abstracts, in-citing and out-citing citations, respectively. We did not compute the MeSH TF-IDF vector due to our preliminary study from which we found that the TF-IDF representation for MeSH terms did not improve the performance. Instead, we included all unique MeSH terms in the collection to represent an author's MeSH profile.

We then derive learning features from two authors' research profiles. Specifically we define features *simText*, *simOutcite*, and *simIncite* as the cosine similarity of two researchers' abstract profiles, out-citing citation profiles, and in-citing citation profiles, respectively. Concretely,

$$simText(x,y) = \frac{abstract(x) \cap abstract(y)}{|abstract(x)||abstract(y)|}$$

where *abstract(.)* is the TF-IDF term vector of the author's publication history. Similarly

$$simOutcite(x,y) = \frac{outcite(x) \cap outcite(y)}{|outcite(x)||outcite(y)|}$$

and

$$simIncite(x,y) = \frac{incite(x) \cap incite(y)}{|incite(x)||incite(y)|}$$

where *outcite(.)* is the TF-IDF term vector of the author's out-citing citations from the publication history, while *incite(.)* is the TF-IDF term vector of the author's in-citing citations of the publication history. We also define *simMeSH* as the Jaccard coefficient of the two researchers' MeSH profiles. Concretely,

$$simMeSH(x,y) = \frac{|MeSH(x) \cap MeSH(y)|}{|MeSH(x) \cup MeSH(y)|}$$

Where *MeSH(.)* is the MeSH terms of author's publication history.

Collective Productivity. The total number of publications of an author was explored in [13] and was shown effectiveness for predicting author collaborations. Here we use the average number of publications per year to measure productivity and *sumPub* as the sum of two researchers' average publications. Formally it is defined as

$$sumPub(x,y) = avgPub(x) + avgPub(y)$$

where *avgPub(.)* is average publication per year. In addition,

similar to age effect [42] we attempted to increase the weight of recent year productivity and defined an author's *recency* as the sum of the inversed publication time distances to the present. This metrics will weigh-in most recent activity. The *recency* of author x is defined as

$$recency(x) = \sum_{i \in papers(x)} 1/t_i$$

where *papers(x)* is the publications of the author x, and t_i is the distance between publication year of paper i and the *present* year. The *sumRecency* is thus the sum of two researcher's *recency* scores.

Seniority. The relations between two researchers include junior–senior relations (e.g., student and advisor) and collegial relations, which may be important for research collaborations. We define an author's *seniority* as the average number of times that the author has been a senior author, which is approximated by the corresponding author in this study. The *seniority* of author x is defined as

$$seniority(x) = \frac{1}{|papers(x)|} \sum_{i \in papers(x)} I(x)$$

where $I(x)$ is the indicator function; it equals one if x is the corresponding author of the particular publication, and is zero otherwise. For example, assume an author had 5 publications and was the corresponding author on 2 of them, thus the author's seniority is 2/5. The feature *diffSeniority* is thus defined as the seniority difference between two researchers. Table 1 shows the formal definitions of features we explored.

Baselines

The first baseline model, called *PreferentialAttachment*, is based on the Barabasi-Albert scale-free model[16]. As described in the background section, preferential attachment is a well-studied network growth pattern. The more existing links a node has, the higher the chance a new node will link to it. We implemented this baseline based on Liben-Nowell and Kleinberg's description [12]. Specifically $score(x,y) = |\Gamma(x)||\Gamma(y)|$, where $\Gamma(.)$ represents the set of neighboring nodes. For each pair of nodes x and y in a testing set, we computed the corresponding *score(x,y)*. The higher the score, the larger the chance that the two nodes x and y will connect (or collaborate).

We also used *JaccardBaseline*, which describes the importance of the common co-author in the author pair, as the third baseline model as it has demonstrated strong performance in previous research [18]. Its definition is the same as for the feature *coauthorJaccard*.

Data

We used the citation/co-authorship network database CiteGraph [43] as the data source. The database comprises of 1.6 million full-text articles, a joint set of the Elsevier database (1899–2011) and the MEDLINE database. Each article entry includes the title, author(s), abstract, full text, year of publication, and the MeSH terms, as well as the in-cites and out-cites. We disambiguated author names and built a co-authorship network.

Figure 2 shows the collaboration frequency distribution in the CiteGraph dataset. As shown in the figure, an 80.5% majority of researcher pairs collaborate only once, while less than 20% collaborate two or more times. The highest number of collaborations for the same researcher pair is 159, spanning 12 years. The percentage, y, of researcher pairs that collaborate x times follows

Table 1. Feature definition.

Category	Feature	Definition
Connectivity	numCommonCoauthor	$\|\Gamma(x)\cap\Gamma(y)\| \Gamma(.)$ is the set of the researcher's co-authors (neighbors in the network)
	coauthorJaccard	$\dfrac{\|\Gamma(x)\cap\Gamma(y)\|}{\|\Gamma(x)\cup\Gamma(y)\|}$
	Adamic	$\displaystyle\sum_{z\in\Gamma(x)\cap\Gamma(y)}\frac{1}{\log\|\Gamma(z)\|}$ The idea is that two nodes are more similar if they share a lot of neighbors that mainly connect to these two nodes.
	sumClusteringCoef	Sum of both researchers' clustering coefficients
	sumCoauthor	avgCoauthor(x) + avgCoauthor(y) avgCoauthor(.) is the researcher's average number of unique co-authors per year.
Research profile similarity	simText	$\dfrac{abstract(x)\circ abstract(y)}{\|abstract(x)\|\|abstract(y)\|}$ Cosine similarity of two researcher's publication history, measured by abstract TF-IDF term vectors
	simOutcite	$\dfrac{outcite(x)\circ outcite(y)}{\|outcite(x)\|\|outcite(y)\|}$ Cosine similarity of two researchers' out-citing citations' TFIDF term vectors
	simIncite	$\dfrac{incite(x)\circ incite(y)}{\|incite(x)\|\|incite(y)\|}$ Cosine similarity of two researchers' in-citing citations' TF-IDF term vectors
	simMeSH	$\dfrac{\|MeSH(x)\cap MeSH(y)\|}{\|MeSH(x)\cup MeSH(y)\|}$ MeSH(.) is the MeSH term set of the researcher's publication history
Collective productivity	sumPub	$avgPub(x)+avgPub(y)$ avgPub(.) is the researcher's average number of publications per year
	sumRecency	$recency(x)+recency(y)$ and $recency(x)=\displaystyle\sum_{i\in papers(x)}1/t_i$ t_i is the time difference between the publication date of paper i to the present, and Paper is the publications of author x.
Seniority	diffSeniority	$seniority(x)-seniority(y)$ seniority(.) is the average number of times a researchers has been a senior author

the power law distribution log $y = -3.59 * \log x + 0.885$, where x refers to the number of collaborations, with statistical significance ($p<0.05$, t-test) for the linear regression.

Training Dataset

We used CiteGraph for both the training and testing data. We selected equal numbers of positive and negative instances for

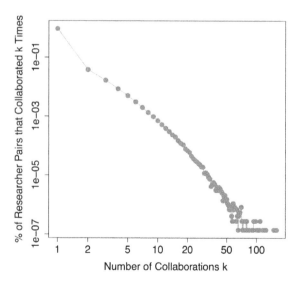

Figure 2. Collaboration frequency distribution for the Cite-Graph dataset. It is a power law distribution log $(y) = -3.59* \log (x)+ 0.885$, where y refers to the percentage of researcher pairs that collaborate x times, with x referring to the number of collaborations.

training and testing as such makes results more comparable to previous work [13]. The positive training instances are author pairs whose first collaborations took place in 2007 or 2008. The negative training instances are author pairs who did not collaborate before 2009. We randomly selected 10,000 positive and 10,000 negative author pairs and extracted each pair's features, and the sampling method is similar with the static graph sampling algorithm proposed in [44]. Since article information, including the abstract, was not available for some authors, we filtered out these pairs, resulting total of 5361 positive instances and 5361 negative instances. The combined group of 10,722 author pairs was used as the training set.

Testing Datasets

We created two sets of testing data. The first set of data, *RandomPairCategory*, was created from a random selection of publications from 2009 and 2010 using the same sampling approach as training set. The positive instances were those in which the author pair first collaborated in 2009 or 2010, while the negative instances were author pairs who never collaborated before 2011. We randomly identified a total of 10,000 positive and 10,000 negative author pairs. Of these, we found that 4726 positive and 4726 negative author pairs had complete features. These 9,452 author pairs were used as the testing set. Note that the selection method of *RandomPairCategory* was utilized for the training data; therefore, the two datasets represent the same distribution.

The second testing dataset, *IndividualAuthorCategory*, was selected based on the collaboration network topology. We randomly selected four authors (target authors) with multiple publications (we set a minimum of 10) in 2009 and 2010. For each author, we built a sub-graph comprising three hops of a breadth-

Table 2. *IndividualAuthorCategory* testing sets.

Author	Number of Publications	Sub-Graph Size	Positives	Negatives Sampled
Jeroen Bax	28	69,487	31	200
Mathew Farrer	10	66,876	13	200
Filippo Marte	59	418	11	200
Christodoulos Stefanadis	30	33,869	16	200

first traversal of the collaboration network established prior to 2011. We thus not only built a sub-graph, but also created the testing set with authors who are close topologically. The positive instances are collaborations established by authors (in the sub-graph) who collaborated with the target author during 2009 and 2010 and the negative instances are those (in the sub-graph) who did not collaborate with the target author before the end of 2010. The statistics of each sub-graph are shown in Table 2. When constructing the testing set for each author, we used all the positive instances and randomly sampled 200 negative instances for each author.

The *IndividualAuthorCategory* evaluation dataset complements the *RandomPairCategory* dataset because the former consists of author pairs who tend to be more similar in research while the latter represents a broader selection of potential collaborators.

We calculate precision (TP/(TP + FP)), recall (TP/(TP + FN)), the receiver operating characteristic—or ROC, the area under the curve of the true positive rate (TPR) over the false positive rate (FPR)—sensitivity (the same as recall), specificity (TN/(FP+TN)), and accuracy ((TP+TN)/ALL), where TP, FP, TN, FN and ALL stand for number of true positives, false positives, true negatives, false negatives, and number of total instances respectively. F1 score is defined as the harmonic mean of recall and precision, specifically 2*recall*precision/(recall+precision). In addition we use log loss[33] to measure the prediction cost of logistic regression model. It is defined as

$$J = \sum_{i=1}^{m} y_t \log y_p + (1 - y_t) \log (1 - y_p)$$

where $y_t \in \{0,1\}$ is class label, and $y_p = P(y_t = 1)$ is the predicted probability of being positive.

Feature and Research Profile Analysis

We analyzed the importance of features using information gain and feature value distributions of true positive (TP), false positive (FP), true negative (TN), and false negative (FN) predictions. We also studied how features' contributions evolve over time, as the authors presumably become more senior.

Results

10-fold Cross Validation on the Training Set

Table 3 shows the 10-fold cross-validation results on the training dataset. The logistic regression and SVM demonstrated the best performance, with a 0.878 ROC and 0.797 F1 for logistic regression and 0.878 ROC and 0.780 F1 for SVM. The naïve Bayes model performs the second best, with an ROC of 0.838. The naïve Bayes multinomial performed the worst among the models. Logistic regression as well as SVM outperformed the naïve

Bayes and naïve Bayes multinomial models with statistical significance (p<0.05, t-test).

Testing Set 1

Table 4 shows the results of models that were trained on the entire training dataset and then tested on the *RandomPairCategory* testing set, which was created by randomly selecting author pairs published during 2009 and 2010. Consistent with the cross-validation results, the logistic regression and SVM outperformed the other models, yielding an ROC of 0.871 and an F1 of 0.789 for logistic regression and 0.871 ROC and 0.769 F1 for SVM. The topology baseline models *PreferentialAttachment* and *JaccardBaseline* yielded ROC values of 0.4583 and 0.278, respectively. All the supervised machine-learning models outperformed the baseline systems. Logistic regression outperformed the naïve Bayes and naïve Bayes multinomial models with statistical significance (p< 0.05, t-test).

Testing Set 2

We evaluated the top-performing supervised machine-learning model, logistic regression, on the *IndividualAuthorCategory* testing set, and the results are shown in Table 5. Our model yielded ROC ranging from 0.766 to 0.980, while the best ROC for the baseline models was 0.634 for the prediction for author Jeroen Bax for the *PreferentialAttachment* model; the *JaccardBaseline* model performed best for predicting collaborators of Mathew Farrer, with an ROC of 0.917. The performance differences between ARCR and the baselines are both statistically significant (p<0.05, t-test).

Inter- vs. Intra-discipline Collaboration

We further examined inter- and intra-disciplinary collaboration predictions separately. Although *simMeSH* can be used as the discipline measure, we assume that the abstract has more detailed information than keywords. We therefore split the training data using different values of *simText* as our threshold in order to approximate inter-discipline and intra-discipline collaboration. The training set and the *RandomPairCateory* testing set were divided into inter-/intra-disciplinary training/testing sets using the threshold. We varied the threshold from *simText* values of 0.01 to 0.30 with 15 evenly distributed data points. For example, when the *simText* threshold was set to 0.01, author-pair instances with *simText* values less than 0.01 were categorized as inter-disciplinary whereas the author-pair instances with *simText* values greater than 0.01 were categorized as intra-disciplinary. For each threshold value, we trained inter-and intra-disciplinary learning models using their respective training sets and then tested them on the corresponding inter-/intra-disciplinary testing sets. Since there was very little data when *simText* is larger than 0.3, we did not explore larger thresholds. Overall the inter-disciplinary collaboration resulted in ROC and F1 ranging from 0.75–0.86 and 0.66–0.77 respectively across different thresholds. The intra-disciplinary

Table 3. 10-fold cross-validation on the training set.

Model	ROC	Precision	Recall	F1	Accuracy
Naïve Bayes	0.838	0.798	0.708	0.684	0.708
Naïve Bayes Multinomial	0.659	0.795	0.655	0.609	0.655
Logistic Regression	**0.878**	**0.803**	**0.797**	**0.796**	**0.797**
SVM	**0.878**	**0.855**	**0.718**	**0.780**	**0.798**
KNN (N = 51)	**0.858**	**0.868**	**0.636**	**0.734**	**0.769**

prediction achieved ROC 0.78–0.87 and F1 0.61–0.77. As shown in Figure 3, when *simText* <0.19, the intra-disciplinary model yielded a better performance according to F1. When *simText* was over 0.19, the inter-disciplinary model outperformed the intra-disciplinary model. The inter-disciplinary collaboration in general under-performed the intra-disciplinary one, suggesting it is more difficult to predict former and there might be other potentially important factors that influence the inter-disciplinary collaboration. Although there is no absolute *simText* value to divide the inter-/intra-diciplinary collaboration, as a reference our preliminary study shows that *simText* 0.1 represent distant topic (research interest) pair such as *diabetes* and *gene regulation*, while more related topic pair *brain* and *Alzheimer* has *simText* 0.3.

Feature Ranking

To identify the features' contributions, we ranked them using information gain [45]. As shown in Table 6, the research interest features *simOutcite* and *simText* are the top-ranked features, both with information gain greater than 0.2. The features *coauthor-Jaccard*, *Adamic*, and *numCommonCoauthor* are the next top ranked, based on the common co-author count. The next features are *simMeSH* and *simIncite*, which also represent research interest. In contrast, the contributions of *sumClusteringCoef* and *sumPub* are considerably smaller and *diffSeniority* shows no contribution.

In order to analyze error patterns across different datasets, we calculated sensitivity and specificity for the logistic regression model evaluation on all the testing sets. As shown in Table 7, the *RandomPairCategory* testing set has lower sensitivity than the *IndividualAuthorCategory* testing set, while the former has higher specificity than the latter.

In our study we found that research interest features are an important feature category (Table 6). In contrast, such features were not studied extensively in other work. We therefore further analyzed their characteristics, such as their relation with author

seniority. As shown in Figure 4, the research profile similarity features *simText*, *simIncite*, and *simOutcite* all increase as author seniority increases. Our results suggested that young researchers are more likely to collaborate with those whose research interests differ while senior or experienced researchers tended to collaborate with those whose research interests are close to them. In addition, the number of author pairs decreases as the authors get more senior, indicating fewer collaborations as the researchers become more senior. We did not show the data in which the authors have over 15 years of research because the data was very sparse.

Discussion

Models

The evaluation results for both the 10-fold cross-validation and the testing data (on the *RandomPairCategory* testing set) show that the logistic regression and SVM models are two best performing supervised machine learning models (logistic regression yielded an F1 score of 0.796 for 10-fold cross-validation and a score of 0.788 for testing while SVM produced 0.780 and 0.769 respectively, as shown in Tables 3 and 4). *IndividualAuthorCategory* is a more challenging evaluation data set as we tried to predict collaborators for individual researchers from the candidates that were collected from their close neighbors in the network. ARCR outperformed all the baselines with statistical significance (Table 5), which further shows that our model has the ability to recommend collaborators for the researcher. KNN model, which learns a non-linear decision boundary, did not perform as well as SVM and Logistic Regression as it only had an F1 score of 0.734 for the cross-validation on the training dataset and 0.726 on the *RandomPairCategory* testing set. This suggests that a linear decision boundary might be preferred.

In contrast, neither the naïve Bayes nor the naïve Bayes multinomial model performed well: an F1 score of 0.684 for 10-

Table 4. *RandomPairCategory* evaluation results.

Model	ROC	Precision	Recall	F1	Accuracy
Naïve Bayes	0.819	0.786	0.694	0.667	0.694
Naïve Bayes Multinomial	0.626	0.790	0.644	0.592	0.644
Logistic Regression	**0.871**	**0.794**	**0.789**	**0.788**	**0.789**
SVM	**0.871**	**0.842**	**0.708**	**0.769**	**0.787**
KNN (n = 51)	0.850	0.854	0.632	0.726	0.762
PreferentialAttachment	0.584	0.574	0.567	0.556	0.567
JaccardBaseline	0.639	0.789	0.639	0.585	0.639

Table 5. *IndividualAuthorCategory* evaluation results.

Author	ARCR ROC	Pref. Attach.* ROC	JaccardBaseline ROC
Jeroen Bax	0.917	0.634	0.620
Mathew Farrer	0.980	0.537	0.917
Filippo Marte	0.800	0.302	0.455
Christodoulos Stefanadis	0.766	0.548	0.313
Macro Average	0.866'	0.505	0.576

*Pref. Attach stands for *PreferentialAttachment*.

fold cross-validation and a score of 0.667 for the *RandomPairCategory* test set with the naïve Bayes model and an F1 score of 0.609 for 10-fold cross validation and 0.592 for the *RandomPairCategory* test set with the naïve Bayes multinomial model. The performance differences between logistic regression and the naïve Bayes and naïve Bayes multinomial models are both statistically significant (p<0.05, t-test). A possible reason for this under-performance is that both models assume conditional independence, which might not hold in our study. For example, *coauthorJaccard* and *numCommonCoauthor* are related, since they both depend on the number of common co-authors.

Feature Analysis

Table 6 shows that the most important feature for collaboration prediction, according to information gain, is the similarity of out-citing citations, which represents an author's knowledge background. True positive instances tend to have a larger *simOutcite* value than negative instances (mean *simOutcite* is 0.305 for *RandomPairCategory* and 0.462 *IndividualAuthorCategory* positive instances while it is 0.159 and 0.264 for negative instances in the two categories respectively), suggesting that common background knowledge increases the chance for collaboration. As for the feature *simOutcite*, collaborating pairs have a higher *simText* score than non-collaborating pairs do. An author's publication history represents the author's research area and *simText* shows the similarity of two researchers' fields. Our results also show that

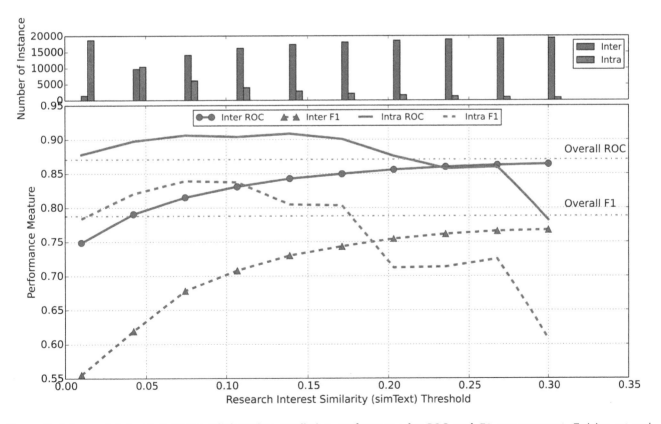

Figure 3. Intra and inter-disciplinary collaboration prediction performance by ROC and F1 measurement. Training set and *RandomPairCateory* test set were divided by the threshold into inter- (<threshold) and intra-disciplinary (>threshold) training/test sets. For each threshold, we trained inter- and intra-disciplinary models and tested them on the corresponding inter-/intra-disciplinary testing sets. The histogram on the top is the number of instances (training+testing) of inter- and intra-disciplinary subset according to the threshold cutoff. The ROC and F1 of overall data are also denoted as the two dotted horizontal lines.

Table 6. Training set feature ranking, by information gain.

Rank	Feature	Information Gain
1	*simOutcite*	0.265
2	*simText*	0.202
3	*coauthorJaccard*	0.173
4	*Adamic*	0.173
5	*numCommonCoauthor*	0.173
6	*simMeSH*	0.145
7	*simIncite*	0.101
8	*sumCoauthor*	0.055
9	*sumRecency*	0.024
10	*sumPub*	0.022
11	*sumClusteringCoef*	0.002
12	*diffSeniority*	0

research field overlap is positively related to potential collaborations.

MeSH terms can be considered the topics of a biomedical article, with the feature *simMeSH* a measure of research interest similarity. Therefore it is not surprising that *simMeSH* contributes to the classification. Keyword overlap was explored in [13] and was a top-ranking feature. In contrast to that study and [14] that did not explore text as features, we found that the feature *simMeSH* ranks below *simText* in information gain. We speculate that although MeSH terms represent an article's semantic content, they are not as robust as the bag of words formulation of *simText* for the task of author collaboration classification, because MeSH terms may not be considered as fine grained as word features in the abstract.

Our results also show that neighborhood structure plays an important role in predicting collaboration. The features *numCommonCoauthor*, *coauthorJaccard*, and *Adamic* all have large information gain. Note that [12] did not find *Adamic* is a useful feature. Positive instances tend to have larger number of common co-authors than negative instances, as for the *IndividualAuthor-Category* testing dataset, where researchers are topologically close (mean *numCommonCoauthor* is 1.0) but negative pairs still tend not to have common collaborators (the mean is closed to 0). Our results suggest that the strength of social ties is important for establishing collaboration. This conclusion is consistent with our hypothesis and previous findings, which show that common neighbors are a very effective predictor in social networks [12,13].

Features that are related to researcher activity level, such as *sumCoauthor*, *sumRecency*, and *sumPub*, are ranked lower than *simOutcite*, *simText*, *coauthorJaccard*, *Adamic*, *numCommonCoauthor*, *simMeSH*, and *simIncite*, as measured by information gain, suggesting that two researchers' specific activities do not have to be closely related to establish a new collaboration. In contrast, the sum of co-authors was found to be among top features in [13], but it is not clear if this was influenced by the normalization by year, as carried out in our study. Consistent with previous findings, the clustering coefficient, which describes the transitivity of a collaboration, is not an effective feature [13]. It is also interesting to note that difference in seniority between collaborators, described by *diffSeniority*, has no impact on establishing a new collaboration in our approach.

Furthermore, we trained classifiers using every single feature individually and analyzed the performance as shown in Figure 5. 1) Research interest features *simText*, *simMesh* and *simOutcite* (Figure 5 panels *a*, *b* and *d*) have large ROC areas, showing that they are informative for the classification. *simIncite* (Figure 5*c*) however is not as large as other features in this category with 0.56 ROC only. 2) Common co-author based features (Figure 5 panels *f*, *k* and *l*) exhibit distinct patterns and are essentially equivalent features as they have large correlation coefficients among each other. For example *Adamic* and *numCommonCoauthor* have a correlation coefficient of 0.96. This suggests that we can use *numCommonCoauthor* as a feature and remove *Adamic*. The ROC curve for them is a straight line due to the fact that most of the author pairs (especially negative instances) don't have any

Table 7. Sensitivity and specificity for all testing sets by logistic regression model.

Testing Set		Sensitivity	Specificity
RandomPairCategory		0.718	0.859
IndividualAuthor Category	Jeroen Bax	0.968	0.345
	Mathew Farrer	1.000	0.125
	Filippo Marte	0.818	0.460
	Christodoulos Stefanadis	0.813	0.352
	Macro Average	0.900	0.321

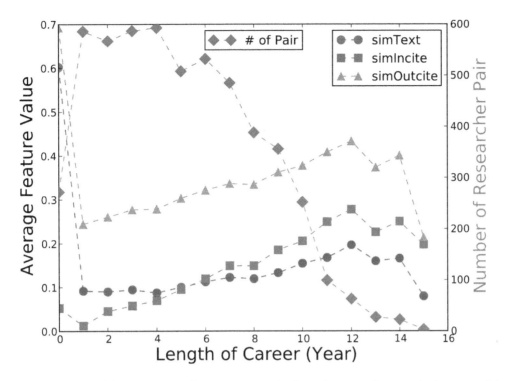

Figure 4. Research interest similarities over researcher career span. In the early stages of a researcher's career, collaborators with less research similarity are found but collaboration between two experienced researchers shows greater research interest similarity.

common coauthors, and only 1/3 of the positive instances have non-zero common co-authors. 3) Other features such as *sumCoauthor* (Figure 5*f*) is also effective for classification with 0.67 ROC. Activity feature *sumRecency* (Figure 5*i*) has 0.60 ROC, and so does *sumPub* (Figure 5*e*). *sumClusteringCoef* and *diffSeniority* (Figure 5 panels *h*, *j*) show only 0.53 and 0.54 ROC respectively The individual ROC is consistent with information gain analysis, which also shows that the research interest features are most informative, followed by common neighbor based features.

There are inconsistency between the single feature logistic regression and information gain, and it is due to the fact that these two ranking mechanisms address feature contribution from slightly different perspectives. Information gain is the entropy difference of before and after splitting the data set by a specific value of this particular feature, and the entropy itself measures the level of impurity of the dataset. Single feature logistic regression, on the other hand, is essentially fitting the data by the particular feature. *simMeSH and sumCoauthor* is ranked lower than connectivity features (*numCommonCoauthor, coauthorJaccard* and *Adamic*) by information gain but higher than them by single feature logistic regression. The reason is that the above connectivity features have skewed distribution (almost all the negative instances don't have any common coauthors, and only 1/3 of the positive instances have non-zero common co-authors). Therefore it is easier to split the data set into two by value zero to yield high information gain. On the other hand *simMeSH* and *sumCoauthor* better fit the overall data due to their less skewed distributions.

In summary, previous work in author collaboration prediction mainly explored topological features. Our results, in contrast, show that in addition to topological features, semantic features are important. For example, we found that research interest is important for establishing a new collaboration. Specifically, research profile similarity features such as *simOutcite* and *simText*,

as shown in Table 6, are the most important features–surpassing any of the topological features–for the classification. Tables 4 and 5 show that the supervised machine learning models that incorporate research similarity features significantly outperformed the baseline systems, which were built upon widely used topological features (*PreferentialAttachment, JaccardBaseline*). Possible interpretation is that knowing the other's work is a form of shared experience and the foundation of trust between two researchers. Their common knowledge, represented by the research similarity features, plays an important role for building collaboration.

As discussed earlier, although seniority plays a limited role in a collaboration, Figure 4 shows that when in their early career stage, researchers are more likely to collaborate with those whose research interests differ from theirs, suggesting that junior faculty are more open to collaborations. In contrast, collaborations between two senior researchers exhibit a higher degree of research interest similarity, suggesting that established researchers are more comfortable in their own fields and are less likely to initiate collaborations.

Error Analysis

In order to determine if our data size or features are sufficient, we analyzed the learning curve for the logistic regression model. The training set was split into two sets: training (66% of total instances) and validation set, and log loss is used for the error metric for the curve. As shown in Figure 6 the training error and validation error converges by the time the dataset reaches the size of 1000 author pairs. Therefore the training set size (5361 positive instances and 5361 negative instances) is sufficient for the task.

We also manually analyzed the prediction errors. We found that authors' publication history is important as many of our semantic features are derived from authors' research profile. If an author has few publications and few co-authors in the past, there is little

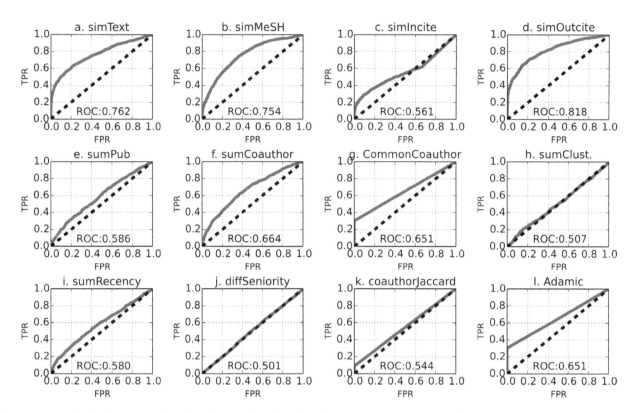

Figure 5. ROC for logistic regression classifiers trained by single feature. 1) Research interest features *simText*, *simMesh* and *simOutcite* (panels a, b and d) have large ROC areas, showing that they are informative for the classification. *simIncite* (panel c) however is not as large as other features in this category with 0.56 ROC only. 2) The ROC curves for common co-author based features (panels f, k and l) are a straight lines due to the fact that most of the author pairs (especially negative instances) don't have any common coauthors, and only 1/3 of the positive instances have non-zero common co-authors. 3) Other features such as *sumCoauthor* (panel f) is also effective for classification with 0.67 ROC. Activity feature *sumRecency* (panel i) has 0.60 ROC, and so does *sumPub* (panel e). *sumClusteringCoef* and *diffSeniority* (panels h, j) show only 0.53 and 0.54 ROC respectively. The individual ROC is consistent with information gain analysis, which also shows that the research interest features are most informative, followed by common neighbor based features.

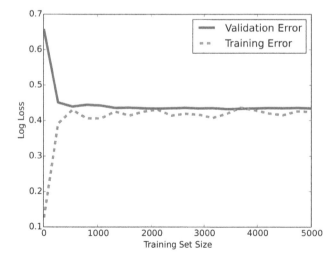

Figure 6. Learning curve for logistic regression with log loss metric. The training error and validation error converges by the time the dataset reaches the size of 1000 author pairs. Therefore the training set size (5361 positive instances and 5361 negative instances) is sufficient for the task.

information we can derive for features such as research interest, network topology, and productivity (or activity) level and therefore will not be able to predict accurately his/her future collaborators.

We found that the data incompleteness is one of the most important reasons for false negatives. The network that we used in this study is a sub-graph of MEDLINE publications only and therefore provides an incomplete picture of the publication history of certain authors. For example, Flaumenhaft R (author of PMID 12837380) has only one publication prior to 2009 with only one co-author. His/her pairing with Laurence RG (author of PMID 18715793) has a *simText* value of 0.010 and a *simOutcite* value of 0.143 (the average value for each feature in the positive training data was 0.134 and 0.325, respectively), although Laurence RG is more prolific in our network with 12 publications and 35 co-authors. In fact, by searching the larger database the MEDLINE we found that Flaumenhaft R has been publishing almost every year from 2003 to present and has many common co-authors with Laurence RG; this information was missing entirely in our network, which was built using the joint MEDLINE and Elsevier data only. As a result, our models predicted Flaumenhaft R was unlikely to collaborate with Laurence RG, which is therefore a false negative. On the other hand, false positive errors can arise due to the fact that these author pairs have features very much like those in the positive training data. These authors, however, might never have had a chance to actually know each other, leading to a false positive.

We also found that the noises in author name disambiguation contribute errors for both false positives and false negatives. Data sparseness arises when one author is mapped to two unique IDs by the author name disambiguation database we used. For example Guida M (author of PMID 17113552) has two IDs. There are only five publications assigned to the ID that we happened to use in our network, while there are 94 publications under the other ID. We are also aware that it is possible for an author to share the same ID with another, unrelated author; this can also cause a disambiguation error and the information from the unrelated author will be wrongly attributed to the original author. However, we did not actually find any such cases in our test sets.

Recall we have built two different testing data sets, and our analyses of true positive, false positive, true negative and false negative of the three testing data sets show interesting results (Table 7). In the *RandomPairCategory* dataset (i.e, positive and negative author pair data were randomly selected) our classification has low sensitivity (0.718) and high specificity (0.859) while *IndividualAuthorCategory* yielded the opposite (0.900 sensitivity and 0.321 specificity). The high specificity of *RandomPairCategory* is due to the fact that the negative instances are "very negative" as they were constructed by the random combination of two authors; therefore, they tend to share few research interests and even fewer common friends. In contrast, the negative instances of *IndividualAuthorCategory* testing set were from the sub-graph of the author, so they do have similar research interests and have a higher chance of sharing a common collaborator. The sensitivity advantage of *IndividualAuthorCategory* can be understood in a similar way, as the positive instances, which were sampled from the sub-graph of the author, are "very positive" and share research interests and common collaborators, which increases the likelihood of the classifiers to classify them as positive.

Limitations

There are several limitations to this study. First, we did not explore learning features of broad social factors, including institutional policies like the status of an IRB application or institution-specific restrictions, because it is difficult to obtain these data. Second, our data are incomplete and contain missing information. We used a sub-graph of the MEDLINE co-author network and therefore the author publication histories may not be complete, as we described in the error analysis. Missing publications indicates missing the important research interest information. It also takes time for an article to accumulate citations, so *simIncite* may be biased to have more citations for older works than for recent ones. Finally, our training and testing period time cutoff is ad-hoc, and we define a negative instance pair as authors who did not collaborate by the time of the training or testing period, which might not be true in reality for every pairs.

Future Work

We identify the following directions for future research. First, we would like to incorporate all the available MEDLINE records to minimize the challenges of missing data. Secondly, we would like to explore additional learning features including the funding status, the collaboration strength, and the impact of an article. Thirdly, it is important to analyze in depth the research collaboration network and its topological characteristics. For example, interdisciplinary collaboration may involve a sub-graph (inter-group collaboration) that may exhibit different characteristics from the overall graph. Finally we may explore other machine learning models, including collaborative filtering.

Conclusions

In this study we applied and evaluated four established supervised machine-learning models, namely naïve Bayes, naïve Bayes multinomial, SVMs, and logistic regression, and explored rich learning features for automatic research collaboration prediction. We found supervised machine learning models can predict research collaboration with a high performance with an ROC ranging from 0.766 to 0.980 on different datasets, and logistic regression and SVMs performed the best. In addition, we identified three key factors for establishing new collaboration: research interest, common collaborators, and research productivity. Our research is important as it not only produces an important tool for automatic author collaboration prediction, but also contributes significantly to the science of evolving network modeling.

Acknowledgments

The authors would like to thank Ricky J. Sethi for the comments and Rebecca Kinney for the proofreading.

Author Contributions

Conceived and designed the experiments: QZ. Performed the experiments: QZ. Analyzed the data: QZ. Contributed reagents/materials/analysis tools: QZ. Wrote the paper: QZ HY.

References

1. Newman ME (2004) Coauthorship networks and patterns of scientific collaboration. Proc Natl Acad Sci U S A 101: 5200.

2. Luo J, Flynn JM, Solnick RE, Ecklund EH, Matthews KRW (2011) International stem cell collaboration: how disparate policies between the United States and the United Kingdom impact research. PLoS ONE 6: e17684. doi:10.1371/journal.pone.0017684.

3. Jones BF, Wuchty S, Uzzi B (2008) Multi-university research teams: shifting impact, geography, and stratification in science. Science 322: 1259–1262.

4. Okeke IN, Wain J (2008) Post-genomic challenges for collaborative research in infectious diseases. Nat Rev Micro 6: 858–864. doi:10.1038/nrmicro1989.

5. Gray B (2008) Enhancing transdisciplinary research through collaborative leadership. Am J Prev Med 35: S124–S132.

6. Ashburner M, Ball CA, Blake JA, Botstein D, Butler H, et al. (2000) Gene Ontology: tool for the unification of biology. Nat Genet 25: 25–29.

7. Harvard Catalyst. Available: http://catalyst.harvard.edu/. Accessed 27 November 2012.

8. Vardell E, Feddern-Bekcan T, Moore M (2011) SciVal Experts: a collaborative tool. Med Ref Serv Q 30: 283–294.

9. COS Pivot: the next generation of funding and research expertise—connected. Available: http://www.refworks-cos.com/pivot/. Accessed 27 November 2012.

10. The VIVO Project. Available: http://vivoweb.org. Accessed 3 February 2012.

11. Elsevier Inc. BiomedExperts – Your scientific professional network. Available: http://www.biomedexperts.com. Accessed 3 February 2012.

12. Liben-Nowell D, Kleinberg J (2003) The link prediction problem for social networks. Proceedings of the Twelfth International Conference on Information and Knowledge Management. CIKM '03. New York, NY, USA: ACM. pp. 556–559.

13. Al Hasan M, Chaoji V, Salem S, Zaki M (2006) Link prediction using supervised learning. SDM'06: Workshop on Link Analysis, Counter-terrorism and Security.

14. Backstrom L, Leskovec J (2011) Supervised random walks: predicting and recommending links in social networks. Proceedings of the fourth ACM international conference on Web search and data mining. pp. 635–644.

15. Erdös P, Rényi A (1959) On random graphs I. Publ Math Debr 6: 290–297.

16. Barabási AL, Albert R (1999) Emergence of scaling in random networks. science 286: 509–512.

17. Newman MEJ (2001) The structure of scientific collaboration networks. Proc Natl Acad Sci 98: 404–409.

18. Adamic LA, Huberman BA (2000) Power-law distribution of the world wide web. Science 287: 2115–2115.

19. Barabási A-L, Oltvai ZN (2004) Network biology: understanding the cell's functional organization. Nat Rev Genet 5: 101–113.

20. Newman MEJ (2009) The first-mover advantage in scientific publication. EPL Europhys Lett 86: 68001.
21. Bianconi G, Barabási A-L (2001) Bose-Einstein condensation in complex networks. Phys Rev Lett 86: 5632.
22. Watts DJ, Strogatz SH (1998) Collective dynamics of "small-world"networks. nature 393: 440–442.
23. Kleinberg JM, others (2000) Navigation in a small world. Nature 406: 845–845.
24. Newman MEJ (2003) Mixing patterns in networks. Phys Rev E 67: 026126.
25. Sun Y, Barber R, Gupta M, Aggarwal CC, Han J (2011) Co-author relationship prediction in heterogeneous bibliographic networks. Advances in Social Networks Analysis and Mining (ASONAM), 2011 International Conference on. pp. 121–128.
26. Wang C, Satuluri V, Parthasarathy S (2007) Local probabilistic models for link prediction. Data Mining, 2007. ICDM 2007. Seventh IEEE International Conference on. pp. 322–331.
27. Newman ME (2001) Clustering and preferential attachment in growing networks. Phys Rev E 64: 025102.
28. Huang J, Zhuang Z, Li J, Giles CL (2008) Collaboration over time: characterizing and modeling network evolution. Proceedings of the international conference on Web search and web data mining. pp. 107–116.
29. Ding Y (2011) Scientific collaboration and endorsement: network analysis of coauthorship and citation networks. J Informetr 5: 187–203. doi:10.1016/j.joi.2010.10.008.
30. Lee K, Brownstein JS, Mills RG, Kohane IS (2010) Does collocation inform the impact of collaboration? PloS One 5: e14279.
31. Pan RK, Kaski K, Fortunato S (2012) World citation and collaboration networks: uncovering the role of geography in science. Sci Rep 2.
32. Choi BC, Pak AW (2007) Multidisciplinarity, interdisciplinarity, and transdisciplinarity in health research, services, education and policy: 2. Promotors, barriers, and strategies of enhancement. Clin Invest Med 30: E224–E232.
33. Hastie T, Tibshirani R, Friedman JJH (2001) The elements of statistical learning. Springer New York.
34. McCallum A, Nigam K (1998) A comparison of event models for naive bayes text classification. AAAI-98 Workshop on Learning for Text Categorization.
35. Manning CD, Raghavan P, Schütze H (2008) Introduction to information retrieval. Cambridge University Press Cambridge. p319.
36. Caruana R, Niculescu-Mizil A (2006) An empirical comparison of supervised learning algorithms. Proceedings of the 23rd international conference on Machine learning. pp. 161–168.
37. Huang J, Lu J, Ling CX (2003) Comparing naive Bayes, decision trees, and SVM with AUC and accuracy. Data Mining, 2003. ICDM 2003. Third IEEE International Conference on. pp. 553–556.
38. Weka3: Data Mining Software In Java. Available: http://www.cs.waikato.ac.nz/ml/weka/. Accessed March 2013.
39. LIBSVM. Available: http://www.csie.ntu.edu.tw/~cjlin/libsvm/. Accessed November 2013.
40. Scikit. Available: http://scikit-learn.org/stable/. Accessed November 2013.
41. Adamic LA, Adar E (2003) Friends and neighbors on the web. Soc Netw 25: 211–230.
42. Dorogovtsev SN, Mendes JFF (2001) Effect of the accelerating growth of communications networks on their structure. Phys Rev E 63: 025101.
43. Zhang Q, Yu H (2013) CiteGraph: A Citation Network System for MEDLINE Articles and Analysis. Studies in Health Technology and Informatics. Vol. 192. pp. 832–836. doi:10.3233/978-1-61499-289-9-832.
44. Ahmed NK, Neville J, Kompella R (2012) Network sampling: from static to streaming graphs. ArXiv Prepr ArXiv12113412.
45. Yang Y, Pedersen JO (1997) A comparative study on feature selection in text categorization. ICML. Vol. 97. pp. 412–420.

11

Identification of the PLK2-Dependent Phosphopeptidome by Quantitative Proteomics

Cinzia Franchin[1,2], **Luca Cesaro**[1], **Lorenzo A. Pinna**[1,3], **Giorgio Arrigoni**[1,2]*, **Mauro Salvi**[1]*

1 Department of Biomedical Sciences, University of Padova, Padova, Italy, **2** Proteomics Center of Padova University, Padova, Italy, **3** CNR Institute of Neurosciences, Padova, Italy

Abstract

Polo-like kinase 2 (PLK2) has been recently recognized as the major enzyme responsible for phosphorylation of α-synuclein at S129 *in vitro* and *in vivo*, suggesting that this kinase may play a key role in the pathogenesis of Parkinson's disease and other synucleinopathies. Moreover PLK2 seems to be implicated in cell division, oncogenesis, and synaptic regulation of the brain. However little is known about the phosphoproteome generated by PLK2 and, consequently the overall impact of PLK2 on cellular signaling. To fill this gap we exploited an approach based on *in vitro* kinase assay and quantitative phosphoproteomics. A proteome-derived peptide library obtained by digestion of undifferentiated human neuroblastoma cell line was exhaustively dephosphorylated by lambda phosphatase followed by incubation with or without PLK2 recombinant kinase. Stable isotope labeling based quantitative phosphoproteomics was applied to identify the phosphosites generated by PLK2. A total of 98 unique PLK2-dependent phosphosites from 89 proteins were identified by LC-MS/MS. Analysis of the primary structure of the identified phosphosites allowed the detailed definition of the kinase specificity and the compilation of a list of potential PLK2 targets among those retrieved in PhosphositePlus, a curated database of in cell/*vivo* phosphorylation sites.

Editor: Kyung S. Lee, National Cancer Institute, NIH, United States of America

Funding: This work was supported by Associazione Italiana per la Ricerca sul Cancro, AIRC (grant number IG10312) (to LAP). The funders had no role in study design, data collection and analysis, decision to publish, or preparation of the manuscript.

* Email: giorgio.arrigoni@unipd.it (GA); mauro.salvi@unipd.it (MS)

Introduction

The Polo like-kinase 2 (PLK2) is a serine/threonine kinase belonging to the POLO like kinase family playing a role in cell cycle progression, mitosis, cytokinesis, and DNA damage response. In mammals, five members of this family have been described: the best characterized PLK1, the closely related PLK3 and PLK2, a distant member PLK4, and PLK5, a protein that lacks the kinase domain in humans. The members of this family share the same domain topology, consisting of a conserved N-terminal kinase domain and one or two POLO box domains at the C-terminus [1,2,3]. PLK2 was initially named Serum inducible kinase (Snk) having been classified as an early response gene as its expression increases following stimulation by growth factors. PLK2 is involved in cell cycle regulation, is required for centriole duplication in mammalian cells [4], regulates mitotic spindle in the mammary gland [5], and is a direct transcriptional target of p53 activating G2-M checkpoint, which prevents mitotic catastrophe following spindle damage [6].

While PLK1 has been pre-clinically validated as a cancer target and is generally overexpressed in different forms of human tumors [7], PLK2 has been initially described as a tumor suppressor gene [3]. However recent works disclose a more complex scenario where also PLK2 inhibition has been suggested as a promising therapeutic strategy against some type of tumors. In this regard PLK2 can bind and phosphorylate the mutant p53, inducing an oncogenic feedback loop in cancer cells [8], or may promote Mcl-1 stabilization, thus providing resistance to cell death induced by TRAIL in Cholangiocarcinoma [9].

Moreover, PLK2 is required for the regulation of the homeostatic synaptic plasticity in the brain: PLK2 acts on Ras and Rap signaling by phosphorylating four Ras and Rap regulators [10]. Recently PLK2 took the center of the stage after being identified as the major kinase responsible for the phosphorylation of Ser-129 of α-synuclein both *in vitro* and *in vivo* [11,12,13,14]. α-Synuclein is constitutively phosphorylated at low levels in normal brain and an accumulation of α-synuclein pS129 in Lewy bodies is observed in Parkinson disease and other synucleinopathies. Although the pathophysiology of the Ser-129 phosphorylation in Parkinson's disease is not completely understood and it has not been clarified whether this phosphorylation is protective or harmful for neurons, PLK2 is considered a very promising target for Parkinson disease treatment [15,16,17].

Despite the fact that the involvement of PLK2 in different biological processes is emerging, the precise functions of this kinase remain elusive as, with few exceptions, its main cellular targets are unknown. Indeed, the PLK2 substrates identified so far are just a dozen or so and the phosphoresidues are often not characterized.

We have here exploited a strategy based on *in vitro* kinase phosphorylation of proteome-derived peptide libraries combined with a mass spectrometry-based quantitative proteomic approach

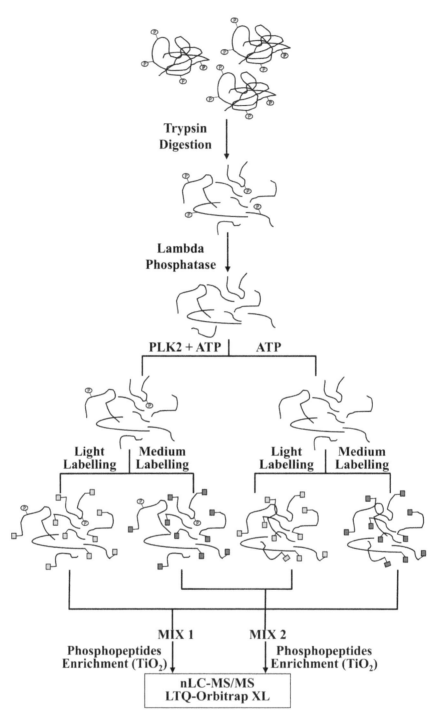

Figure 1. Workflow for PLK2 peptide substrate identification.

to identify the PLK2-dependent phosphopeptidome. A similar approach was successfully applied by Zou's group to identify putative substrates of the protein kinase CK2 [18]. Our analysis allowed for the detailed definition of the PLK2 kinase specificity and the compilation of a list of its potential targets to gain a deeper understanding of the involvement of this kinase in signal transduction pathways.

Materials and Methods

Materials

Recombinant human Dopa decarboxylase, Annexin A2 and Prostaglandin E Synthase 3 were purchased from ProSpec (Tany TechnoGene Ltd.). All chemicals and solvents were of MS-grade.

c-DNA constructs and production of recombinant proteins

Plasmids encoding human GST-HDGF [19] and human PLK2-PGEX4TI [20] were previously described. GST-PLK2 T210D constitutively active mutant and GST-HDGF T225A were produced by PCR site-directed mutagenesis and mutations were confirmed by sequencing analysis.

Recombinant GST-HDGF, GST-CK2, and GST-PLK2 T210 D, have been expressed in *E. coli* BL-21 pLysS and purified as described in [19] and [20], respectively.

Cell culture

Human neuroblastoma SK-N-BE cells [21] were maintained in 5% CO_2 in DMEM supplemented with 10% FBS, 2 mM l-glutamine, 100 U/ml penicillin and 100 mM streptomycin, in an atmosphere containing 5% CO_2.

Cell lysate dephosphorylation and in vitro assay

Undifferentiated cells were detached, centrifuged, extensively washed with PBS and lysed by the addition of ice-cold buffer containing 8 M urea in 25 mM Hepes (pH 8.0), protease inhibitor cocktail Complete (Roche) and ultrasonicated in an ice-bath. After 40 min, the lysate was centrifuged 15 min at $10000 \times g$ at 4°C. The supernatant was collected and protein concentration was measured by BCA method.

Extracted proteins (2 mg) were reduced with 20 mM dithiothreitol for 1 h at 56°C and alkylated with 40 mM iodoacetamide for 45 min at room temperature in the dark. The sample was diluted 8 times with 25 mM Hepes pH 8.0 to reach a concentration of urea compatible with trypsin activity. Sequencing grade modified trypsin (45 µg) (Promega) was added to the sample and the protein mixture was digested at 37°C overnight.

Tryptic peptides were acidified with formic acid and desalted on SepPak Vac 1cc C18 Cartridges (Waters) following the manufacturer's instructions. Eluted peptides were dried under vacuum and then dissolved in 0.5 mL of dephosphorylation reaction buffer containing 50 mM Hepes pH 7.5, 2 mM $MnCl_2$, 0.1 mM EGTA, 5 mM DTT and 0.01% BRIJ35. Dephosphorylation of peptides was carried out by adding 2000 U of lambda phosphatase (Santa Crutz). After 7 h at 37°C, other 2000 U of lambda phosphatase was added. This second dephosphorylation reaction was carried out overnight at 37°C. Finally the solution was heated at 95°C for 15 min to inactivate the phosphatase and subjected to *in vitro* phosphorylation. PLK2 phosphorylation conditions are described in [22]. Briefly, the sample was divided into two identical aliquots of 250 µl and each of them was diluted to 500 µL with a solution 2× containing 20 mM $MgCl_2$, 10 mM DTT, and 200 µM ATP. One of the aliquots was supplemented with PLK2-GST T210D (1 µg) and both aliquots were incubated for 2h at 30°C. After incubation the samples were frozen and dried.

Dimethyl labeling and phosphopeptides enrichment

Samples were labeled according to the dimethyl labeling method described in [23] and following the scheme reported in Figure 1. 400 µg of each peptide solution (control sample and PLK2 phosphorylated sample) was diluted to 500 µl of 5% formic acid. Each sample was then divided into two identical aliquots of 250 µl to perform a "forward" and a "reverse" experiment. Two isotopic forms of formaldehyde were used: the "light" form (CH_2O) and the "medium" form (CD_2O). Labeling was performed on-column using SepPak Vac 1cc C18 Cartridges, as described in [23]. Samples were mixed in a 1:1 ratio as described in Figure 1 and dried under vacuum.

Peptides from each of the two samples were dissolved in 100 µl of 80% acetonitrile, 6% of trifluoroacetic acid and phosphopeptides enrichment was performed using home-made micro columns packed with 400 µg of TiO_2 (Titansphere) as described in [20]. Eluted peptides were acidified with formic acid, dried under vacuum, and samples were finally dissolved in 45 µl of 3% acetonitrile 0.1% formic acid just prior to LC-MS/MS analysis.

Mass Spectrometry analysis

Mass spectrometry analyses were performed on an LTQ-Orbitrap XL mass spectrometer (Thermo Fisher Scientific) coupled with an on-line nano-HPLC Ultimate 3000 (Dionex – Thermo Fisher Scientific). Peptides were loaded onto a Trap column (300 mm I.D., 300 Å, C18, 3 mm; SGE Analytical Science) using a flow rate of 8 µL/min of 0.1% formic acid (solvent A), transferred into a homemade pico-frit column packed with C18 material (Aeris Peptide 3.6 µm XB-C18, Phenomenex), and separated using a linear gradient of acetonitrile/0.1% formic acid (solvent B) from 3% to 50% in 90 minutes at a flow rate of 250 µL/min. Ion source capillary temperature was set at 200°C, and spray voltage at 1.5 kV. To increase the number of identified phosphopeptides, each sample was analyzed three times with the same chromatographic conditions but using different fragmentation methods as described in [24].

Data analysis

For each of the two final samples, MS/MS data derived from the different analyses were analyzed with a MudPIT protocol using Proteome Discoverer 1.4 software (Thermo Fisher Scientific) interfaced to a Mascot server (version 2.2.4, Matrix Science, London, UK). Searches were performed against the Uniprot Human protein database (version 2014.01.22, 88479 sequences). Enzyme specificity was set to trypsin and a maximum of two missed cleavages were allowed. The precursor and fragment mass tolerances were set to 10 ppm and 0.6 Da respectively. Light-marked dimethylation (+28.0313 Da) and medium-marked dimethylation (+32.0564 Da) were selected as variable modifications at N-terminus and lysine residues. Phosphorylation of serine, threonine, and tyrosine were also inserted as variable modifications, while carbamidomethylation of cysteines was set as static modification. The search was done also against a randomized database and the confidence level of all the identified peptides was assessed using the Percolator algorithm, and only peptides with a q-value <0.05 were considered as correctly identified. For quantification, all data were reported as "PLK2-treated" over control, with a maximum ratio of 100.

In vitro phosphorylation

In vitro PLK2 phosphorylation assays were performed as described in [22]. Briefly, recombinant proteins were incubated at the indicated concentrations in a radioactive mixture consisting in 50 mM Tris (pH 7.5), 100 µM ATP ([γ-^{33}P]ATP ~ 2000 cpm/pmol), 10 mM $MgCl_2$, and 5 mM DTT, in absence (control) or with GST-PLK2 T210D (20 ng) at 37°C for 10 min. For CK2 *in vitro* phosphorylation assay, protein substrate was incubated in the same radioactive mixture, without DTT and in presence of the GST-CK2 kinase (20 ng). The reaction was stopped with the addition of 2× Laemmli sample buffer and samples were subjected to SDS-PAGE. Gels were stained with colloidal coomassie, dried, exposed overnight to a multipurpose storage phosphor screen, and analyzed using a Cyclone storage phosphor system (Packard).

A

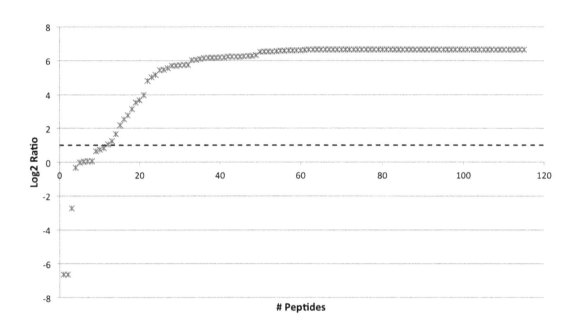

B

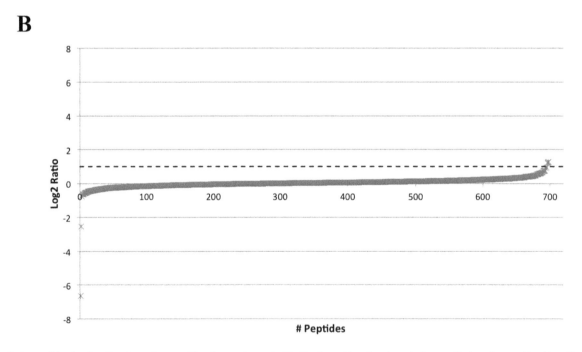

Figure 2. Logarithmic distribution of quantification values. A. Distribution of Log2 ratios relative to all phosphopeptides identified in this study. B. Distribution of Log2 ratios relative to all non-phosphopeptides.

Two-sample logo analysis and molecular dynamics simulations

Sequence motif analysis was performed with a Two-Sample logo tool (t-test) [25] using up to a +7,−7 residue window around each modified phospho- Ser/Thr identified. These data were compared with the +7, −7 residue window surrounding Ser/Thr residues randomly extracted from the human proteome obtained from the Swiss Prot database using a homemade script and unix

text processing commands. Non-redundant sequences have been randomized using unix command shuff.

Bona fide CK2, PLK1, and CK1δ substrates (+7, −7 residue window) were collected from PhosphositePlus database [26] and analysed using Two-Sample Logo vs random Ser/Thr peptides as described above.

Molecular dynamics (MD) simulations of peptide, PLK2, and ATP inserted manually in the active site, was studied using

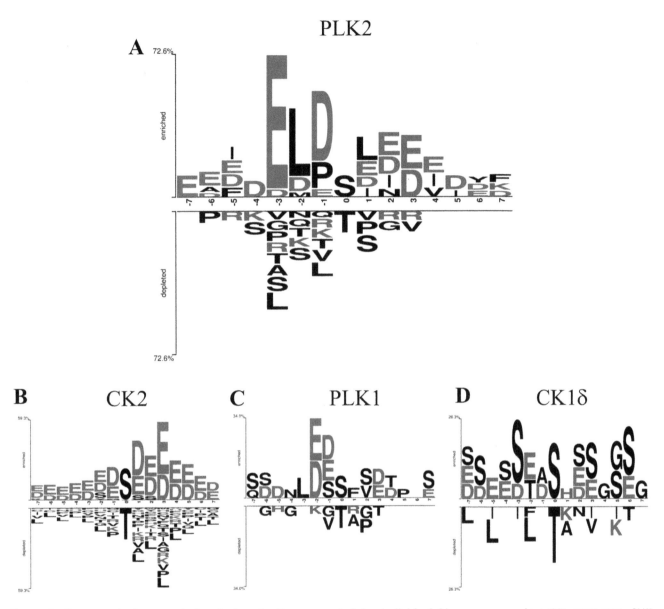

Figure 3. Two-sample logo analysis of phosphosites generated by individual kinases vs. random S/T proteome. PLK2 phosphopeptides identified in this paper (A) or *bona fide* CK2 (B), PLK1 (C), and CK1δ (D) substrates collected from PhosphositePlus database, have been analyzed as described in Materials and Methods.

Desmond-Maestro. MD simulations of the minimized complexes (parameterized with OPLS 2005) were performed in order to verify their stability over time; in particular a 70 ns of NPT (1 atm, 300 K) MD simulation was performed.

Results and Discussion

Identification of the PLK2 phosphopeptidome

The workflow utilized for the identification of PLK2 peptide substrates is shown in Figure 1. We have generated a peptide library from undifferentiated human neuronal SK-NB-E cells that has been subjected to extensive dephosphorylation by lambda phosphatase. After phosphatase inactivation, the sample has been divided in two equal aliquots. One was incubated with recombinant PLK2 and the other was incubated in the same buffer but without the kinase, as detailed in the methods section. After the reaction, each of the two samples was further split in two identical aliquots. Each aliquot was then separately labeled with the dimethyl labeling reagents, combined (as schematized in Figure 1), subjected to TiO₂ phosphopeptides enrichment, and finally analyzed by LC-MS/MS. With this approach, we performed a "forward" experiment where the light-labeled sample incubated with PLK2 was mixed with the not phosphorylated medium-labeled sample, and a "reverse" experiment where the medium-labeled sample incubated with PLK2 was mixed with the not phosphorylated light-labeled sample. The stable isotope-based quantification was used to differentiate phosphosites generated by PLK2 from background phosphorylation that could be still present due to an incomplete dephosphorylation reaction. Moreover, for each of the experiments ("forward" and "reverse") we performed 3 technical replicates, by analyzing the same samples with 3 different fragmentation methods. With this approach we have identified in total 98 unique, PLK2-dependent phosphosites from 89 proteins (Table S1, supplementary material). These phospho-

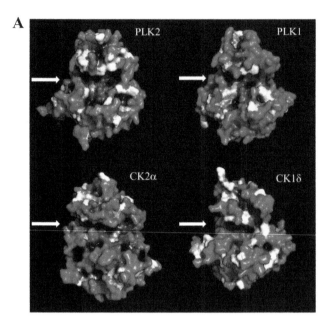

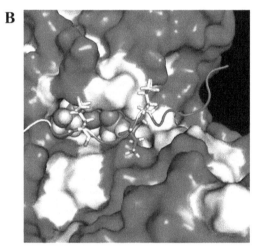

Figure 4. *In silico* **analysis of substrate binding zone of PLK2.** A. Hydrophobic surface calculation of acidophilic kinases PLK2, PLK1, CK2α, CK1δ. In yellow the hydrophobic areas. Kinase active sites have been indicated by an arrow. B. Interaction between PLK2 and the phosphopeptide EAIAELDtLNEESYK (P31946). −3 and +1 leucine residues are shown in yellow, threonine in blue. ATP is shown in spheres.

peptides were divided in two categories: the first comprises all phosphopeptides quantified both in the "forward" and in the "reverse" experiment. The reported PLK2-treated/control ratios were calculated as the average value obtained from the technical replicates of each experiment (class 1 phosphopeptides). The second category comprises phosphopeptides that were identified in only one of the experiments (class 2 phosphopeptides) and whose quantification was calculated as the average value obtained from the technical replicates, either in the "forward" or in the "reverse" experiment. All data regarding peptide identifications (protein accession number, peptide sequence, modifications, quantification values, Mascot scores, PEP values, q-values, chromatographic- and MS-relevant information) are reported in Tables S2 and S3, supplementary material.

Figure 2 shows the logarithmic distribution of dimethyl label ratios for phosphorylated and non-phosphorylated peptides. In particular, panel A shows the distribution of Log2 ratios relative to phosphorylated peptides, where it is evident that, except for few cases, the very large majority of identified phosphopeptides is present almost exclusively in the sample treated with recombinant PLK2 (the maximum ratio was set at 100, as specified in the methods section). To assess a threshold above which we could consider the fold change as significant, we plotted Log2 ratios for all quantified non-phosphorylated peptides (panel B). As it is possible to see, the Log2 ratio for these peptides never exceeds the value of 1 (dashed line), equivalent to a PLK2-treated/control of 2. Hence this was chosen as the threshold above which the differences between PLK2-treated samples and untreated samples were considered as significant.

Phosphosites primary structure analysis

The identification of a relatively large number of peptides phosphorylated by PLK2 *in vitro* allowed us to perform a primary structure analysis to define the kinase consensus sequence. Primary structure strongly contributes to the process of substrate recognition, making the determination of the consensus sequence a primary aim for the characterization of a protein kinase. However, it should be borne in mind that other factors may influence the kinase specificity such as tertiary and quaternary structures, and conditions that favor substrate recruitment (for example docking sites not involving the catalytic domain, or the presence of scaffolding and adaptor proteins). Therefore the conformity of a specific substrate to the consensus sequence may be variable [27,28].

The Two-sample logo is here utilized to obtain a detailed analysis of positive and negative selection of individual residues at given positions around the target site [25]. More in details, this logo provides a graphical representation of the differences between two sets of sequence alignment, i.e. sequences surrounding identified phosphorylated Ser/Thr vs sequences randomly selected from human proteome surrounding Ser/Thr: the upper section displays residues over-represented at a given position in the identified phosphosites as compared to the random one; the lower section displays residues under-represented at a given position in the identified phosphosites.

Several considerations can be made observing the Two-sample logo of Figure 3A. Foremost this analysis confirms the acidophilic nature of PLK2 (initially observed by Johnson *et al.* [29]), showing an enrichment of acidic residues in all positions considered. Positions upstream from the site of phosphorylation (in particular from −3 to −1) display a higher selection consistent with previous observations that the specific determinants of PLK2 are mostly located on the N-terminal side of the target residue [13,20,30]. Moreover the main determinants in PLK2 target selection here identified correlate well with previous observations [13,20,30].

Particularly remarkable is the striking overrepresentation of glutamic acid at position n-3, present at a frequency of 75% in the identified phosphosites, followed by leucine at −2 and aspartic acid at −1 present at 62,5% and 59%, respectively.

The Two-sample logo generated on PLK2-phosphorylated peptides can be compared with those generated using *bona fide* substrates of the most common acidophilic kinases, i.e. CK2α, CK1δ, and PLK1 (Figure 3). This comparative analysis shows that the four acidophilic kinases present a distinct substrate specificity. Even if all these kinases show an acidophilic nature in substrate recognition, the main acidic determinants are indeed observed at different positions: −3 and −1 for PLK2, +1 and +3 for CK2α, −2 and −1 for PLK1 (Figure 3). In the case of CK1δ the picture

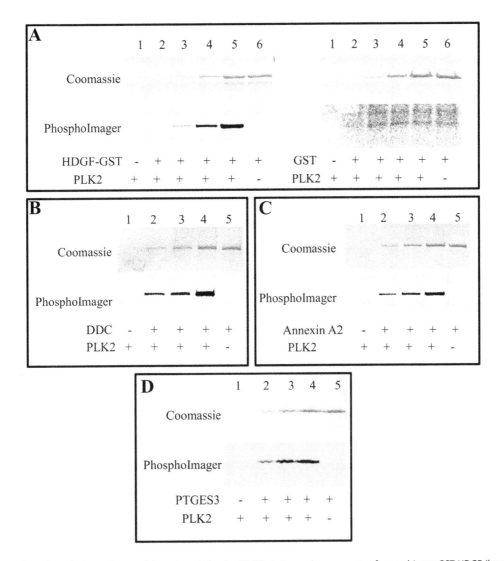

Figure 5. In vitro phosphorylation of recombinant proteins by PLK2. A. Increasing amounts of recombinant GST-HDGF (lane 2, 50 ng; lane 3, 100 ng; lane 4 250 ng; lane 5 and 6, 500 ng) were incubated in radioactive mixture in presence (lanes 1–5) of absence (lane 6) of PLK2 recombinant kinase as described in Materials and Methods. B–D Increasing amounts of purified proteins (lane 2, 100 ng; lane 3, 250 ng; lane 4 and 5, 500 ng) were incubated in radioactive mixture in presence (lanes 1–4) of absence (lane 5) of PLK2 recombinant kinase as described in Materials and Methods. Samples were loaded on SDS-PAGE, stained with colloidal coomassie and ^{33}P incorporation was analyzed by PhopshorImager. A- Hepatoma-derived growth factor. B- Aromatic L-amino acid decarboxylase (Dopa decarboxylase). C- Annexin A2. D- Prostaglandin E synthase 3 (PTGES3).

is less clear, revealing, besides a "background" of acidic residues at all nearby positions (especially upstream), the recurrent selection of seryl residues reflecting the canonical primed consensus of CK1 (pS-X-X-S) [31]. It is noteworthy that the two-sample logo of PLK2 displays a significant preference for an acidic residue at +3 position that corresponds to the major acidic determinant for CK2 phosphorylation. Moreover about 10% of the identified PLK2 phosphosites presents the strict CK2 consensus sequence s/t [DE]x[DE], thus suggesting a partial target overlap between these two kinases.

Of special interest is the enrichment in hydrophobic residues close to the PLK2 target residue, at −2 (the above-mentioned leucine) and at +1 position. The preference for hydrophobic residues is uncommon among acidophilic kinases even if this feature is shared with PLK1 [29]. Therefore we decided to further investigate this aspect. To provide a structural basis for this enrichment in hydrophobic residues at −2 and +1 position, an *in silico* analysis of the substrate binding zone of PLK2 was performed. Analyzing the hydrophobic amino acid distribution of PLK2 (Figure 4A) it is possible to observe the presence of hydrophobic regions in the active site (yellow areas). These hydrophobic regions, albeit less pronounced, are also present in the active site of PLK1 that also displays a preference for hydrophobic residues at −3 and +1 position (Figure 3C). By sharp contrast, these two hydrophobic regions are absent in the acidophilic kinases CK2 and CK1δ active sites (Figure 4A) consistent with the aminoacid preference observed in Figure 3.

To better analyze this interaction a series of protein-protein docking experiments between PLK2 and one of the phosphopeptides identified in this study EAIAELDtLNEESYK (P31946) were performed. From this analysis it is possible to observe that these hydrophobic regions are responsible for the interaction with the leucine at position −2 and with the hydrophobic residue at position +1, thus further supporting this peculiar feature of PLK2 specificity (Figure 4B).

Table 1. List of phosphosites identified in this study as PLK2 substrates that are present in Phosphosite database.

Acc. Number	Name	P-Site	Kinase
P31946	14-3-3 protein beta/alpha	T207	No
P62258	14-3-3 protein epsilon	T208	PLK2/PLK3
P63104	14-3-3 protein zeta/delta	T205	No
Q02952	A-kinase anchor protein 12	S381	No
Q9H4A4	Aminopeptidase B	T408	No
Q9Y2×7	ARF GTPase-activating protein GIT1	S643	No
Q07021	Complement component 1 Q subcomponent-binding protein	S201	No
Q14566	DNA replication licensing factor MCM6	S762	No
P55265	Double-stranded RNA-specific adenosine deaminase	S481	No
P24534	Elongation factor 1-beta	S95	No
P14625;P08238	Endoplasmin/Heat shock protein HSP 90-beta	S106/S45	No
Q9H501	ESF1 homolog	S663	No
P55884	Eukaryotic translation initiation factor 3 subunit B	S152	No
P56537	Eukaryotic translation initiation factor 6	S175	CK1δ
P35269	General transcription factor IIF subunit 1	S218	No
O60763	General vesicular transport factor p115	S942	CK2/GCK
P08238	Heat shock protein HSP 90-beta	S365	No
P51858	Hepatoma-derived growth factor	T225	No
P31943/P55795	hnRNA H1/hnRNP H2	S63	No
P17096	High mobility group protein HMG-I/HMG-Y	S99	No
P46821	Microtubule-associated protein 1B	S1156	No
Q14978	Nucleolar and coiled-body phosphoprotein 1	S637	No
Q9NR30	Nucleolar RNA helicase 2	S84	No
Q9NR30	Nucleolar RNA helicase 2	S121	No
P19338	Nucleolin	S28	No
P09874	Poly [ADP-ribose] polymerase 1	S785	No
Q99623	Prohibitin-2	S119	No
Q15185	Prostaglandin E synthase 3	S113	CK2
Q15084	Protein disulfide-isomerase A6	S428	CK2
P13521	Secretogranin-2	S104	No
Q13813	Spectrin alpha chain, non-erythrocytic 1	S391	No
Q96I25	Splicing factor 45	T224	No
Q13428	Treacle protein	S270	No
P40939	Trifunctional enzyme subunit alpha, mitochondrial	S669	No
P60174	Triosephosphate isomerase	S260	No
G3V1U9;P68363	Tubulin alpha-1A chain/Tubulin alpha-1B chain	S48	No
Q9BVA1;Q13509;P07437	Tubulin beta-2B/Tubulin beta-3/Tubulin beta	T72	No
P68371;P07437	Tubulin beta-4B chain/Tubulin beta chain	S126	No
P15374	Ubiquitin carboxyl-terminal hydrolase isozyme L3	S161	No
Q15942	Zyxin	S150	No

Potential novel substrates of PLK2

Having used tryptic peptides derived from undifferentiated human neuronal cells as PLK2 *in vitro* substrates, the identified phosphopeptides may help to predict putative PLK2 substrates *in vivo*. Although a residue phosphorylated within a peptide not necessarily undergoes phosphorylation in the full length protein, some observations suggest a good correlation between the phosphopeptidome and the phosphoproteome: two of the substrates identified in fact, i.e. 14-3-3 epsilon and endoplasmin,

have been previously identified as *in vitro* protein substrates [20], moreover we have also randomly selected from this list four proteins that have been subjected to *in vitro* phosphorylation by PLK2. All four proteins, GST-HDGF but not GST alone, Annexin A2, Aromatic L-amino acid decarboxylase (Dopa decarboxylase), and Prostaglandin E Synthase 3, were efficiently phosphorylated *in vitro* by PLK2 recombinant kinase (Figure 5). Two of these substrates were further analysed to confirm that the

A

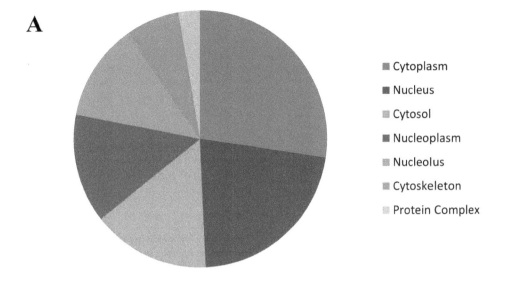

- Cytoplasm
- Nucleus
- Cytosol
- Nucleoplasm
- Nucleolus
- Cytoskeleton
- Protein Complex

B

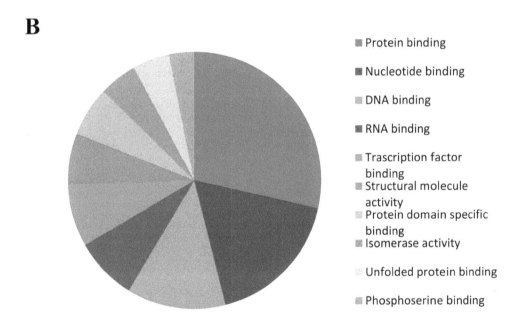

- Protein binding
- Nucleotide binding
- DNA binding
- RNA binding
- Trascription factor binding
- Structural molecule activity
- Protein domain specific binding
- Isomerase activity
- Unfolded protein binding
- Phosphoserine binding

Figure 6. Putative PLK2-substrate localization (A) and functional (B) analysis. Subcellular localization (A) and functional analysis (B) for each protein have been assigned using GeneCoDis3 webserver [35,36].

site phosphorylated within the intact proteins corresponds to that identified in the phosphopeptidome (see Figure S1).

These observations strongly support the idea that the newly identified phosphosites are physiologically relevant and can provide new insights into the role of PLK2 in cells. In this connection, we have checked if the phosphosites here identified are already annotated in PhosphositePlus database (www. phosphosite.org) [26]. About 40% of the phosphosites identified in this study have been reported as phosphorylated in cell/*in vivo*. The list of these proteins is shown in Table 1, together with the indication of the phosphosites and, if known, of the kinase/s responsible for their generation. About 90% of these phosphosites

are "orphan", meaning that the kinase/s responsible for their generation are not known.

Figure 6 shows the analysis of subcellular localization (A) and molecular functions (B) of putative PLK2 substrates identified in this study. Identified proteins localize both in cytoplasmic and nuclear compartments and participate to several processes where the involvement of PLK2 kinase has not been described yet. As mentioned above the number of *bona fide* PLK2 substrates identified so far is low and includes not only cytosolic proteins, but also plasma membrane [32] and nuclear [33] substrates. The localization of PLK2 at centrosomes where it regulates centriole duplication, has been deeply investigated [4]. However PLK2 has been identified also in different subcellular compartments, such as cytoplasm, nucleus (PLK2 contains a nuclear localization signal [34]), and membranes in HEK 293T cells [12], while in primary hippocampal neurons PLK2 shows primarily a nuclear localization [12]. Co-localization between the kinase and its putative substrates suggests unanticipated regulatory roles for PLK2 in nuclear functions.

Finally, given the known role of PLK2 in synaptic remodeling, it would be interesting to extend the analysis also to a model of differentiated neuronal cells, such as human cortex or primary neuron cultures. This approach could reveal substrates of PLK2 that are only expressed at the synapse and that were not identified in the present study. This will increase the panel of putative substrates of PLK2 and, on the other hand, will allow to identify substrates correlated to specific neuronal functions.

Supporting Information

Figure S1 Confirmation of PLK2 phosphorylation sites in intact proteins. A. 200 ng (lane 1) or 400 ng (lane 3) of GST-HDGF wild type and 200 ng (lane 2) or 400 ng (lane 4) of GST-HDGF T225A were incubated for 10 minutes in the radioactive mixture as described in the Material and Methods section in presence of PLK2 (left panel) or CK2 (right panel), loaded in SDS-PAGE gel, coomassie stained and analyzed by PhosphorImager. B. Prostaglandin E Synthase 3 (400 ng) was phosphorylated by recombinant PLK2 as in Figure 5, loaded in SDS-PAGE gel, coomassie stained, and trypsin digested. Phosphopeptides were enriched and identified as described in Material and Methods. The annotated MS/MS spectrum relative to the phosphopeptide DWEDDpSDEDMSNFDR is displayed together with all relevant information regarding peptide identification.
(TIF)

Table S1 List of phosphopeptides specifically phosphorylated by PLK2. The Table lists all phosphopeptides

identified in this study with a PLK2-treated/control ratio above 2. The ratios were obtained as the average values from all technical replicates. Class 1 phosphopeptides were quantified both in the "forward" and in the "reverse" experiment, while class 2 phosphopeptides were quantified only in one of the experiments. Stretches of sequences in brackets indicate that the same phosphosite was found in peptides with different number of missed-cleavages.
(XLSX)

Table S2 Relevant information relative to the peptides identified in the "forward" experiment. The table lists the sequences of all identified peptides, together with protein accession numbers, modifications, quantification values, Mascot scores, PEP values, q-values, chromatographic- and MS-relevant information.
(XLSX)

Table S3 Relevant information relative to the peptides identified in the "reverse" experiment. The table lists the sequences of all identified peptides, together with protein accession numbers, modifications, quantification values, Mascot scores, PEP values, q-values, chromatographic- and MS-relevant information.
(XLSX)

Acknowledgments

The authors wish to thank the "Cassa di risparmio di Padova e Rovigo" (Cariparo) holding, for funding the acquisition of the LTQ-Orbitrap XL mass spectrometer.

Author Contributions

Conceived and designed the experiments: MS GA. Performed the experiments: CF MS GA. Analyzed the data: GA MS CF LC. Contributed reagents/materials/analysis tools: GA MS LAP. Wrote the paper: MS GA LAP.

References

1. Archambault V, Glover DM (2009) Polo-like kinases: conservation and divergence in their functions and regulation. Nat Rev Mol Cell Biol 10: 265–275.
2. de Carcer G, Manning G, Malumbres M (2011) From Plk1 to Plk5: functional evolution of polo-like kinases. Cell Cycle 10: 2255–2262.
3. Strebhardt K (2010) Multifaceted polo-like kinases: drug targets and antitargets for cancer therapy. Nat Rev Drug Discov 9: 643–660.
4. Warnke S, Kemmler S, Hames RS, Tsai HL, Hoffmann-Rohrer U, et al. (2004) Polo-like kinase-2 is required for centriole duplication in mammalian cells. Curr Biol 14: 1200–1207.
5. Villegas E, Kabotyanski EB, Shore AN, Creighton CJ, Westbrook TF, et al. (2014) Plk2 regulates mitotic spindle orientation and mammary gland development. Development 141: 1562–1571.
6. Burns TF, Fei P, Scata KA, Dicker DT, El-Deiry WS (2003) Silencing of the novel p53 target gene Snk/Plk2 leads to mitotic catastrophe in paclitaxel (taxol)-exposed cells. Mol Cell Biol 23: 5556–5571.
7. Cholewa BD, Liu X, Ahmad N (2013) The role of polo-like kinase 1 in carcinogenesis: cause or consequence? Cancer Res 73: 6848–6855.
8. Valenti F, Fausti F, Biagioni F, Shay T, Fontemaggi G, et al. (2011) Mutant p53 oncogenic functions are sustained by Plk2 kinase through an autoregulatory feedback loop. Cell Cycle 10: 4330–4340.
9. Fingas CD, Mertens JC, Razumilava N, Sydor S, Bronk SF, et al. (2013) Polo-like kinase 2 is a mediator of hedgehog survival signaling in cholangiocarcinoma. Hepatology 58: 1362–1374.
10. Lee KJ, Lee Y, Rozeboom A, Lee JY, Udagawa N, et al. (2011) Requirement for Plk2 in orchestrated ras and rap signaling, homeostatic structural plasticity, and memory. Neuron 69: 957–973.
11. Inglis KJ, Chereau D, Brigham EF, Chiou SS, Schobel S, et al. (2009) Polo-like kinase 2 (PLK2) phosphorylates alpha-synuclein at serine 129 in central nervous system. J Biol Chem 284: 2598–2602.
12. Mbefo MK, Paleologou KE, Boucharaba A, Oueslati A, Schell H, et al. (2010) Phosphorylation of synucleins by members of the Polo-like kinase family. J Biol Chem 285: 2807–2822.
13. Salvi M, Trashi E, Marin O, Negro A, Sarno S, et al. (2012) Superiority of PLK-2 as alpha-synuclein phosphorylating agent relies on unique specificity determinants. Biochem Biophys Res Commun 418: 156–160.
14. Bergeron M, Motter R, Tanaka P, Fauss D, Babcock M, et al. (2014) In vivo modulation of polo-like kinases supports a key role for PLK2 in Ser129 alpha-synuclein phosphorylation in mouse brain. Neuroscience 256: 72–82.
15. Lashuel HA, Overk CR, Oueslati A, Masliah E (2013) The many faces of alpha-synuclein: from structure and toxicity to therapeutic target. Nat Rev Neurosci 14: 38–48.
16. Looyenga BD, Brundin P (2013) Silencing synuclein at the synapse with PLK2. Proc Natl Acad Sci U S A 110: 16293–16294.
17. Oueslati A, Schneider BL, Aebischer P, Lashuel HA (2013) Polo-like kinase 2 regulates selective autophagic alpha-synuclein clearance and suppresses its toxicity in vivo. Proc Natl Acad Sci U S A 110: E3945–3954.
18. Wang C, Ye M, Bian Y, Liu F, Cheng K, et al. (2013) Determination of CK2 specificity and substrates by proteome-derived peptide libraries. J Proteome Res 12: 3813–3821.
19. Salvi M, Sarno S, Cesaro L, Nakamura H, Pinna LA (2009) Extraordinary pleiotropy of protein kinase CK2 revealed by weblogo phosphoproteome analysis. Biochim Biophys Acta 1793: 847–859.
20. Salvi M, Trashi E, Cozza G, Franchin C, Arrigoni G, et al. (2012) Investigation on PLK2 and PLK3 substrate recognition. Biochim Biophys Acta 1824: 1366–1373.
21. Massimino ML, Ballarin C, Bertoli A, Casonato S, Genovesi S, et al. (2004) Human Doppel and prion protein share common membrane microdomains and internalization pathways. Int J Biochem Cell Biol 36: 2016–2031.
22. Salvi M, Trashi E, Cozza G, Negro A, Hanson PI, et al. (2012) Tools to discriminate between targets of CK2 vs PLK2/PLK3 acidophilic kinases. Biotechniques: 1–5.
23. Boersema PJ, Raijmakers R, Lemeer S, Mohammed S, Heck AJ (2009) Multiplex peptide stable isotope dimethyl labeling for quantitative proteomics. Nat Protoc 4: 484–494.

24. Venerando A, Franchin C, Cant N, Cozza G, Pagano MA, et al. (2013) Detection of phospho-sites generated by protein kinase CK2 in CFTR: mechanistic aspects of Thr1471 phosphorylation. PLoS One 8: e74232.

25. Vacic V, Iakoucheva LM, Radivojac P (2006) Two Sample Logo: a graphical representation of the differences between two sets of sequence alignments. Bioinformatics 22: 1536−1537.

26. Hornbeck PV, Kornhauser JM, Tkachev S, Zhang B, Skrzypek E, et al. (2012) PhosphoSitePlus: a comprehensive resource for investigating the structure and function of experimentally determined post-translational modifications in man and mouse. Nucleic Acids Res 40: D261−270.

27. Ubersax JA, Ferrell JE Jr (2007) Mechanisms of specificity in protein phosphorylation. Nat Rev Mol Cell Biol 8: 530−541.

28. Toppo S, Pinna L, Salvi M (2010) Matching up Phosphosites to Kinases: A Survey of Available Predictive Programs. Current Bioinformatics 5: 141−152.

29. Johnson EF, Stewart KD, Woods KW, Giranda VL, Luo Y (2007) Pharmacological and functional comparison of the polo-like kinase family: insight into inhibitor and substrate specificity. Biochemistry 46: 9551−9563.

30. Kettenbach AN, Wang T, Faherty BK, Madden DR, Knapp S, et al. (2012) Rapid determination of multiple linear kinase substrate motifs by mass spectrometry. Chem Biol 19: 608−618.

31. Venerando A, Ruzzene M, Pinna LA (2014) Casein kinase: the triple meaning of a misnomer. Biochem J 460: 141−156.

32. Schwarz J, Schmidt S, Will O, Koudelka T, Kohler K, et al. (2014) Polo-like kinase 2, a novel ADAM17 signaling component, regulates tumor necrosis factor alpha ectodomain shedding. J Biol Chem 289: 3080−3093.

33. Krause A, Hoffmann I (2010) Polo-like kinase 2-dependent phosphorylation of NPM/B23 on serine 4 triggers centriole duplication. PLoS One 5: e9849.

34. Zimmerman WC, Erikson RL (2007) Finding Plk3. Cell Cycle 6: 1314−1318.

35. Tabas-Madrid D, Nogales-Cadenas R, Pascual-Montano A (2012) GeneCodis3: a non-redundant and modular enrichment analysis tool for functional genomics. Nucleic Acids Res 40: W478−483.

36. Nogales-Cadenas R, Carmona-Saez P, Vazquez M, Vicente C, Yang X, et al. (2009) GeneCodis: interpreting gene lists through enrichment analysis and integration of diverse biological information. Nucleic Acids Res 37: W317−322.

12

Proteomic-Coupled-Network Analysis of T877A-Androgen Receptor Interactomes Can Predict Clinical Prostate Cancer Outcomes between White (Non-Hispanic) and African-American Groups

Naif Zaman[1], Paresa N. Giannopoulos[2], Shafinaz Chowdhury[1,2], Eric Bonneil[3], Pierre Thibault[3], Edwin Wang[1], Mark Trifiro[2,4], Miltiadis Paliouras[2,4]*

1 Biotechnology Research Institute - National Research Council, Montréal, QC, Canada, 2 Lady Davis Institute for Medical Research, Segal Cancer Centre - Jewish General Hospital, Montréal, QC, Canada, 3 Institut de recherche en immunologie et en cancérologie, Université de Montréal, Montréal, QC, Canada, 4 Department of Medicine, Department of Oncology and Division of Experimental Medicine, McGill University, Montréal, QC, Canada

Abstract

The androgen receptor (AR) remains an important contributor to the neoplastic evolution of prostate cancer (CaP). CaP progression is linked to several somatic AR mutational changes that endow upon the AR dramatic gain-of-function properties. One of the most common somatic mutations identified is Thr877-to-Ala (T877A), located in the ligand-binding domain, that results in a receptor capable of promiscuous binding and activation by a variety of steroid hormones and ligands including estrogens, progestins, glucocorticoids, and several anti-androgens. In an attempt to further define somatic mutated AR gain-of-function properties, as a consequence of its promiscuous ligand binding, we undertook a proteomic/ network analysis approach to characterize the protein interactome of the mutant T877A-AR in LNCaP cells under eight different ligand-specific treatments (dihydrotestosterone, mibolerone, R1881, testosterone, estradiol, progesterone, dexamethasone, and cyproterone acetate). In extending the analysis of our multi-ligand complexes of the mutant T877A-AR we observed significant enrichment of specific complexes between normal and primary prostatic tumors, which were furthermore correlated with known clinical outcomes. Further analysis of certain mutant T877A-AR complexes showed specific population preferences distinguishing primary prostatic disease between white (non-Hispanic) vs. African-American males. Moreover, these cancer-related AR-protein complexes demonstrated predictive survival outcomes specific to CaP, and not for breast, lung, lymphoma or medulloblastoma cancers. Our study, by coupling data generated by our proteomics to network analysis of clinical samples, has helped to define real and novel biological pathways in complicated gain-of-function AR complex systems.

Editor: Mohammad Saleem, Hormel Institute, University of Minnesota, United States of America

Funding: This study was supported by the Canadian Institutes for Health Research (http://www.cihr-irsc.gc.ca) operating grant MOP-106477 (M.T., M.P, E.W.). The funders had no role in study design, data collection and analysis, decision to publish, or preparation of the manuscript.

Competing Interests: The authors have declared that no competing interests exist.

* Email: miltiadis.paliouras@mcgill.ca

Introduction

Significant advances in genomic sequencing methodology have allowed a better assessment of the extent of somatic mutations accrued in common neoplasms [1,2]. More important is the realization that tumors significantly vary genetically from one patient to another and within a singular patient there exists extensive inter-tumoral heterogeneity and intra-tumoral heterogeneity [3,4,5]. A significant number of these genetic alterations are missense mutations that provoke new gain-of-function properties that render a particular gene proactive to tumoral evolution and are referred to as driver mutations. A better understanding of these new properties would lead to a better interpretation of oncogenesis, but this is difficult due to a large number of different mutations, the unpredictable nature of gain-of-function properties associated with somatic mutations, the possible extensive interplay of different somatic mutants and the ensuing selection processes initiated by the microenvironment or by therapy itself. Such complex "systems" require a more global "omics" approach and more network analysis, rather than the classical single gene approach, to garner more critical information related to neoplastic evolution.

In keeping with the newly defined mutational landscape of tumors, prostate cancer (CaP) also has extensive genetic alterations that range from single missense mutations, copy number variation, splicing variants, genetic rearrangements and short DNA alterations in a large number of genes [1,2,6,7], including the androgen receptor (AR) gene. It is not unexpected that AR mutations can add to the protein's repertoire of powerful new functions [8,9] and these gain-of-function attributes may allow the AR to function in an aberrant manner. A number of somatic CaP AR mutants, especially the most commonly occurring CaP AR mutation, Thr877Ala (T877A), have unique gain-of-function properties: they can bind several classes of steroids promiscuously (e.g. estrogens, progestins, glucocorticoids) with subsequent transactivation, or be hyperactivated by normal ligands [10]. Classic anti-androgen treatments [e.g. flutamide, cyproterone acetate (CPA) or bicalutamide] have generated, through selection pressure, specific somatic AR mutations, e.g. Trp741Cys (W741C) and His874Tyr, resulting in subversive ARs that are fully active with these drugs [11]. Even the next generation of anti-androgen drugs exemplified by enzalutamide (MDV-3100) has provoked specific AR mutations [12,13]. This observation also correlates with a dramatic fall in PSA levels subsequent to anti-androgen withdrawal [11]. The T877A-AR mutations, which is also present in prostate cancer cell line LNCaP, has been reported by various individuals to occur in 25 to 33% of androgen-independent or castrate-resistant tumors [11,14,15,16].

Recently, our own work strongly suggests that the AR function extends beyond its classical role as a transcription factor and includes the novel properties of RNA splicing, DNA methylation, proteasomal interaction and protein translation at the polyribosomes themselves [17]. Furthermore, the great functional diversity of the components of AR complexes exemplifies the intricate nature of protein-protein interactions associated with generating the appropriate AR biological output. These novel AR functions may mediate cellular processes and offer new areas in which somatic AR CaP mutants might "indulge" and promote CaP oncogenesis.

In an attempt to describe novel gain-of-function properties associated with mutant CaP ARs, a proteomic-coupled network analysis was performed. Multiple proteomics-mass spectroscopy investigations were carried out in order to fully characterize the protein composition of T877A-AR "complexes" (interactome) under different classes of hormone/ligand conditions reflecting the promiscuity of ligand binding associated with T877A. Critically, mutant CaP ARs may have their own unique ability to undergo and define new interactions. The coupling of the data generated by our proteomics screen to system biology analysis has been helpful in defining real and novel biological endpoints in AR complex systems, within a clinical disease perspective.

Materials and Methods

Cell lines

The LNCaP prostate cancer cell lines was obtained from the American Type Culture Collection (ATCC), Rockville MD.

Cell culture, steroid hormones, ligands and stimulation experiments

LNCaP cell line was cultured in RPMI 1640 media supplemented with FBS (10%), at 37°C, 5% CO_2 in T-75 plastic culture flasks. Once confluent, the medium was changed to RMPI supplemented with 10% charcoal–dextran stripped FBS and incubated for an additional 24 h. The following day, the medium was changed to fresh RMPI/charcoal–dextran stripped FBS for overnight hormone/ligand stimulation studies for an 18 hour period. Steroids hormones were used at the following final concentrations, 10 nM dihydrotestosterone (DHT), 10 nM mibolerone (MB), 10 nM R1881, 10 nM testosterone+10 μM finasteride, 10 nM 17β-estradiol, 10 nM progesterone, 10 nM dexamethasone, and 100 nM cyproterone acetate (CPA).

Affinity purification and Western blotting

LNCaP whole cell lysates were prepared by freeze-thaw method with 1X PDG buffer containing the appropriate hormone/ligand [18]. Lysates were then carried over for α-AR co-immunoprecipitations [AR(N20), Santa Cruz Biotechnology, Santa Cruz, CA] overnight at 4°C. A 50% Protein A Sepharose slurry was added to each sample and incubated at room temperature for 90 min. Beads were washed three times with wash buffer (50 mM Tris-HCl pH 8.0, 150 mM NaCl, 1% Tween 20) and resuspended in 100 μL 1X SDS gel loading buffer. Samples were denatured by boiling, and resolved on a 10% SDS-polyacrylamide gel before silver staining according to manufacturer's guidelines (BioRad) or transfer to a nitrocellulose membrane for Western blot analysis using monoclonal antibody AR(441) (NeoMarkers, Fremont, CA).

Mass spectrometry and peptide comparison

TCEP (tris(2-carboxyethyl)phosphine) was added to the protein samples to reach the concentration of 5 mM. Samples were incubated at 37°C for 30 min. One μg of trypsin was added and the samples digested overnight at 37°C, then dried down in a SpeedVac and resolubilized in 50 μl of ACN 5%/formic acid (FA) 0.2%.

All MS analyses were performed using an LTQ-Orbitrap hybrid mass spectrometer with a nanoelectrospray ion source (ThermoFisher, San Jose, CA) coupled with an Eksigent nano-LC 2D pump (Dublin, CA) equipped with a Finnigan AS autosampler (Thermo Fisher, San Jose, CA). Twenty μl of each sample was injected on a C18 precolumn (0.3 mm i.d.×5 mm) and samples separated on a C18 analytical column (150 μm i.d.×100 mm) using an Eksigent nanoLC-2D system. A 76-min gradient from (A/B) 10–60% (A: formic acid 0.2%, B: acetonitrile/0.2% formic acid) was used to elute peptides with a flow rate set at 600 nL/min. The conventional MS spectra (survey scan) were acquired in profile mode at a resolution of 60,000 at m/z 400. Each full MS spectrum was followed by three MS/MS spectra (four scan events), where the three most abundant multiply charged ions were selected for MS/MS sequencing. Tandem MS experiments were performed using collision-induced dissociation in the linear ion trap.

The comparisons of peptide abundance across the different experimental paradigms were achieved using label-free quantitative proteomics [19,20]. Briefly, raw data files from the Xcalibur software was converted into peptide map files representing all ions according to their corresponding m/z values, retention time, intensity and charge state. Peptide abundance was then assessed using the "peak top" intensity values. Intensities of peptides eluting across several fractions were summed together, and only a coefficient of variance (CV) allowing the maximal ion transmission was considered to calculate peptide intensity. Clustering of peptide maps across different sample sets was performed on the peptide-associated Mascot entry using hierarchical clustering with specific tolerances (+/−15 ppm of peptide mass and +/−1 min of peptide retention time). Normalization of retention time was performed on the initial peptide cluster using a dynamic and nonlinear correction that confines the retention time distribution to less than 0.1 min on average. Reproducibility changes in abundance across conditions was determined using a two-tail homoscedastic t-test on sample replicates to identify peptide clusters with p-values <0.1 with fold changes greater than 7 standard deviations. Peptide clusters fulfilling these selection criteria was inspected manually to validate identification and changes in abundance. Expression analyses were performed on proteins identified by at least two different peptide sequences. Expression values and relative standard deviation were gained by averaging the intensity differences and standard deviations of the four most intense peptide triplets after removing outlying peptide clusters. Normalized proteomic data can be found in **Table S1**.

Datasets for network construction and gene expression analysis

All AR interactors were given NCBI gene IDs. Human protein interaction information was compiled from diverse data resources and annotation databases such as Biomolecular Interaction Network Database (BIND), the Database of Interacting Proteins (DIP), Human Protein Reference Database (HPRD), IntAct, and Molecular INTeraction database (MINT), most of which contain curated interaction data and high-throughput data. We generated a metadata of protein interactions by merging these data with our own manually curated human signaling network containing 4,000 proteins and 22,000 signaling relations [21,22,23].

A protein interaction network was constructed for each hormone using our manually curated human signaling network and protein-to-protein interaction network. We only considered proteins that were ≥2.5 times the hormone condition abundance (signal) vs. the vehicle control abundance from the MS dataset, to be considered significantly present. In the network, a node and link represent the protein and interaction, respectively. To find the highly interconnected regions of the network, for each hormone, we scored each pair of interaction based on the number of neighboring nodes they have in common. This gave us a matrix with all the scores between all the interaction pairs. Hierarchical clustering was applied to the matrix and a threshold score calculated based on the partition density allowing us to identify the highly interconnected regions (clusters) for each of the hormone networks.

Next, we used these protein clusters and identified GO-terms (http://www.geneontology.org/) that are significantly associated with each of the protein clusters (p-value<0.05, hyper-geometric). We also used Gene Set Enrichment Analysis (GSEA) defined pathways and performed GSEA if 25% of the protein cluster's genes were in the pathway. For the pathways and GO-terms that were significant from the GSEA results (p-value<0.05), we also did survival analysis. We used genes associated with these GO-terms and performed GSEA using GSE21034 [24].

RNA extractions and microarray analysis

LNCaP cells were stimulated with panel of ligands as described above, and total RNA was extracted using TRIZOL (Invitrogen, Carlsbad, CA). RNA samples were then processed by Quebec Genome Innovation Centre (McGill University, Montreal, Canada), for microarray analysis with Illumina Human HT-12 Expression Beadchip v4 (Illumina, San Diego, CA). Raw data was processed using R [22,25].

Two global heatmaps were made. The first heatmap includes the most differentially expressed gene for each hormone using t-test (p-value<0.05), where each hormone is compared against all others. The second heatmap includes the top 10, 20, 50 and 100 genes with the highest variance. The t-test finds the most differentially expressed genes that are specific to each hormone, whereas the variance helps us observe genes that are differentially expressed across multiple hormones. GSEA analysis was performed using GSE21034 [24], of the 10, 20, 50 and 100 genes that showed the highest variance.

Progression and Survival Analysis

Gene expression profiles, patient survival data, and demographic information for the 267 clinical prostate samples (29 normal, 181 primary and 37 metastatic tumors) were obtained from GSE21034 [24]. Breast, lung, lymphoma and medulloblastoma datasets were obtained from the Broad Institute (http://www.broadinstitute.org/cgi-bin/cancer/datasets.cgi) [26]. We examined the post-radical prostatectomy prognostic values of a subnetwork based on gene expression profiles of primary tumors, and performed Kaplan-Meier analysis by implementing the Cox-Mantel log-rank test using R as described previously [22,25]. If the p-value is less than 0.05, the subnetwork was treated as statistically significant to classify the tumors into non-metastatic and metastatic tumors. We stratified recurrent vs. non-recurrent CaP cancer based on the following criteria: PSA (≥4 ng/mL), Gleason (≥7), Tumor Stage (≥3) and combined (PSA+Gleason+Tumor).

Structure Preparation for MD simulation

The Crystal structure of DHT (1I38) and CPA (2OZ7) bound to T877A mutant AR-LBD and testosterone (2AM9) and R1881 (1E3K) bound to wild type AR-LBD are available in the Protein Data Bank (PDB). The complexes of different ligands bound to the T877A mutant AR-LBDs were further prepared for MD simulations using Xleap [27] in AMBER10 [28]. The generalized AMBER force field (GAFF) and ff99SB [29] parameters were used for eight different ligands. The complex was solvated in a truncated octahedron TIP3P [30] water box. The distance between the wall of the box and the closest atom of the solute was 12.0 Å, and the closest distance between the solute and solvent atoms was 0.8 Å. Counterions (Cl^-) were added to maintain electroneutrality of the system. Each system was minimized, first by applying harmonic restraints with force constants of 10 kcal/mol/$Å^2$ to all solute atoms; second, by heating from 100 to 300 K over 25 ps in the canonical ensemble (NVT); and lastly by equilibrating to adjust the solvent density under 1 atm pressure over 25 ps in the isothermal–isobaric ensemble (NPT) simulation. The harmonic restraints were then gradually reduced to zero with four rounds of 25-ps NPT simulations. After additional 25-ps simulation, a 15-ns production run was obtained with snapshots collected every 1 ps. For all simulations, 2 fs time step and 9 Å non-bonded cut-off were used. The particle mesh Ewald method [31] was used to treat long-range electrostatics and bond lengths involving bonds to hydrogen atoms were constrained by SHAKE [32].

Results

Comparative network characterization of T877A-AR complexes: interactome and gene expression studies

Our ability to capture both ligand-bound and unliganded full-length wild-type AR by affinity chromatography under physiological conditions has previously allowed us to pursue a proteomics approach in order to characterize the components of wild-type AR complexes. This was done by subjecting such complexes to tryptic digestion followed by mass spectrometry (MS) to assign protein identification, in effect creating AR interactomes initiated by ligand binding [33]. To our MS data, a label-free quantitative method has now been applied across the different experimental paradigms (see Materials and Methods) [19,20], which allowed us to obtain data related to protein identification, along with abundance, thus allowing for direct comparisons between stimulation conditions.

The aim of differential hormone stimulation conditions will allow us to determine whether disease etiology of the T877A-AR mutation is dependent upon ligand and co-factor status. Therefore, we used LNCaP cells that endogenously express the T877A-AR mutation to characterize the ligand promiscuous protein interactome complexes under different hormonal conditions, in order to highlight the possible distinct complexes that may be linked to disease progression. LNCaP cells were stimulated with the following hormones, four androgens: DHT, mibolerone (MB), R1881, or testosterone (in the presence of finasteride (to prevent the conversion of testosterone to DHT), or 17β-Estradiol, progesterone, dexamethasone, or the anti-androgen cyproterone actetate (CPA), alone or with no-ligand/alcohol-vehicle control. Hormone stimulated T877A-AR complexes were immunopurified with an N-terminal specific AR antibody, which would not interfere with hormone binding to the AR ligand-binding domain. Eluates from all experimental conditions were analyzed by LC-MS/MS. In order to increase our sample frequency for peptide detection in our MS analysis, each experimental condition was performed four times. Our proteomics data is a compilation of only fully characterized proteins, with full gene ontology and function.

Quantitative MS data, for each of the eight hormone stimulation conditions, was used to create a protein interaction map (**Figure 1A**). The protein interaction network map, allows for a visual analysis of the relationship of the interaction of each protein with the mutant AR. It is most likely that not all proteins interact directly with the mutant AR, but can through intermediate proteins. Further ontological function classification (see Materials and Methods) is based on this interaction network map, by discerning significant clusters of interacting proteins based on the number of protein-protein interaction connections. Of the eight hormone stimulation T877A-AR protein lists, we then systemically applied hierarchal clustering analysis to the experimental conditions. Hierarchical clustering heat-maps representing the grouping of between whole T877A-AR agonist and antagonist experimental treatments were then generated (**Figure 1B**). Between the different experimental conditions (hormone treatments), a comparative network analysis was applied [21,22,23], and although four different androgens were used (DHT, testosterone, MB and R1881), the proteomic profiles of these androgen ligands do not segregate together, and we observed that progesterone and dexamethasone AR complexes have proteomic profiles that look like R1881 and MB, respectively. Moreover, the protein interaction complex for AR-estradiol-stimulated complexes was most similar to the AR-DHT response interactome.

We also proceeded to characterize the gene expression patterns of the multi-panel hormone stimulated LNCaP cells. Analysis of most variably expressed genes between the hormone conditions gave a hierarchical clustering pattern that was much different to the T877A-AR protein-interaction profile (**Figure 1C**). Quite clearly, we again observed that the different androgens used in these stimulation profiles do not segregate together, and that synthetic androgens, like R1881 and MB, do not have the same AR-stimulated transactivation profiles as the natural ligands like testosterone and DHT. Moreover, the functional ontological properties between protein-interaction vs. gene expression profiles of our differential ligand stimulated cells also appear to be very different. Discerning the impact on disease progression from these profiles was of particular interest.

To establish statistically significant biological functions, we implemented the incorporation of Gene Ontological (GO)/ pathways terms using DAVID (Database for Annotation, Visualization and Integrated Discovery, http://david.abcc.ncifcrf.gov/). Using the protein interaction data from all the ligand stimulation conditions, the major ontological functions are: RNA pol II-dependent transcription, protein biosynthesis, with components of the translational machinery (translation initiation, elongation factors, ribosomal proteins and other regulatory proteins), RNA

metabolism (specifically RNA splicing), DNA repair (through an interaction with members of the DNA repair complex), and the proteasome/ubiquitination pathways (see **Table S2**). In contrast, the ligand-dependent gene expression pattern ontological classes included pathways involved in DNA replication, steroid/sterol biosynthesis, and apoptosis (see **Table S3**). Thus, although the AR is classically described as a transcription factor, its proteome profile would suggest that the AR is capable of functions beyond what has initially been described as a gene activator.

Structural analysis of hormone binding to the T877A-AR

To explore the possible mechanisms by which ligands binding to the mutant AR create different sets of AR interacting proteins, we obtained the detailed conformation of the receptor, using 15 ns molecular dynamic (MD) simulation studies (see Materials and Methods) of the eight different ligands used. Using the docking program WILMA (**Figure 2**), we obtained structural data of the mutant AR bound with progesterone, estrogen, dexamethasone and MB. The structural data of the mutant AR with testosterone, DHT, R1881 and cyproterone acetate are already available and as such, these structures served as the starting points for the MD simulation studies.

Mutant AR MD simulations were performed over 15 ns production runs. To inspect the local flexibility of each protein/ hormone complex, we calculated the root-mean-squared deviation (RSMD) fluctuations of backbone atoms of each amino acid residue for each mutant AR complex. In the study of globular protein conformations, one customarily measures the similarity in the three-dimensional structure by the RMSD of the central carbon atoms in amino acids, after optimal rigid body superposition. The most significant fluctuations correspond to loop regions between α3/α4, α9, α10/α11 and α11/α12 and the helices α11 and α12 themselves of the LBD of AR (**Figure 3**). It can be seen that the loop between α9, α10 and α11 is the most flexible region in all the complexes (**Figure 3B**). Among the eight different AR ligand bound complexes, the testosterone- and estradiol-bound complexes demonstrate the highest flexibility, with some loop residues having RMSD fluctuations as high as 2.4 Å. Although these loops are distant from the hormone binding pocket, they are exposed to the surface and may serve a role as potential binding sites to other protein partners.

The other major differences observed between the mutant AR complexes are the positions of α11 and 12, which are known to be critical for dictating hormone binding and co-activator interactions. Residues of α11 and α12 and the loop between them showed higher flexibility (**Figure 3C**). The T877A-AR mutation located in α11 allows for a more spacious hormone-binding pocket and will accommodate steroids with different extensions within the D ring.

The examination of the nature and size of the solvent accessible surface area (SASA) of proteins is an important tool to measure potential interaction propensity with neighboring proteins. We calculated the average SASA of each AR complex from the 15 ns MD trajectory. No large differences were observed among the calculated SASAs of different AR mutant complexes, which ranged from 12,124 to 12,390 Å^2. These results show that the differences are mostly associated with the α11–α12 regions of the AR-LBD, where the T877A mutation is located. Therefore, we decided to compare the dynamics among the eight different complexes, specifically in this region, by using RMSD matrices (**Table 1**). The matrices, which essentially capture extreme movements, reveal regions of high flexibility, especially for CPA and dexamethasone, compared to other AR-ligand complexes.

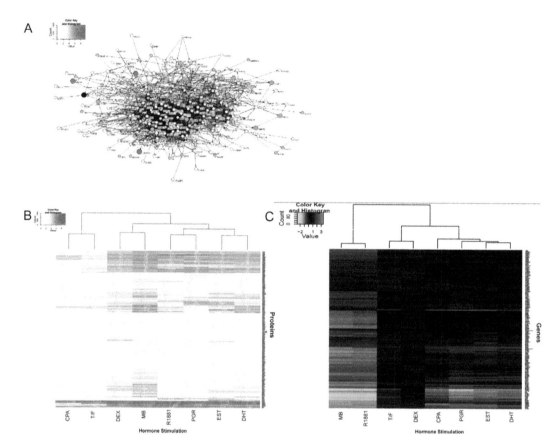

Figure 1. Characterization of T877A-AR protein interaction and gene expression profiles. A. Quantified protein interaction network of DHT stimulated LNCaP cells. The value of each protein, defined by our label-free quantitative MS, is distinguished by both color and size, and subsequent protein-protein interactions. The AR is designated by the black circle. To determine the relationship between protein interaction and gene expression patterns, hierarchical clustering of multi-panel hormone-stimulated LNCaP cells was carried out. **B.** Interacting proteins identified by mass spectrometry. **C.** Most variably expressed genes upon ligand stimulation. Although it has been suggested that all hormones used are able to activate androgen-dependent gene transcription with the T877A-AR mutant receptor, there are differences between AR protein complexes and AR gene expression patterns. This cluster analysis illustrates that even the synthetic androgens like Mibolerone (MB) and R1881, have similar gene expression profiles, their protein-interaction complexes are more similar to dexamethasone (DEX) and progesterone (PGR), respectively, than to either natural androgens testosterone and DHT. Moreover, even natural androgens (DHT and testosterone) can be segregated by their protein complexes. This would suggest that there are functions for the AR beyond gene transactivation. Protein and gene expression values are given as a ratio of quantified protein of each stimulation vs. vehicle control stimulation.

Computational modeling has also been carried out for a number of other AR somatic mutations in the ligand-binding domain (LBD), and show sensitivity to a broad range of hormone ligands, including, AR-W741L[34], -L701H [35,36,37,38], -H874Y [39,40,41] and -F876L [13]. Determination of the LBD structure between AR-WT and -H874Y (present in 22Rv1 cells), when bound to Testosterone in the presence of the N-terminal FXXLF motif peptide or the TIF2 coactivator peptide, found that all structures conformed to the canonical nuclear receptor LBD fold [41]. Moreover, the -H874Y DHT and R1881 structures conformed to -T877A and -W741L LBD bound to steroid and nosteroid ligands [34,42]. The double AR-mutant cell lines MDA-PCa-2a and MDA-PCa-2b, possess the L701H and T877A somatic mutations, these cells shows similar AR transactivation and LBD structural properties to the single AR-T877A mutant LNCaP cells to a broad spectrum of steroid ligands and anti-androgens, but also show an increased sensitivity cortisol steroids [35,36]. Most recently, the structure AR-F876L mutations has been investigated that allows the cell to use the new anti-androgen enzalutamide as an agonist. Similarly the Leu876 mutation allows for the antagonist-agonist switch, by accommodating of enzaluta-

mide to the ligand binding pocket [39]. Although only LBD structural data is available for the AR-F876L mutant, we don't believe that enzalutamide vs. androgen binding would significantly alter helices movement to drastically affect overall global AR structure, as has been observed with other AR LBD mutants.

AR protein interaction functional clusters, but not clinically derived gene expression profiles, correlate with CaP progression and survival outcome

From each of the ligand mutant AR protein interaction network, we identified specific sub-network modules. These sub-networks suggest hormone-specific activated pathways involved in either tumor initiation or progression. The characterization of ontological functions across the stimulation conditions is important as these differences or similarities in interacting proteins within the T877A-AR mutant complex may account for unique or shared cellular properties contributing to disease progression and outcomes. Therefore, we annotated significant GO-terms directly on sub-network modules extracted from the eight different hormone-protein interaction networks to highlight functions that

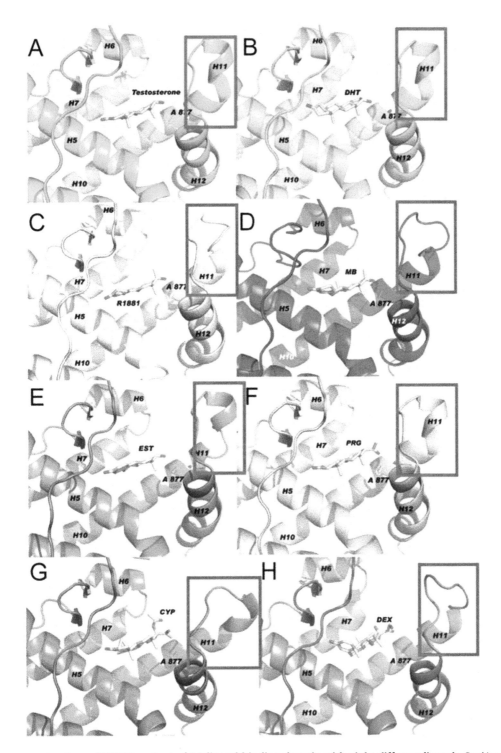

Figure 2. MD average structure of T877A mutant of AR ligand-binding domain with eight different ligands. Docking program WILMA was used to examine structural changes of the ligand-binding domain of the T877A mutation, upon binding to ligand binding. Illustrated is the average structure of T877A-AR mutation ligand binding domain with eight different ligands. **A.** testosterone, **B.** DHT, **C.** R1881, **D.** MB, **E.** estrogen (EST), **F.** progesterone (PGR), **G.** CPA, **H.** dexamethasone (DEX). A red box highlights the changes in the helix α 11 loop upon binding to each hormone ligand.

may be unique to each of stimulation condition, and extracted the list of genes corresponding to those GO-terms.

Using the lists of genes from the annotated GO-terms (now to be referred to as "gene-sets"), we determined whether these gene-sets are enriched in the publicly available clinical prostatic tumor microarray dataset, by applying Gene Set Enrichment Analysis (GSEA). We used the clinical data set GSE21034 [24], containing 247 clinical specimens (29 normal, 181 primary and 37 metastatic tumors). Initial analysis of all primary tumors from this data-set did not yield obvious enrichment of any gene-sets. However, after

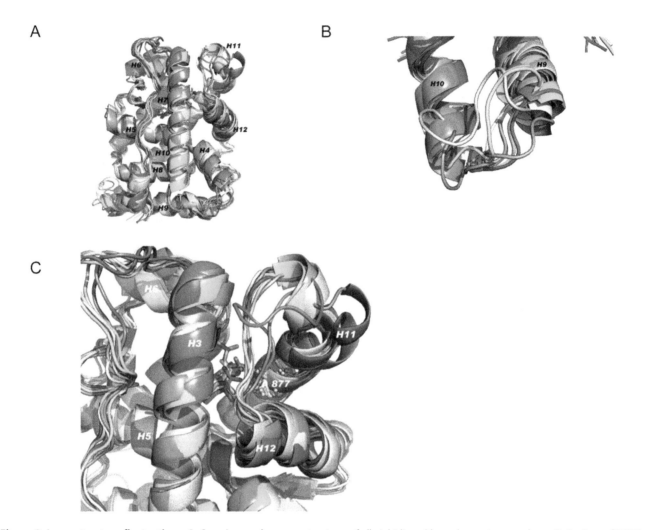

Figure 3. Loop structure fluctuations. A. Superimposed average structures of all eight ligand bound receptor complexes. **B.** Regions of T877A AR-LBD loop between Helixα9 and Helixα10 showed maximum flexibility. **C.** Expanded view of Helix α11 and Helix α12 regions of T877A AR-LBD bound to eight different ligands studied in this study. Residues of α11 and α12 and loop between them showed higher flexibility compared to other regions of the receptor where T877A is located. Color corresponds to the following ligands: Green - testosterone, Cyan- DHT, Yellow-R1881, Pink- MB, Rose- EST, Grey- PGR, Orange-CPA, Purple-DEX.

further inspection of the data, unique features were noted in certain tumor samples. Thus, upon returning to the patient pathology information that accompanied the clinical data-set, two diverse patient populations could be immediately discerned, and thus we resegregated our data-sets between 142 White (non-Hispanic) and 25 African-American samples [other population groups (Hispanic and Asian) in the 181 available primary samples were too few to perform GSEA]. By segregating the dataset along available ethnic demographical information, we immediately were able to distinctly differentiate gene-sets between White (non-Hispanic) vs. African-American populations. From these results, we identified 138 T877A-AR -interacting protein sub-network modules (gene-sets) that show significant (p≤0.05) enrichment of the T877A-AR -interacting partners that discerned CaP primary tumors vs. normal samples. Two gene-set examples are shown in **Figure. 4**.

Figure 4A represents a gene-set (gene-set 1) that is significantly enriched among primary tumors of the African-American population, however subsequent GSEA of the same gene-set did not show significant enrichment in primary tumors of White (non-Hispanic) group. One of the unique features of the African-

American gene-set 1 is that is enriched in normal rather than tumor samples. This suggests that these genes may have anti-tumorgenic properties or subsequent loss of the expression may contribute detrimentally to prostate disease. Of particular note, the genes represented in gene-set 1 are part of the transcription-dependent DNA repair pathway. **Figure 4B** shows another gene-set (gene-set 2) that is significantly enriched in primary tumor samples of White (non-Hispanic) males and not in the corresponding African-American cohort. Furthermore, the GSEA results of these two T877A-AR interacting protein gene-sets are also distinct, and not shared, between primary tumor data of population groups of White (non-Hispanic) or African-American males. However, African-American gene-set 1 was significantly enriched in metastatic tumor data of White (non-Hispanic) males (data not shown). Due to limited data available on metastatic tumors from the African-American demographic (2 data sets), and reciprocal analysis could not be performed. This data highlights that we can identify gene-sets that show population-specific distinctions from primary tumors, but also suggests that the molecular characteristics of AR function underlying disease

Table 1. Two dimensional RMSD matrices of the residues of Helix α11 and Helix α12 and loop between them without fitting.

	DHT	TEST	MB	R1881	CPA	DEX	EST	PRG
DHT	0	1.72	1.83	1.79	3.11	3.23	1.79	2.03
TEST	1.72	0	1.69	2.05	3.17	3.19	1.73	2.2
MB	1.83	1.69	0	2.19	2.93	2.93	1.88	2.26
R1881	1.79	2.05	2.19	0	3.54	3.38	1.98	2.11
CPA	3.11	3.16	2.93	3.54	0	3.28	3.14	3.4
DEX	3.23	3.19	2.93	3.38	3.38	0	3.43	3.72
EST	1.79	1.73	1.88	1.98	3.12	3.46	0	1.82
PRG	2.01	2.18	2.26	2.11	3.41	3.72	1.83	0

etiology in metastatic tumors may be common between African-American and White (non-Hispanic) men.

We next analyzed whether or not these distinct population T877A-AR gene-sets were predictive of 10-year survival outcomes, available from the GSE21034 data set [24], based on the following criteria; PSA (≥4 ng/mL), Gleason (≥7), Tumor stage (≥T3) or Combined (PSA + Gleason + Tumor stage). From our initial 138 characterized gene-sets, we identified 10 AR-interacting protein complexes that are indeed predictive for disease outcomes, 8 representing the White (non-Hispanic) population and the remaining the African-American group. From these 10 gene-sets (see **Table S4**), a single gene-set from the White (non-Hispanic) population (gene-set 2) was able to predict disease outcomes across all scoring criteria, (PSA, Gleason, Tumor Stage, and combined (**Figure 4C**). This particular gene set did not exhibit any predictive value when analyzed against the African-American cohort of samples. The African-American gene-set (gene-set 1) was only able to predict disease outcome using the combined criteria (PSA+Gleason+Tumor Stage) (**Figure 4D**). All other gene-sets, from all population groups, were able to predict survival outcomes for only one scoring criteria (**see Figure S1**). Furthermore, these predictive clusters were shared between all the hormones, supporting the results from our dynamic modeling study, where the T877A mutation accommodates all steroid hormones to and exhibits very subtle structural differences, although the overall structures appear to elicit the same functional interaction platform.

A similar analysis to that performed with our proteomic data was performed using our LNCaP multi-panel hormone micro-array gene expression data across our hormone stimulation conditions. We selected 10, 20, 50 and 100 of the most variably expressed genes from our microarray data set to assess the ability to predict disease progression and outcome between White (non-Hispanic) or African-American men. We identified two gene-sets of 10 and 50 genes respectively that were able to distinguish between normal *vs.* tumor in White (non-Hispanic) men, but not African-American men. There appears to be no predictive value associated with each different class of hormone stimulation, irrespective of whether the hormones act through T877A-AR or through their cognate receptor. It is also apparently clear that the two different data-sets (protein interactome *vs.* gene expression), result in two different capabilities of predicting disease outcomes. This is a direct result of linking ontological function to a specific protein sub-network vs. arbitrarily selecting a defined number of genes linked solely to expression profiles.

Finally, although LNCaP cells are derived from a metastatic CaP lymph-node biopsy from a White (non-Hispanic) male, other cell lines also possessing the T877A-AR mutation, MDA-PCa-2a and MDA-PCa-2b, from bone metastatic CaP from African-American also exist. However, extensive genome-wide gene expression characterization between these cell lines and LNCaP, have found them to be most similar to one another vs. other androgen sensitive cell lines, LAPC4 (possessing a wild-type AR) or 22Rv1 (H874Y-AR) or vs. AR-null cell lines PC3 and DU145 [43]. Thus, if this differential hormone stimulation experiment were to be performed using MDA-PCa-2a or -2b cell lines, we would identify the same AR protein complexes *in vitro*, as LNCaP, and would also predict disease survival *in vivo* from clinical population data.

Characterized T877A-AR protein gene-sets cannot predict survival outcomes in four other non-CaP cancers

We subsequently determined whether these predictive gene-sets were cancer specific and extracted gene expression datasets with available clinical outcome profiles for breast, lymphoma, lung and

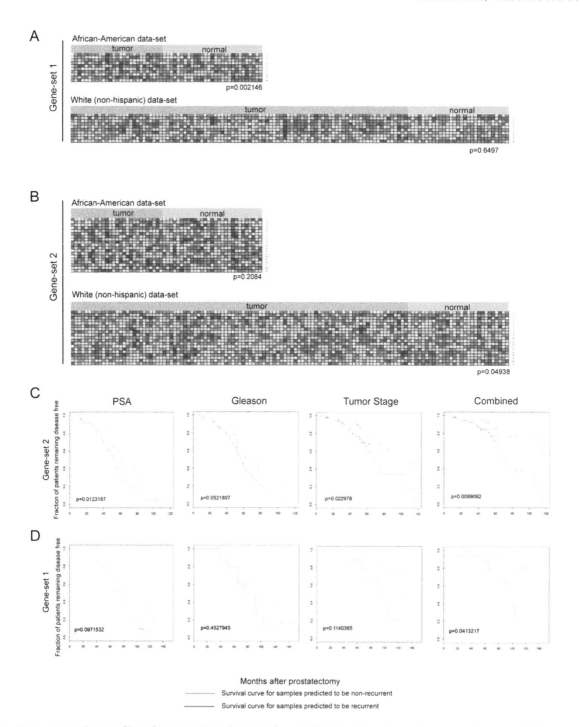

Figure 4. Gene expression profiles of AR protein sub-networks correlate to CaP progression and outcomes. GSEA was applied to determine the correlation of transcript expression profiles of AR subnetwork protein clusters from raw microarray data of normal and primary tumor datasets of clinical prostatic samples (GSE21034). **A.** Illustrated is an enrichment profile of a single AR subnetwork which has significant differential expression between normal and tumor datasets, and was only significant for a population of White (non-Hispanic) but not African-American men. **B.** Enrichment profile for African-American men gene-set 1. Survival outcome analysis was followed-up on these gene-sets. Disease-free survival of CaP patients stratified by serum PSA, Gleason score, Tumor stage, and combined risk (PSA, Gleason and Tumor Stage) with respect to the AR subnetwork cluster. Kaplan-Meier analysis was used to plot the fraction of at-risk patients remaining free of disease (y-axis) at the indicated time after radical prostatectomy (x-axis). Patient stratification is based on serum PSA (≥4 ng/mL), Gleason score (≥7), Tumor Stage (≥T3) and combination of serum PSA, Gleason score and Tumor Stage score values. **C.** White (non-Hispanic) functional cluster gene-set 2. **D.** African-American functional clusters gene-set 1.

medulloblastoma clinical samples [26]. The gene-sets illustrated in **Figure 4** did not give significant outcome values for any of the four other cancers (**Figure 5**). Furthermore, the remaining 8

gene-sets, described above, also lacked significant predictive outcomes for the same 4 non-prostate cancers analyzed (see **Figure S2**). In hormone-dependent breast cancer, certain ethnic population differences have been observed, with higher incidences of breast cancer occurring in African-American woman vs. other groups [44,45,46], but similar population demographic data used for the CaP cohort analyzed within our study was not available for the non-CaP cancers used in this analysis. Such genetic expression data would have been useful for further confirmatory follow-up studies. However, the expression of the AR has been described in a number of non-CaP cancers, especially breast cancer [47,48,49,50,51,52,53]. It has also been shown that several cell lines from these non-CaP cancers show androgen sensitivity and androgen-dependent gene expression profiles similar to CaP cell lines. However, we have identified AR interactome gene-sets that can that can differentiate between CaP and non-CaP disease survival which suggests that there are unique molecular characteristics of AR function in CaP and part of a CaP-specific pathway in neoplastic development, and also that these gene-sets can be used to predict CaP disease outcomes between genetically diverse groups.

White (non-Hispanic) gene-set can predict CaP disease outcome based on gene copy number

Finally, to deduce a molecular mechanism to account for differential gene expression patterns for each gene-set, we analyzed copy number variation of a CGH array dataset (GSE21035) [24], for the patients used for our GSEA and survival prediction outcome. For the White (non-Hispanic) gene-set 2, we were able to confirm that patient survival based on the criteria of PSA value, Gleason score and combined (PSA+Gleason), was dependent on copy number variation (**Figure 6**). This was the only gene-set that

was predictive for survival outcomes based on copy number variation.

Discussion

Knowledge of various molecular mechanisms of action contribute to our understanding of wild type AR function. Most mechanisms require the involvement of ligand binding and interacting partners. Examples of this include AR interactors involved in gene transactivation, including HSP70, HSP90, p300 and components of the RNA polII complex [54]. Gain-of-function somatic mutations, abundant in cancerous tissues, typically add new functions, adding to the complexity of physiological and disease outcomes. We investigated the T877A-AR mutation, as it represents the most common AR mutation in clinical CaP specimens, and is the AR mutation found in the most studied prostatic cancer cell line, LNCaP. Mutations like T877A-AR, and several others in the ligand-binding domain of the receptor, allow the AR to bind to other classes of steroid ligands such as estradiol, dexamethasone and progesterone, including anti-androgens such as CPA, resulting in subsequent AR dependent gene transactivation [55]. It is also now clear that as cancers evolve through many somatic mutations [56,57] and undergo selection processes induced by classical drug therapies themselves. Futhrermore, a number of studies have used LNCaP cells as a model for studying the progression from androgen-dependent to –independent/ castrate resistant prostate cancer (AIPC/CRPC) state and support the hypothesis that continuous AR activity and signaling continues to be one of the most important mechanisms in CRPC [58,59,60,61,62]. These studies have substantiated extensive

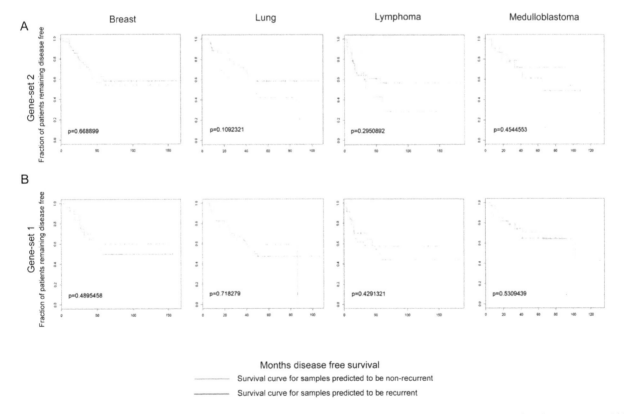

Figure 5. Disease-free survival of 4 non-CaP cancers. Datasets were retrieved from the Broad Institute (http://www.broadinstitute.org/cgi-bin/cancer/datasets.cgi) representing expression and disease outcomes for the following cancers: breast, lymphoma, lung and medulloblastoma, to determine survival outcomes for the gene-sets 1 and 2, as described in Figure 4.

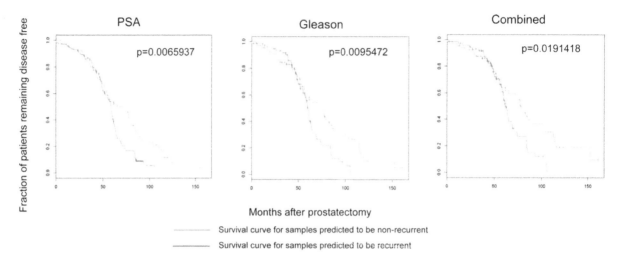

Figure 6. Survival outcomes reflected by Copy Number Variation. Survival outcomes based on copy number variation (GSE21035) of gene-set 2 and same patients described in Figure 4, was applied. Patient stratification is based on serum PSA ($\geq$4 ng/mL) and Gleason score ($\geq$7) and combination of serum PSA and Gleason score values.

genetic alterations that range from single missense mutations, to copy number variations, splicing variants, genetic rearrangements and short DNA alterations in a large number of genes and AR co-factor interactions to reproduce androgen-independent scenario [1,2,6,7,36,63,64]. Modeling AIPC using LNCaP cell lines and actual tissue from AIPC patients, Wang et al., 2009 [64] found that the of gene expression regulated by the AR in the absence of hormone is distinct from androgen-regulated program and can selectively and directly upregulate M-phase genes found in androgen-independent CaP and may explain why maximal androgen deprivation (AR antagonists and LHRH inhibitors), and cannot prolong androgen-independent survival. Most recently, it was found that overexpression of AR was a result to prolonged exposure of LNCaP-derived xenografts with the anti-androgen enzalutamide, and was similar to chronic androgen depletion, as a significant mechanism for drug resistance and CRPC development [65].

In this study we characterized whole mutant AR protein complexes with several classes of steroids and ligands known to bind T877A-AR. The specific interactomes were dependent on the ligand utilized; so too were the specific gene expression profiles associated with each ligand. Thus mutant AR gain-of-function properties are not singular but multiple, dictated by the class of steroid hormones used. Further exploration of other adrenal androgens such as DHEA or androstenedione or other anti-androgens such as flutamide and bicultamide were not examined in this study, however, we did select a diverse class of ligands known to bind to the T877A-AR variant.

High-throughput gene expression microarray approaches described in CaP cells have identified hundreds of androgen-regulated genes and also characterized genome-wide AR recruitment sites [66,67,68,69]. The classical AR complex contains general transcription factors, coregulators and specific transcription factors that associate either directly or indirectly with the AR to enhance or repress its transcriptional activity function without themselves necessarily binding to DNA. As shown in our recent proteomic studies [17], including this one, AR complexes may also include a larger number of functionally diverse proteins involved in a multitude of "non-classical" AR cellular processes such as histone acetylation, DNA methylation, ubiquitination, RNA

splicing, apoptosis, and protein synthesis, with all pathways found to be dependent on hormone stimulation conditions.

From our data, there are several clusters of AR-interacting proteins that are worth exploring to understand their role in disease progression. The first cluster of AR-interacting proteins is unique to the African-American population group and consists of the following proteins: **ERCC1, ERCC2, ERCC3, ERCC5** and **FEN1**. The function of these proteins is required for mediating DNA damage excision repair. However, they are known components of the RNA polymerase II transcriptional complex [70] and ERCC2 and ERCC3 have been previously published as AR interactors [71]. Experimental evidence show that over-activity of this complex can lead to instability of CAG/CTG triplet repeats, resulting in a shortening of the repeat [72,73].

The second cluster of AR-interacting proteins is unique for the White (non-Hispanic) population and is involved in chromatin remodeling and histone deacetylation activity. From this cluster we identified the following proteins: **KDM1A, KDM4C, SIRT1, CTBP1, NR2C1** and **SMARCD3**. NR2C1 has been previously described as an AR interactor [74]. The selection of this cluster of proteins for further analysis would be interesting because of the increased attention histone deacetylase inhibitors that are garnering in cancer biology [75].

A final pathway for further investigation, and unique to the White (non-Hispanic) group of men, is the role of the AR in participating in the negative regulation of apoptosis via its interaction with **BCL2, RELA, FAS, EEF1A2** and **NR4A2**. The role of these proteins have been well described in apoptotic pathways, and RELA (NFκB p65) is a well characterized interactor of AR [76] and BCL2 [77]. A proposed mechanism by which these proteins may facilitate regulating apoptosis would be via a signal transduction cascade that would negatively regulate pro-apoptotic genes and proteins [77].

Using the T877A-AR hormone-specific interaction complexes as a basis for a novel systems biology network analysis exercise established gene-sets with clear predictive CaP clinical outcome value. In doing so, we confirmed the critical importance of the genetic backgrounds of the CaP individuals in the clinical dataset. Without segregating the microarray expression data of CaP patients along White (non-Hispanic) and African-American datasets, no defined gene-sets with predictive clinical values could be identified. Once the data had been segregated, gene-sets with very powerful clinical outcome parameters were discovered. African–American men have long been considered to have clinically different CaP from White (non-Hispanic) men, based on their genetic background [78,79]. This is not entirely surprising, as African–American descent has long been associated with higher incidence and more aggressive disease, characterized by greater tumor volume for each clinical stage, have greater PSA levels and a more aggressive cancer for Gleason score of 8 or greater with compare to White (non-Hispanics) males [80,81]. Investigations, excluding socio-economic disparities that would limit individuals to health care [82], to explain racial differences between African-American vs. White (non-Hispanic) males, have excluded hormones levels, as serum testosterone levels (later in life) have similar levels at the time of prostate biopsy and in the their prostate biopsy tissue [83,84]. However, analysis of AR expression in malignant vs. benign prostate tissue African-American males found to be 27% more likely to stain positive for AR and the nuclear localization of the AR was 81% greater than White (non-Hispanics) [85]. It was also obersved that there was significantly expression of CaP biomarkers that those of white men, one of which was the AR [86]. Interestingly, African-American males vs. other ethnic groups display shorter *AR* CAG repeat lengths which code for the polyglutamine tract of the AR [87]. AR with shorter polyglutamine tracts exhibits higher AR activity and represents a potential risk factor for CaP [88,89,90]. Therefore, dysregulation of DNA repair function, specifically loss of expression, and the link to contraction of CAG tract length in African-American males is a mechanism to investigate as a mechanism to explain racial differences in CaP. Furthermore, to gain more insight into CaP molecular/genetic etiology, our approach, encompassing proteomics, expression studies and network analysis, would benefit from investigating even more populations of diverse genetic origin backgrounds that have very low rates of CaP including Chinese and Middle Eastern men. By identifying such protective pathways of AR function, along with identifying powerful prognostic tools to predict disease, now we can assess pathways to also offer novel therapeutic targets.

In conclusion, our unique approach of using an important gain-of-function AR mutation, has generated gene-sets along functional organization lines showing that we can distinguish prostatic disease between White (non-Hispanic) and African-American men. Moreover, in our study for classification of AR function based on interaction profiles was a much more powerful tool predicting disease and survival outcomes than analyzing androgen-dependent gene expression patterns. The identification of these new functional properties of the AR and somatic mutations of the AR, further suggests that there is a role for the AR beyond that as a transcription factor and also implicates the ability of non-androgenic hormones to activate CaP disease-linked pathways.

Supporting Information

Figure S1 CaP 10-year survival outcomes forT877A-AR Gene-sets 3–10.
(PDF)

Figure S2 Disease-free survival outcomes of 4 non-CaP cancers for Gene-sets 3–10.
(PDF)

Table S1 Normalized Proteomic Data.
(XLS)

Table S2 GO-Term definitions of Proteomic data.
(XLS)

Table S3 GO-Term definitions of microarray data.
(XLS)

Table S4 GO-Term definitions of T877A-AR Gene-sets 1–10.
(XLS)

Author Contributions

Conceived and designed the experiments: MP MT EW. Performed the experiments: NZ PG SC EB. Analyzed the data: SC EB PT MP. Contributed reagents/materials/analysis tools: PT EW MT MP. Contributed to the writing of the manuscript: EB MT MP.

References

1. Kumar A, White TA, MacKenzie AP, Clegg N, Lee C, et al. (2011) Exome sequencing identifies a spectrum of mutation frequencies in advanced and lethal prostate cancers. Proc Natl Acad Sci U S A 108: 17087–17092.
2. Berger MF, Lawrence MS, Demichelis F, Drier Y, Cibulskis K, et al. (2011) The genomic complexity of primary human prostate cancer. Nature 470: 214–220.
3. Barbieri CE, Tomlins SA (2014) The prostate cancer genome: Perspectives and potential. Urol Oncol 32: 15–22.
4. Xu X, Zhu K, Liu F, Wang Y, Shen J, et al. (2013) Identification of somatic mutations in human prostate cancer by RNA-Seq. Gene 519: 343–347.
5. Baca SC, Prandi D, Lawrence MS, Mosquera JM, Romanel A, et al. (2013) Punctuated evolution of prostate cancer genomes. Cell 153: 666–677.
6. Wu C, Wyatt AW, Lapuk AV, McPherson A, McConeghy BJ, et al. (2012) Integrated genome and transcriptome sequencing identifies a novel form of hybrid and aggressive prostate cancer. J Pathol 227: 53–61.
7. Hieronymus H, Sawyers CL (2012) Traversing the genomic landscape of prostate cancer from diagnosis to metastasis. Nat Genet 44: 613–614.
8. Bielas JH, Loeb KR, Rubin BP, True LD, Loeb LA (2006) Human cancers express a mutator phenotype. Proc Natl Acad Sci U S A 103: 18238–18242.
9. Venkatesan RN, Bielas JH, Loeb LA (2006) Generation of mutator mutants during carcinogenesis. DNA Repair (Amst) 5: 294–302.
10. Vogelstein B, Kinzler KW (2002) The genetic basis of human cancer. New York: McGraw-Hill, Medical Pub. Division. xv, 821 p.
11. Taplin ME, Bubley GJ, Ko YJ, Small EJ, Upton M, et al. (1999) Selection for androgen receptor mutations in prostate cancers treated with androgen antagonist. Cancer Res 59: 2511–2515.
12. Joseph JD, Lu N, Qian J, Sensintaffar J, Shao G, et al. (2013) A clinically relevant androgen receptor mutation confers resistance to second-generation antiandrogens enzalutamide and ARN-509. Cancer discovery 3: 1020–1029.
13. Korpal M, Korn JM, Gao X, Rakiec DP, Ruddy DA, et al. (2013) An F876L mutation in androgen receptor confers genetic and phenotypic resistance to MDV3100 (enzalutamide). Cancer discovery 3: 1030–1043.
14. Gaddipati JP, McLeod DG, Heidenberg HB, Sesterhenn IA, Finger MJ, et al. (1994) Frequent detection of codon 877 mutation in the androgen receptor gene in advanced prostate cancers. Cancer Res 54: 2861–2864.
15. Taplin ME, Bubley GJ, Shuster TD, Frantz ME, Spooner AE, et al. (1995) Mutation of the androgen-receptor gene in metastatic androgen-independent prostate cancer. N Engl J Med 332: 1393–1398.
16. Veldscholte J, Ris-Stalpers C, Kuiper GG, Jenster G, Berrevoets C, et al. (1990) A mutation in the ligand binding domain of the androgen receptor of human LNCaP cells affects steroid binding characteristics and response to anti-androgens. Biochem Biophys Res Commun 173: 534–540.
17. Paliouras M, Zaman N, Lumbroso R, Kapogeorgakis L, Beitel LK, et al. (2011) Dynamic rewiring of the androgen receptor protein interaction network correlates with prostate cancer clinical outcomes. Integrative biology : quantitative biosciences from nano to macro 10: 1020–1032.
18. Beitel LK, Sabbaghian N, Alarifi A, Alvarado C, Pinsky L, et al. (1995) Characterization of normal and point-mutated human androgen receptors expressed in the baculovirus system. J Mol Endocrinol 15: 117–128.
19. Saba J, Bonneil E, Pomies C, Eng K, Thibault P (2009) Enhanced sensitivity in proteomics experiments using FAIMS coupled with a hybrid linear ion trap/Orbitrap mass spectrometer. J Proteome Res 8: 3355–3366.
20. Kearney P, Thibault P (2003) Bioinformatics meets proteomics–bridging the gap between mass spectrometry data analysis and cell biology. J Bioinform Comput Biol 1: 183–200.
21. Awan A, Bari H, Yan F, Moksong S, Yang S, et al. (2007) Regulatory network motifs and hotspots of cancer genes in a mammalian cellular signalling network. IET Syst Biol 1: 292–297.
22. Cui Q, Ma Y, Jaramillo M, Bari H, Awan A, et al. (2007) A map of human cancer signaling. Mol Syst Biol 3: 152.
23. Cui Q, Yu Z, Purisima EO, Wang E (2006) Principles of microRNA regulation of a human cellular signaling network. Mol Syst Biol 2: 46.

24. Taylor BS, Schultz N, Hieronymus H, Gopalan A, Xiao Y, et al. (2010) Integrative genomic profiling of human prostate cancer. Cancer Cell 18: 11–22.

25. Li J, Lenferink AE, Deng Y, Collins C, Cui Q, et al. (2010) Identification of high-quality cancer prognostic markers and metastasis network modules. Nature communications 1: 34.

26. Ramaswamy S, Ross KN, Lander ES, Golub TR (2003) A molecular signature of metastasis in primary solid tumors. Nat Genet 33: 49–54.

27. Case DA, Cheatham TE 3rd, Darden T, Gohlke H, Luo R, et al. (2005) The AMBER biomolecular simulation programs. J Comput Chem 26: 1668–1688.

28. Gotz AW, Williamson MJ, Xu D, Poole D, Le Grand S, et al. (2012) Routine microsecond molecular dynamics simulations with AMBER on GPUs. 1. Generalized Born. Journal of Chemical Theory and Computation 8: 1542–1555.

29. Hornak V, Abel R, Okur A, Strockbine B, Roitberg A, et al. (2006) Comparison of multiple AMBER force fields and development of improved protein backbone parameters. Proteins 65: 712–725.

30. Jorgensen WL, Chandrasekhar J, Madura JD, Impey RW, Klein ML (1983) Comparison of simple potential functions for simulating liquid water. Journal of Chemical Physics 79: 926–935.

31. Darden T, York D, Pedersen L (1993) Particle mesh ewald - an N.Log(N) method for ewald sums in large systems. Journal of Chemical Physics 98: 10089–10092.

32. Ryckaert JP, Ciccotti G, Berendsen HJC (1977) Numerical-integration of cartesian equations of motion of a system with constraints - molecular-dynamics of N-alkanes. Journal of Computational Physics 23: 327–341.

33. Parker CE, Warren MR, Loiselle DR, Dicheva NN, Scarlett CO, et al. (2005) Identification of components of protein complexes. Methods Mol Biol 301: 117–151.

34. Bohl CE, Gao W, Miller DD, Bell CE, Dalton JT (2005) Structural basis for antagonism and resistance of bicalutamide in prostate cancer. Proceedings Of The National Academy Of Sciences Of The United States Of America 102: 6201–6206.

35. Zhao XY, Boyle B, Krishnan AV, Navone NM, Peehl DM, et al. (1999) Two mutations identified in the androgen receptor of the new human prostate cancer cell line MDA PCa 2a. J Urol 162: 2192–2199.

36. Zhao XY, Malloy PJ, Krishnan AV, Swami S, Navone NM, et al. (2000) Glucocorticoids can promote androgen-independent growth of prostate cancer cells through a mutated androgen receptor. Nat Med 6: 703–706.

37. Matias PM, Carrondo MA, Coelho R, Thomaz M, Zhao XY, et al. (2002) Structural basis for the glucocorticoid response in a mutant human androgen receptor (AR(ccr)) derived from an androgen-independent prostate cancer. J Med Chem 45: 1439–1446.

38. van de Wijngaart DJ, Molier M, Lusher SJ, Hersmus R, Jenster G, et al. (2010) Systematic structure-function analysis of androgen receptor Leu701 mutants explains the properties of the prostate cancer mutant L701H. J Biol Chem 285: 5097–5105.

39. Mahmoud AM, Zhu T, Parray A, Siddique HR, Yang W, et al. (2013) Differential effects of genistein on prostate cancer cells depend on mutational status of the androgen receptor. PLoS One 8: e78479.

40. Zhou J, Geng G, Shi Q, Sauriol F, Wu JH (2009) Design and synthesis of androgen receptor antagonists with bulky side chains for overcoming antiandrogen resistance. J Med Chem 52: 5546–5550.

41. Askew EB, Gampe RT Jr, Stanley TB, Faggart JL, Wilson EM (2007) Modulation of androgen receptor activation function 2 by testosterone and dihydrotestosterone. J Biol Chem 282: 25801–25816.

42. Bohl CE, Wu Z, Miller DD, Bell CE, Dalton JT (2007) Crystal structure of the T877A human androgen receptor ligand-binding domain complexed to cyproterone acetate provides insight for ligand-induced conformational changes and structure-based drug design. J Biol Chem 282: 13648–13655.

43. Zhao H, Kim Y, Wang P, Lapointe J, Tibshirani R, et al. (2005) Genome-wide characterization of gene expression variations and DNA copy number changes in prostate cancer cell lines. Prostate 63: 187–197.

44. Wu AH, Gomez SL, Vigen C, Kwan ML, Keegan TH, et al. (2013) The California Breast Cancer Survivorship Consortium (CBCSC): prognostic factors associated with racial/ethnic differences in breast cancer survival. Cancer causes & control: CCC 24: 1821–1836.

45. Long J, Zhang B, Signorello LB, Cai Q, Deming-Halverson S, et al. (2013) Evaluating genome-wide association study-identified breast cancer risk variants in African-American women. PLoS One 8: e58350.

46. Lindner R, Sullivan C, Offor O, Lezon-Geyda K, Halligan K, et al. (2013) Molecular phenotypes in triple negative breast cancer from African American patients suggest targets for therapy. PLoS One 8: e71915.

47. Park HS, Bae JS, Noh SJ, Kim KM, Lee H, et al. (2013) Expression of DBC1 and androgen receptor predict poor prognosis in diffuse large B cell lymphoma. Translational oncology 6: 370–381.

48. Mikkonen L, Pihlajamaa P, Sahu B, Zhang FP, Janne OA (2010) Androgen receptor and androgen-dependent gene expression in lung. Mol Cell Endocrinol 317: 14–24.

49. Recchia AG, Musti AM, Lanzino M, Panno ML, Turano E, et al. (2009) A cross-talk between the androgen receptor and the epidermal growth factor receptor leads to p38MAPK-dependent activation of mTOR and cyclinD1 expression in prostate and lung cancer cells. Int J Biochem Cell Biol 41: 603–614.

50. Kaiser U, Hofmann J, Schilli M, Wegmann B, Klotz U, et al. (1996) Steroid-hormone receptors in cell lines and tumor biopsies of human lung cancer. International journal of cancer Journal international du cancer 67: 357–364.

51. McGhan LJ, McCullough AE, Protheroe CA, Dueck AC, Lee JJ, et al. (2014) Androgen receptor-positive triple negative breast cancer: a unique breast cancer subtype. Annals of surgical oncology 21: 361–367.

52. Vera-Badillo FE, Templeton AJ, de Gouveia P, Diaz-Padilla I, Bedard PL, et al. (2014) Androgen receptor expression and outcomes in early breast cancer: a systematic review and meta-analysis. Journal of the National Cancer Institute 106: djt319.

53. Ren Q, Zhang L, Ruoff R, Ha S, Wang J, et al. (2013) Expression of androgen receptor and its phosphorylated forms in breast cancer progression. Cancer 119: 2532–2540.

54. Heemers HV, Tindall DJ (2007) Androgen receptor (AR) coregulators: a diversity of functions converging on and regulating the AR transcriptional complex. Endocr Rev 28: 778–808.

55. Ngan S, Stronach EA, Photiou A, Waxman J, Ali S, et al. (2009) Microarray coupled to quantitative RT-PCR analysis of androgen-regulated genes in human LNCaP prostate cancer cells. Oncogene 28: 2051–2063.

56. Lee W, Jiang Z, Liu J, Haverty PM, Guan Y, et al. (2010) The mutation spectrum revealed by paired genome sequences from a lung cancer patient. Nature 465: 473–477.

57. Pleasance ED, Cheetham RK, Stephens PJ, McBride DJ, Humphray SJ, et al. (2010) A comprehensive catalogue of somatic mutations from a human cancer genome. Nature 463: 191–196.

58. Marques RB, Dits NF, Erkens-Schulze S, van Weerden WM, Jenster G (2010) Bypass mechanisms of the androgen receptor pathway in therapy-resistant prostate cancer cell models. PLoS One 5: e13500.

59. Debes JD, Tindall DJ (2004) Mechanisms of androgen-refractory prostate cancer. N Engl J Med 351: 1488–1490.

60. Feldman BJ, Feldman D (2001) The development of androgen-independent prostate cancer. Nat Rev Cancer 1: 34–45.

61. Heinlein CA, Chang C (2004) Androgen receptor in prostate cancer. Endocr Rev 25: 276–308.

62. Ueda T, Mawji NR, Bruchovsky N, Sadar MD (2002) Ligand-independent activation of the androgen receptor by interleukin-6 and the role of steroid receptor coactivator-1 in prostate cancer cells. J Biol Chem 277: 38087–38094.

63. Baca SC, Garraway LA (2012) The genomic landscape of prostate cancer. Front Endocrinol (Lausanne) 3: 69.

64. Wang Q, Li W, Zhang Y, Yuan X, Xu K, et al. (2009) Androgen receptor regulates a distinct transcription program in androgen-independent prostate cancer. Cell 138: 245–256.

65. Loriot Y, Wyatt A, Al Nakouzi N, Beraldi E, Toren P, et al. (2013) Mechanisms of resistance to enzalutamide in LNCaP models. Cancer Res 73.

66. Massie CE, Adryan B, Barbosa-Morais NL, Lynch AG, Tran MG, et al. (2007) New androgen receptor genomic targets show an interaction with the ETS1 transcription factor. EMBO Rep 8: 871–878.

67. Chen H, Libertini SJ, George M, Dandekar S, Tepper CG, et al. (2010) Genome-wide analysis of androgen receptor binding and gene regulation in two CWR22-derived prostate cancer cell lines. Endocr Relat Cancer 17: 857–873.

68. Takayama K, Kaneshiro K, Tsutsumi S, Horie-Inoue K, Ikeda K, et al. (2007) Identification of novel androgen response genes in prostate cancer cells by coupling chromatin immunoprecipitation and genomic microarray analysis. Oncogene 26: 4453–4463.

69. Takayama K, Tsutsumi S, Katayama S, Okayama T, Horie-Inoue K, et al. (2011) Integration of cap analysis of gene expression and chromatin immunoprecipitation analysis on array reveals genome-wide androgen receptor signaling in prostate cancer cells. Oncogene 30: 619–630.

70. van Brabant AJ, Stan R, Ellis NA (2000) DNA helicases, genomic instability, and human genetic disease. Annu Rev Genomics Hum Genet 1: 409–459.

71. Chymkowitch P, Le May N, Charneau P, Compe E, Egly JM (2011) The phosphorylation of the androgen receptor by TFIIH directs the ubiquitin/proteasome process. The EMBO journal 30: 468–479.

72. Hubert L Jr, Lin Y, Dion V, Wilson JH (2011) Xpa deficiency reduces CAG trinucleotide repeat instability in neuronal tissues in a mouse model of SCA1. Hum Mol Genet 20: 4822–4830.

73. Lin Y, Wilson JH (2009) Diverse effects of individual mismatch repair components on transcription-induced CAG repeat instability in human cells. DNA repair 8: 878–885.

74. Mu XM, Chang CS (2003) TR2 orphan receptor functions as negative modulator for androgen receptor in prostate cancer cells PC-3. Prostate 57: 129–133.

75. Russo D, Durante C, Bulotta S, Puppin C, Puxeddu E, et al. (2013) Targeting histone deacetylase in thyroid cancer. Expert opinion on therapeutic targets 17: 179–193.

76. Palvimo JJ, Reinikainen P, Ikonen T, Kallio PJ, Moilanen A, et al. (1996) Mutual transcriptional interference between RelA and androgen receptor. Journal of Biological Chemistry 271: 24151–24156.

77. de Moissac D, Zheng H, Kirshenbaum LA (1999) Linkage of the BH4 domain of Bcl-2 and the nuclear factor kappaB signaling pathway for suppression of apoptosis. J Biol Chem 274: 29505–29509.

78. Abern MR, Bassett MR, Tsivian M, Banez LL, Polascik TJ, et al. (2013) Race is associated with discontinuation of active surveillance of low-risk prostate cancer: Results from the Duke Prostate Center. Prostate Cancer Prostatic Dis 16: 85–90.

79. Fukagai T, Namiki T, Carlile RG, Namiki M (2009) Racial differences in clinical outcome after prostate cancer treatment. Methods in molecular biology 472: 455–466.

80. Powell IJ (1998) Prostate cancer in the African American: is this a different disease? Seminars in urologic oncology 16: 221–226.

81. Chornokur G, Dalton K, Borysova ME, Kumar NB (2011) Disparities at presentation, diagnosis, treatment, and survival in African American men, affected by prostate cancer. The Prostate 71: 985–997.

82. Ellis SD, Blackard B, Carpenter WR, Mishel M, Chen RC, et al. (2013) Receipt of National Comprehensive Cancer Network guideline-concordant prostate cancer care among African American and Caucasian American men in North Carolina. Cancer 119: 2282–2290.

83. Kubricht WS 3rd, Williams BJ, Whatley T, Pinckard P, Eastham JA (1999) Serum testosterone levels in African-American and white men undergoing prostate biopsy. Urology 54: 1035–1038.

84. Marks LS, Hess DL, Dorey FJ, Macairan ML (2006) Prostatic tissue testosterone and dihydrotestosterone in African-American and white men. Urology 68: 337–341.

85. Gaston KE, Kim D, Singh S, Ford OH 3rd, Mohler JL (2003) Racial differences in androgen receptor protein expression in men with clinically localized prostate cancer. J Urol 170: 990–993.

86. Kim HS, Moreira DM, Jayachandran J, Gerber L, Banez LL, et al. (2011) Prostate biopsies from black men express higher levels of aggressive disease biomarkers than prostate biopsies from white men. Prostate Cancer Prostatic Dis 14: 262–265.

87. Lange EM, Sarma AV, Ray A, Wang Y, Ho LA, et al. (2008) The androgen receptor CAG and GGN repeat polymorphisms and prostate cancer susceptibility in African-American men: results from the Flint Men's Health Study. Journal of human genetics 53: 220–226.

88. Bratt O, Borg A, Kristoffersson U, Lundgren R, Zhang QX, et al. (1999) CAG repeat length in the androgen receptor gene is related to age at diagnosis of prostate cancer and response to endocrine therapy, but not to prostate cancer risk. British Journal of Cancer 81: 672–676.

89. Ding D, Xu L, Menon M, Reddy GP, Barrack ER (2004) Effect of a short CAG (glutamine) repeat on human androgen receptor function. The Prostate 58: 23–32.

90. Rodriguez-Gonzalez G, Cabrera S, Ramirez-Moreno R, Bilbao C, Diaz-Chico JC, et al. (2009) Short alleles of both GGN and CAG repeats at the exon-1 of the androgen receptor gene are associated to increased PSA staining and a higher Gleason score in human prostatic cancer. J Steroid Biochem Mol Biol 113: 85–91.

13

Adaptive Firefly Algorithm: Parameter Analysis and its Application

Ngaam J. Cheung[1⑨], **Xue-Ming Ding**[2⑨], **Hong-Bin Shen**[1*]

1 Institute of Image Processing and Pattern Recognition, Shanghai Jiao Tong University, and Key Laboratory of System Control and Information Processing, Ministry of Education of China, Shanghai, China, **2** School of Optical-Electrical and Computer Engineering, University of Shanghai for Science and Technology, Shanghai, China

Abstract

As a nature-inspired search algorithm, firefly algorithm (FA) has several control parameters, which may have great effects on its performance. In this study, we investigate the parameter selection and adaptation strategies in a modified firefly algorithm — adaptive firefly algorithm (AdaFa). There are three strategies in AdaFa including (1) a distance-based light absorption coefficient; (2) a gray coefficient enhancing fireflies to share difference information from attractive ones efficiently; and (3) five different dynamic strategies for the randomization parameter. Promising selections of parameters in the strategies are analyzed to guarantee the efficient performance of AdaFa. AdaFa is validated over widely used benchmark functions, and the numerical experiments and statistical tests yield useful conclusions on the strategies and the parameter selections affecting the performance of AdaFa. When applied to the real-world problem — protein tertiary structure prediction, the results demonstrated improved variants can rebuild the tertiary structure with the average root mean square deviation less than 0.4Å and 1.5Å from the native constrains with noise free and 10% Gaussian white noise.

Editor: Haipeng Peng, Beijing University, China

Funding: This work was supported by the China Scholarship Council (NJC, No. 201406230154), the National Natural Science Foundation of China (HBS, No. 61222306, 91130033, 61175024), Shanghai Science and Technology Commission (HBS, No. 11JC1404800), and A Foundation for the Author of National Excellent Doctoral Dissertation of PR China (HBS, No. 201048). The funders had no role in study design, data collection and analysis, decision to publish, or preparation of the manuscript.

* Email: hbshen@sjtu.edu.cn

⑨ These authors contributed equally to this work.

Introduction

Firefly algorithm (FA) is a simple yet quite efficient nature-inspired search technique for global optimization. Since FA was developed, it has attracted a lot of attentions and becomes more popular in solving various real-world problems [1–6]. FA is a swarm-based intelligence algorithm, which mimics the flashing behavior of fireflies [7]. A firefly flashes as a signal to attract others for some purposes, e.g. predation or mating. Accordingly, this biological phenomenon is formulated as a meta-heuristic algorithm depending on following three rules [7,8]:

- All fireflies are attracted by each other without respect to their sex;
- Attractiveness is proportional to its brightness, that is, the less bright one will move towards the brighter one;
- If there are no brighter fireflies than a particular firefly, it will move randomly in the space.

Similar to other heuristic algorithms, FA also has several control parameters to be tuned, for instance, the light absorption coefficient, the randomization control factor, and the population size, for good performance on different problems. The values of these control parameters greatly determine the qualities of the achieved solutions and the efficiency of the FA algorithm. Generally, it is a problem dependent task to select suitable control parameters for FA. As to a complex problem, it will be a hard case to deal with if the problem with numerous local optima, in which most optimization algorithms will be trapped [9–11]. Although it is highly important, there is no consistent methodology for determining the control parameters of a FA variant. Mostly, the parameters are fixed experientially or set arbitrarily within some predefined ranges.

In FA algorithm, there are two important issues: (1) the variation of light intensity; and (2) formulation of attractiveness [7]. However, these parameters are either set constants or fixed empirically in the traditional FA [7], which may make the algorithm inefficient for the problems with complex landscapes [3]. Researchers have made numerous contributions to the improvement of FA considering the alteration of the control parameters. For example, Gandomi et al. applied several chaos mechanisms to tune light absorption coefficient and attractiveness coefficient [12]. In ref. [1], a geometric progression reduction scheme for the randomization parameter was introduced in FA to enhance the solution quality. Coelho and Mariani adopted Gaussian distribution probability functions to tune the light absorption coefficient and the randomization parameter [3]. In ref. [2], a chaotic mapping operator and a chaotic component were used to generate initial solutions and replace the random component of the standard FA, respectively. The former was to improve the quality of the initial population, while the latter was to perform ergodic search of the solution space.

Although much progress has been achieved on the FA-based algorithms since 2008, more efforts are required to further improve their performance:

- Providing the sufficient analysis for the control parameter settings;
- Efficient strategies or mechanisms for the selections of the control parameters;
- Employing heterogeneous search rules to enhance the performance of FA.

In this paper, the contributions are to develop several mechanisms and strategies for improving standard firefly algorithm, involving: (1) distance-based light absorption coefficient and gray coefficient for adaptively altering the attractiveness and enhancing difference information sharing, respectively; (2) five strategies for controlling the randomization parameter; and (3) employing heterogeneous search rules supported by the adaptively altering control parameters, to enhance the search ability of the original FA. The control parameters are adjusted over time or depending on heuristic rules, which employ the evolutionary information among the fireflies.

Methods

Firefly algorithm

The firefly algorithm (FA) is a nature-inspired optimization method [7], which maintains a population of fireflies to find the global optimum of an optimization problem.

In FA, the distance between any two fireflies i and j at $\mathbf{x}_i$ and $\mathbf{x}_j$, respectively, can be defined as the Euclidean distance r_{ij}, which is formulated as follows,

$$r_{ij} = \|\mathbf{x}_i - \mathbf{x}_j\| = \sqrt{\sum_{d=1}^{D}(x_{i,d} - x_{j,d})^2} \tag{1}$$

where D is the dimension of an optimization problem.

Indeed, the larger the distance r_{ij} is, the less light the fireflies can see from each other. Accordingly, it is necessary to define

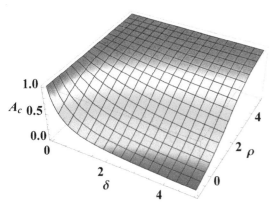

Figure 1. The relationship among A_c, σ and ρ. As σ and ρ increase, A_c will decrease while the value of A_c will sharply increase when σ and ρ are very small (or large).

monotonically decreasing functions for light intensity and attractiveness, respectively. They are presented in Eqs. (2) and (3).

$$I(r) = I_0 e^{-\gamma r^2} \tag{2}$$

where I_0 is initial light intensity, and γ is the light absorption coefficient, which controls the decrease of light intensity. Accordingly, the attractiveness β of a firefly is defined as shown in Eq. (3).

$$\beta(r) = \beta_0 e^{-\gamma r^2} \tag{3}$$

where β_0 is a constant, which is the attractiveness at $r = 0$.

The step of a firefly i is attracted to move to another more attractive (brighter) firefly j is determined by

$$\Delta \mathbf{x}_i(t) = \beta \cdot (\mathbf{x}_i(t) - \mathbf{x}_j(t)) + \alpha(\mathbf{e}_t - \mathbf{c}) \tag{4}$$

where $\mathbf{c}$ is a constant vector $[0.5, 0.5, \ldots, 0.5]^D$ and t is the time step, $\mathbf{e}_t$ is drawn from a normal distribution $N(0,1)$. $\Delta \mathbf{x}$ is the step size of the ith firefly moving. The first term is the attraction from the jth firefly, while the second term is randomization controlled by α, which is a constant in the range of $(0,1)$. Therefore, the update of the ith firefly is formulated as follows,

$$\mathbf{x}_i(t+1) = \mathbf{x}_i(t) + \Delta \mathbf{x}_i(t) \tag{5}$$

The Eqs. (4) and (5) show that the ith firefly will move towards the jth firefly, which is a more attractive one.

The procedure of FA algorithm is summarized as follows (*Algorithm S1*):

The selections of parameters are crucially important in FA, such as the light absorption coefficient γ in Eq. (3) and the randomization parameter α in Eq. (4), but the values of the control parameters are chosen in predefined ranges dogmatically [13]. During the search process, traditional FA does not alter the values of the control parameters or only use constant parameters throughout the whole process. Also the information of the search or the knowledge achieved by the fireflies are not taken into account in the selections of parameters. All these static designs may be optimal for one problem, but not efficient or even fail to guarantee convergence for another one [13]. Proper selections of these parameters highly determine the quality of the solution and the search efficiency. Although researchers have proposed many improved FA variants [1,14–16], premature convergence can still occur in the original firefly algorithm and its variants. FA may be easily trapped in local regions when it is used to deal with the complex problems with numerous local optima if the randomness is reduced to quickly [3]. To overcome these weaknesses in FA, we develop five variants of FA based on two main mechanisms, which will be described in details in following sections.

Adaptive firefly algorithm

The standard FA employs three parameters for solving the optimizations, and the parameters may result in significantly different performance of FA, such as the absorption coefficient γ and the randomization parameter α. Proper selections of these parameters can be a useful way to improve the search ability of FA. However, considering different problems with distinguish features, it is difficult to manually tune the parameters. To enhance FA, two main mechanisms and five strategies are employed to avoid the premature convergence of the classical FA. Accordingly, five variants of FA are yielded to balance the exploitation (local search) and exploration (global search), which are denoted as AdaFa-$\mathcal{S}_s$ ($s = 1,2,\cdots,5$). The two main mechanisms are distance-based adaptive mechanism for different information sharing and gray-based coefficients for efficiently enhancing the heterogeneous search. All the two mechanisms are adaptive to exchange messages and applied to tune the control parameters in FA. Additionally, we also propose five different strategies for the selection of the randomization parameter α in Eq. (4).

Distance-based adaptive strategy. The motivation for us to investigate a distance-based strategy is that the adaptive absorption coefficient can efficiently deal with different problems whatever their landscape are, while the traditional FA uses a constant one throughout the search process.

It is obvious that there are two limited cases for a constant light absorption coefficient γ [4], which can be concluded as follows:

- The attractiveness of other fireflies will be a constant when γ approaches 0. That is, a firefly can be seen by all the other ones. In this case, FA is the same as a classical PSO.

- If $\gamma \rightarrow \infty$, the attractiveness will be equal to 0. All the fireflies cannot take their bearings to move but in random flight. In this case, FA becomes a pure random search algorithm.

As can be seen, the parameter γ is crucially important in characterizing the variations of the attractiveness, and the speed of the convergence is also dependent on γ [12]. As a result, the performance of FA will be significantly constrained when a constant γ is used to solve the optimization problems as done in traditional FA. As is well known, the attractiveness should be linked with the distance among the fireflies, and it should also vary with the different distances among the population during the search process. The information of the distances is useful for promising search adaptively. Hence, we propose to use the distance to adaptively adjust the trajectories of the fireflies. The mean distance of the ith firefly to the other fireflies is calculated as follows,

$$dis_i = \frac{1}{N-1} \sum_{j=1, j \neq i}^{N} \sqrt{\sum_{k=1}^{D} (x_i^k - x_j^k)^2} \qquad (6)$$

where N is the number of the fireflies.

Based on the distance calculated in Eq. (6), we then define the distance ratio as follows,

$$D_r = \frac{dis_{opt} - dis_{\min}}{dis_{\max} - dis_{\min}} \qquad (7)$$

where dis_{opt} is the distance from the ith firefly to the global best firefly. $dis_{\max}$ and $dis_{\min}$ are the maximum and the minimum distances from the i_{th} firefly to other fireflies, respectively.

Accordingly, we define an adaptive absorption coefficient based on the distance ratio, which is to adaptively track the promising flight direction. It is defined as follows,

$$A_c = \frac{1}{1 + \sigma e^{-\rho \cdot D_r}} \qquad (8)$$

where σ denotes as amplitude factor, which controls the amplitude of A_c. ρ is called contraction index. The relationship between A_c and the two factors is illustrated in Fig. 1.

In Fig. 1, the value of D_r is set to 1, which does not have effect on the lower and upper boundaries of A_c. As can be seen, the larger the values of σ and ρ are, the smaller the value of A_c is; vice versa. If we remove the effect of D_r, that is $D_r = 1$, then A_c will be a constant. As a result, it is greatly difficult to deal with the balance between exploration and exploitation. Because the light absorption coefficient is determined by the distance information adaptively, the fireflies are able to adjust their flight directions for promising search.

Then, the constant light absorption coefficient γ in traditional FA is replaced by the adaptive coefficient A_c for efficient search. The new attractiveness is obtained in Eq. (9).

$$\tau = D_r \cdot e^{-A_c r^2} \qquad (9)$$

Gray-based coefficients. Gray relational analysis (GRA) is a similarity measure for finite sequences with incomplete information [17]. Let $\mathbf{y} = (y_1, y_2, y_3, \ldots, y_D)$ be the reference sequence, and $\mathbf{x}_i = (x_{i1}, x_{i2}, \ldots, x_{iD}), i = 1, 2, 3, \ldots, N$ be the ith comparative sequence, then the gray relational coefficient between $\mathbf{y}$ and $\mathbf{x}_i$ can be defined as

$$GRC(y_j, x_{ij}) = \frac{\delta_{\min} + \xi \cdot \delta_{\max}}{\delta_{ij} + \xi \cdot \delta_{\max}} \qquad (10)$$

where $\delta_{ij} = |y_j - x_{ij}|$, $\delta_{\min} = \min_i \quad \min_j \delta_{ij}$, $\delta_{\max} = \max_i \quad \max_j \delta_{ij}$, and $\xi \in (0,1]$ is employed to distinguish $\delta_{\max}$ from $\delta_{\min}$ and called as distinguishing constant. Based on Eq. (10), we can calculate the corresponding gray relational grade as follows

$$GRG(\mathbf{y}, \mathbf{x}_i) = \sum_{j=1}^{N} \left[\lambda_j \cdot GRC(y_j, x_{ij}) \right] \qquad (11)$$

where λ_j is a weighted constant of the gray relational coefficient $GRC(y_j, x_{ij})$ and subjects to $\sum_{j=1}^{n} \lambda_j = 1$.

To properly use the gray relational analysis, the new best firefly $\mathbf{x}_{opt}$ is employed as the reference sequence, while all the fireflies are the comparative ones. Base on these assumptions, we can use the gray relational analysis to measure the similarity between them. Let the gray relational grade between the new best firefly $\mathbf{x}_{opt}$ and the ith firefly $\mathbf{x}_i$ be GRG_i. As can be seen from Eq. (11), the closer the new best firefly $\mathbf{x}_n$ and the ith firefly are, the larger the GRG_i is. Accordingly, the relational grade GRG_i can be used to control the diversity of the firefly population. Since the gray relational grade involves the information of population distribution, we define a gray coefficient η to satisfy the requirement of diversity as follows,

$$\eta = \frac{(\eta_{\min} - \eta_{\max}) \cdot GRG_i + \eta_{\max} GRG_{\max} - \eta_{\min} GRG_{\min}}{(GRG_{\max} - GRG_{\min})} \qquad (12)$$

where $\eta_{\max}$ and $\eta_{\min}$ are the upper and the lower boundaries, which ensure the population can converge in finite time. $GRG_{\max} = \max\{GRG_{i|i=1,2\ldots,N}\}$ and $GRG_{\min} = \min \{GRG_{i|i=1,2\ldots,N}\}$.

Update rules. Fireflies in the traditional FA and most of its modifications, follow the same search law and share similar information throughout the search process. Due to the same search characteristics, the fireflies cannot always exhibit diverse and useful information for promising search. In an optimization

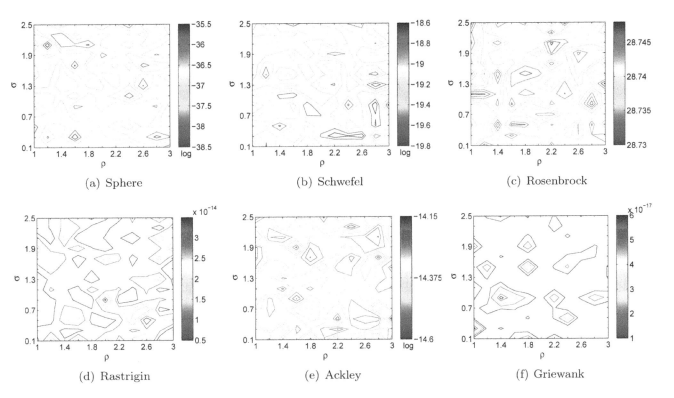

(a) Sphere   (b) Schwefel   (c) Rosenbrock

(d) Rastrigin   (e) Ackley   (f) Griewank

Figure 2. The relationship between the two factors (σ and ρ) and the performance of AdaFa.

algorithm, it is important to balance the exploration and exploitation. The algorithm should make contribution to exploration initially, while focusing on exploitation and convergence in the later search process [3,18]. To accomplish the task, we introduce heterogeneous update rules to improve the search abilities of the fireflies. Two updating equations are defined and selected randomly in the search process, which are presented in Eq. (13).

$$\mathbf{x}_i(t+1) = \begin{cases} (1-\tau)\cdot\mathbf{x}_i(t) + \tau\cdot\mathbf{x}_j(t) + \mathbf{x}_r(t+1) & rand > 0.5, \\ \dfrac{N_G - t}{N_G}(1-\eta)\cdot\mathbf{x}_i(t) + \eta\cdot\mathbf{x}_{opt} & \text{else,} \end{cases} \quad (13)$$

where N_G is the maximum number of generations. $\mathbf{x}_r$ is the randomization term, which is defined based on new α-strategies as follows,

$$\mathbf{x}_r(t+1) = \alpha(t+1)\cdot(\mathbf{rand} - \mathbf{c})\cdot|\mathbf{U}_b - \mathbf{L}_b| \quad (14)$$

where $\mathbf{U}_b$ and $\mathbf{L}_b$ are the upper and lower boundaries, respectively.

In FA, the strategy ($\mathcal{S}_0$) for α is

$$\alpha(t+1) = \alpha(t)\cdot\left(\frac{10^{-4}}{0.9}\right)^{\frac{1}{N_G}} \quad (15)$$

As can be seen from Eq. (15), α decreases linearly depending on the generation number, and it does not always work well for different problems. As an important parameter controlling the randomization step, α has significant effect on the performance of

FA. To enhance its search ability, we propose five different strategies for α, which all decrease dynamically with the generation number, population size and the size of the optimized problem. These designs promise AdaFa to deal with different types of problems, and all of them increase the range of α. The strategies are presented in Eq. (16)–Eq. (20).

- strategy $\mathcal{S}_1$

$$\alpha(t) = \left(\frac{e^{\delta}(N_G - t)}{t\cdot\epsilon}\right)^M, \quad (16)$$

where δ is constant defined by user to clamp α, and ϵ is also a constant, which is 10^{-6}.

- strategy $\mathcal{S}_2$

$$\alpha(t) = \exp\left(\delta\left(\frac{N_G - t}{t}\right)^{\frac{1}{t}}\right)^M, \quad (17)$$

- strategy $\mathcal{S}_3$

$$\alpha(t) = \left(e^{\delta}t^{-\frac{N}{D}}\right)^{\frac{M(N_G - t)}{N_G}}, \quad (18)$$

where N and D are the population size and the size of an optimization problem, respectively.

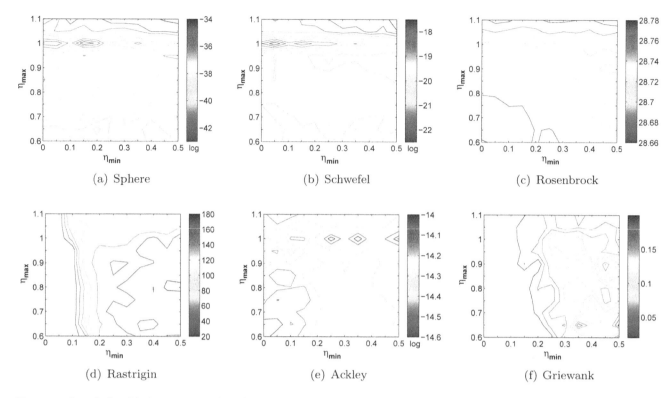

Figure 3. The relationship between η and performance of AdaFa.

- strategy $\mathcal{S}_4$

$$\alpha(t) = \left[t^{(N^P - 1)} \cdot (N_G - t) \right]^M, \qquad (19)$$

where P is the power of population size N.

- strategy $\mathcal{S}_5$

$$\alpha(t) = t^{\frac{M(N+D)(N_G - t)}{N_G}}. \qquad (20)$$

In Eq. (16)–Eq. (20), $M = -\exp\left(-\dfrac{N_G - t}{t} \right)$. From Eq. (16)–Eq.

(20), α varies with time t non-linearly, and the analysis of the parameters in the strategies are presented in following section.

Based on the designs of different adaptive mechanisms and strategies discussed above, the fireflies are allowed to learn more useful information from others and adjust the flight directions adaptively. The AdaFa algorithm is summarized in (*Algorithm S2*).

Results

In this section, we demonstrate the performance of AdaFa variants over twelve benchmark functions F_1–F_{12} summarized in Table S1 of Supplementary Materials (more details can be referred to the study [19,20]) and apply AdaFa variants to rebuild protein tertiary structure. Firstly, we conduct numerical experiments for parameters analysis and then compare AdaFa with standard particle swarm optimization (SPSO) [21], adaptive particle swarm optimization (APSO) [19], grey particle swarm optimization (GPSO) [20] and FA [8]. In numerical experiments, we analyze

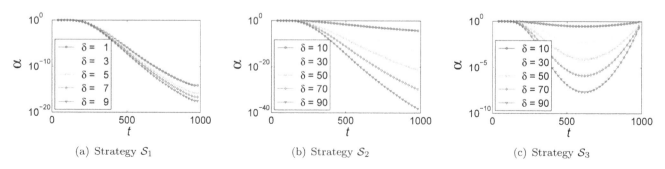

Figure 4. Different strategies for α: (a) strategy S_1, (b) strategy S_2, (c) strategy S_3 varying with different δ.

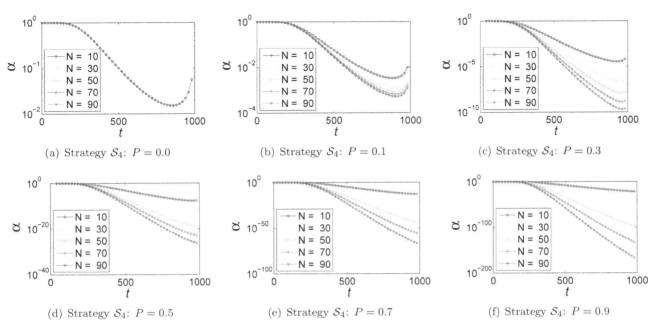

Figure 5. In strategy S_4, the relationship between α and the population size N varying with different power value P.

the parameter sensitivity, provide the parameter settings of each algorithm, present the results of the numerical experiment, and analyze the numerical results.

Parameter sensitivity analysis

In this section, we analyze four parameters in AdaFa including the amplitude factor σ, the contraction index ρ, η_{min} and η_{max}. The different selections of the four parameters may have influence on its performance. To investigate the impacts of these parameters, we conducted the experiments on functions F_1–F_2, F_5, and F_9–F_{11}, which are Sphere, Schwefel, Rosenbrock, Rastrigin,

Ackley and Griewank functions, respectively. The maximum number of generations was set to 1,000, and the population size was 10 (This is similar to ref. [8], where the authors analyzed the effect of population size on the optimization problems). The averages of best-so-far values were used to measure the performance of AdaFa.

To investigate the amplitude factor σ and the contraction index ρ, η_{min} and η_{max} were set to $\eta_{min} = 0.05$ and $\eta_{max} = 0.95$, respectively. AdaFa was performed 20 times with different σ and ρ. σ was varied from 0.1 to 2.5 with an increment of 0.2, while ρ was changed from 1 to 3 in the same increment. Fig. 2 reveals that

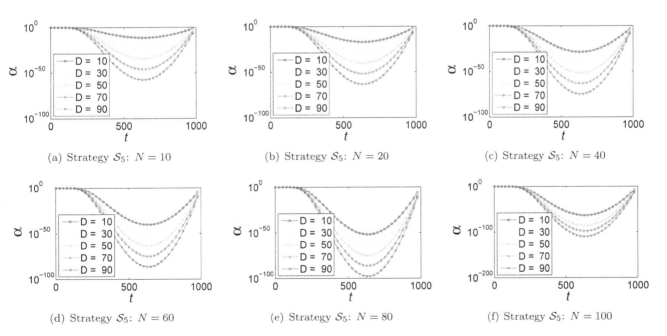

Figure 6. In strategy S_5, the relationship between α and the size of problem D with different population size N.

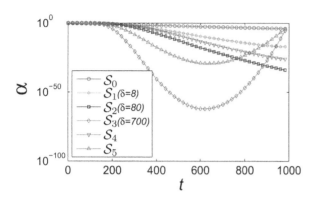

Figure 7. The comparison of different strategies for the randomization parameter α.

the relationship between the best-so-far and the two factors. As can be seen, the optimal σ and ρ were around 1.5 and 2.5, respectively. Hence, in this study $\sigma = 1.5$ and $\rho = 2.5$ were used for all the following experiments.

In the gray strategy, we employ a gray coefficient η to trade-off between the ith firefly and the best firefly. As shown in Eq. (12), it is necessary to clamp η in a fixed interval, such as $\eta_{min} \leq \eta \leq \eta_{max}$. To fix η_{min} and η_{max}, we conducted trial tests on the same six benchmark functions, and the maximum number of generations was also set to 1,000. η_{min} and η_{max} were set in $\eta_{min} \in [0,0.5]$ and $\eta_{max} \in [0.6,1.1]$, respectively, with the same increment of 0.05. Fig. 3 shows the relationship between η and performance of AdaFa. From Fig. 3, AdaFa will perform well when $\eta_{min} \approx 0.1$ and $\eta_{max} \approx 1.0$. Hence, in this study, we use $\eta_{min} = 0.1$ and $\eta_{max} = 0.95$ to conduct the experiments.

In strategies S_1 S_2 and S_3 of α (Eq. (16)–Eq. (18)), δ can be defined by user before the optimization. How to select δ is important to α. We thus analyze the relationship between δ and α, which is illustrated in Fig. 4. As shown in Fig. 4, the larger δ is, the larger the range of α is. It is interesting to note that there is a tail in strategy S_3 at the latter of generations, which can enhance the search abilities of the fireflies in exploitation region at latter search process.

In strategy S_4 and strategy S_5, the values of α are dependent on the population size N and the size D of an optimization problem, and we also analyze the relationship among them. As illustrated in Fig. 5, in strategy S_4 the smaller P is, the more similar the trajectories of α are. As well as strategy S_1–S_3, the range of α is enlarged with the increment of the value of P and population size N. In strategy S_5, as shown in Fig. 6, although the trajectories of α are different from each other varying with D, they all converge to similar points. These points are independent of the population size N.

In Fig. 7, different strategies of α are compared, where the ranges of α of all proposed strategies are larger than the strategy in standard FA. The larger ranges allow the fireflies with stronger exploration and exploitation abilities throughout the search process.

Parameters settings

Experiments were conducted to compare different algorithms on 12 benchmark functions with 30 dimensions. In these experiments, the maximum number of generation was set to 2,000, and the population of each algorithm was set to 40. According to previous analysis, the parameter settings are listed as follows:

- SPSO [21]: $\omega = \dfrac{1}{2log(2)}$, $c_1 = c_2 = 0.5 + log(2)$.
- APSO [19]: $\omega_{start} = 0.9$, $c_1 = c_2 = 2.0$.
- GPSO [20]: $\xi = 1$, $\omega_{min} = 0.4$, $\omega_{max} = 0.9$, $c_{min} = 1.5$, $c_{max} = c_{final} = 2.5$
- FA [8]: $\alpha = 0.5$, $\beta_{min} = 0.2$, $\gamma = 1.0$.
- IFA: $\phi = 0.05$, $\alpha = 0.5$, $\beta_0 = 1.0$, $\beta_{min} = 0.2$, $\gamma = 1.0$.
- AdaFa-S_1: $\delta = 8$, $\sigma = 1.5$, $\rho = 2.5$, $\eta_{min} = 0.1$, $\eta_{max} = 0.95$.
- AdaFa-S_2: $\delta = 80$, $\sigma = 1.5$, $\rho = 2.5$, $\eta_{min} = 0.1$, $\eta_{max} = 0.95$.
- AdaFa-S_3: $\delta = 700$, $\sigma = 1.5$, $\rho = 2.5$, $\eta_{min} = 0.1$, $\eta_{max} = 0.95$.
- AdaFa-S_4: $P = 0.75$, $\sigma = 1.5$, $\rho = 2.5$, $\eta_{min} = 0.1$, $\eta_{max} = 0.95$.
- AdaFa-S_5: $\sigma = 1.5$, $\rho = 2.5$, $\eta_{min} = 0.1$, $\eta_{max} = 0.95$.

Numerical Results

The experimental results over the twelve benchmark functions are presented in this section. Table 1 shows the mean values and the standard deviation of the results over 30 independent trials of each algorithm, and the success rate. The best are highlighted in bold, and the mean values are also illustrated in Fig. S1 in File S1 to compare the convergence of each algorithm (more details can be found in Supplementary Materials).

As illustrated in Table 1 and Fig. S1 in File S1, the five versions of AdaFa achieved better results than the other four algorithms on eight benchmark functions (F_2–F_5, F_7, and F_9–F_{11}), which can be also observed from Fig. S1 in File S1. SPSO achieved the target solution on function F_1, although the five versions of AdaFa did not as well as SPSO, they obtained the comparable results to each other, and the results were better than APSO, GPSO and FA. AdaFa-S_3 was better than AdaFa with other four different strategies on function F_2–F_4, while other algorithms were all trapped and failed to obtain good results. It is easy to solve function F_6 by the FA-based methods as shown in Table 1, as shown FA and five variants of AdaFa all found the target solutions to the problem. Due to the complex landscape of function F_8, GPSO was the sole algorithm that found a most approaching results to the target. Similarly, AdaFa-S_1 and AdaFa-S_4 were the only ones that achieved sharp result than the compared algorithms on functions F_{11} and F_9, respectively. Although AdaFa-S_2, AdaFa-S_3 and AdaFa-S_5 were all trapped into the local regions on function F_9, they performed so well that they all highly better than the three variants of PSO and FA on functions F_{10} and F_{11}. FA was good for the function F_{12}, on which it was superior to others, but the rest possessed similar convergent characteristic as illustrated in Fig. S1 in File S1.

Comparison of CPU efficiency

To compare the computational efficiency, we use CPU time to measure the complexity of each algorithm. For each algorithm, the computational efficiency is formulated as follows,

$$p = \frac{T_e(f)}{T_{tot}(f)} \times 100\% \tag{21}$$

where T_e is the computational time of an algorithm on the benchmark function f, while the T_{tot} is the total time of all the algorithms on function f.

In Fig. 8, the computational efficiency is illustrated. As can be seen, APSO is the fastest among all the algorithms, and it is followed by FA. Their computational efficiency are quite similar to each other over each benchmark function. The costs of the proposed five versions of AdaFa are also similar to each other, and they are all comparable to those of SPSO. However, all the five

Table 1. The results achieved by different algorithms on the benchmark functions.

Algorithm	F_1	F_2	F_3	F_4
SPSO	**0.00E+00** ± 0.00E+00	1.54E+00 ± 8.56E-01	3.55E-04 ± 3.68E-04	1.98E+00 ± 1.36E+00
APSO	2.47E-37 ± 7.47E-37	3.02E-20 ± 1.56E-19	5.88E+00 ± 7.02E+00	6.62E-01 ± 3.30E-01
GPSO	2.83E-23 ± 8.63E-23	6.15E-10 ± 1.65E-09	2.15E-01 ± 1.67E-01	1.20E+00 ± 7.70E-01
FA	1.22E-03 ± 2.53E-04	4.80E-02 ± 3.53E-02	3.67E+01 ± 2.83E+01	4.38E-02 ± 1.27E-02
AdaFa-S_1	1.56E-55 ± 4.84E-55	3.45E-28 ± 1.05E-27	1.93E-54 ± 5.11E-54	1.60E-28 ± 2.95E-28
AdaFa-S_2	2.36E-87 ± 9.56E-87	5.09E-45 ± 8.84E-45	4.19E-86 ± 9.73E-86	6.85E-45 ± 1.68E-44
AdaFa-S_3	1.26E-123 ± 3.17E-123	**7.44E-63** ± 6.60E-63	**2.16E-122** ± 7.92E-122	**1.30E-62** ± 1.44E-62
AdaFa-S_4	3.51E-78 ± 1.15E-77	8.13E-40 ± 2.33E-39	8.95E-77 ± 3.37E-76	1.11E-39 ± 2.06E-39
AdaFa-S_5	3.67E-74 ± 4.93E-74	4.75E-38 ± 3.59E-38	2.47E-73 ± 4.25E-73	7.16E-38 ± 6.57E-38

Algorithm	F_5	F_6	F_7	F_8
SPSO	3.56E+01 ± 2.84E+01	7.67E-01 ± 1.01E+00	3.38E-03 ± 1.36E-03	-7.28E+03 ± 1.04E+03
APSO	2.94E+01 ± 2.71E+01	6.67E-02 ± 2.54E-01	1.17E-02 ± 3.88E-03	-2.10E+04 ± 2.96E+03
GPSO	3.92E+01 ± 2.51E+01	6.67E-02 ± 2.54E-01	1.27E-02 ± 5.15E-03	**-1.30E+04** ± 1.13E+03
FA	8.28E+01 ± 1.31E+02	**0.00E+00** ± 0.00E+00	3.50E-02 ± 3.76E-02	-7.04E+03 ± 6.91E+02
AdaFa-S_1	**2.87E+01** ± 1.82E-02	**0.00E+00** ± 0.00E+00	3.00E-05 ± 3.05E-05	-4.83E+03 ± 7.70E+02
AdaFa-S_2	**2.87E+01** ± **1.74E-02**	**0.00E+00** ± 0.00E+00	3.61E-05 ± 4.76E-05	-4.87E+03 ± 5.70E+02
AdaFa-S_3	2.88E+01 ± 2.08E-02	**0.00E+00** ± 0.00E+00	**2.20E-05** ± **1.78E-05**	-4.96E+03 ± 6.57E+02
AdaFa-S_4	**2.87E+01** ± 2.01E-02	**0.00E+00** ± 0.00E+00	4.59E-05 ± 4.18E-05	-4.90E+03 ± **5.40E+02**
AdaFa-S_5	2.88E+01 ± 2.26E-02	**0.00E+00** ± 0.00E+00	3.55E-05 ± 2.96E-05	-4.94E+03 ± 6.55E+02

Algorithm	F_9	F_{10}	F_{11}	F_{12}
SPSO	3.48E+01 ± 3.03E+01	1.04E+00 ± 8.29E-01	6.56E-03 ± 1.12E-02	2.42E-02 ± 8.47E-02
APSO	5.00E+01 ± 1.37E+01	3.15E-01 ± 5.88E-01	1.42E-02 ± 1.73E-02	1.63E-01 ± 2.34E-01
GPSO	3.30E+01 ± 6.63E+00	8.41E-13 ± 1.06E-12	1.05E-02 ± 1.10E-02	3.11E-02 ± 8.67E-02
FA	3.49E+01 ± 1.31E+01	8.82E-03 ± 1.32E-03	2.95E-03 ± 2.24E-03	**5.87E-06** ± **1.56E-06**
AdaFa-S_1	5.55E+00 ± 3.04E+01	**4.56E-15** ± **6.49E-16**	**0.00E+00** ± 0.00E+00	8.80E-02 ± 1.69E-01
AdaFa-S_2	7.20E+00 ± 2.74E+01	8.47E-15 ± 5.00E-15	2.96E-17 ± 9.64E-17	9.33E-02 ± 2.06E-01
AdaFa-S_3	3.23E+00 ± 1.77E+01	2.78E-14 ± 2.09E-14	9.25E-16 ± 1.31E-15	6.91E-03 ± 2.63E-02
AdaFa-S_4	**3.03E-14** ± **5.33E-14**	6.10E-15 ± 1.80E-15	3.33E-17 ± 8.82E-17	1.73E-02 ± 5.50E-02
AdaFa-S_5	4.21E+00 ± 2.30E+01	1.14E-14 ± 3.55E-15	2.22E-17 ± 6.12E-17	3.80E-02 ± 1.07E-01

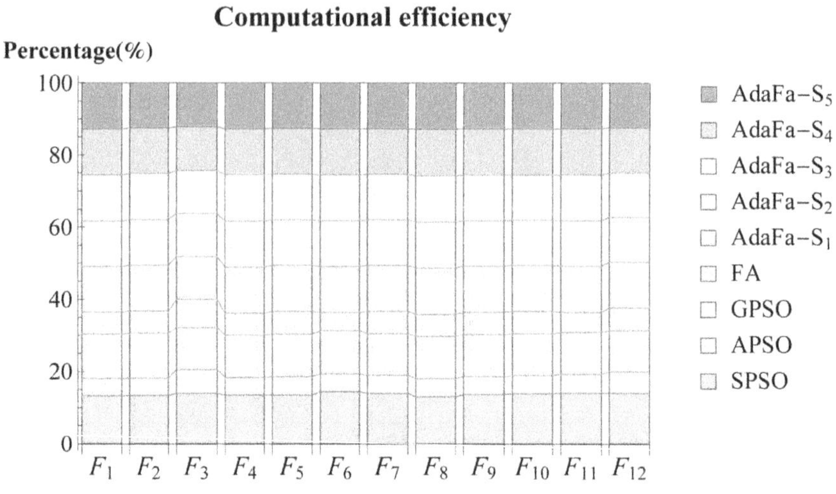

Computational efficiency

Figure 8. The comparison of computational efficiency of each algorithm on the test functions.

version of AdaFa are worse than FA in terms of computational time due to several new mechanisms have been employed.

Statistical analysis

Generally, it is necessary to use the non-parametric tests to analyze the experimental results. In this section, the Friedman, Aligned Friedman and Quade tests [22] were used to validate the performances of all the algorithms.

Table 2 presents the average rankings calculated from the Friedman, Aligned Friedman, and Quade tests. Each algorithm and its score are listed in ascending order. The statistics and the corresponding p-values are shown at the bottom of the table. In terms of computed p-values, there exists significant differences among the algorithms with the 5% significance level.

Simulations on trans-membrane protein helix

Optimization approaches have been widely used in computational biology, for instance, immune algorithm (IA) was applied to discover a protein conformation with minimal energy based on lattice models [23]. Estimation of distribution algorithms (EDAs)

[24] was used to solve the PSP in simplified models. Also based on simple protein models, Islam and Chetty employed memetic algorithm (MA) with several features to accomplish the structure prediction [25]. One important step of *ab initio* PSP methods is to reconstruct the tertiary structure of a protein by some optimization algorithms. Many methods reconstruct the tertiary structure of a protein depending on the residue contact maps, which can be solved in the framework of optimizing NP-hard problem [26], for example, in [27], a stochastic method based on simulated annealing (SA) was developed to derive a three-dimensional structure from a contact map. Vassura et al. used the a heuristic method and contact map to accelerate the process [28]. As we know, the aim of PSP is to obtain the Cartesian coordinates of all the atoms, which are bonded together by inter-atomic forces called chemical bonds. It has been observed that the bond lengths subject to a Gaussian distribution with a small standard deviation in high resolution protein structural data [29], and contain the essential information to determine the backbone structure of a protein [30]. Hence, given the torsion angle and bond length constrains, conformation of the geometry of the global protein structure is also

Table 2. Average Ranks of all compared algorithms over all benchmark functions.

Average Rank	Friedman		Aligned Friedman		Quade	
	Algorithm	**Score**	**Algorithm**	**Score**	**Algorithm**	**Score**
1	AdaFa-S_4	3.3333	AdaFa-S_4	43.1667	AdaFa-S_4	3.2756
2	AdaFa-S_3	3.375	AdaFa-S_3	43.2083	AdaFa-S_3	3.5833
3	AdaFa-S_1	3.75	AdaFa-S_5	43.8333	AdaFa-S_5	3.891
4	AdaFa-S_5	3.8333	AdaFa-S_1	46	AdaFa-S_1	3.9167
5	AdaFa-S_2	4.3333	AdaFa-S_2	46.5833	AdaFa-S_2	4.4551
6	GPSO	6.5	APSO	59.2917	APSO	6.109
7	APSO	6.5417	GPSO	66	GPSO	6.3462
8	FA	6.625	FA	69.4583	SPSO	6.7115
9	SPSO	6.7083	SPSO	72.9583	FA	6.7115
Statistic	30.35556		10.088976		3.022701	
p-value	0.000183		0.2588364		0.004785	

Table 3. Information of Trans-membrane protein helix.

PDB ID	Res. No.	PDB ID	Res. No.	PDB ID	Res. No.
1afoA	40	2c0xA	50	2kdrA	28
1bzkA	42	2gofA	19	2l0eA	31
1eq8A	23	2h95A	18	2l6wA	39
1javA	19	2hacA	33	2lk9A	24
1lb0A	13	2htgA	27	2lx0A	32
1lcxA	13	2jtwA	25	2xkmA	46
1pjdA	15	2jwaA	44	3c9jA	25
1wazA	46	2k9jA	42	3e86A	70
1z65A	30	2k9yA	41	3hroA	37
2beqA	36	2ka1A	35	3mraA	25

an optimization problem, where the final structure resolution is significantly dependent on the algorithms.

In this study, we applied the developed AdaFa variants on the protein structure prediction problem with focus on the transmembrane helix proteins (TMHP). Membrane proteins account for 30% of the whole genome and more than 70% known drug targets. However, because of the hydrophobic environment of TMHP, they are the extreme difficult targets for the experimental structure biology studies [31,32].

In high resolution protein structural data, it has been investigated that the bond lengths and angles subject to Gaussian distribution with a small standard deviation [33]. Here, we try to rebuild the tertiary structure directly from the bond lengths and angles with AdaFa. The data set of TMHP is filtered from the RCSB protein data bank (PDB) benchmark with 30% sequence identity cutoff, and it is released after Jan. 1st, 2010. We use 30 TMHPs for the tests, where the number of the residues (Res. No.) is less than or equal to 100, and the atomic resolutions of all the proteins are better than 2.5Å, without missed internal residues and sequence redundancies. These benchmark proteins are presented in Table 3.

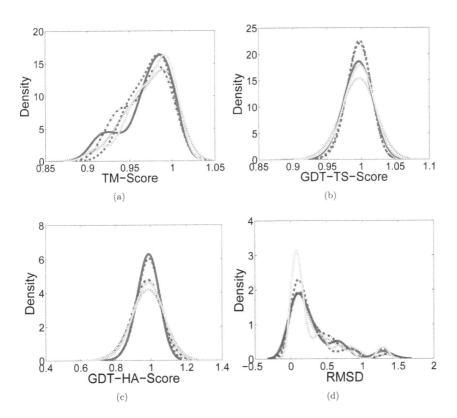

Figure 9. The kernel smoothing density estimate of (a) TM-Score, (b) GDT-TS-Score, (c) GDT-HA-Score, and (d) RMSD achieved over the native constrains. AdaFa-S_1–AdaFa-S_5 were represented by red solid line, black dotted line, blue dotted dashed line, magenta dashed line, and green solid line, respectively.

From the 30 PDB structural data set, we extract only the angles of protein backbone including bond angles and torsion angles, which will be used as the constrains of the construction by the proposed AdaFa. In the experiments, the coordinate of the first backbone C_α atom was set randomly, the second one was determined by the standard bond length, the third one was fixed by two standard bond lengths and a bond angle, and the fourth one was calculated by the former three bond lengths, two bond angles and a dihedral angle. From the fifth C_α atom, the coordinates were determined by the same constrains involving four bond lengths, three bond angles and two dihedral angles. When considering the whole process as an optimization problem, we define the objective function as an energy, which is to calculate the distance between the undetermined C_α atom and its former four fixed C_α atom. For example, there are four determined coordinates of C_α atoms (C_i, $i \geq 4$), and we need to determine the coordinate of the $(i+4)$th C_α atom using the bond lengths, bond angles, and dihedral angles among the C_α atom C_i, C_{i+1}, C_{i+2} and C_{i+3} in the backbone of a chain. We transform the obtained knowledge into distance-based constrains in geometrical respect, and the distances can be denoted as $d_{i,j}$, $d_{i+1,j}$, $d_{i+2,j}$ and $d_{i+3,j}$ ($j = 1, 2, 3, 4$). Accordingly, the energy can be formulated as $E = \sum_{j=1}^{4} (d_{i,j} - d_{i,j}^{o})$, where $d_{i,j}^{o}$ is transformed from the native constrains. The energy is used as the criterion of the proposed AdaFa in the optimization process. Hence, the backbone of a protein can be achieved by AdaFa iteratively. We then use the entire backbone of a protein chain as input of the PULCHRA [34] to determine the other atoms, such as hydrogen atoms and nitrogen atoms.

As illustrated in Fig. S2 in File S1, the tertiary structures of the proteins listed in Table 3 were predicted by AdaFa with high accuracy from the native constrains of each protein. In the experiments over 30 proteins, the averaged RMSDs of AdaFa-S_1–AdaFa-S_5 are 0.3074Å, 0.2960Å, 0.3043Å, 0.3079Å, and 0.2838Å, respectively. To validate the robustness of AdaFa, the native constrains with 10% Gaussian white noise were used to predict the tertiary structures. According to the results in Fig. 9 and Fig. S3 in File S1, it can be seen that AdaFa variants (AdaFa-S_1–AdaFa-S_5) are able to construct the tertiary structure of the protein with high accuracy and good robustness, where the averaged RMSDs of AdaFa-S_1–AdaFa-S_5 are 1.4425Å, 1.164Å, 1.2946Å, 1.0354Å, and 1.3508Å, respectively.

The different scores (TM-Score [35], GDT-TS-Score [36] and GDT-HA-Score [37]) and RMSD are illustrated by kernel smoothing density estimate [38] in Fig. 9 (More details can be found in Supporting Information). From Fig. 9 (a), the proposed AdaFa with five various strategies achieved different TM-Scores, in which AdaFa-S_1, AdaFa-S_4 and AdaFa-S_5 were better than the rest two variants of AdaFa. AdaFa-S_4 was superior to the other one in terms of GDT-TS-Score and GDT-HA-Score as shown in Fig. 9 (b) and Fig. 9 (c), while the AdaFa-S_2 was the worst one among the variants of AdaFa over the protein benchmark dataset. On the other hand, AdaFa-S_1 and AdaFa-S_3 exhibited a little different and competed with each other in the two score items. In respect of RMSD as shown in Fig. 9 (d), AdaFa-S_5 occupied an overwhelming position, which was far better than the other four AdaFa variants, among which AdaFa-S_1–AdaFa-S_3 possessed similar density estimation.

Conclusion

In this paper, we develop an adaptive firefly algorithm (AdaFa) and its five variants to enhance the search ability of the original FA. In AdaFa, we propose a distance-based technique to overcome the two main drawbacks in using a constant light absorption coefficient, which tunes the light among the fireflies dynamically to control the sharing distance information leading to the variation of the attractiveness. Simultaneously, the differences among the fireflies can be adequately used to enhance the local search ability of each firefly, hence we employ the gray relational analysis to design a gray coefficient as another self-adaptively altering parameter. According to the designed parameters, AdaFa uses heterogeneous update laws to accomplish the balance between the exploitation and the exploration throughout the search process.

In the numerical experiments, we compared the performances of all five proposed AdaFa variants with FA and other three PSO-based algorithms, and the statistical results demonstrated the five AdaFa variants were significantly better than the other four algorithms with the 5% significance level. The experiments on the reconstruction of the tertiary structure of the protein showed that AdaFa had the potential ability in predicting the helix structures with high accuracy and good robustness. Although AdaFa exhibited good performance on either the numerical experiments or real-world application on the prediction of the proteins' tertiary structures, it is still a challenging problem to deal with the cooperation among the fireflies for further improving the performance of FA, which is our future efforts. The codes of AdaFa is available upon request at http://www.csbio.sjtu.edu.cn/bioinf/AdaFa-PAA/.

Supporting Information

Algorithm S1   The FA algorithm.
(TXT)

Algorithm S2   The AdaFa algorithm.
(TXT)

File S1   Combined file of supporting figures and tables.
Figure S1: The mean value over the benchmark functions with 30-dimensions. Figure S2: Simulation results over thirty proteins. Figure S3: The kernel smoothing density estimates of different measurement metrics. Table S1: Benchmark Functions.
(ZIP)

Author Contributions

Conceived and designed the experiments: NJC XMD HBS. Performed the experiments: NJC. Analyzed the data: NJC XMD. Wrote the paper: NJC XMD HBS. Method development and implementation: NJC HBS.

References

1. Gandomi AH, Yang XS, Alavi AH (2011) Mixed variable structural optimization using firefly algorithm. Computers & Structures 89: 2325–2336.
2. Kazem A, Sharifi E, Hussain FK, Saberi M, Hussain OK (2013) Support vector regression with chaos-based firefly algorithm for stock market price forecasting. Applied Soft Computing 13: 947–958.
3. dos Santos Coelho L, Mariani VC (2013) Improved firefly algorithm approach applied to chiller loading for energy conservation. Energy and Buildings 59: 273–278.
4. Miguel LFF, Lopez RH, Miguel LFF (2013) Multimodal size, shape, and topology optimisation of truss structures using the firefly algorithm. Advances in Engineering Software 56: 23–37.
5. Amiri B, Hossain L, Crawford JW, Wigand RT (2013) Community detection in complex networks: Multiobjective enhanced firefly algorithm. Knowledge-Based Systems.
6. Su Z, Li L, Peng H, Kurths J, Xiao J, et al. (2014) Robustness of interrelated traffic networks to cascading failures. Sci Rep 4.

7. Yang XS (2008) Nature-Inspired Metaheuristic Algorithms. Luniver Press.

8. Yang XS (2009) Firefly algorithms for multimodal optimization. In: Stochastic Algorithms: Foundations and Applications, SAGA 2009. volume 5792, pp. 169–178.

9. Liang JJ, Qin AK, Suganthan PN, Baskar S (2006) Comprehensive learning particle swarm optimizer for global optimization of multimodal functions. IEEE Transactions on Evolutionary Computation 10: 281–295.

10. Cheung NJ, Ding XM, Shen HB (2014) OptiFel: A convergent heterogeneous particle swarm optimization algorithm for Takagi-Sugeno fuzzy modeling. Fuzzy Systems, IEEE Transactions on 22: 919–933.

11. Li L, Peng H, Kurths J, Yang Y, Schellnhuber HJ (2014) Chaos-order transition in foraging behavior of ants. Proceedings of the National Academy of Sciences 111: 8392–8397.

12. Gandomi A, Yang XS, Talatahari S, Alavi A (2013) Firefly algorithm with chaos. Communications in Nonlinear Science and Numerical Simulation 18: 89–98.

13. Brest J, Greiner S, Boskovic B, Mernik M, Zumer V (2006) Self-adapting control parameters in differential evolution: a comparative study on numerical benchmark problems. IEEE Transactions on Evolutionary Computation 10: 646–657.

14. Yang XS (2010) Firefly algorithms for multimodal optimization. ArXiv e-prints.

15. Yang XS, Deb S (2010) Eagle strategy using Lévy walk and firefly algorithms for stochastic optimization. In: Gonzlez J, Pelta D, Cruz C, Terrazas G, Krasnogor N, editors, Nature Inspired Cooperative Strategies for Optimization (NICSO 2010), Springer Berlin Heidelberg, volume 284 of *Studies in Computational Intelligence*. pp. 101–111.

16. Yang XS (2010) Firefly algorithm, stochastic test functions and design optimisation. International Journal of Bio-inspired Computation 2: 78–84.

17. Deng JL (1989) Introduction to grey system theory. J Grey Syst 1: 1–24.

18. Engelbrecht AP (2010) Heterogeneous particle swarm optimization. In: Swarm Intelligence, Springer Berlin Heidelberg, volume 6234 of *Lecture Notes in Computer Science*. pp. 191–202.

19. Zhan ZH, Zhang J, Li Y, Chung HH (2009) Adaptive particle swarm optimization. IEEE Transactions on Systems, Man, and Cybernetics, Part B: Cybernetics 39: 1362–1381.

20. Leu MS, Yeh MF (2012) Grey particle swafrm optimization. Applied Soft Computing 12: 2985–2996.

21. Clerc M (2012) Standard particle swarm optimisation. http://clerc.maurice.free.fr/pso/SPSO_descriptions.pdf.

22. Derrac J, García S, Molina D, Herrera F (2011) A practical tutorial on the use of nonparametric statistical tests as a methodology for comparing evolutionary and swarm intelligence algorithms. Swarm and Evolutionary Computation 1: 3–18.

23. Cutello V, Nicosia G, Pavone M, Timmis J (2007) An immune algorithm for protein structure prediction on lattice models. IEEE Transactions on Evolutionary Computation 11: 101–117.

24. Santana R, Larranaga P, Lozano J (2008) Protein folding in simplified models with estimation of distribution algorithms. IEEE Transactions on Evolutionary Computation 12: 418–438.

25. Islam M, Chetty M (2013) Clustered memetic algorithm with local heuristics for *ab initio* protein structure prediction. IEEE Transactions on Evolutionary Computation 17: 558–576.

26. Breu H, Kirkpatrick DG (1998) Unit disk graph recognition is NP-hard. Computational Geometry 9: 3–24.

27. Vendruscolo M, Kussell E, Domany E (1997) Recovery of protein structure from contact maps. Folding and Design 2: 295–306.

28. Vassura M, Margara L, Lena PD, Medri F, Fariselli P, et al. (2008) Reconstruction of 3D structures from protein contact maps. IEEE/ACM Transactions on Computational Biology and Bioinformatics 5: 357–367.

29. Engh RA, Huber R (1991) Accurate bond and angle parameters for X-ray protein structure refinement. Acta Crystallographica Section A 47: 392–400.

30. Wu S, Zhang Y (2008) ANGLOR: A composite machine-learning algorithm for protein backbone torsion angle prediction. PLoS ONE 3: e3400.

31. Shen HB, Chou JJ (2008) MemBrain: Improving the accuracy of predicting transmembrane helices. PLoS ONE 3: e2399.

32. Yang J, Jang R, Zhang Y, Shen HB (2013) High-accuracy prediction of transmembrane inter-helix contacts and application to GPCR 3D structure modeling. Bioinformatics 29: 2579–2587.

33. Engh RA, Huber R (2006) Structure quality and target parameters, John Wiley & Sons, Ltd. doi:10.1107/97809553602060000695.

34. Rotkiewicz P, Skolnick J (2008) Fast procedure for reconstruction of full-atom protein models from reduced representations. Journal of Computational Chemistry 29: 1460–1465.

35. Zhang Y, Skolnick J (2004) Scoring function for automated assessment of protein structure template quality. Proteins: Structure, Function, and Bioinformatics 57: 702–710.

36. Zemla A, MJFK, Venclovas Č (1991) Processing and analysis of casp3 protein structure predictions. Proteins S3: 22–29.

37. Zemla A (2003) LGA: a method for finding 3D similarities in protein structures. Nucleic Acids Research 31: 3370–3374.

38. Bowman AW, Azzalini A (1997) Applied Smoothing Techniques for Data Analysis. Oxford University Press, USA.

14

Protein Interaction Networks Reveal Novel Autism Risk Genes within GWAS Statistical Noise

Catarina Correia[1,2,3], Guiomar Oliveira[4,5,6], Astrid M. Vicente[1,2,3]*

1 Departamento de Promoção da Saúde e Doenças não Transmissíveis, Instituto Nacional de Saúde Doutor Ricardo Jorge, 1649-016 Lisboa, Portugal, **2** Center for Biodiversity, Functional & Integrative Genomics, Faculty of Sciences, University of Lisbon, 1749-016 Lisboa, Portugal, **3** Instituto Gulbenkian de Ciência, 2780-156 Oeiras, Portugal, **4** Unidade Neurodesenvolvimento e Autismo, Centro de Desenvolvimento, Hospital Pediátrico (HP) do Centro Hospitalar e Universitário de Coimbra (CHUC), 3000-602 Coimbra, Portugal, **5** Centro de Investigação e Formação Clínica do HP-CHUC, 3000-602 Coimbra, Portugal, **6** Faculdade de Medicina da Universidade de Coimbra, 3000-548 Coimbra, Portugal

Abstract

Genome-wide association studies (GWAS) for Autism Spectrum Disorder (ASD) thus far met limited success in the identification of common risk variants, consistent with the notion that variants with small individual effects cannot be detected individually in single SNP analysis. To further capture disease risk gene information from ASD association studies, we applied a network-based strategy to the Autism Genome Project (AGP) and the Autism Genetics Resource Exchange GWAS datasets, combining family-based association data with Human Protein-Protein interaction (PPI) data. Our analysis showed that autism-associated proteins at higher than conventional levels of significance ($P<0.1$) directly interact more than random expectation and are involved in a limited number of interconnected biological processes, indicating that they are functionally related. The functionally coherent networks generated by this approach contain ASD-relevant disease biology, as demonstrated by an improved positive predictive value and sensitivity in retrieving known ASD candidate genes relative to the top associated genes from either GWAS, as well as a higher gene overlap between the two ASD datasets. Analysis of the intersection between the networks obtained from the two ASD GWAS and six unrelated disease datasets identified fourteen genes exclusively present in the ASD networks. These are mostly novel genes involved in abnormal nervous system phenotypes in animal models, and in fundamental biological processes previously implicated in ASD, such as axon guidance, cell adhesion or cytoskeleton organization. Overall, our results highlighted novel susceptibility genes previously hidden within GWAS statistical "noise" that warrant further analysis for causal variants.

Editor: Branko Aleksic, Nagoya University Graduate School of Medicine, Japan

Funding: The AGP study was funded by Autism Speaks (USA), the Health Research Board (HRB, Ireland; AUT/2006/1, AUT/2006/2, PD/2006/48), The Medical Research Council (MRC, UK), Genome Canada/Ontario Genomics Institute and the Hilibrand Foundation (USA). Additional support for individual groups was provided by the US National Institutes of Health (NIH Grants: HD055751, HD055782, HD055784, MH52708, MH55284, MH061009, MH06359, MH066673, MH080647, MH081754, MH66766, NS026630, NS042165, NS049261), the Canadian Institutes for Health Research (CIHR), Assistance Publique - Hôpitaux de Paris (France), Autism Speaks UK, Canada Foundation for Innovation/Ontario Innovation Trust, Deutsche Forschungsgemeinschaft (Grant: Po 255/17-4) (Germany), EC Sixth FP AUTISM MOLGEN, Fundação Calouste Gulbenkian (Portugal), Fondation de France, Fondation FondaMental (France), Fondation Orange (France), Fondation pour la Recherche Médicale (France), Fundação para a Ciência e Tecnologia (Portugal), the Hospital for Sick Children Foundation and University of Toronto (Canada), INSERM (France), Institut Pasteur (France), the Italian Ministry of Health (convention 181 of 19 October 2001), the John P. Hussman Foundation (USA), McLaughlin Centre (Canada), Ontario Ministry of Research and Innovation (Canada), the Seaver Foundation (USA), the Swedish Science Council, The Centre for Applied Genomics (Canada), the Utah Autism Foundation (USA) and the Wellcome Trust core award 075491/Z/04 (UK). The Autism Genetic Resource Exchange is a program of Autism Speaks and is supported, in part, by grant 1U24MH081810 from the National Institute of Mental Health to Clara M. Lajonchere (PI). Catarina Correia is supported by grant SFRH/BPD/64281/2009 from the Fundação para a Ciência e Tecnologia. The funders had no role in study design, data collection and analysis, decision to publish, or preparation of the manuscript.

Competing Interests: The authors have declared that no competing interests exist.

* Email: astrid.vicente@insa.min-saude.pt

Introduction

Autism Spectrum Disorder (ASD) is a complex neurodevelopmental illness with significant clinical and genetic heterogeneity. Family and twin studies demonstrated that ASD is one of the most heritable neuropsychiatric disorders, but there is yet no consensus on the underlying genetic architecture [1,2]: while single-gene disorders, metabolic disorders and Copy Number Variants (CNVs) account for approximately 30% of the etiology of ASD [1,3–7], the contribution of common risk variants to the remaining heritability is still unclear. Thus far, each large genome-wide association study (GWAS) carried out for ASD highlighted a single, non-overlapping locus [8–11], which frequently was not replicated by subsequent independent replication studies [12].

Devlin et al. (2011) have recently predicted that common variants having an odds ratio of 1.5 or more are very unlikely to exist; few, if any, common variants with an impact on risk exceeding 1.2 may still await discovery, but require much larger sample sizes, while variants with modest impact may range from zero to many thousands [13]. The small effect of common risk variants for ASD represents a challenge for their individual detection using conventional single-marker association analysis, which likely allows many true *loci* to remain hidden within the GWAS statistical "noise". Evidence from classical quantitative genetic analysis further suggests that most of the heritability missing in complex diseases is rather hidden below the threshold for genome-wide significant associations [14,15].

New strategies are therefore needed to increase the power of GWAS analysis. The use of molecular networks, which is not limited by *a priori* sorting the genes into incompletely annotated predefined gene sets, is emerging as an appealing unbiased alternative to pathway analysis. Network-based approaches have been widely applied in the analysis of high-throughput expression data from a wide range of diseases [16] and have proven successful in the identification of subnetwork markers more reproducible and with a higher prediction performance than individual markers [17]. More recent studies incorporated protein networks into the analysis of genome-wide association data, using networks to search for interacting *loci* in human GWAS data [18,19] or to identify genome wide-enriched pathways [20–24]. However, an unsupervised global network analysis of ASD GWAS data that includes all signals without arbitrary significance thresholds has not been performed, and may lead to the identification of many risk variants of small effect below the accepted threshold for statistical significance.

Based on the premise that disease-causing genes are likely to be functionally related, in the present study we applied a network-based approach to two ASD GWAS datasets, the AGP consortium GWAS and the GWAS carried out in the Autism Genetic Resource Exchange (AGRE) dataset [10]. For this purpose we integrated genome wide association data with Human Protein-Protein interaction data and examined topological network properties indicative of connectivity at various levels of association, confirming our hypothesis that genes associated to ASD at a "statistical noise" level are functionally connected beyond random expectation. We compared the enrichment in known ASD candidates of network genes versus top GWAS genes, and the overlap of network genes vs the overlap at gene or SNP level between the two ASD datasets. The network obtained was further tested for ASD specificity using networks derived from six unrelated diseases GWAS, and explored for biological processes associated with ASD.

Materials and Methods

A workflow of the strategy for network definition, validation and identification of the most relevant candidate genes is shown in Figure 1.

Ethics statement

All the data used is previously published and publicly available. Written informed consent has been previously obtained from all families and procedures had approval from institutional review boards from all the institutions involved in recruitment and research, following national and international ethical and legal regulations and the principles of the Declaration of Helsinki.

Datasets

The AGP dataset included 2818 trios consisting of autistic patients and both parents collected as part of the AGP Consortium. Patients were diagnosed and genotyped as previously reported [8]. Written informed consent was obtained from all families and procedures had approval from institutional review boards [8]. A total of 723 423 SNPs meeting the QC criteria [9], genotyped in 8491 individuals, were tested for association using the Transmissions Disequilibrium Test (TDT) implemented in PLINK v1.07 [25].

The GWAS replication dataset from the Autism Genetic Resource Exchange (AGRE) included 943 ASD families (4,444 subjects) from the AGRE cohort [10]. SNP genotyping data was obtained from AGRE [10]. Analysis in this study was limited to

SNPs in common with the AGP GWAS and meeting the same QC criteria (425 587 SNPs).

Summary SNP association results were obtained from the database of Genotype and Phenotype (dbGAP) repository for 6 case-control GWAS for other pathologies, including Parkinson's Disease (PD) [26], Systemic Lupus Erythematosus (SLE) [27], Multiple Sclerosis (MS) [28], Type 1 Diabetes (T1D) [29], Breast Cancer (BC) [30] and Neuroblastoma (NB) [31] (Table S1). All individuals included were of European ancestry and the sample size was as similar as possible to the replication ASD dataset (AGRE).

Integration of gene association data with Protein-Protein interaction data

Genotyped SNPs from the AGP and AGRE GWAS were assigned to specific genes if they were located within or up to 10 kb from the gene, using the GRCh37/hg19 genome build (Step 1). Each gene was assigned a gene score using MAGENTA (Meta-analysis Gene-set Enrichment of variant associations) [32], which allocates to each gene the most significant *P*-value among the TDT *P*-values of all individual SNPs mapped to that gene. MAGENTA then uses step-wise multivariate linear regression analysis to regress out of this *P*-value the confounding effects of gene size, number of SNPs per kilobase (kb), number of independent SNPs, number of recombination hotspots and the number of linkage disequilibrium units per kb.

Genes selected at various gene-wise *P*-value cutoffs ($0.5 < -LogP < 5$) were superimposed onto their corresponding protein on a large human protein-protein interaction (PPI) network, converting Entrez gene IDs to Uniprot IDs (release 2010_04) (Step 2). This PPI network, covering 12372 proteins and 58365 interactions, was previously built compacting data from six public PPI databases: BIND, BioGRID, HPRD, IntAct, MINT and MPPI [33–40].

PPI network analysis

Topological properties from the resulting network were analyzed to select the gene-wise *P*-value for which corresponding proteins were functionally connected beyond random expectation, thus the lowest gene-wise *P*-value for which there is still relevant biological data in the GWAS that can be captured through network analysis (step 3). Three metrics indicative of this functional coherence were estimated for various association gene-wise *P*-value thresholds, for the two ASD datasets, and compared with those determined for 1000 equal size sets of randomly selected proteins from the human PPI network. The metrics evaluated were 1) the percentage of proteins directly interacting; 2) the percentage of isolated nodes, which represents the fraction of selected proteins with no interactions with any other selected protein; and 3) the size of the largest connected component (LCC), the largest group of selected proteins that are reachable from each other in the network. An empirical *P*-value was obtained computing the fraction of random samples where the value of the network metric is greater (or smaller in the case of isolates) than the observed one. Network analysis was performed using python module Network X.

Performance against a candidate gene list and overlap between datasets

To evaluate the performance of the proteins included in the LCC in retrieving known ASD candidate genes, the precision and recall against a curated list of ASD candidate genes were calculated (step 4). This list was obtained from SFARIGene and

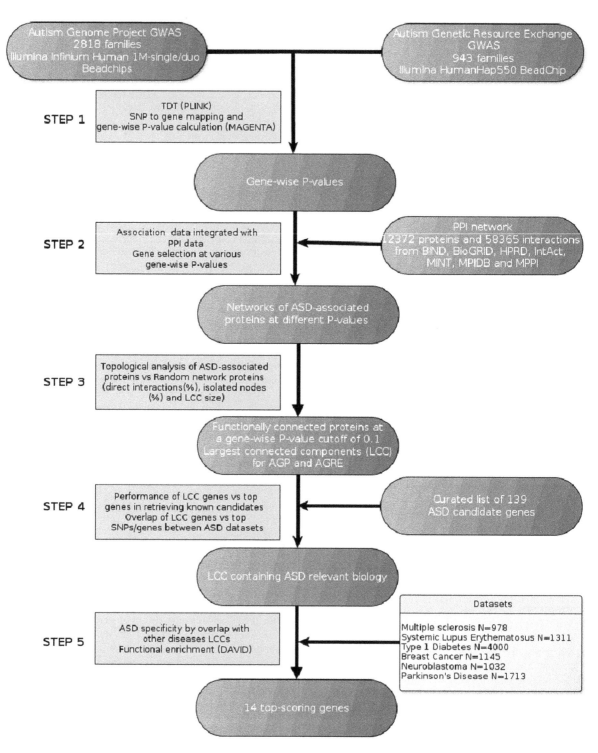

Figure 1. Workflow of the strategy for network definition, validation and identification of most relevant candidate genes.

includes 236 genes having at least minimal evidence of association with ASD (categories 1 to 4) or categorized as syndromic (https://gene.sfari.org/autdb/Welcome.do).

Precision (Positive Predictive value) is the proportion of known candidate genes among the selected genes, while recall (Sensitivity) is the proportion of known candidate genes retrieved by the selection. The precision and recall calculated for the genes encoding LCC proteins were compared to those determined using two other gene selection criteria: a) all genes selected at the same gene P-value cutoff used to derive LCC; b) the same number of top genes (ranked according to gene-wise P-values) as those included in the LCC.

Overlap between the AGP and AGRE datasets at SNP, gene and LCC levels was determined using the Jaccard index, defined as the size of the intersection divided by the size of the union of the datasets. For comparison purposes the size of each dataset LCC was used to select from each GWAS dataset an equal number of top SNPs (ranked by their TDT P-value) and top genes (ranked by their gene-wise P-value).

Gene ranking and functional enrichment

To rank ASD-associated proteins included in the AGP LCC by ASD specificity and reproducibility, a prioritization system was created, assigning a score to each protein based on their presence in the LCC derived from the AGRE ASD replication dataset and from each of the six unrelated disease datasets (step 5). Each protein included in the AGP LCC had an initial score of 0.5. If the protein was present in the AGRE ASD dataset LCC, 0.5 was added to the initial protein score, whereas for each unrelated disease dataset LCC where the protein was present, one sixth of 0.5 was subtracted from the score. Therefore, protein scores vary between 0 and 1, with zero representing a protein present in the LCCs of the AGP dataset and the 6 unrelated diseases, while a score of 1 is attributed to a protein present only in the LCCs of both ASD datasets.

Functional enrichment was tested by DAVID (The Database for Annotation, Visualization and Integrated Discovery 2008_version6[th]; http://david.abcc.ncifcrf.gov) [41,42], a publicly available bioinformatics tool that identifies functionally related groups of genes. Overrepresentation of mouse-mutant phenotypes was evaluated using the web tool MamPhea [43]. The complete list of the genes in the PPI network was used as background and P-values were corrected by the Benjamini correction. Top-scoring genes were further investigated using NextBio platform (Cupertino, CA, USA), a curated and correlated repository of experimental data derived from an extensive set of public resources (eg. ArrayExpress and GEO) [44]. Protein-protein networks were visualized in Cytoscape [45].

Results

Genes associated to ASD at $P<0.1$ are functionally related

Transmission Disequilibrium Tests were initially carried out in parallel for the AGP and AGRE datasets to identify small effect risk variants. In the sample of 2818 AGP families, single SNP transmission disequilibrium tests of the 723423 SNPs meeting the QC criteria showed no SNPs reaching the threshold for genome-wide significance. Two SNPs showed association signals at $P<1\times10^{-6}$ and very few exceeded $P<1\times10^{-5}$. In the AGRE dataset, after a similar quality control protocol and using only SNPs common to both datasets, three SNPs located in regions with no overlap with the AGP top findings showed association at $P<1\times10^{-6}$. Given the dearth of meaningful results from these two GWAS efforts, we proceeded with a network analysis strategy.

The first step involved calculating gene-wise association P-values corrected for gene size and linkage disequilibrium, taking into account only the SNPs mapping within 10kb from each gene (403360 SNPs), followed by the integration of GWAS data onto protein interaction data. Then, we determined the lowest gene-wise P-value threshold for which genes encoding the network proteins were functionally related, inferred by their proximity in the network. Statistical noise is expected to have random connections in the network, while disease proteins are more likely to establish direct interactions between them and more rarely be isolated in the network, translating into a larger group of proteins that are all interconnected. For both ASD datasets, proteins encoded by genes selected at a gene-wise $-\text{Log}_{10}P$ cutoff between 0.5 and 1.5 were found to establish significantly more direct interactions than equal sized sets of randomly selected proteins (Empirical P values $0.001<P<0.043$), with the significance maintained up to $-\text{Log}_{10}P = -2.0$ in the case of AGRE dataset (Figure S1, Figure 2A). The number of isolated nodes was found to be significantly smaller in sets of ASD-associated proteins at the same range of gene-wise $-\text{Log}_{10}P$-values than in random sets (Empirical P values $0.001<P<0.038$), again with significant differences maintained for lower gene-wise P-values in the AGRE dataset (Figure S1, Figure 2A). When compared to the same number of random proteins from the network, proteins encoded by genes selected at a gene-wise $-\text{Log}_{10}P<1$ from either ASD dataset are interconnected in a significantly larger LCC (Empirical P values $0.001<P<0.007$) (Figure S1, Figure 2B). The large size of the largest connected components, 416 and 367 proteins for the AGP and AGRE datasets, respectively, indicates the existence of several small effect risk genes reinforcing the high genetic heterogeneity in ASD.

Based on the lowest gene-wise P-value for which the percentage of direct interactions was significantly higher, the percentage of isolated nodes significantly smaller and the size of the LCC significantly larger than random expectation (Figure 2A and B), we established gene-wise $-\text{Log}_{10}P = 1$ as the cutoff value to infer functional coherence from the two ASD datasets.

The overall results indicate that, as hypothesized, genes associated with ASD at the range of GWAS statistical noise encode proteins that are functionally related and preferentially directly interact, confirming our expectation that there is indeed unexplored relevant biology at this statistical level.

Functionally coherent sub networks associated with ASD contain relevant ASD biology

To test whether the identified groups of functionally connected proteins captured by the largest connected components indeed contain ASD-relevant biology, we compared the performance of the genes selected through the LCC against a list of known candidates, [5] with the performance of all genes selected from the GWAS at the same gene-wise P-value cutoff or the performance of a number of GWAS top genes equal to the number of genes encoding LCC proteins. Genes implicated in ASD are largely unknown, thus low precision values are expectable given the incompleteness and noise in the available knowledge in the field.

Table 1 shows that, for both datasets, genes encoding proteins included in the LCC presented a 2 to 2.5 fold higher precision against the list of known genes than all the GWAS genes selected at the same statistical level cutoff. In other words, genes included in the LCC, and thus encoding functionally related proteins, are enriched in known candidates compared with the set of genes selected from the GWAS at the same statistical level, demonstrating that our filtering approach of association results based on PPIs more specifically captures ASD-relevant genes. A 1.3 to 3.3 fold increase is observed when comparing LCC genes with the same number of GWAS top genes, showing that a protein interaction-based selection was more accurate than selecting only the most strongly associated genes.

A.

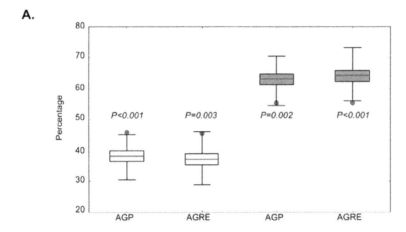

B.

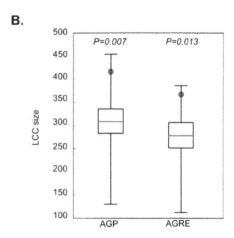

Figure 2. Network properties of proteins selected at gene-wise *P*<0.1 in each ASD. a) Comparison of percentage of direct interactions and isolated nodes between proteins selected at gene-wise *P*<0.1 in each GWAS dataset (red circles) vs 1000 random samples of network proteins (represented by light gray and dark gray box plots, for direct interactions and isolated nodes, respectively). The bottom and top of the box represent the 25th and 75th percentile and the extremity of the whiskers the maximum and minimum of the random samples data. **b)** Same comparison for the largest connected component (LCC) size.

Concerning the proportion of known genes that are retrieved by our selection, or recall, LCC encoding genes had a lower recall compared with all genes selected at the same cutoff, as expected since LCC genes are a subset of this selection (Table 1). However, compared with the top-gene selection, the 1.4 to 3 fold increase in the recall achieved by LCC encoding genes, indicates that additional relevant low effect genes are being captured. Further inspection of the known genes present in the top gene set and the LCC encoding genes confirmed that LCCs capture not only larger effect genes overlapping with top genes, such as *MET* (Uniprot P08581)(in AGP dataset), but additionally capture low effect genes, such as *TSC2* (Uniprot P49815), which single gene association analysis alone does not have the power to detect.

One of the major problems in ASD GWAS and GWAS in general is the low reproducibility of results between different datasets. Indeed, we found only one SNP (rs11837890 in *TBK1* gene) and 10 genes in common between the two datasets, when comparing the same number of SNPs or genes (ranked by *P*-values) than genes included in the LCCs from each dataset. Remarkably, we observed a 25 and 2.5-fold increase in the overlap between the two ASD datasets (AGP and AGRE) at PPI network level when compared to SNP or gene level, respectively (Figure S2).

Taken together, these results showed that our selection of functionally connected genes based on the largest connected component is an effective approach to capture ASD-relevant disease candidate genes, which might escape detection in an analysis based only on association evidence, even at gene-level.

Functionally connected genes in ASD suggest novel susceptibility genes

Given the observation that the largest connected component contains ASD-relevant proteins, we further explored this network for biological processes implicated in ASD (step 5). The largest connected components generated by genes selected at $-\text{Log}_{10}P<1$ from the AGP and AGRE datasets comprised 416 and 367 proteins, respectively. A first look into the biological processes represented in these networks, using functional enrichment analysis, revealed an enrichment in pathways related to regulation of apoptosis and cell cycle. Additionally, intersection of the protein network data with knockout mice phenotypes from the Mouse Genome Informatics Database, showed that these proteins are primarily involved in aberrant embryogenic and developmental processes and anomalous immune system phenotypes.

A closer inspection of these LCCs at the gene level showed that around 30 (7–8%) of the encoding genes were implicated in neuropsychiatric or neurodegenerative disorders (Table S2). More interestingly, 20 (5–6%) of the LCC encoding genes were found to carry *de novo* mutations in ASD described in at least one of the three whole exome sequencing studies recently published [4,7,46], with 3 genes overlapping between the two datasets (*CSDE1* (Uniprot O75534), *PGD* (Uniprot P52209), *TSC2*). In addition, 80

Table 1. Precision and recall were consistently higher for LCC genes relative to top GWAS genes or genes selected at *P*<0.1.

Gene subset	Precision (%)		Recall (%)	
	AGP dataset	**AGRE dataset**	**AGP dataset**	**AGRE dataset**
LCC genes	2.16	2.74	3.81	4.24
GWAS Top genes	1.68	0.82	1.27	2.97
Genes selected at *P*<0.1	0.96	1,11	8.47	9.43

Precision and Recall (Percentage), by ASD dataset, of three sets of genes (genes selected at a gene wise *P*-value cutoff of 0.1, genes included in the LCC and the same number of GWAS top genes) against a list of known disease candidates.

(~19%) of the AGP LCC-encoding genes were deleted or duplicated by CNVs identified by the AGP whole genome analysis as potentially pathogenic (with less than 50% of length overlap with control datasets) (Table S2).

To further examine the specificity of the proteins in the AGP LCC for ASD, this network was compared with LCCs generated from six unrelated diseases GWAS (MS, SLE, T1D, BC, NB, PD). Based on the presence of each protein in the LCC of each unrelated disease and in the AGRE LCC, we derived a highly stringent ASD-specificity protein score, allowing the prioritization of encoding genes for follow-up. Low scoring proteins were not replicated in the AGRE dataset, and were present in one or more unrelated diseases, whereas the highest scoring proteins were present in both ASD LCCs, but in none of the LCCs generated from the unrelated diseases. This analysis revealed that the majority of proteins (~63%) were present only in the AGP network, while 31% of the proteins were present in at least one additional non-ASD network, and thus were not specific. From the 25 proteins identified in both ASD networks, the majority (56%) was not present in any ASD-unrelated network and 28% were present in one of the ASD-unrelated networks.

Using this gene scoring system, based on gene reproducibility and specificity for ASD, we built a network with the 14 top scoring genes and their first neighbors in the LCC network (Figure 3). The largest component of this network, although approximately 7 times smaller than the original LCCs, showed a similar overlap (~5%) with genes reported to have *de novo* mutations in ASD (*PGD, SYNE1* (Uniprot Q8NF91), *TSC2*) and an increased overlap with known candidate genes (*SYNE1, TSC2* and *SHANK3* (Uniprot Q9BYB0)) and with genes contained in potentially relevant CNVs identified by the AGP analysis (~26%). Enrichment in mouse phenotypes was also similar but, in addition, an enrichment in abnormal nervous system phenotype became significant, and in abnormal behavior/neurological phenotype borderline significant.

The genes encoding the 14 top scoring proteins were considered the best candidates for harboring common variants associated with ASD risk (Table 2). These genes are involved in various biological processes, such as NGF signaling, axon guidance, cell adhesion and migration, cytoskeleton regulation, apoptosis and DNA repair. A *de novo* mutation in the phosphogluconate dehydrogenase gene (*PGD*) has recently been reported in ASD [4], while potentially pathogenic CNVs deleting or duplicating the *ABL1* (Uniprot P00519), *RPS6KA1* (Uniprot Q15418) and *PPP1CB* (Uniprot P62140) genes were identified in ASD patients from the AGP study. A query of our genes in the NEXTBIO platform, a data mining framework that integrates and correlates global public datasets with several normal and disease phenotypes, revealed correlations of six genes with ASD. For instance, deletions within the *NASP* (Uniprot P49321) gene were identified in ASD patients from the Simons Simplex Collection (SSC) [47]. An altered expression of this gene, as well as of the *NR4A1* (Uniprot P22736), *ABI1* (Uniprot Q8IZP0), *BBS4* (Uniprot Q96RK4), *LMNA* (Uniprot P02545) and *ABL1* genes, was found in postmortem brain tissue [48] or lymphoblastoid cells [49] of ASD patients. Some of the 14 top-scoring genes, namely the *CTSB* (Uniprot P07858), *BBS4, LMNA* and *ABL1* genes, were associated with abnormal nervous system phenotypes in animal models. The most strongly associated genes to ASD, using the AGP data, were the peroxiredoxin 1 gene (*PRDX1* (Uniprot Q06830)) and cathepsin B gene (*CTSB*).

Discussion

In this study we have conducted a network-based analysis of two ASD GWAS datasets, hypothesizing that small effect ASD risk variants hidden at the level of GWAS statistical noise can be discovered from networks of genes with related biological functions. Mapping of association data to a PPI network indeed revealed that, in both datasets, ASD-associated genes at $P < 0.1$ encoded proteins that directly interact beyond random expectation, are more rarely found isolated in the network and are connected in significantly larger LCCs than expected by chance, suggesting a functional connection. These results support recent findings from the AGP consortium, showing that stronger association of allele scores with case status was generally achieved when those scores were based on markers associated at significance thresholds higher than 0.2 [8]. The International Schizophrenia GWAS consortium had similar results of optimal discrimination between cases and controls only after the inclusion of markers with P-values as high as 0.2, [14] using this allele scoring approach.

The relevance to ASD of these networks was further illustrated by their higher performance in retrieving known ASD candidates compared to top GWAS genes, and the increased similarity between the two ASD datasets, when compared to SNP or gene level overlap. Remarkably, the AGP and AGRE LCCs included 20 genes, respectively, in which *de novo* mutations have been described in whole-exome sequencing studies of nearly a thousand ASD patients [4,7,50]. A large overlap of our results with the published data of these sequencing studies was not expected, because the LCCs encoding genes are likely to harbor variants transmitted by unaffected parents, whereas these sequencing studies mainly focused, and reported only, *de novo* variants which do not explain the heritability of the disorder, but support recent observations that common and rare variants associated with ASD disturb common neuronal networks [51]. Moreover, around 20% of the AGP LCC encoding genes were deleted or duplicated by potentially pathogenic CNVs detected in the AGP whole genome CNV screening of 2446 ASD patients.

As an additional filter for meaningful ASD biology, we derived an ASD candidate gene prioritization system ranking the genes encoding proteins included in the AGP LCC for ASD reproducibility and specificity. The scoring system used was very stringent, in particular since some of the control disorders are neurological (Parkinson's, multiple sclerosis or neuroblastoma) and may share susceptibility genes and pathways with autism [52–56]. While we may have discarded relevant autism risk genes that are ubiquitous and common to these disorders, we believe that we enriched our list of genes in true positive results with a higher chance of experimental validation. In fact, the enrichment analysis performed with the top-scoring genes and their first neighbors showed a high content in mouse genes associated with nervous system or neurological phenotypes and a similar or higher overlap with candidate genes or genes reported with *de novo* mutations or potentially pathogenic CNVs in ASD.

This approach generated a list of 14 top-scoring genes, present in the two ASD networks and none of the other disorders, which were considered strong candidates to harbor common variants associated with ASD risk. These genes are mostly novel candidates for ASD, and are involved in nervous system pathways or other more fundamental biological processes which have been widely associated to ASD, such as ubiquitination [4,9,57,58], cytoskeleton organization and regulation [5,47,59] and cell adhesion [10,60]. For instance, the *CTSB, BBS4, LMNA* and *ABL1* genes have been associated with neurobiological phenotypes identified in an enrichment analysis of mouse neurobiological phenotypes from a list of 112 ASD candidate genes [61], with *CTSB* and *ABL1* associated with cerebellum morphological and development abnormalities. The AGP genome-wide analysis identified potentially pathogenic CNVs spanning *ABL1, RPS6KA1* and

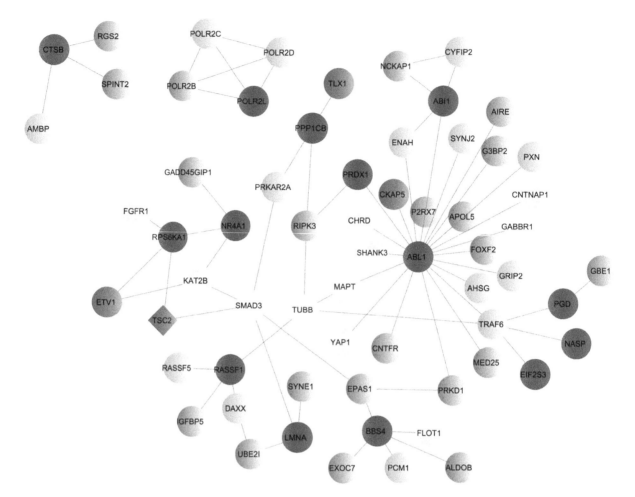

Figure 3. ASD top scoring gene network. This network illustrates the 14 top scoring genes included in the ASD LCC and their first neighbors. Nodes are colored based on a score reflecting their presence in the second ASD dataset and in the 6 unrelated diseases LCCs. A darker color represents a higher score, which means a higher specificity for ASD.

PPP1CB, whose relevance needs to be further established. Likewise, in the phosphogluconate dehydrogenase gene (PGD), a *de novo* mutation has recently been reported in a patient with ASD [4], although with an uncertain deleterious effect. This gene plays a critical role in protecting cells from oxidative stress [62] and, together with *PRDX1*, which also has an important antioxidant protective role in cells [63,64] and shows the strongest association with ASD, supports emerging evidence for a role of oxidative stress in ASD pathophysiology [65,66].

Thus far the use of protein networks to address common risk variants in ASD was limited to enrichment analysis of GWAS top hit genes in co-expressed or differentially expressed networks [51,67]. In contrast, this study incorporated protein interaction data into GWAS analysis, without *a priori* assumptions of association thresholds. The present results have shown that autism-associated genes at higher than conventional levels of significance are functionally related, and were used to extract relevant disease biology and uncover small effect variants contributing to the disorder. The study highlighted a group of novel susceptibility genes relevant for CNS function with a high probability of bearing common variants associated with autism, which have been elusive thus far, and warranting further analysis for identification of causal variants.

Supporting Information

Figure S1   Network properties per gene-wise *P*-value for each ASD dataset. For each $-\text{Log}_{10}$ gene wise association *P*-value cutoff in the x-axis, the percentage of direct interactions (A) and isolated nodes (B) and the logarithm of the LCC size (C) were plotted for proteins encoded by disease-associated genes (red line) and for the mean of 1000 equal sized random samples of proteins (blue line). Dark grey areas represent the range between the 25th and 75th quartiles and light gray areas indicate the range between the minimum and maximum values of the random data. Empirical *P*-values are indicated for each gene wise association *P*-value comparison. Values are plotted until the $-\text{Log}_{10}$ for which the percentage of direct interactions and isolated nodes reaches 0 and 100%, respectively.
(TIF)

Figure S2   Overlap between the two ASD datasets at SNP, gene or network level. Venn diagrams showing the overlap between the two ASD datasets (AGP and AGRE) at SNP, gene or network level.
(TIF)

Table S1   GWAS datasets used in the analysis and genotyping platforms.
(XLSX)

Table 2. Top scoring ASD network genes.

Gene (Uniprot ID)	Description	Location	Relevant biological processes	Gene-wise association P-value (MAGENTA)	Published studies in autism	Neurological and behavioral features in mouse models
NASP(P49321)	nuclear autoantigenic sperm protein (histone-binding)	1p34.1	blastocyst development, cell proliferation, cell cycle	$1.470e^{-02}$	NEXTBIO: deletion in idiopathic females (Sakai et al, 2011), significantly downregulated in brain samples (Chow et al. 2011)	NA
PRDX1(Q06830)	peroxiredoxin 1	1p34.1	redox regulation, cell proliferation	$1.760e^{-04}$	-	NA
RPS6KA1(Q15418)	ribosomal protein S6 kinase, 90kDa, polypeptide 1	1p36.11	protein kinase, synaptic transmission, axon guidance, long-term potentiation, toll-like and NGF receptor signaling pathway	$2.422e^{-02}$	-	NNP
PGD(P52209)	phosphogluconate dehydrogenase	1p36.22	cell redox regulation	$7.851e^{-02}$	De novo mutation in autistic patient (O'Roak et al. 2012)	NNP
LMNA(P02545)	lamin A/C	1q22	regulation of cell migration, regulation of apoptotic process, spermatogenesis	$8.226e^{-02}$	Nextbio:altered expression in Lymphoblastoid cells from males with autism (15q11–13 duplication) and brain samples (Chow et al. 2011; Nishimura et al. 2007)	abnormal axon morphology, abnormal myelination
PPP1CB(P62140)	protein phosphatase 1, catalytic subunit, beta isozyme	2p23	regulation of cell cycle, focal-adhesion, long-term potentiation, regulation of actin cytoskeleton	$3.749e^{-02}$	-	NA
RASSF1(Q9NS23)	Ras association (RalGDS/AF-6) domain family member 1	3p21.3	Cell cycle, response to DNA damage stimulus, positive regulation of protein ubiquitination	$1.949e^{-02}$	-	NA
CTSB(P07858)	cathepsin B	8p23.1	regulation of apoptotic process, cellular response to thyroid hormone stimulus	$2.343e^{-03}$		Purkinje cell degeneration, abnormal neuron apoptosis (details) neuron degeneration
ABL1(P00519)	c-abl oncogene 1, non-receptor tyrosine kinase	9q34.1	axon guidance, regulation of cell adhesion, motility, cycle, actin cytoskeleton organization, response to DNA damage stimulus	$9.560e^{-02}$	NextBio:altered expression in autistic brain samples (Chow et al. 2011)	abnormal cerebellum morphology, small cerebellum, abnormal cerebellum development, abnormal cerebellar foliation, ectopic Purkinje cell, abnormal cerebellar lobule formation, absent cerebellar lobules, abnormal neuron differentiation
ABI1(Q8IZP0)	Abl-interactor 1 previously known as spectrin SH3 domain binding protein 1	10p12.1	transmembrane receptor protein tyrosine kinase signaling pathway, negative regulation of cell proliferation	$7.352e^{-02}$	NextBio:downregulation in autistic brain samples (Chow et al. 2011)	NA
POLR2L(P62875)	polymerase (RNA) II (DNA directed) polypeptide L, 7.6kDa	11p15.5	DNA repair, regulation of transcription	$7.821e^{-02}$	-	NNP

Table 2. Cont.

Gene (Uniprot ID)	Description	Location	Relevant biological processes	Gene-wise association P-value (MAGENTA)	Published studies in autism	Neurological and behavioral features in mouse models
NR4A1 (P22736)	nuclear receptor subfamily 4, group A, member 1 also known as nerve Growth factor IB (NGFIB)	12q13.13	nuclear transcription factor, epidermal and fibroblast growth factor receptor signaling pathway, nerve growth factor receptor signaling pathway	$4.783e^{-02}$	NextBio:downregulation in autistic brain samples (Chow et al. 2011)	NA
BBS4 (Q96RK4)	Bardet-Biedl syndrome 4	15q22.3–q23	centrosome organization, microtubule cytoskeleton organization, neural tube closure, dendrite, striatum, hippocampus, cerebral cortex development	$7.511e^{-02}$	Nextbio:altered expression in lymphoblasts and brain samples (Chow et al. 2011; Nishimura et al. 2007)	abnormal neural tubemorphology/development, thincerebral cortex, abnormal basal ganglion morphologyabnormal olfactory neuron morphology, small hippocampusenlarged lateral ventriclesenlarged third ventricle
EIF2S3 (P41091)	eukaryotic translation initiation factor 2, subunit 3 gamma, 52kDa	Xp22.11	cellular protein metabolic process	$7.322e^{-02}$	-	NNP

List of the 14 top scoring ASD network genes, present in both ASD networks and in none of the other disorders (ASD specificity score = 1), with information on gene-wise association P-value and biological processes relevant for ASD.

NNP - No neurological phenotypes]; NA - No mouse model available.

Table S2 AGP LCC network genes. List of the 416 genes included in the AGP LCC with information on gene-wise association *P*-value, specificity score for ASD and previous findings regarding implication in ASD and other neurological disorders.
(XLSX)

Acknowledgments

We gratefully acknowledge the children with ASD and their families enrolling at the AGP and AGRE participating study sites. We are thankful to José Pereira Leal, Yoan Dieckmann and the remaining members of the Computational Genomics group at Instituto Gulbenkian de Ciência for helpful discussions and data sharing. We are grateful to all the AGP investigators, and particularly to Richard Anney, for sharing data, resources and scientific discussions. We gratefully acknowledge the resources provided by the Autism Genetic Resource Exchange (AGRE) Consortium and the participating AGRE families.

We acknowledge the NIH GWAS repository. Public available analysis data for Multiple Sclerosis, Systemic Lupus Erythematosus, Type 1 diabetes, Breast Cancer, Neuroblastoma and Parkinson's Disease were obtained from dbGaP at http://www.ncbi.nlm.nih.gov/gap through dbGaP accession numbers pha002861, pha002848, pha002862, pha002853, pha002845, pha002868, respectively.

Author Contributions

Conceived and designed the experiments: CC AMV. Performed the experiments: CC. Analyzed the data: CC AMV. Contributed reagents/materials/analysis tools: CC GO AMV. Wrote the paper: CC AMV.

References

1. Devlin B, Scherer SW (2012) Genetic architecture in autism spectrum disorder. Curr Opin Genet Dev 22: 229–237.
2. Geschwind DH (2011) Genetics of autism spectrum disorders. Trends Cogn Sci 15: 409–416.
3. Betancur C (2010) Etiological heterogeneity in autism spectrum disorders: more than 100 genetic and genomic disorders and still counting. Brain Res 1380: 42–77.
4. O'Roak BJ, Deriziotis P, Lee C, Vives L, Schwartz JJ, et al. (2012) Exome sequencing in sporadic autism spectrum disorders identifies severe de novo mutations. Nat Genet 44: 471.
5. Pinto D, Pagnamenta AT, Klei L, Anney R, Merico D, et al. (2010) Functional impact of global rare copy number variation in autism spectrum disorders. Nature 466: 368–372.
6. Sanders SJ, Ercan-Sencicek AG, Hus V, Luo R, Murtha MT, et al. (2011) Multiple recurrent de novo CNVs, including duplications of the 7q11.23 Williams syndrome region, are strongly associated with autism. Neuron 70: 863–885.
7. Sanders SJ, Murtha MT, Gupta AR, Murdoch JD, Raubeson MJ, et al. (2012) De novo mutations revealed by whole-exome sequencing are strongly associated with autism. Nature 485: 237–241.
8. Anney R, Klei L, Pinto D, Almeida J, Bacchelli E, et al. (2012) Individual common variants exert weak effects on the risk for autism spectrum disorderspi. Hum Mol Genet 21: 4781–4792.
9. Anney R, Klei L, Pinto D, Regan R, Conroy J, et al. (2010) A genome-wide scan for common alleles affecting risk for autism. Hum Mol Genet 19: 4072–4082.
10. Wang K, Zhang H, Ma D, Bucan M, Glessner JT, et al. (2009) Common genetic variants on 5p14.1 associate with autism spectrum disorders. Nature 459: 528–533.
11. Weiss LA, Arking DE, Daly MJ, Chakravarti A (2009) A genome-wide linkage and association scan reveals novel loci for autism. Nature 461: 802–808.
12. Curran S, Bolton P, Rozsnyai K, Chiocchetti A, Klauck SM, et al. (2011) No association between a common single nucleotide polymorphism, rs4141463, in the MACROD2 gene and autism spectrum disorder. Am J Med Genet B Neuropsychiatr Genet 156B: 633–639.
13. Devlin B, Melhem N, Roeder K (2011) Do common variants play a role in risk for autism? Evidence and theoretical musings. Brain Res 1380: 78–84.
14. Purcell SM, Wray NR, Stone JL, Visscher PM, O'Donovan MC, et al. (2009) Common polygenic variation contributes to risk of schizophrenia and bipolar disorder. Nature 460: 748–752.
15. Yang J, Benyamin B, McEvoy BP, Gordon S, Henders AK, et al. (2010) Common SNPs explain a large proportion of the heritability for human height. Nat Genet 42: 565–569.
16. Barabasi AL, Gulbahce N, Loscalzo J (2011) Network medicine: a network-based approach to human disease. Nat Rev Genet 12: 56–68.
17. Chuang HY, Lee E, Liu YT, Lee D, Ideker T (2007) Network-based classification of breast cancer metastasis. Mol Syst Biol 3: 140.
18. Emily M, Mailund T, Hein J, Schauser L, Schierup MH (2009) Using biological networks to search for interacting loci in genome-wide association studies. Eur J Hum Genet 17: 1231–1240.
19. Pan W (2008) Network-based model weighting to detect multiple loci influencing complex diseases. Hum Genet 124: 225–234.
20. Akula N, Baranova A, Seto D, Solka J, Nalls MA, et al. (2011) A network-based approach to prioritize results from genome-wide association studies. PLoS One 6: e24220.
21. Baranzini SE, Galwey NW, Wang J, Khankhanian P, Lindberg R, et al. (2009) Pathway and network-based analysis of genome-wide association studies in multiple sclerosis. Hum Mol Genet 18: 2078–2090.
22. Jensen MK, Pers TH, Dworzynski P, Girman CJ, Brunak S, et al. (2011) Protein interaction-based genome-wide analysis of incident coronary heart disease. Circ Cardiovasc Genet 4: 549–556.
23. Jia P, Zheng S, Long J, Zheng W, Zhao Z (2011) dmGWAS: dense module searching for genome-wide association studies in protein-protein interaction networks. Bioinformatics 27: 95–102.
24. Lee I, Blom UM, Wang PI, Shim JE, Marcotte EM (2011) Prioritizing candidate disease genes by network-based boosting of genome-wide association data. Genome Res 21: 1109–1121.
25. Purcell S, Neale B, Todd-Brown K, Thomas L, Ferreira MA, et al. (2007) PLINK: a tool set for whole-genome association and population-based linkage analyses. Am J Hum Genet 81: 559–575.
26. Simon-Sanchez J, Schulte C, Bras JM, Sharma M, Gibbs JR, et al. (2009) Genome-wide association study reveals genetic risk underlying Parkinson's disease. Nat Genet 41: 1308–1312.
27. Hom G, Graham RR, Modrek B, Taylor KE, Ortmann W, et al. (2008) Association of systemic lupus erythematosus with C8orf13-BLK and ITGAM-ITGAX. N Engl J Med 358: 900–909.
28. Baranzini SE, Wang J, Gibson RA, Galwey N, Naegelin Y, et al. (2009) Genome-wide association analysis of susceptibility and clinical phenotype in multiple sclerosis. Hum Mol Genet 18: 767–778.
29. Barrett JC, Clayton DG, Concannon P, Akolkar B, Cooper JD, et al. (2009) Genome-wide association study and meta-analysis find that over 40 loci affect risk of type 1 diabetes. Nat Genet 41: 703–707.
30. Hunter DJ, Kraft P, Jacobs KB, Cox DG, Yeager M, et al. (2007) A genome-wide association study identifies alleles in FGFR2 associated with risk of sporadic postmenopausal breast cancer. Nat Genet 39: 870–874.
31. Maris JM, Mosse YP, Bradfield JP, Hou C, Monni S, et al. (2008) Chromosome 6p22 locus associated with clinically aggressive neuroblastoma. N Engl J Med 358: 2585–2593.
32. Segre AV, Groop L, Mootha VK, Daly MJ, Altshuler D (2010) Common inherited variation in mitochondrial genes is not enriched for associations with type 2 diabetes or related glycemic traits. PLoS Genet 6: e1001058.
33. Bader GD, Betel D, Hogue CWV (2003) BIND: the Biomolecular Interaction Network Database. Nucleic Acids Research 31: 248–250.
34. Ceol A, Chatr Aryamontri A, Licata L, Peluso D, Briganti L, et al. (2010) MINT, the molecular interaction database: 2009 update. Nucleic acids research 38: D532–539.
35. Kerrien S, Aranda B, Breuza L, Bridge A, Broackes-Carter F, et al. (2011) The IntAct molecular interaction database in 2012. Nucleic Acids Research 40: D841–D846.
36. Keshava Prasad TS, Goel R, Kandasamy K, Keerthikumar S, Kumar S, et al. (2009) Human Protein Reference Database—2009 update. Nucleic acids research 37: D767–772.
37. Mishra GR, Suresh M, Kumaran K, Kannabiran N, Suresh S, et al. (2006) Human protein reference database—2006 update. Nucleic acids research 34: D411–414.
38. Pagel P, Kovac S, Oesterheld M, Brauner B, Dunger-Kaltenbach I, et al. (2005) The MIPS mammalian protein–protein interaction database. Bioinformatics 21: 832–834.
39. Peri S, Navarro JD, Amanchy R, Kristiansen TZ, Jonnalagadda CK, et al. (2003) Development of human protein reference database as an initial platform for approaching systems biology in humans. Genome Res 13: 2363–2371.
40. Stark C (2006) BioGRID: a general repository for interaction datasets. Nucleic Acids Research 34: D535–D539.
41. Huang DW, Sherman BT, Lempicki RA (2009) Systematic and integrative analysis of large gene lists using DAVID bioinformatics resources. Nat Protoc 4: 44–57.
42. Huang DW, Sherman BT, Lempicki RA (2009) Bioinformatics enrichment tools: paths toward the comprehensive functional analysis of large gene lists. Nucleic Acids Res 37: 1–13.
43. Weng MP, Liao BY (2010) MamPhEA: a web tool for mammalian phenotype enrichment analysis. Bioinformatics 26: 2212–2213.

44. Kupershmidt I, Su QJ, Grewal A, Sundaresh S, Halperin I, et al. (2010) Ontology-based meta-analysis of global collections of high-throughput public data. PLoS One 5.

45. Shannon P, Markiel A, Ozier O, Baliga NS, Wang JT, et al. (2003) Cytoscape: a software environment for integrated models of biomolecular interaction networks. Genome Res 13: 2498–2504.

46. Neale BM, Kou Y, Liu L, Ma'ayan A, Samocha KE, et al. (2012) Patterns and rates of exonic de novo mutations in autism spectrum disorders. Nature 485: 242–245.

47. Sakai Y, Shaw CA, Dawson BC, Dugas DV, Al-Mohtaseb Z, et al. (2011) Protein interactome reveals converging molecular pathways among autism disorders. Sci Transl Med 3: 86ra49.

48. Chow ML, Li HR, Winn ME, April C, Barnes CC, et al. (2011) Genome-wide expression assay comparison across frozen and fixed postmortem brain tissue samples. BMC Genomics 12: 449.

49. Nishimura Y, Martin CL, Vazquez-Lopez A, Spence SJ, Alvarez-Retuerto AI, et al. (2007) Genome-wide expression profiling of lymphoblastoid cell lines distinguishes different forms of autism and reveals shared pathways. Hum Mol Genet 16: 1682–1698.

50. Neale BM, Kou Y, Liu L, Ma'ayan A, Samocha KE, et al. (2012) Patterns and rates of exonic de novo mutations in autism spectrum disorders. Nature 485: 242–245.

51. Ben-David E, Shifman S (2012) Networks of neuronal genes affected by common and rare variants in autism spectrum disorders. PLoS Genet 8: e1002556.

52. Diskin SJ, Hou C, Glessner JT, Attiyeh EF, Laudenslager M, et al. (2009) Copy number variation at 1q21.1 associated with neuroblastoma. Nature 459: 987–991.

53. Eijkelkamp N, Linley JE, Baker MD, Minett MS, Cregg R, et al. (2012) Neurological perspectives on voltage-gated sodium channels. Brain 135: 2585–2612.

54. Hollander E, Wang AT, Braun A, Marsh L (2009) Neurological considerations: autism and Parkinson's disease. Psychiatry Res 170: 43–51.

55. Scheuerle A, Wilson K (2011) PARK2 copy number aberrations in two children presenting with autism spectrum disorder: further support of an association and possible evidence for a new microdeletion/microduplication syndrome. Am J Med Genet B Neuropsychiatr Genet 156B: 413–420.

56. Crespi B (2011) Autism and cancer risk. Autism Res 4: 302–310.

57. Glessner JT, Wang K, Cai G, Korvatska O, Kim CE, et al. (2009) Autism genome-wide copy number variation reveals ubiquitin and neuronal genes. Nature 459: 569–573.

58. Yaspan BL, Bush WS, Torstenson ES, Ma D, Pericak-Vance MA, et al. (2011) Genetic analysis of biological pathway data through genomic randomization. Hum Genet 129: 563–571.

59. Gilman SR, Iossifov I, Levy D, Ronemus M, Wigler M, et al. (2011) Rare de novo variants associated with autism implicate a large functional network of genes involved in formation and function of synapses. Neuron 70: 898–907.

60. Hussman JP, Chung RH, Griswold AJ, Jaworski JM, Salyakina D, et al. (2011) A noise-reduction GWAS analysis implicates altered regulation of neurite outgrowth and guidance in autism. Mol Autism 2: 1.

61. Buxbaum JD, Betancur C, Bozdagi O, Dorr NP, Elder GA, et al. (2012) Optimizing the phenotyping of rodent ASD models: enrichment analysis of mouse and human neurobiological phenotypes associated with high-risk autism genes identifies morphological, electrophysiological, neurological, and behavioral features. Mol Autism 3: 1.

62. He W, Wang Y, Liu W, Zhou CZ (2007) Crystal structure of Saccharomyces cerevisiae 6-phosphogluconate dehydrogenase Gnd1. BMC Struct Biol 7: 38.

63. Hofmann B, Hecht HJ, Flohe L (2002) Peroxiredoxins. Biol Chem 383: 347–364.

64. Immenschuh S, Baumgart-Vogt E (2005) Peroxiredoxins, oxidative stress, and cell proliferation. Antioxid Redox Signal 7: 768–777.

65. Frustaci A, Neri M, Cesario A, Adams JB, Domenici E, et al. (2012) Oxidative stress-related biomarkers in autism: systematic review and meta-analyses. Free Radic Biol Med 52: 2128–2141.

66. Ghanizadeh A, Akhondzadeh S, Hormozi M, Makarem A, Abotorabi-Zarchi M, et al. (2012) Glutathione-related factors and oxidative stress in autism, a review. Curr Med Chem 19: 4000–4005.

67. Voineagu I, Wang X, Johnston P, Lowe JK, Tian Y, et al. (2011) Transcriptomic analysis of autistic brain reveals convergent molecular pathology. Nature 474: 380–384.

15

Community Structure Detection for Overlapping Modules through Mathematical Programming in Protein Interaction Networks

Laura Bennett[1], Aristotelis Kittas[2], Songsong Liu[1], Lazaros G. Papageorgiou[1], Sophia Tsoka[2]*

1 Centre for Process Systems Engineering, Department of Chemical Engineering, UCL (University College London), Torrington Place, WC1E 7JE, London, United Kingdom,
2 Department of Informatics, King's College London, Strand, WC2R 2LS, London, United Kingdom

Abstract

Community structure detection has proven to be important in revealing the underlying properties of complex networks. The standard problem, where a partition of *disjoint* communities is sought, has been continually adapted to offer more realistic models of interactions in these systems. Here, a two-step procedure is outlined for exploring the concept of *overlapping* communities. First, a hard partition is detected by employing existing methodologies. We then propose a novel mixed integer non linear programming (MINLP) model, known as OverMod, which transforms disjoint communities to overlapping. The procedure is evaluated through its application to protein-protein interaction (PPI) networks of the rat, *E. coli*, yeast and human organisms. *Connector* nodes of hard partitions exhibit topological and functional properties indicative of their suitability as candidates for multiple module membership. OverMod identifies two types of connector nodes, *inter* and *intra-connector*, each with their own particular characteristics pertaining to their topological and functional role in the organisation of the network. Inter-connector proteins are shown to be highly conserved proteins participating in pathways that control essential cellular processes, such as proliferation, differentiation and apoptosis and their differences with intra-connectors is highlighted. Many of these proteins are shown to possess multiple roles of distinct nature through their participation in different network modules, setting them apart from proteins that are simply 'hubs', i.e. proteins with many interaction partners but with a more specific biochemical role.

Editor: Dongxiao Zhu, Wayne State University, United States of America

Funding: The funding of the EU (to ST; HEALTH-F2-2011-261366), the Leverhulme Trust (to ST and LGP; RPG- 2012-686) and the UK Engineering & Physical Sciences Research Council (to LGP; EPSRC Centre for Innovative Manufacturing in Emergent Macromolecular Therapies) is gratefully acknowledged. The funders had no role in study design, data collection and analysis, decision to publish, or preparation of the manuscript.

Competing Interests: The authors have declared that no competing interests exist.

* Email: sophia.tsoka@kcl.ac.uk

Introduction

Community structure detection is widely accepted as a means of elucidating the underlying properties of complex networks. In the standard community structure detection problem, the aim is to partition a network into disjoint communities, also known as modules, which are generally regarded as semi-independent units. In protein interaction networks, disjoint community structure detection methods have served to propose functionally coherent modules [1,2]. However, in reality proteins may carry out more than one task or belong to more than one protein complex [3], corresponding to membership of more than one module. If disjoint communities are assumed to correspond to functional units, then overlapping communities offer a means of expressing the coordination of these functions within the context of the entire system. Consequently, relaxing the constraint of strictly non-overlapping communities in models of community structure may represent a more true to life abstraction, thus leading to a more accurate representation of cellular interactions.

The overlapping community structure detection problem is less well-defined than the standard problem and can be formulated in various ways depending on analytical requirements or user interpretation. As a result, existing approaches vary to a large

degree. Methods exist based on clique percolation (CFinder [4]), local expansion and optimisation (OSLOM [5], OCG [6] and ClusterONE [7]), agent-based and dynamical algorithms (GAN-XiSw [8]), Approximate Minimum Degree Ordering (MOFinder [9]), normalised cut calculation (Graclus [10]), regularized sparse random graph model (RSRGM [11]), modularity optimisation (OMIM [12]), consensus clustering [13], hub duplication [14] and Markov clustering (R-MCL [15]). The first challenge is therefore to decide how to interpret the problem and subsequently to define a suitable solution procedure in relation to biological systems. Overlapping communities in protein interaction networks have been studied to derive the functional cohesion of overlapping communities compared with disjoint communities, according to enrichment of GO terms and correspondence with protein complexes [6,7,9,11,12,15,16]. In [6], proteins belonging to many modules in the human PPI network were found to have on average a higher node degree and node betweenness, to contain more protein domains and to be annotated with more GO terms than proteins belonging to only one module. Similarly, in [9] proteins participating in many modules in the human PPI network were found to be enriched for druggable targets according to the Druggable Genome [17]. The method proposed in [18] has been applied to a structural brain network, where each node

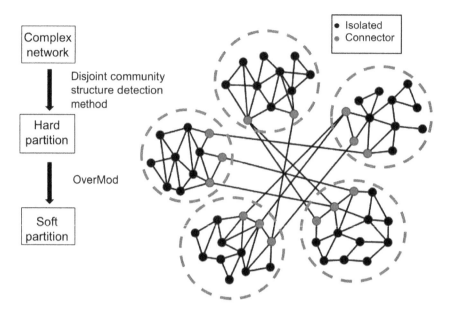

Figure 1. Outline of the two-stage procedure for detecting overlapping community structure. Black nodes (isolated nodes) have their module membership fixed, whereas red nodes (connector nodes) are free to be assigned to one or more modules when solving OverMod.

corresponds to a brain region [19] and nodes assigned to more than one module were found to have a higher degree and higher nodal efficiency. Analogous ideas have been found in a social network modelling the spread of disease, where nodes bridging communities were found to be potential immunisation targets [20].

These studies introduce the idea that nodes with multiple module membership may play an important role within a system, either in topological or functional terms. However, despite such previous investigations, there is still no concrete definition or understanding of the nature of nodes with multiple module membership, leaving much scope for (i) the development of a solution procedure that clearly demarcates overlapping modules and (ii) the systematic investigation of links between node clustering properties and their functional potential. We address these issues by first extending previous work on network module detection based on modularity optimisation to allow for multiple module membership. We then use the clustering features of protein nodes as means to explore their topological and functional significance on system properties.

Modularity is a measure that expresses how well-defined the community structure of a network is by comparing the fraction of edges that lie within modules minus the expected value in a null model, i.e. a network with same degree distribution but edges placed at random [21]. The metric provides a natural, intuitive description of community structure for a wide range of biological applications [1,2,22–24]. Optimisation of modularity is one of the most popular methods for community structure detection (e.g. [25–28]) and has been applied in various solution procedures, including simulated annealing [26,29], greedy algorithms [27,30] and spectral methods [28]. Mathematical programming has also been used to solve modularity maximisation problems, achieving globally optimal solutions in small to medium networks [31,32] and competitive results for larger networks [33–37]. Here, we extend our previous

mathematical programming approaches to modularity optimisation ([31,34,36]) to detect overlapping modules as outlined below.

Given a *hard partition* of a network, i.e. a partition of non-overlapping communities, if each module relates to a semi-independent functionally cohesive unit, then nodes that form edges across the borders of the communities can be thought of as bridges between different functional groups. We define these nodes as *connector* nodes and we distinguish them from *isolated* nodes, that only possess links with nodes of the same community. Consequently, our proposed method takes existing modularity optimisation methodologies one step further by considering the dynamics at the borders of communities. We pose the following questions: what effect do connector nodes have on the modularity of their neighbouring modules, what parameters determine whether a connector node has a multiple module membership, what effect do these connector nodes have on the cohesion of the network and do they exhibit some biological relevance?

Our proposed two-stage procedure is outlined in Figure 1. First, a hard partition is detected by defining the disjoint communities of a network using existing well-established and tested methodologies for modularity optimisation. Through this step, connector nodes are identified and distinguished from isolated nodes. In stage two, the association of connector nodes with their neighbouring modules is considered by allowing them to be allocated to any of the modules that they interact with. Finally, connector nodes will be assigned to a module if modularity is increased and therefore they become either *inter-connectors* if they are assigned to multiple communities, or *intra-connectors* if they remain a member of a single community, resulting in a *soft partition* of the network. The transformation from disjoint to overlapping communities is achieved by a mixed integer non linear programming (MINLP) model, known as OverMod.

Our method is evaluated through the investigation of the disjoint and overlapping community structures of the rat, *E. coli*, yeast and human PPI networks. Properties of connector against isolated nodes are first examined and OverMod then determines which of the connector nodes remain assigned only to their original module (intra-connectors), and which are distributed across many modules (inter-connectors). Analysis of each category of nodes reveals their own particular characteristics pertaining to their topological and functional role in the organisation of the network. Finally, a comparative analysis of related methodologies from the literature is presented, where method performance is discussed in relation to synthetic networks and the PPI networks.

Materials and Methods

A Mathematical Programming Model for Transforming Disjoint to Overlapping Communities

In our previous work, modularity optimisation has been formulated as a mixed integer quadratic programming (MIQP) model [31] and a mixed integer non linear programming (MINLP) model [34,36] to detect disjoint communities. In this work, modularity is again used as the objective function in the optimisation problem, but here node-module allocations for nodes that have no connections outside their community, known as isolated nodes, are fixed, leaving only connector nodes free to be assigned to one or more modules. The new model, OverMod, transforms a disjoint partition of a network into a partition with overlapping communities. The input required for OverMod is an undirected, weighted or unweighted network together with a hard partition of the network. The output is a set of overlapping communities. The indices, sets, parameters and variables associated with OverMod are defined below.

Indices

n,e	nodes
m	modules

Sets

CN	set of all connector nodes
C_m	connector nodes for module m
IS_m	isolated nodes for module m

Parameters

β_{ne}	weight of the edge between nodes n and e
α_n	weight of the edge node n makes with itself, i.e. a loop
d_n	strength (weighted degree) of node n
L	sum of the weights of all edges in the network

Continuous variables

L_m	sum of weights of all edges among nodes within module m
D_m	sum of strengths of nodes in module m

Binary variables

YS_{nm}	node membership in the soft partition; equal to 1 if node n is allocated to module m; 0 otherwise

If β_{ne} is non-zero, then an edge exists between nodes n and e and $\beta_{ne} = \beta_{en}$. The sets IS_m and C_m are defined according to each module, m, in the hard partition. IS_m is the set of isolated nodes in module m; nodes which belong to module m and do not interact with nodes outside of module m. C_m is the set of connector nodes associated with module m; nodes in module m that interact with nodes in neighbouring modules or, nodes outside module m that are connected to nodes within module m.

We adopt modularity as our objective function, defined as follows:

$$Q_{ov} = \sum_m \left[\frac{L_m}{L} - \left(\frac{D_m}{2L} \right)^2 \right] \tag{1}$$

where D_m is the sum of the strengths (weighted degrees) of all nodes in module m, L_m is the sum of the weights of the edges with both associated nodes belonging to module m and L is the sum of the weights of all edges in the network. We label modularity Q_{ov} instead of simply Q in order to distinguish it from modularity values where each node is assigned to only one module. The idea behind the approach is that since isolated nodes do not connect with nodes in other modules, they would make little or no contribution to the modularity of modules other than their own. Consequently, their module membership remains fixed and only connectors have the possibility of belonging to multiple modules in the course of the conversion procedure. In other words, for all $n \in IS_m$, YS_{nm} is fixed to 1, and for all $n \in C_m$, YS_{nm} is assigned a random initial value of 0 or 1. The number of variables in the optimisation problem is therefore reduced and in turn, so is computational cost.

Equation 1 is optimised subject to the following constraints. First, L_m and D_m are defined as:

$$L_m = \sum_{n \in IS_m} \alpha_n + \sum_{n \in C_m} \alpha_n YS_{nm} + \sum_{\substack{n,e \in IS_m \\ n < e}} \beta_{ne}$$
$$+ \sum_{\substack{n \in C_m \\ e \in IS_m}} \beta_{ne} YS_{nm} + \sum_{\substack{n,e \in C_m \\ n < e}} \beta_{ne} YS_{nm} YS_{em}, \quad \forall m \tag{2}$$

and

$$D_m = \sum_{n \in IS_m} d_n + \sum_{n \in C_m} d_n YS_{nm}, \quad \forall m \tag{3}$$

where d_n is the strength of node n and is defined as $d_n = 2\alpha_n + \sum_{e,n<e} \beta_{ne}$. Note that OverMod accommodates self-interactions, also known as loops, α_n.

In order to account for the overlapping aspect of the model, the following constrains each connector to belong to at least one module:

$$\sum_{m:n \in C_m} YS_{nm} \geq 1, \quad \forall n \in CN \tag{4}$$

The resulting MINLP model comprises a non-linear objective function with a combination of integer and continuous variables, summarised as:

Maximise:

$$Q_{ov} = \sum_m \left[\frac{L_m}{L} - \left(\frac{D_m}{2L} \right)^2 \right] \tag{5}$$

Subject to:
Constraints (2-4)

$$L_m, D_m \geq 0, \quad \forall m \tag{6}$$

$$YS_{nm} \in \{0,1\}, \quad \forall m, n \in C_m \tag{7}$$

Implementation

OverMod was implemented in GAMS (General Algebraic Modelling System) [38], where the MINLP is solved using the SBB (standard branch and bound method) mixed integer optimisation solver and CONOPT as the NLP solver. Due to the non-convex nature of the model, globally optimal solutions cannot be guaranteed. Thus, the MINLP is solved iteratively 100 times, each time with a different random initial solution, giving a approximate representation of solution space. The largest value of Q corresponds to the best soft partition. The GAMS binaries of the MINLP algorithm are available on request.

Networks

The proposed procedure is evaluated through its application to protein-protein interaction (PPI) networks of the rat, *E. coli*, yeast

and human organisms. The rat PPI network was downloaded from BioGRID (version 3.1.86, July 2012) [39]. Only interactions where both nodes were proteins of rat were retained. We consider only the main component of the network, which has 487 nodes and 572 interactions. The protein interaction network of *E. coli* was downloaded from the IntAct database (July 2013) [40], comprising exclusively interactions with a relation of type 'direct interaction' or 'physical association' that have been experimentally verified. The main connected component has 668 nodes and 846 interactions. We also consider the yeast PPI network of Collins et al. [41], from the BioGRID database. The main component of the yeast network comprises 1002 nodes and 8313 interactions. Finally, we include the main component of the human PPI network, as used in [6] and made available by the authors, which comprises 6160 nodes and 24014 interactions.

Statistical Analysis

Various comparisons are made where the average degree, betweenness, eigenvector centrality, number of GO terms and number of protein domains of groups of nodes were calculated. Betweenness and eigenvector centrality were found using the igraph library [42] in the statistical computing environment R [43]. GO annotations for rat, *E. coli*, yeast and human were downloaded from [44], [45], [46] and [47], respectively. Each protein was mapped to all possible GO terms. Parent terms were removed to keep the most specific GO annotations. Protein domains for each organism were downloaded from the Pfam database [48]. Only distinct domain annotations for each protein were retained. The population means of each group of nodes for each property were determined statistically significantly different or not using the Mann-Whitney-Wilcoxon U test (two sided) as implemented in R. A p-value of less than 0.01 indicates a statistically significant difference. Finally, essential genes were downloaded from the the Online GEne Essentiality (OGEE) Database [49]. Enrichment of essential genes in multi-clustered node was determined according to the Fisher's Exact test as implemented in R.

Node Removal

We investigated the effect of node removal using Monte Carlo simulations. At each step a random node is removed from the sets of isolated, inter and intra-connector nodes respectively. We then calculate, s, the size (number of nodes) of the largest connected component, over the initial component size. Each step is the average of 100 independent runs.

Results and Discussion

In this section, the disjoint and overlapping community structures of the PPI networks of rat, *E. coli*, yeast and human are investigated. Hard partitions of the networks are first detected by three different modularity optimisation methods. Based on the hard partitions, each node is classified as either an isolated node or a connector node. Characteristics of the connector nodes are investigated in order to determine whether they possess some topological and/or functional relevance relating to their position in the network. The hard partitions are subsequently transformed to overlapping communities by the proposed mathematical programming method, OverMod. The effect of node removal and the functional significance of the inter and intra-connectors are then explored. Finally, the proposed procedure is discussed in the context of other existing methodologies.

Detection of Hard Partitions

We employ three of the most well known methods of modularity optimisation to detect hard partitions of the PPI networks: iMod [34], Louvain [27] and QCUT [28]. Each method has been shown to perform well in applications on various sizes of complex networks. In particular, iMod outperformed several other well-known methods on medium to large sized networks, including finding known globally optimal solutions for small sized networks [34] and Louvain, a heuristic method, is known for its low computational cost and high quality results on very large networks [50]. Employing these three methods allows us first to explore the effect of the choice of hard partition on the final results found by OverMod and in turn the stability of the algorithm, and second, to combine the information from them to determine the most robustly multi-clustered proteins. In doing this, we take advantage of the information offered by a range of results from three of the best available methods.

Table 1 gives the value of modularity, the corresponding number of modules and the number of connector nodes found by iMod, Louvain and QCUT for each of the PPI networks. For the rat and E. coli networks, iMod finds partitions with marginally larger values of modularity, for the yeast network, Louvain performs best and for the human network QCUT performs best. These differences in modularity are minimal and generally for each network, the methods perform similarly in terms of value of modularity, number of modules and number of connectors. The difference between the three sets of connector nodes produced for each network is quantified by employing the Jaccard index (results not shown). The Jaccard index measures similarity between finite sample sets, and is defined as the number of nodes in the intersection of the sets divided by the number of nodes in the union of the sets. The average of the Jaccard values between pairwise sets of connectors for Rat, E. coli, Yeast and Human are 0.78, 0.80, 0.88 and 0.80, respectively. The difference between the sets of connectors found by each method across the networks is relatively small. It is investigated later how these differences are reflected in the output of OverMod, i.e. how stable is the algorithm to perturbations in the input.

Finally, as an illustrative example of the hard-partitioning step, Figure 2 shows the rat PPI network before partitioning, followed by the clustered network, where the modules found by iMod are identified by an individual colour. Hub nodes, UBC (54 interaction partners) and SUMO3 (187 interaction partners), are identified by yellow circles. The hard partition is dominated by one large module (red). It comprises 168 nodes and corresponds to the SUMO3 gene and its surrounding nodes, while the second largest module (blue), comprises the UBC gene and 63 surrounding nodes.

Properties of Connector Nodes. We investigate the nature of the two types of nodes that are defined according to the hard partitions: isolated and connector nodes. If connector nodes are interpreted as bridges between functional units, one expects them to exhibit properties that reflect such activity. Here, various topological and functional measures are employed to characterise protein nodes and investigate whether connectors can be distinguished from isolated nodes in terms of these properties.

The following topological properties are employed as measures of a node's structural features. First, node degree is used as it has been shown that complex networks are vulnerable to targeted removal of nodes with a high degree, also known as hubs [51,52]. Hubs therefore are considered to possess particular topological characteristics that may also link to relevant functional properties. Second, node betweenness is used to indicate the number of shortest paths that traverse a particular node. It has been suggested that nodes of high betweenness usually lie between communities, according to the betweenness clustering method of Girvan and Newman [53], potentially indicating connector properties. Finally, eigenvector centrality is considered, which is a method of computing the centrality of a node based on the centrality values of the nodes that it is connected to.

Where the above properties offer topological measures for illustrating structural importance, we also consider descriptive features based on protein function. We use the number of GO annotations as a measure of functional importance of a node, both in combined form (ALL GO) as well in terms of the individual GO categories (molecular function, MF, biological process, BP and cellular compartment, CC). Additionally, the number of domains that a protein contains are used as measure of its multifunctionality. Finally, the Online GEne Essentiality (OGEE) Database [49] that contains genes that have been tested experimentally for essentiality, is employed here and we investigate whether the connector nodes are enriched for essential proteins.

For all four networks node degree, node betweenness and eigenvector centrality were on average significantly higher for connector nodes when compared to isolated nodes (Table S1 in File S1), thus demonstrating their distinct topological properties. We now reinforce these results by showing that topological features correspond to particular functional properties. First, it has been shown that signalling domains are found more often in intermodular hub proteins, which were also more frequently associated with oncogenesis than their intramodular hub counterparts [54]. Our analysis therefore demonstrates that the high connectivity of these nodes is correlated with their multiple roles and their potential to act as bridges between functional modules.

In the human network, connectors have a significantly higher average number of GO annotations and protein domains (Table S2 in File S1). While in the rat network, this is true for GO terms. For the yeast and E. coli networks these effects are less pronounced, perhaps owing to the fact that, as yeast and E. coli are unicellular organisms, they have much simpler biochemistry than multicellular species. The difference in organismal complexity may be therefore reflected, in this case, in the significance of GO term enrichment. Additionally, with respect to essentiality properties, we find significant enrichment of essential genes in the connector nodes of E. coli, yeast and human (Table S3 in File S1). For the rat network, only a small number of essential genes have been identified, so statistical evaluation was not attempted.

The p-values for each property do not vary greatly between the different hard partitioning methods employed, i.e the overall properties of connector nodes are generally the same regardless of the hard partitions used. The above results indicate then that in general, connector nodes have not only distinct topological properties, but also a wider functional repertoire than isolated proteins. Our next task was to investigate the properties of connector nodes through the detection of overlapping communities (soft partition), which we discuss in the following section.

Converting to Soft Partitions

We denote the application of OverMod to each hard partition as the following three methods: (i) iMod for the hard partition followed by OverMod, (ii) hard partition by Louvain followed by OverMod and (iii) hard partition by QCUT followed by OverMod. Applying OverMod to the hard partitions results in connector nodes being determined as either inter or intra-connector. Table 2 summarises the number of inter and intra-connectors.

Figure 3 shows the number of modules that protein nodes belong to across the reference organisms tested. Nodes belonging

176

Bioinformatics, Proteomics and Genomics

to only one module include both isolated and intra-connector nodes. For rat and *E. coli*, the inter-connector nodes belong to at most 3 modules, whereas inter-connector nodes participate in up to 7 and 13 modules in the yeast and human networks respectively.

The stability of OverMod is now investigated, i.e. it is determined how much the output is perturbed by changes to the input. In this case the input is the hard partition found by either iMod, Louvain or QCUT and the output is the classification of connectors as either inter or intra-connectors. Of particular interest is how the sets of inter-connectors vary depending on which hard partition is used as input. The commonality between inter-connector nodes across methods (i) – (iii) is illustrated in Figure 4. Inter-connector proteins found by all three methods were 36 in the rat network, 40 in *E. coli*, 233 in yeast, and 3002 in the human network. The Jaccard index is again employed as a means of quantifying the similarity between the inter-connector sets returned by OverMod for methods (i) – (iii). The Jaccard index has previously been used a measure of cluster stability [55]. For each PPI network, the Jaccard index is calculated for iMod + OverMod vs. Louvain + OverMod, iMod + OverMod vs. QCUT + OverMod and Louvain + OverMod vs. QCUT + OverMod. Results are presented in Table S4 in File S1 and here the average Jaccard index for each network is reported: 0.72, 0.75, 0.85 and 0.74 for the rat, *E. coli*, yeast and human networks, respectively. The similarity between inter-connector sets is relatively stable across all networks. Overall, the results show a good degree of stability regarding individual inter-connector proteins, indicating that OverMod exhibits a good level of robustness to small perturbations in the input.

Topological Features and the Effect of Node Removal. Node degree, betweenness and eigenvector centrality of the two types of connector nodes were calculated. Statistical analysis showed a significant difference in node degree for the rat, *E. coli* and human networks, with the intra-connectors having consistently higher values (Table S5 in File S1). For the yeast network, although inter-connectors have a higher average degree than intra-connectors, the difference is not significant. Less consistent results are found for betweenness and eigenvector centrality.

On first inspection, the observation that nodes characterised as intra-connectors by OverMod had higher degree was counter-intuitive, we therefore sought to look more closely at inter-module connections to determine if inter- and intra-connectors can be characterised according to their communication with neighbouring modules. A measure known as participation coefficient is adopted, which measures how uniformly distributed the edges of a node are among the communities of a partition [26]:

$$P_n = 1 - \sum_{m=1}^{M} \left(\frac{d_{nm}}{d_n}\right)^2 \tag{8}$$

where M is the number of modules in the partition, d_{nm} is the number of links node n has with nodes in module m and d_n is the degree of node n. The larger the participation coefficient of a connector, the more evenly distributed its connections are with different modules. We found that for all four networks studied, inter-connectors have a significantly higher average participation coefficient than intra-connectors (Table S6 in File S1), indicating that despite lower node degree on average, inter-connector genes distribute their edges across communities more widely than intra-connectors.

The topological properties of inter-connector, intra-connector, as well as isolated nodes were also established through simulating

Table 1. Hard partition summary.

Network	iMod Q	iMod Mod	iMod Con	Louvain Q	Louvain Mod	Louvain Con	QCUT Q	QCUT Mod	QCUT Con
Rat	0.7734	18	54	0.7729	17	54	0.7701	18	54
E. coli	0.8537	24	59	0.8523	23	59	0.8522	24	59
Yeast	0.7572	20	310	0.7612	22	299	0.7608	21	290
Human	0.5382	18	4529	0.5462	19	4372	0.5527	16	4246

Summary of the hard partitions of the rat, *E. coli*, yeast and human PPI networks, showing modularity value (*Q*), corresponding number of modules (*Mod*) and the number of connectors (*Con*). For each network, the hard partition results appear to be relatively stable. These hard partitions are the input for OverMod.

the effect of their removal on network integrity. Figure 5 shows the relative size of the largest component *s* against the number of nodes *n* removed from the network. In all cases the removal of isolated nodes has the smallest effect on network structure, in line with our results of centrality of isolated vs connector nodes, where degree and betweenness centrality are always significantly lower in isolated nodes. This illustrates that the isolated nodes are the least important in maintaining network integrity.

Our simulations also show that the removal of intra-connector nodes breaks the network consistently faster than inter-connectors, in accordance with their significantly higher node degree. The effect can also be seen in the yeast and human networks, however it is less pronounced due to the density and size of these networks, i.e. a large portion of the nodes need to be removed in order to reduce the largest component size. Alternatively, in rat and *E. coli* where the effect was more marked is attributed to the fact that the integrity of these smaller networks is largely maintained by a few intra-connectors (Table 2), which for both networks contain the most highly connected nodes in the entire network. These two cases are discussed in more detail below.

The most highly connected node in the rat PPI network (degree 187) corresponds to the SUMO3 gene and the most highly connected node in the *E. coli* networks is the CH60 (GroEL) protein (degree 128). Both are intra-connector nodes despite their large degree and the fact that they interact with nodes in 12 and 9 modules respectively. In other words, there is a large number of possible modules that OverMod could assign them to. We investigate why these two nodes are intra-connectors, when the highest degree nodes in the yeast and human networks are inter-connectors.

We consider the application of OverMod to the hard partitions found by iMod in the following discussion regarding the rat and *E. coli* networks. SUMO3 is not only the node with the highest degree in the rat network, it is also the node with the largest number of possible modules it can be assigned to. However, OverMod, assigns SUMO3 to only one module. On further investigation, we find that only 20 out of 187 connections link to nodes in other modules, indicating (i) a strong connection to the

module where it was assigned during the hard partition stage and (ii) a rather low participation coefficient of 0.2. Bearing in mind that OverMod simultaneously optimises the modularity of all modules in the soft partition by either assigning the connector nodes to one or multiple modules, we calculate the modularity for each of the 12 possible modules, with SUMO3 present and absent. It is found that SUMO3 decreases the modularity of all modules other than its own. Therefore, to optimise the total modularity across the soft partition, OverMod assigns SUMO3 to its own module only.

Similarly, in the *E. coli* network, the CH60 (GroEL) protein, also has the highest number of possible modules it can be assigned to. However, of its 128 interactions, only 12 are among the 9 neighbouring modules, corresponding to a participation coefficient of 0.18. Again, CH60 was found to decrease the modularity of all neighbouring modules and only increase the modularity of its own module, despite its high centrality and potential to belong to many modules. These two cases illustrate that high degree is not necessarily the principal driving force when OverMod allocates a connector node to more than one module. In fact, these two examples are representative of high degree connector nodes that become intra-connectors and thus why so many high degree connectors remain in a single module.

We compare these cases with the most highly connected nodes in the yeast and human networks, which are both multi-clustered by OverMod. These have degrees 127 and 182 respectively. However, the difference is their much higher participation coefficient, 0.5 and 0.7 respectively. Therefore these high centrality nodes have a more even distribution of edges, whereas the corresponding nodes in the rat and *E. coli* have most edges concentrated in their original module. It is worth noting that if one calculated the average degree of the intra-connectors for the rat and *E. coli* networks without SUMO3 and CH60 respectively, the intra-connectors would still show a significantly higher average degree than the inter-connectors (with no significant difference for betweenness and eigenvector centrality as before). Similarly, inter-connectors continue to have a significantly higher participation coefficient than intra-connectors.

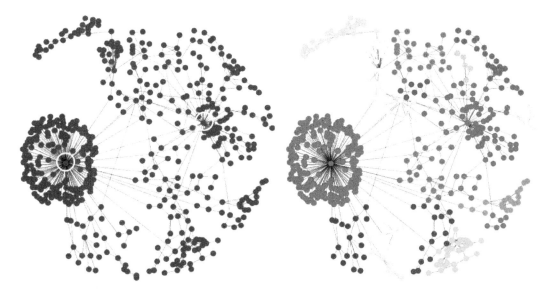

Figure 2. The left hand side of the figure shows the rat PPI network and on the right hand side, each community found by iMod is given an individual colour. Hub nodes (SUMO3 and UBC) are highlighted by yellow circles in the network on the left. Visualisation of the network was done using Cytoscape [82].

Overall, these results show that OverMod reveals the structural significance of a portion of the connector nodes, i.e. those with high connectivity which end up assigned to only one of their possible modules. While the majority of connectors maximise modularity by becoming inter-connectors ($\approx 90\%$ in most cases), a small number of highly connected nodes optimise modularity by remaining in their original module. Thus our method discovers a small subset of intra-connectors that are topologically important for the integrity of the network. OverMod also identifies the set of inter-connectors which may generally have a lower degree, but exhibit a higher participation coefficient, signifying that their edges are more evenly distributed among various modules. We now investigate the functional properties of inter- and intra-connectors and in particular we look into the biological evidence that corresponds to the association of multi-clustered nodes with multiple functional units.

Functional Comparison of Inter and Intra-connector Proteins. A PPI network is a collection of interactions that take place across time and space, no matter whether they happen simultaneously or not, or whether they are exclusive or not. Therefore, a PPI network may not capture a precise representation of protein interactions *in vivo*. While high centrality has been traditionally correlated with essentiality for survival [56–58], we investigate below whether this is consistently the case by comparing the most robust inter-connectors and intra-connectors with the highest connectivity.

First, we determine which nodes are considered as 'strong' inter-connectors. For the rat and *E. coli* networks, since all inter-connector proteins belong to either 2 or 3 modules, we consider proteins belonging to 3 modules as strong inter-connectors. For the yeast and human networks, the number of modules that inter-connector nodes belong to is higher and so strong inter-connectors are defined by finding the range of number of modules to which the top ten strongest inter-connectors are assigned. For example, for the yeast network, the top ten strongest inter-connectors for iMod+OverMod belong to between 5 and 6 modules. However, there are actually 19 nodes that belong to 5 and 6 modules and therefore we consider all of them as strong inter-connector proteins. It follows that any inter-connector defined as strong by two or more methods is described as being robust (Table 3). We now examine the biological functionality of robust inter-connector proteins and high degree intra-connector proteins with high and low inter-modular degree (shown in tables 3 and 4 respectively).

We investigate connectors by considering cases of proteins in the literature with significant functionality. CH60 (GroEL) is a molecular chaperone and belongs to a group of proteins that assist in the folding, translocation and assembly of proteins in the cell [59] and are the subject of significant research interest. Another example is Ubiquitin, which is conjugated to target proteins via an isopeptide bond either as a monomer (monoubiquitin - UBQ), a polymer linked via different Lys residues of the ubiquitin (polyubiquitin chains - UBC). The linkage type of the ubiquitin chain determines whether a modified protein is degraded by the proteasome or serves to attract proteins to initiate signalling cascades or be internalised [60]. UBC is assigned as intra while UBQ as inter-connector. The various types of Ub modifications are linked to distinct physiological functions in cells. UBQ, for example, regulates DNA repair and receptor endocytosis, whereas lysine 48-linked Ub chains label proteins for proteasomal degradation [61]. Since less is known about the functionality of UBC chains than the UBQ monomer, it is possible that UBQ rather than UBC becomes an inter-connector because of its associated multi-functionality, as the molecular mechanisms involving specificity in UBC chain synthesis and recognition are still incompletely understood [62] and thus less information about UBC interactions exists in PPI databases.

Genes in the RPS family encode approximately 80 different ribosomal proteins, which in conjunction with rRNA make up the ribosomal subunits. RPS4A, RPS5, RPS8A etc. are such proteins. RPS5 is a component of the small ribosomal subunit. Mature ribosomes consist of a small (40S) and a large (60S) subunit. Because the ribosome is such a vital component of the translational machinery and therefore of all cellular life, ribosomal proteins (RPs) have been highly conserved throughout evolution [63,64]. The RSM and MRP genes encode proteins of the 37S small subunit of mature mitochondrial ribosomes [65]. Surprisingly, only a minority of MRPs that have been characterised show significant sequence similarities to known ribosomal proteins from other sources [66]. With respect to our analysis, ribosomes in the cytosol are found to be inter-connectors while mitochondrial ribosomes are intra-connector proteins. We hypothesise that this is the case because cytosol ribosomes have a broader functionality, while mitochondrial ribosomes participate in a more limited spectrum of functions, since their main role is to synthesise proteins of these organelles.

The TGF-beta type I receptor is a transmembrane kinase which transduces TGF signalling from the cell surface to the cytoplasm and thus regulates a plethora of physiological and pathological processes [67–70]. Although TGF-beta is important in regulating crucial cellular activities, the full mechanism behind the suggested activation pathways is not yet well understood. Some of the known activating pathways are cell or tissue-specific, while some are seen in multiple cell types and tissues [71,72]. TGF beta receptor was the intra-connector node with the highest inter-modular degree (equal to 12), so it links to twelve different modules, yet remains assigned in its original module. Because it acts as a signalling

Table 2. Soft partition summary: inter and intra-connectors.

	(i)		(ii)		(iii)	
	Inter	Intra	Inter	Intra	Inter	Intra
Rat	45	9	48	8	45	9
E. coli	51	8	51	7	50	7
Yeast	274	36	274	25	260	30
Human	4207	322	3952	420	3647	559

Summary of the number of connectors that become inter and intra-connectors in the corresponding soft partitions. For Rat, *E. coli* and Yeast in particular, we see that the number of inter-connectors, i.e. nodes belonging to more than one community, found by each method (i) – (iii) are relatively similar. This stability is reinforced in Figure 4, where we see that the number of common inter-connectors is high for each PPI network.

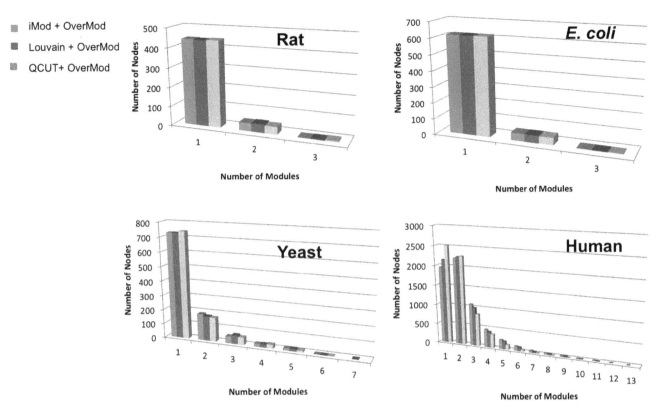

Figure 3. Bar charts showing the number of modules and the corresponding number of proteins for each soft partition for the rat, *E. coli*, yeast and human PPI networks.

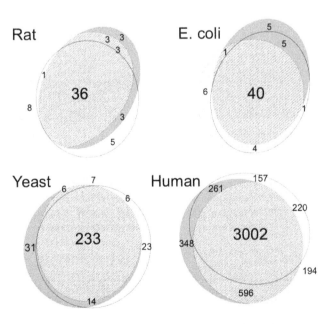

Figure 4. Venn diagrams illustrating the number of common inter-connector proteins across the methods for each network. The pink ellipse represents iMod + OverMod, the green striped ellipse represents Louvain + OverMod and the empty ellipse is QCUT + OverMod.

molecule activating many pathways but not directly participating in the biochemical processes (i.e. further interacting with molecules in the pathway), it is possible that while it's a very central protein, it remains assigned to a particular module rather than to multiple modules.

GBLP (Guanine nucleotide-binding protein), a component of the small (40S) ribosomal subunit, interacts with a wide variety of proteins and is involved in the recruitment, assembly and regulation of a variety of signalling molecules. CSK21 is a kinase complex that phosphorylates a large number of substrates containing acidic residues C-terminal to the phosphorylated serine or threonine and regulates numerous cellular processes, such as cell cycle and apoptosis [73]. CDC2 encodes the protein CDK1 (Cyclin dependent kinase 1), which is a highly conserved cell cycle protein that forms complexes that phosphorylate a variety of target substrates, leading to cell cycle progression [74]. RAF1 is a proto-oncogene which encodes the c-RAF enzyme. It functions as a switch determining cell fate decisions by acting as a regulatory link between the membrane-associated Ras GTPases and the MAPK/ERK cascade [75,76]. Mouse double minute 2 homolog (MDM2) is an important negative regulator of the p53 tumor suppressor, functioning both as an E3 ubiquitin ligase that binds to the p53 tumor suppressor and an inhibitor of p53 transcriptional activation [77]. Cell cycle and cancer related-proteins are thus often candidates for inter-connector molecules, as they regulate key processes (such as cell cycle, proliferation and apoptosis) that are interlinked for the cell's survival and reproduction.

Therefore, we see some similarity in the role of connector nodes, but also many differences. We have discovered that heat shock proteins in the rat network can be classified as both inter and intra-connector (HSP74 and HSP90, Tables 3 and 4). In general, intra-

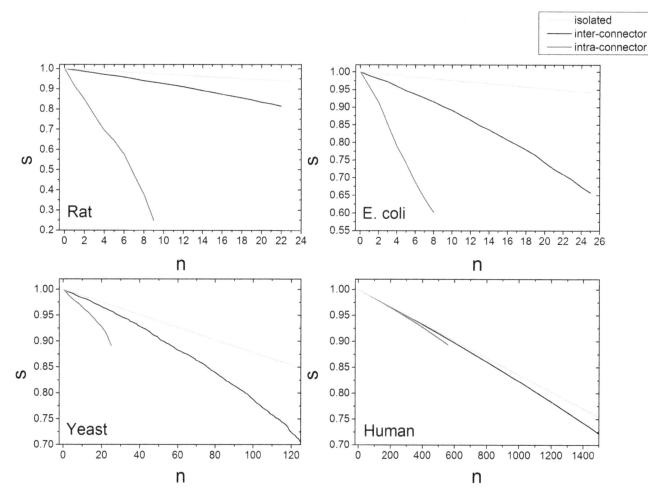

Figure 5. Relative size of largest component s vs number of nodes removed n for human, yeast, *E. coli* and rat networks. At each step, a random node is removed from the set of isolated, inter-connector and intra-connector nodes respectively. Results are an average of 100 runs. From each organism, the hard partitioning method which yielded the best Q was selected, namely iMod for the rat and *E. coli* networks, Louvain for yeast and QCUT for human

connectors do not participate directly in biochemical pathways, but may act as signalling molecules providing multiple connections between the modules (e.g TGF-beta receptor). Cytosol ribosomal proteins in yeast are consistently classified as inter-connectors in contrast to mitochondrial ribosomal proteins which end up in a single module. Overall, many of the robust inter-connectors exhibit the following characteristics:

1. Proteins which are major regulators of the cell cycle and therefore proliferation and apoptosis

2. Proto-oncogenes, oncogenes and regulators of tumour suppressors

3. Proteins which markedly affect cell growth

4. Ribosomal proteins which are essential for the survival and function of the organism and other highly conserved proteins

In a PPI network a protein usually has multiple copies, each acting as a specific molecular entity. These copies may interact with different groups of molecules in the cell [78]. A PPI network would then contain a single vertex that actually represents a collection of that kind of protein, rather than individual protein

Table 3. Robust inter-connectors.

Rat	MDM2, BRDT, *HSP90AA1, SUMO1*
E. coli	EF-Tu1, MutL, *DNA-Pol III*
Yeast	RPL31A, NOP1, RPS4A, RPL7B, RPS7A, RPS8A-B, RPS5, RPS11A-B, XRN1, RPS22A, *RPS13,RPS9B, SRO9, PRP43,RPL8B, CBF5*
Human	GBLP, CSK21, HS90A, ANDR, CDC2, *RAF1, CTNB1, RB, A4, NPM, UBIQ, 1433Z*

Connector proteins that were found to be strong inter-connectors by OverMod when applied to: (i) all three hard partitions, (ii) two out of three hard partitions (italics).

Table 4. Intra-connector proteins.

Rat	UBC, SUMO3, **UBC9, NTRK1, TBA1A, HSP74**
E. coli	CH60, DNAK, **HLDD, KPRS, ODP1, RPE, RS2, SYP**
Yeast	**CLP1, CKA2, CKB1-2, YEF3, FKS1, VID24, GCD11, NAP1, PAP1, PCF11, PTI1, REF2, MRP4, MRP7, RNA14, RSM19M RSMM22, RSM23M, MRP51, MRPS9, MRPS18, SWD2, YSH1**
Human	TGFR1, EF1G, ATN1, MDFI, PLS1, TRIP6, CACO2, KR412, AP2M1, MCM7, PABP1, ARI2, FHL3

Proteins with inter-modular degree that is (i) greater than 5 and (ii) equal to or less than 5 (bold). Only the former are shown for human, as the number of low inter-modular degree proteins is too large in this network.

copies [79]. This is why a hub vertex can bind to hundreds of interactors in a PPI network, while this is unrealistic in biological cells. Assuming that modules correspond to functional units, with this method we suggest candidate proteins of multiple distinct roles (inter-connectors), by their participation in different distinct modules, rather than hubs which are regulators of highly specific pathways.

Related Methodologies

The overlapping community structure problem has been subject to multiple interpretations due to lack of formalisation of the underlying problem statement. The great variation in existing methodology is reflected in the results, which can even be seen on a small network such as the benchmark Zachary karate network [36]. Here we present a comparative analysis of OverMod with various alternative overlapping community structure detection methods from the literature, namely, CFinder [4], OSLOM [5], OCG [6], ClusterONE [7], GANXiSw [8], MOFinder [9],

RSRGM [11] and R-MCL [15]. The performance of the above methods will be evaluated on a series of synthetic networks and the PPI networks described previously.

First, we note Wang et al. [12] have proposed a similar method to OverMod involving the two stage optimisation of modularity. Here we have addressed the problem more comprehensively and in a more rigorous mathematical manner. By employing mathematical programming and optimising the sum of modularity across all modules in the soft partition, we avoid sequence dependent results, as well as offering the possibility for a connector node to leave its original module in the hard partition. Furthermore, in our evaluation, we focus on the nature of multi-clustered nodes in the context of PPI networks, instead of the functional cohesion of modules. Since, a method implementation is not publicly available, we do not provide results.

A series of synthetic networks of the type described in [80] was generated. Each network comprises 500 nodes, with average degree equal to 10 and with either 75, 150 or 250 (ON) multi-

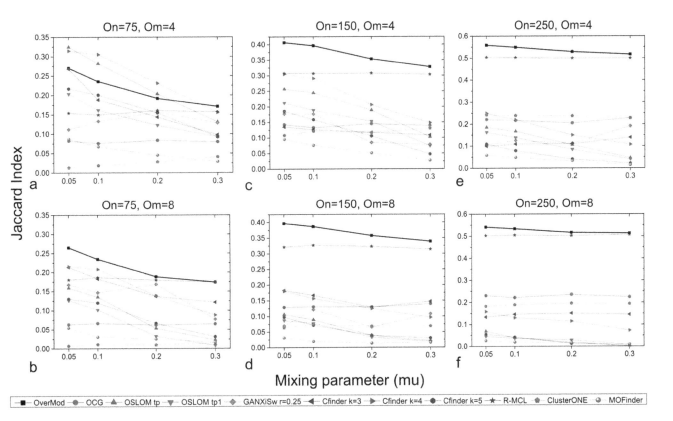

Figure 6. Method comparison on synthetic networks. The average Jaccard index between the sets of 'known' multi-clustered nodes and those predicted to be multi-clustered by OverMod and other overlapping community structure detection methods from the literature.

Table 5. Method comparison on the PPI networks.

	Rat				E. coli				Yeast				Human			
	M	C	MC	Max	M	C	MC	Max	M	C	MC	Max	M	C	MC	Max
CFinder k=3	20	74	4	4	17	97	5	2	N/A	N/A	N/A	N/A	364	2847	477	6
CFinder k=4	3	12	0	N/A	6	37	3	2	N/A	N/A	N/A	N/A	126	808	173	9
CFinder k=5	N/A	N/A	N/A	N/A	8	16	0	N/A	N/A	N/A	N/A	N/A	31	267	28	5
OSLOM tp	8	487	1	2	9	668	1	2	57	1002	65	3	157	6160	657	6
OSLOM tp1	N/A	N/A	N/A	N/A	N/A	N/A	N/A	N/A	16	1002	216	3	48	6160	1048	5
OCG	328	487	86	172	401	668	50	125	252	1002	585	41	393	6160	2104	53
ClusterONE	27	91	11	3	38	140	30	3	115	910	172	4	572	2202	338	4
GANXiSw r=0.25	36	487	22	2	42	668	16	2	58	1002	25	2	302	6160	1120	3
MOFinder	12	37	12	4	13	52	13	4	143	695	316	8	95	338	86	6
RSRGM	43	425	55	3	64	572	60	3	78	837	112	4	651	4688	1028	4
R-MCL	205	476	267	22	360	660	503	19	222	957	588	12	5297	6160	6112	28

Summary of results found by applying the methods from the literature to the four PPI networks in terms of number of modules (M), number of nodes that are assigned at least one community (C), number of nodes that are multi-clustered (MC) and the maximum number of modules the multi-clustered nodes belong to (Max). The results highlight the great variation between each method. Please note that CFinder did not complete within 24 hours for the yeast PPI network, there was no $k=5$ result for CFinder for the rat PPI network and for the rat and E. coli networks, OSLOM did not produce a partition for level tp.

clustered nodes belonging to either 4 or 8 (OM) modules. Therefore there are 6 sets of synthetic networks: (i) $ON = 75$, $OM = 4$, (ii) $ON = 75$, $OM = 8$, (iii) $ON = 150$, $OM = 4$, (iv) $ON = 150$, $OM = 8$, (v) $ON = 250$, $OM = 4$ and (vi) $ON = 250$, $OM = 8$. For (i)–(vi), 10 networks were generated for each of the following mixing parameters (the fraction of all links that lie between modules): 0.05, 0.1, 0.2 and 0.3. OverMod, CFinder, OSLOM, OCG, ClusterONE, GANXiSw, MOFinder and R-MCL were applied to the synthetic networks. Note that RSRGM was not evaluated on the synthetic networks as an upper bound for the number of modules is required to be selected in advance. Louvain was used to find the hard partition for OverMod, levels tp and $tp1$ of the OSLOM results were considered, for GANXiSw, $r = 0.25$ and for CFinder, k was set to 3, 4 and 5. Otherwise, all default parameters were selected.

The aim of OverMod is to identify multi-clustered nodes that adopt the important role of bridging multiple modules. OverMod and the alternative methods are therefore evaluated based on their ability to identify those multi-clustered in synthetic networks. The Jaccard index is employed here to quantify the similarity between the set of 'known' multi-clustered nodes (as given by the synthetic network generating software) and those predicted by the clustering methods being evaluated. Figure 6 shows the average Jaccard index against the mixing parameter for (i)–(iv) above. For (i), OSLOM tp and CFinder $k = 4$ identify the multi-clustered nodes more accurately than OverMod, however for (ii) to (vi) OverMod performs best. Note that there are no results for GANXiSw for (vi) as for many networks the method only found one module. Overall, OverMod identifies the known multi-clustered nodes well for a range of synthetic networks with varying connectivity properties.

Synthetic networks offer a means of benchmarking community structure detection methods, however, determining parameters such as average degree, ON and OM, do not represent real life complex networks. As such, each method, including RSRGM, is now applied to the PPI networks analysed previously. For RSRGM the upper bound on the number of modules was taken to be the average of the results for all other methods for each network. Table 5 summarises the results in terms of number of modules (M), number of nodes that are assigned at least one community (C), number of nodes that are multi-clustered (MC) and the maximum number of modules the multi-clustered nodes belong to (Max). Clearly, the results vary to a large degree for all of these factors. In particular, the range of number of modules in each soft partition for each network is vast, e.g. for the rat network partitions have from between 3 and 328 modules. Note also that for the rat PPI network, OSLOM finds only 1 multi-clustered protein, CFinder ($k = 3$) finds 4 multi-clustered proteins but only 74 our of 487 nodes are assigned at least one module and CFinder ($k = 4$) clusters only 12 nodes with no multi-clustered nodes. Similarly for the *E. coli* network, OSLOM finds only 1 multi-clustered node. Therefore, for some applications, these methods may not offer satisfactory or relevant solutions.

The Jaccard index is employed to carry out a pair-wise comparison between the sets of multi-clustered proteins found by all methods for each of the four PPI networks (results not shown). For the rat PPI network, the two methods with the most similar set of multi-clustered nodes are OCG and R-MCL. Furthermore, these are the two methods with results that are most similar to OverMod. The same is true for the *E. coli* network. For the Yeast network, MOFinder, closely followed by OCG and R-MCL, generate sets of multi-clustered nodes most similar to those found by OverMod. For the human network, R-MCL finds the most similar multi-clustered nodes to OverMod. Furthermore, OCG and R-MCL multi-cluster all of the robust inter-connectors

(Table 3) for the Rat, *E. coli* and Human networks, while the same methods, and additionally MOFinder, multi-cluster all of the robust inter-connectors for the Yeast network.

Evaluating method performance on the PPI networks is difficult. Unlike the synthetic networks, a benchmark of known multi-clustered nodes is not available. Each method's approach varies to such a large degree and for many of the above methods, choice of parameter values can greatly affect the final results, with often no way of determining the 'correct' values. Furthermore, as our results show, and has also previously been reported [12], CFinder generally leaves a large portion of the network un-clustered. Similarly, ClusterONE, MOFinder, RSRGM and R-MCL do not always assign each node to at least one module. Each of these factors makes a fair comparison very difficult. Ultimately, one must choose a method that is suited to their application and user requirements. Based on the assumption that modularity optimisation is meaningful in biological networks, we have chosen our proposed two-stage approach and in light of the results presented in the previous sections, we believe OverMod to be a reasonable and successful approach to finding overlapping communities and in particular identifying 'important' nodes in PPI networks.

Conclusions

In this work, a two-stage procedure for identifying overlapping community structure is outlined. Stage one involves detecting disjoint communities of a network using existing methodologies. In stage two, we propose, OverMod, a novel mixed integer non-linear programming (MINLP) model to convert disjoint to overlapping network communities. We extend the use of modularity optimisation and mathematical programming in community detection and present a thorough investigation into the relevance of this methodology in PPI networks. Connector proteins exhibited a range of topological and functional properties indicative of their role in these networks, thus demonstrating their suitability as candidate nodes for multiple module membership. OverMod was then shown to identify two types of connector nodes: inter and intra-connector, each with their own distinguishing topological features. In general, intra-connectors have a higher node degree than inter-connectors, their removal breaks down the network faster, while inter-connectors exhibit a larger dispersion of their connections across modules, thereby acquiring a higher average participation coefficient. Further investigation suggested characteristics that differentiated these nodes in terms of functionality.

Through the above discussion of our results and comparison to other methods, the comparative advantages of the two-stage procedure become apparent. In particular, OverMod can be applied to any hard partitioning method that is deemed suitable to the problem. Owing to the nature of the mathematical programming framework used, modelling can be flexible enough to allow additional constraints and parameters to be easily implemented, as relevant according to user requirements. Prior knowledge on a particular system can be incorporated, for example in the form of nodes with similar functional annotations that may be constrained to be allocated in the same community. Modelling and solution procedure enhancements can also be investigated in order to improve the efficiency of OverMod (e.g. symmetry breaking constraints [31], column generation techniques [32] solution post processing [81]).

Overall, the development of overlapping community detection procedures has the potential to uncover the principles of communication across distinct functional modules through the investigation of nodes which connect cellular processes, providing a greater understanding of system properties. OverMod identifies two types of connector proteins that may play different but central roles in linking distinct processes, distinguishing them from single hubs that may connect a large number of possibly functionally homogeneous interacting proteins.

We have demonstrated the potential of inter-connectors through their participation in different modules, providing a reasonable interpretation of proteins with multiple interactors in a PPI network, since this is unrealistic in biological cells. Many of these proteins are major regulators of proliferation and apoptosis, including oncogenes and regulators of tumour suppressors. The application of our method in disease networks may therefore be relevant in prioritising OMICs results, especially when looking for disease biomarkers, suggesting potential drug targets and regulators of disease related pathways. These results demonstrate the potential of the proposed method in future functional genomics applications and especially in discovering important proteins in less well-characterised systems.

Supporting Information

File S1   Supporting tables. Table S1, The significance values for the comparison of isolated nodes with connector nodes (topological features). Table S2, The significance values for the

comparison of isolated nodes with connector nodes (functional features). Table S3, Essentiality results summary, where p-values less than 0.01 indicate that connector nodes in the corresponding organism are significantly enriched for essential genes. Table S4, The Jaccard index for each pair-wise comparison of sets of inter-connector nodes. Table S5, Summary of the significance values for the comparison between inter and intra-connectors based on topological measures. Table S6, Average participation coefficient for inter and intra-connectors with corresponding significance values.

(PDF)

Acknowledgments

We would like to thank Jierui Xie, Yu-Keng Shih and Xiao-Fei Zhang for providing the software and for GANXiSw, R-MCL and RSRGM, respectively. We would also like to thank the two anonymous reviewers for their helpful suggestions.

Author Contributions

Conceived and designed the experiments: LB SL ST LGP. Performed the experiments: LB AK. Analyzed the data: LB AK. Wrote the paper: LB AK LGP ST. Designed the computational model: SL LGP LB.

References

1. Lewis A, Jones N, Porter M, Deane C (2010) The function of communities in protein interaction networks at multiple scales. BMC Systems Biology 4: 100.
2. Voevodski K, Teng SH, Xia Y (2009) Finding local communities in protein networks. BMC Bioinformatics 10: 297.
3. Kühner S, van Noort V, Betts MJ, Leo-Macias A, Batisse C, et al. (2009) Proteome Organization in a Genome-Reduced Bacterium. Science 326: 1235–1240.
4. Palla G, Dernyi I, Farkas I, Vicsek T (2005) Uncovering the overlapping community structure of complex networks in nature and society. Nature 435: 814–818.
5. Lancichinetti A, Radicchi F, Ramasco JJ, Fortunato S (2011) Finding Statistically Significant Communities in Networks. PLoS ONE 6.
6. Becker E, Robisson B, Chapple CE, Guénoche A, Brun C (2012) Multifunctional proteins revealed by overlapping clustering in protein interaction network. Bioinformatics 28: 84–90.
7. Nepusz T, Yu H, Paccanaro A (2012) Detecting overlapping protein complexes in protein-protein interaction networks. Nature Methods 9: 471–472.
8. Xie J, Szymanski BK (2012) Towards linear time overlapping community detection in social networks. CoRR abs/1202.2465.
9. Yu Q, Li GHH, Huang JFF (2012) MOfinder: a novel algorithm for detecting overlapping modules from protein-protein interaction network. Journal of biomedicine & biotechnology 2012: 103702.
10. Dhillon IS, Guan Y, Kulis B (2007) Weighted graph cuts without eigenvectors a multilevel approach. IEEE Trans Pattern Anal Mach Intell 29: 1944–1957.
11. Zhang RC, Lin Y, Yue M, Li Q, Zhang XF, et al. (2012) Exploring overlapping functional units with various structure in protein interaction networks. PLoS One 7.
12. Wang X, Li L, Cheng Y (2012) An overlapping module identification method in protein-protein interaction networks. BMC Bioinformatics 13: S4.
13. Asur S, Ucar D, Parthasarathy S (2007) An ensemble framework for clustering proteinprotein interaction networks. Bioinformatics 23: i29–i40.
14. Ucar D, Asur S, Catalyurek U, Parthasarathy S (2006) Improving functional modularity in protein-protein interactions graphs using hub-induced subgraphs. In: Proceedings of the 10th European Conference on Principle and Practice of Knowledge Discovery in Databases. Berlin, Heidelberg: Springer-Verlag, PKDD'06, pp. 371–382.
15. Shih YK, Parthasarathy S (2012) Identifying functional modules in interaction networks through overlapping markov clustering. Bioinformatics 28: i473–i479.
16. Liu G, Wong L, Chua HN (2009) Complex discovery from weighted PPI networks. Bioinformatics 25: 1891–1897.
17. Russ AP, Lampel S (2005) The druggable genome: an update. Drug Discovery Today 10: 1607–1610.
18. Shen HW, Cheng XQ, Guo JF (2009) Quantifying and identifying the overlapping community structure in networks. Journal of Statistical Mechanics: Theory and Experiment 2009: P07042.
19. Wu K, Taki Y, Sato K, Sassa Y, Inoue K, et al. (2011) The overlapping community structure of structural brain network in young healthy individuals. PLoS ONE 6: e19608.

20. Salath M, Jones JH (2010) Dynamics and control of diseases in networks with community structure. PLoS Comput Biol 6: e1000736.
21. Newman MEJ, Girvan M (2004) Finding and evaluating community structure in networks. Physical Review E 69: 026113.
22. Chen J, Yuan B (2006) Detecting functional modules in the yeast proteinprotein interaction network. Bioinformatics 22: 2283–2290.
23. Zinman G, Zhong S, Bar-Joseph Z (2011) Biological interaction networks are conserved at the module level. BMC Systems Biology 5: 134.
24. Lee J, Gross SP (2013) Improved network community structure improves function prediction.
25. Newman MEJ (2004) Fast algorithm for detecting community structure in networks. Physical Review E 69: 066133.
26. Guimera R, Amaral L (2005) Functional cartography of complex metabolic networks. Nature 433: 895–900.
27. Blondel VD, Guillaume JL, Lambiotte R, Lefebvre E (2008) Fast unfolding of communities in large networks. Journal of Statistical Mechanics: Theory and Experiment 2008: P10008.
28. Ruan J, Zhang W (2008) Identifying network communities with a high resolution. Physical Review E 77: 1–12.
29. Medus A, Acuna G, Dorso C (2005) Detection of community structures in networks via global optimization. Physica A: Statistical Mechanics and its Applications 358: 593–604.
30. Wakita K, Tsurumi T (2007) Finding community structure in mega-scale social networks: [extended abstract]. In: Proceedings of the 16th international conference on World Wide Web. New York, NY, USA: ACM, WWW '07, pp. 1275–1276.
31. Xu G, Tsoka S, Papageorgiou LG (2007) Finding community structures in complex networks using mixed integer optimisation. The European Physical Journal B 60: 231–239.
32. Aloise D, Cafieri S, Caporossi G, Hansen P, Perron S, et al. (2010) Column generation algorithms for exact modularity maximization in networks. Phys Rev E 82: 046112.
33. Agarwal G, Kempe D (2008) Modularity-maximizing graph communities via mathematical programming. Eur Phys J B 66: 409–418.
34. Xu G, Bennett L, Papageorgiou LG, Tsoka S (2010) Module detection in complex networks using integer optimisation. Algorithms for Molecular Biology 5: 36.
35. Cafieri S, Hansen P, Liberti L (2011) Locally optimal heuristic for modularity maximization of networks. Phys Rev E 83: 056105.
36. Bennett L, Liu S, Papageorgiou LG, Tsoka S (2012) Detection of disjoint and overlapping modules in weighted complex networks. Advances in Complex Systems 15: 11500.
37. Aloise D, Caporossi G, Hansen P, Liberti L, Perron S, et al. (2013). Modularity maximization in networks by variable neighborhood search. Bader, David A. (ed.) et al., Graph partitioning and graph clustering. Proceedings of the 10th DIMACS implementation challenge workshop, Atlanta, GA, USA, February 13–14, 2012. Providence, RI: American Mathematical Society (AMS). Contemporary Mathematics 588, 113–127 (2013). doi:10.1090/conm/588.

38. Rosenthal R (2008) GAMS - A user's guide. Washington D.C., USA: GAMS Development Corporation.

39. Stark C, Breitkreutz BJ, Reguly T, Boucher L, Breitkreutz A, et al. (2006) Biogrid: a general repository for interaction datasets. Nucleic Acids Research 34: D535–D539.

40. Hermjakob H, Montecchi-Palazzi L, Lewington C, Mudali S, Kerrien S, et al. (2004) IntAct: an open source molecular interaction database. Nucleic acids research 32.

41. Collins SR, Kemmeren P, Zhao XC, Greenblatt JF, Spencer F, et al. (March 2007) Toward a comprehensive atlas of the physical interactome of saccharomyces cerevisiae. Molecular and Cellular Proteomics 6: 439–450.

42. Csardi G, Nepusz T (2006) The igraph software package for complex network research. InterJournal Complex Systems: 1695.

43. R Core Team (2013) R: A Language and Environment for Statistical Computing. R Foundation for Statistical Computing, Vienna, Austria. Available: http://www.R-project.org.

44. Dwinell MR, Worthey EA, Shimoyama M, Bakir-Gungor B, DePons J, et al. (2009) The Rat Genome Database 2009: variation, ontologies and pathways. Nucleic Acids Research 37: D744–D749.

45. Ashburner M, Ball CA, Blake JA, Botstein D, Butler H, et al. (2000) Gene ontology: tool for the unification of biology. The Gene Ontology Consortium. Nature genetics 25: 25–29.

46. Cherry JM, Hong EL, Amundsen C, Balakrishnan R, Binkley G, et al. (2012) Saccharomyces Genome Database: the genomics resource of budding yeast. Nucleic acids research 40.

47. Dimmer EC, Huntley RP, Alam-Faruque Y, Sawford T, O'Donovan C, et al. (2012) The UniProt-GO Annotation database in 2011. Nucleic acids research 40: D565–D570.

48. Punta M, Coggill PC, Eberhardt RY, Mistry J, Tate J, et al. (2012) The pfam protein families database. Nucleic Acids Research 40: D290–D301.

49. Chen WHH, Minguez P, Lercher MJ, Bork P (2012) OGEE: an online gene essentiality database. Nucleic acids research 40: D901–D906.

50. Newman MEJ (2011) Communities, modules and large-scale structure in networks. Nature Physics 8: 25–31.

51. Cohen R, Erez K, ben Avraham D, Havlin S (2001) Breakdown of the internet under intentional attack. Physical Review Letters 86: 3682–3685.

52. Dartnell L, Simeonidis E, Hubank M, Tsoka S, Bogle IDL, et al. (2005) Robustness of the p53 network and biological hackers. FEBS Letters 579: 3037–3042.

53. Girvan M, Newman MEJ (2002) Community structure in social and biological networks. Proceedings of the National Academy of Sciences 99: 7821–7826.

54. Taylor IW, Linding R, Warde-Farley D, Liu Y, Pesquita C, et al. (2009) Dynamic modularity in protein interaction networks predicts breast cancer outcome. Nature Biotechnology 27: 199–204.

55. Hennig C (2007) Cluster-wise assessment of cluster stability. Computational Statistics and Data Analysis 52: 258–271.

56. Jeong H, Mason SP, Barabsi AL, Oltvai ZN (2001) Lethality and centrality in protein networks. Nature 411: 41–42.

57. Hahn MW, Kern AD (2005) Comparative genomics of centrality and essentiality in three eukaryotic protein-interaction networks. Molecular Biology and Evolution 22: 803–806.

58. He X, Zhang J (2006) Why do hubs tend to be essential in protein networks? PLoS Genet 2: e88.

59. Gething MJ, Sambrook J (1992) Protein folding in the cell. Nature 355: 33–45.

60. Komander D (2009) The emerging complexity of protein ubiquitination. Biochemical Society transactions 37: 937–953.

61. Ikeda F, Dikic I (2008) Atypical ubiquitin chains: new molecular signals. 'Protein modifications: Beyond the usual suspects' review series. EMBO Reports 9: 536–542.

62. Pickart CM, Fushman D (2004) Polyubiquitin chains: polymeric protein signals. Current Opinion in Chemical Biology 8: 610–616.

63. Wool IG (1979) The structure and function of eukaryotic ribosomes. Annual Review of Biochemistry 48: 719–754.

64. Yoshihama M, Nakao A, Nguyen HD, Kenmochi N (2006) Analysis of ribosomal protein gene structures: Implications for intron evolution. PLoS Genet 2: e25.

65. Saveanu C, Fromont-Racine M, Harington A, Ricard F, Namane A, et al. (2001) Identification of 12 new yeast mitochondrial ribosomal proteins including 6 that have no prokaryotic homologues. The Journal of biological chemistry 276: 15861–15867.

66. Graack HR, Wittmann-Liebold B (1998) Mitochondrial ribosomal proteins (MRPs) of yeast. Biochemical Journal 329: 433–448.

67. Wieser R, Wrana JL, Massagu J (1995) GS domain mutations that constitutively activate t beta r-i, the downstream signaling component in the TGF-beta receptor complex. The EMBO journal 14: 2199–2208.

68. Chen YG, Liu F, Massague J (1997) Mechanism of TGFbeta receptor inhibition by FKBP12. The EMBO journal 16: 3866–3876.

69. Massagu J (2000) How cells read TGF- signals. Nature Reviews Molecular Cell Biology 1: 169–178.

70. Shi Y, Massagu J (2003) Mechanisms of TGF-beta signaling from cell membrane to the nucleus. Cell 113: 685–700.

71. Annes JP, Munger JS, Rifkin DB (2003) Making sense of latent TGF activation. Journal of Cell Science 116: 217–224.

72. Dijke Pt, Hill CS (2004) New insights into TGF-Smad signalling. Trends in Biochemical Sciences 29: 265–273.

73. Sayed M, Pelech S, Wong C, Marotta A, Salh B (2001) Protein kinase CK2 is involved in g2 arrest and apoptosis following spindle damage in epithelial cells. Oncogene 20: 6994–7005.

74. Enserink JM, Kolodner RD (2010) An overview of cdk1-controlled targets and processes. Cell Division 5: 11.

75. Chen J, Fujii K, Zhang L, Roberts T, Fu H (2001) Raf-1 promotes cell survival by antagonizing apoptosis signal-regulating kinase 1 through a MEKERK independent mechanism. Proceedings of the National Academy of Sciences 98: 7783–7788.

76. O'Neill E, Rushworth L, Baccarini M, Kolch W (2004) Role of the kinase MST2 in suppression of apoptosis by the proto-oncogene product raf-1. Science 306: 2267–2270.

77. Pei D, Zhang Y, Zheng J (2012) Regulation of p53: a collaboration between mdm2 and mdmx. Oncotarget 3: 228–235.

78. Chen B, Fan W, Liu J, Wu FX (2013) Identifying protein complexes and functional modulesfrom static PPI networks to dynamic PPI networks. Briefings in Bioinformatics.

79. Tsai CJ, Ma B, Nussinov R (2009) Proteinprotein interaction networks: how can a hub protein bind so many different partners? Trends in Biochemical Sciences 34: 594–600.

80. Lancichinetti A, Fortunato S, Radicchi F (2008) Benchmark graphs for testing community detection algorithms. Phys Rev E 78: 046110.

81. Cafieri S, Hansen P, Liberti L (2014) Improving heuristics for network modularity maximization using an exact algorithm. Discrete Applied Mathematics 163, Part 1: 65–72.

82. Shannon P, Markiel A, Ozier O, Baliga NS, Wang JT, et al. (2003) Cytoscape: A software environment for integrated models of biomolecular interaction networks. Genome Research 13: 2498–2504.

Identification of Kernel Proteins Associated with the Resistance to *Fusarium* Head Blight in Winter Wheat (*Triticum aestivum* L.)

Dawid Perlikowski[1,9], **Halina Wiśniewska**[1,9], **Tomasz Góral**[2], **Michał Kwiatek**[1], **Maciej Majka**[1], **Arkadiusz Kosmala**[1]*

1 Institute of Plant Genetics of the Polish Academy of Sciences, Poznan, Poland, 2 Plant Breeding and Acclimatization Institute – National Research Institute, Radzikow, Blonie, Poland

Abstract

Numerous potential components involved in the resistance to *Fusarium* head blight (FHB) in cereals have been indicated, however, our knowledge regarding this process is still limited and further work is required. Two winter wheat (*Triticum aestivum* L.) lines differing in their levels of resistance to FHB were analyzed to identify the most crucial proteins associated with resistance in this species. The presented work involved analysis of protein abundance in the kernel bulks of more resistant and more susceptible wheat lines using two-dimensional gel electrophoresis and mass spectrometry identification of proteins, which were differentially accumulated between the analyzed lines, after inoculation with *F. culmorum* under field conditions. All the obtained two-dimensional patterns were demonstrated to be well-resolved protein maps of kernel proteomes. Although, 11 proteins were shown to have significantly different abundance between these two groups of plants, only two are likely to be crucial and have a potential role in resistance to FHB. Monomeric alpha-amylase and dimeric alpha-amylase inhibitors, both highly accumulated in the more resistant line, after inoculation and in the control conditions. *Fusarium* pathogens can use hydrolytic enzymes, including amylases to colonize kernels and acquire nitrogen and carbon from the endosperm and we suggest that the inhibition of pathogen amylase activity could be one of the most crucial mechanisms to prevent infection progress in the analyzed wheat line with a higher resistance. Alpha-amylase activity assays confirmed this suggestion as it revealed the highest level of enzyme activity, after *F. culmorum* infection, in the line more susceptible to FHB.

Editor: Joshua L. Heazlewood, The University of Melbourne, Australia

Funding: The study was supported by the Polish Ministry of Agriculture and Rural Development (HOR hn-078/PB/18/14 no. 6). The funders had no role in study design, data collection and analysis, decision to publish, or preparation of the manuscript.

Competing Interests: The authors have declared that no competing interests exist.

* Email: akos@igr.poznan.pl

9 These authors contributed equally to this work.

Introduction

Fusarium species are widespread necrotrophic pathogens of small grain cereals, e.g. oat (*Avena sativa* L.), wheat (*Triticum aestivum* L.) and triticale (×*Triticosecale* Wittm.). Three of these species – *F. avenaceum* (Corda ex Fries) Sacc., *F. culmorum* (W.G. Smith) Sacc. and *F. graminearum* (Schwabe.) are considered to be the most important in central European countries [1]. Severity of *Fusarium* head blight (FHB) depends on several agronomic, climatic and genetic factors [2]–[4]. This disease can result in *Fusarium*-damaged kernels (FDK), which are smaller, compared to healthy-looking kernels (HLK), discoloured and shriveled [5]–[7]. Accumulation of *Fusarium* toxins such as deoxynivalenol (DON), nivalenol (NIV), zearalenone and many others in infected chaff, kernels and rachises is also often observed [8]–[10]. Contamination of the harvested grain with toxic fungal secondary metabolites (mycotoxins) may cause mycotoxicoses in humans and domestic animals [11], [12]. Observations of FHB occurrence revealed a high susceptibility of cultivars and breeding lines of spring wheat and oat to most *Fusarium* pathogens [13], [14]. Most of the published papers on triticale situate this species in terms of resistance between wheat and rye (*Secale cereale* L.). However, there are results available showing that susceptibility of triticale to FHB may be higher and even equal to wheat [15]–[17]. Under conditions of artificial inoculation with *F. culmorum* most winter wheat cultivars proved to be susceptible or highly susceptible to FHB, when compared to the known resistant winter wheat, e.g. 'Arina' or 'SVP' lines [18], [19]. Moreover, high yielding winter wheat cultivars that are best adapted to environmental conditions are often susceptible to FHB. The development of cultivars resistant to FHB plays a key role in disease control and the prevention of kernel contamination with mycotoxins [20], [21].

The resistance of wheat to FHB has a relatively complex nature. Five types of physiological resistance have been described [5]: type I or resistance to the initial infection, type II or resistance to spread within the spike, type III or resistance to kernel infection, type IV or tolerance to infection and type V or resistance to DON accumulation. However, the detailed defense mechanisms against FHB infection remain poorly characterized. An interaction between the pathogen and the host causes a defense response involving: hypersensitive reactions, deposition of cell wall reinforcing materials and synthesis of a wide range of antimicrobial compounds, such as pathogenesis-related (PR) proteins [22]. Gene expression studies revealed that the transcripts of defense response genes, coding peroxidase and PR-1-5, accumulated as early as six to 12 hours after inoculation of wheat spikes with *F. graminearum* [23]. Gottwald et al. [24] assumed that jasmonate and ethylene dependent defense and suppression of fungal virulence factors could provide major mechanisms of FHB resistance in wheat. In addition, proteomic studies have been carried out in *F. graminearum* infected wheat, barley (*Hordeum vulgare* L.) and their wild relatives [25]–[28]. Zhou et al. [29], [30] performed research on the interaction between *F. graminearum* and wheat to identify FHB infection response proteins by comparing protein profiles of *F. graminearum*-inoculated with mock-inoculated wheat spikelets of 'Ning7840', *Fhb1* resistance gene carrier. Gel-based proteomic analysis of the resistant cultivar revealed accumulation of plant proteins involved in oxidative stress, PR responses, and nitrogen metabolisms. The results showed up-regulation of proteins in the antioxidant and jasmonic acid-signaling pathway, PR responses and amino acid synthesis after three days of inoculation [29], [30].

Although, numerous potential components involved in the resistance to FHB have been indicated, our knowledge regarding this process in cereals is still limited and further work required. Here, we present comprehensive research on winter wheat, performed to recognize the crucial proteins associated with the resistance. Thus, the current work involved two main proteomic steps: (1) the analysis of protein abundance in the FDK of more resistant and more susceptible wheat lines using two-dimensional gel electrophoresis (2-DE) and (2) mass spectrometry (MS) identification of proteins which were differentially accumulated between the analyzed lines. It is hypothesized here that between the FDK derived from the lines with distinct levels of resistance, differentially accumulated proteins will be identified. Moreover, it is also suggested that the proteins highly accumulated in the more resistant line after *Fusarium* infection, crucial for the resistance, will have also higher abundance in this line in the control conditions, without infection. The procedure of proteome profiling was further followed by alpha-amylase activity assays performed on the FDK and on the kernels in the control conditions, both in the more resistant and more susceptible wheat lines to decipher involvement of this enzyme activity and its inhibition in the resistance of winter wheat to FHB.

Materials and Methods

The study presented here does not require an ethics statement. The field plots described below, in Cerekwica and Radzikow, belong to the Institute of Plant Genetics of the Polish Academy of Sciences and to the Plant Breeding and Acclimatization Institute – National Research Institute, respectively. No special permissions were required to perform trials on these field plots and the trials did not involve endangered or protected species.

Field experiments

The plant materials for the proteomic research were two lines of winter wheat (*Triticum aestivum* L.) – STH 0290, a line with a higher level of resistance to FHB (abbreviated here as 'the resistant line', RL), developed by Plant Breeding Strzelce Ltd., Co. (Poland)

and AND 775/09, a line with a lower level of resistance (abbreviated here as 'the susceptible line', SL), developed by Danko Plant Breeding Ltd., Co. (Poland). The level of resistance of these two lines was estimated under the field conditions in 2013, in two locations: Cerekwica (western Poland; GPS coordinates: N 52.521012, E 16.692005) characterized by poor, sandy-clay soil and Radzikow (central Poland; GPS coordinates: N 52.211754, E 20.631954) with rich sandy-clay soil. The rainfalls and mean temperature during the experiments performed in Cerekwica and Radzikow, are presented in Table S1. The experiments in both locations were carried out according to the same design. The experimental field in each location consisted of four plots for each tested line. The seeds were sown in plots of 1 m^2 size with the sowing rate 300 seeds (September 2012). The fungal material for inoculation was a mixture of three isolates of *F. culmorum* (W.G.Sacc.): KF 846 (DON chemotype) and KF 350 (NIV chemotype) derived from the collection of Institute of Plant Genetics, Polish Academy of Sciences (Poznan, Poland), and ZFR 112, producing zearalenone, derived from the collection of Plant Breeding and Acclimatization Institute – National Research Institute (Radzikow, Poland). In the case of each analyzed line, one plot, used as a control, was not treated with *Fusarium* isolates, and three others were inoculated by spraying flowering heads (stage of development: full flowering, 50% of anthers mature; stage 65 in commonly used BBCH scale) with the spore suspension at a rate of approximately 100 ml m^{-2} (June 2013). Conidia concentration was adjusted to 5×10^4 ml^{-1}. During two days after inoculation micro-irrigation was applied to maintain moisture levels. Two weeks after inoculation, progress of the disease was visually evaluated. The presence of FHB (percentage of heads infected per plot) and percentage of head infection were determined. On the basis of the obtained results, FHB index (FHBi), associated with the resistance type I and II, as described by Mesterházy [5], was calculated, separately for each line (the RL and SL) and location (Cerekwica and Radzikow), according to the formula: FHBi (%) = (% of head infection × % of heads infected per plot)/100.

At the harvest (August 2013) 20 randomly selected heads from each plot (one control plot and three inoculated plots, in each location) were collected for the RL and the SL and threshed manually. Further, within a group of each 20 heads derived from three inoculated plots for each line mature kernels were visually scored and divided into two categories: HLK and FDK. In each category kernel weight [g] and number were recorded. Percentage of FDK (% FDK) was estimated as a percent of infected kernels per head, taking into account both kernel weight and number. Mean values and standard deviations of this parameter calculated on the base of data obtained from three inoculated plots were described in the paper, separately for each experimental location and each analyzed line. The percentage of FDK parameter is associated with the level of resistance type III, as described by Mesterházy [5]. Additionally, total kernel number and weight per head on the base of 20 randomly selected head from one control plot, separately for each analyzed line and location, were also calculated.

Proteome profiling and identification of differentially accumulated proteins

The plant materials derived from one location (Cerekwica) were used for further molecular research – proteome profiling and alpha amylase activity assays. The FDK derived from 20 heads were pooled, separately for each inoculated plot, giving three separate pooled samples (bulks) for each analyzed line, the RL and the SL. The kernels derived from 20 heads of the control plot were

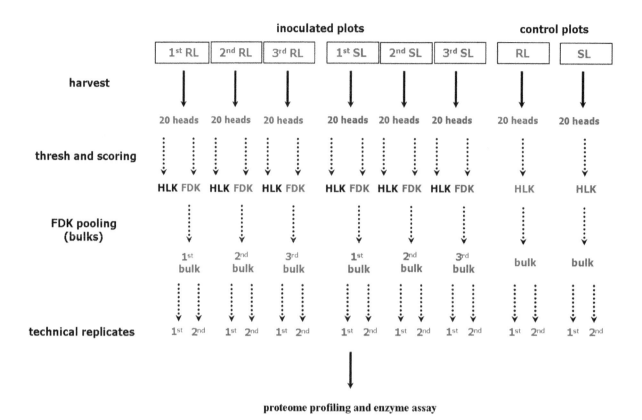

Figure 1. Diagram demonstrating a workflow of sample preparation for proteome analysis. Abbreviations: FDK, *Fusarium*-damaged kernels; HLK, healthy-looking kernels; RL, line of winter wheat more resistant to *Fusarium* head blight; SL, line of winter wheat more susceptible to *Fusarium* head blight.

also pooled for each analyzed line. The pooled samples (bulks) were used for proteomic research – each one in two technical replicates. Thus, the final proteomic survey resulted in 16 2-D gels. A diagram outlining the workflow of sample preparation for proteome analysis is shown in the Fig. 1.

The proteomic protocol used, including 2-DE and MS to identify differentially accumulated kernel proteins between the RL and SL of wheat, was the same as that described in detail by Masojć and Kosmala [31] and Masojć et al. [32]. Protein extraction was performed according to the method of Hurkman and Tanaka [33] using 0.7 M sucrose, 0.5 M TRIS, 30 mM HCl, 50 mM EDTA, 2% DTT, and 0.1 M KCl as components of the extraction buffer. In the first dimension, isoelectrofocusing (IEF), strip gels with linear pH range 4–7 (24 cm) were used to focus the aliquots of proteins extracted from 25 mg of wheat flour. In the second dimension (sodium dodecyl sulphate-polyacrylamide gel electrophoresis) the proteins were separated using 13% polyacrylamide gels (1.5×255×196 mm). Following electrophoresis the gels were stained with colloidal coomassie brilliant blue G-250, using the modified method of Neuhoff et al. [34]. The protein gels were scanned by Image scanner III (GE Healthcare, Buckinghamshire, UK) and subjected to Lab scan 6.0 program (GE Healthcare, Buckinghamshire, UK) processing. Spot detection and image analyses were performed with Image Master 2-D *Platinum* software (GE Healthcare, Buckinghamshire, UK). The abundance of each protein spot was normalized as a relative volume (% Vol) and it was calculated by Image Master software as a ratio of the volume of particular spot to the total volume of all the spots present on the gel. To consider a spot as 'present', it had to be detected in all the gel replicates used in the analysis. The

significance of the differences was assessed using the Student's t-test. The protein spots which showed at least two fold differences (p≤0.05) in protein abundance between the FDK of two analyzed lines together with protein spots present only in the FDK of one of the analyzed lines (the RL or SL), were subjected to MS analyses. However, to be sure that a given protein spot originated from wheat but not from *Fusarium*, one more condition had to be fulfilled by the spot. It had to be also present on the 2-D protein maps obtained for the control conditions, before inoculation.

Gel spots of interest were harvested, and after sequential washing with ammonium bicarbonate and acetonitrile were reduced with 10 mM dithiothreitol and alkylated with 55 mM iodoacetamide. The in-gel protein digestion was performed using trypsin solution (Sequencing Grade Modified Trypsin; Promega, Fitchburg, WI, USA) in 25 mM ammonium bicarbonate (25 ng μl^{-1}). Samples were concentrated and desalted on a RP-C18 precolumn (Waters), and further peptide separation was achieved on a nano-Ultra Performance Liquid Chromatography (UPLC) RP-C18 column (Waters, BEH130 C18 column, 75 µl i.d., 250 mm long) of a nanoACQUITY UPLC system, using a 45 min linear acetonitrile gradient. The proteins were analyzed by liquid chromatography coupled to the Orbitrap Velos type mass spectrometer (Thermo), working in the regime of data dependent MS to MS/MS switch, in the Laboratory of Mass Spectrometry, Institute of Biochemistry and Biophysics, Polish Academy of Sciences (Warsaw, Poland) as described by Kosmala et al. [35] and Perlikowski et al. [36]. The data were analyzed with Mascot Distiller software (version 2.3, MatrixScience) with standard settings for the Orbitrap low resolution measurements (available at http://www.matrixscience.com/distiller.html) to extract MS/

Table 1. The components of the resistance to Fusarium head blight in the more resistant (RL) and more susceptible (SL) winter wheat (*Triticum aestivum*) lines and their yields under control conditions.

Location	Winter wheat line	Conditions after inoculation					Control conditions	
		FHBi	% FDK (weight [g])	% FDK (number)	Total kernel number/head	Total kernel weight [g]/head	Total kernel number/head	Total kernel weight [g]/head
Cerekwica	RL	23.7±0.58	45.5±5.14	54.9±5.70	27.9±4.23	0.7±0.25	39.9	2.0
	SL	38.3±2.08	83.5±4.89	91.1±4.90	16.6±1.89	0.3±0.06	38.4	1.9
Radzikow	RL	29.5±2.65	19.5±1.37	30.3±10.01	29.1±4.63	0.8±0.13	42.3	1.6
	SL	53.3±10.07	34.1±3.68	51.5±6.13	21.9±1.26	0.7±0.07	44.5	1.8

FHBi – Fusarium head blight index, FDK – Fusarium-damaged kernels, RL – more resistant line, SL – more susceptible line; mean values and standard deviations of each parameter calculated after inoculation (three plots) and data from one plot calculated for the control conditions, are shown.

MS peak-lists from the raw files. The obtained fragmentation spectra were matched to the National Center Biotechnology Information (NCBI) non-redundant database (37425594 sequences; 13257553858 residues), with a *Viridiplantae* filter (1760563 sequences) using the Mascot search engine (Mascot Daemon v. 2.3.0, Mascot Server v. 2.4.0, MatrixScience). The following search parameters were applied: enzyme specificity was set to trypsin, peptide mass tolerance to ±40 ppm and fragment mass tolerance to ±0.8 Da. The protein mass was left as unrestricted and mass values as monoisotopic with one missed cleavage being allowed. Alkylation of cysteine by carbamidomethylation as fixed and oxidation of methionine as variable modifications, were set. Ion score was $-10*Log(P)$, where P was the probability that the observed match was a random event. Only peptides with Mascot expect value over 0.05 were accepted as valid identifications. The proteins characterized by the highest Mascot-assigned protein score – Multidimensional Protein Identification Technology-type (MudPIT-type) and/or the highest number of peptide sequences, were selected. When the protein was identified as "a predicted protein" (primary identification), its amino acid sequence was blasted using *blastp* algorithm (http://blast.ncbi.nlm.nih.gov). The protein with the highest score was then selected as the functional homolog of the "predicted protein".

Alpha-amylase activity assays

Alpha-amylase activity in wheat kernels was evaluated according to the Ceralpha method [37] using Megazyme reagents (Ceralpha α-Amylase Assay Kit) and a detail protocol available at the company website: www.megazyme.com. The same biological and technical sample replicates as for proteome profiling, were applied (Fig. 1). Each technical replicate involved 0.5 g of wheat flour. The enzyme activity was expressed in Ceralpha Units (CU) per gram of flour – one unit of activity was defined as the amount of enzyme, in the presence of excess α-glucosidase and glucoamylase, required to release one micromole of *p*-nitrophenol from blocked *p*-nitrophenyl maltoheptaoside (BPNPG7) in one minute under the defined assay conditions. The significance of the differences in amylase activity between the RL and SL was assessed using the Student's t-test (p≤0.05).

Results and Discussion

Field experiments

The two analyzed winter wheat lines differed significantly in their levels of resistance to FHB as manifested by the values of FHBi and % FDK (with respect to kernel weight and number), after *F. culmorum* inoculation in Cerekwica and Radzikow (Table 1). Thus, it is clearly visible that these two groups of plants (the RL and SL) could serve as excellent models to recognize the most crucial proteins associated with the resistance to FHB in winter wheat. In the control conditions (without the artificial inoculation) both lines, the RL and SL, demonstrated similar yield levels (Table 1). The differences observed between both locations might be due to different soil quality and weather conditions (Table S1).

Proteome profiles and identities of differentially accumulated proteins

Linear strips for isoelectrofocusing with pH 4–7 range were selected to resolve the proteins derived from FDK (after inoculation) and kernels (in the control conditions) of both analyzed wheat lines to achieve the best compromise between the number (656 protein spots in the RL and 658 in the SL were present within all the replicate gels) and resolution of the spots.

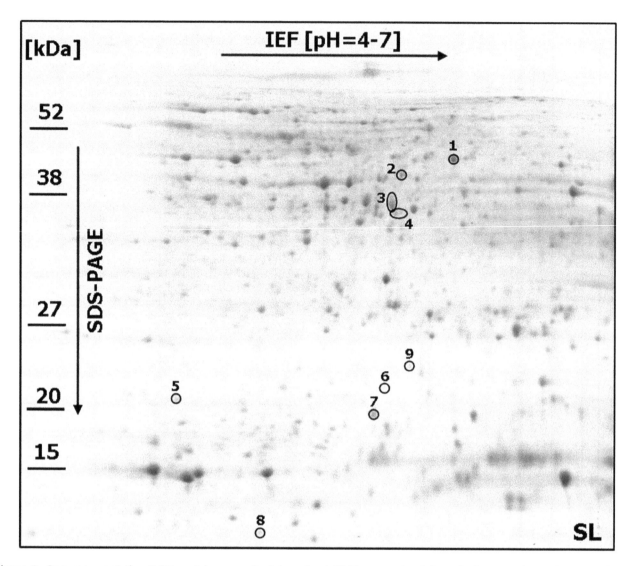

Figure 2. One representative 2-DE protein map of winter wheat (*Triticum aestivum*) kernel after *Fusarium culmorum* infection (*Fusarium*-damaged kernels) for the line more susceptible (SL) to *Fusarium* head blight. The spots with differentially accumulated (p≤0.05) proteins (1–9) identified in the SL, are circled with a solid line. Molecular weight (MW) scale is shown.

Only the representative proteome maps were presented in the paper (Fig. 2 and 3; Fig. S1 and S2). All the analyzed maps showed a relatively high level of similarity with respect to the number of detected spots and their distribution in the gels. This advantage was, in our opinion, mainly due to the scientific approach applied into protein sample preparation – pooling of all the FDK into a single sample (bulk) for a given field plot. This strategy could help to identify the differences between the analyzed wheat lines, which could be associated with their level of resistance to FHB. On the other hand, the spot intensities cannot be compared directly between the gels as a single raw image is not suitable to reveal the protein abundance directly. For each spot the normalized volume of mean derived from gel replicates, to calculate the level of protein accumulation, was used here. The comparative analyses indicated a total of 11 spots that showed differences in protein abundance (according to the criteria described in the materials and methods section) between the more resistant and more susceptible wheat lines after infection (Fig. 2 and 3). Four protein spots (no. 4, 5, 6 and 8) were demonstrated to be specific for the SL (Fig. 2 and 4; Fig. S3) and two other protein

spots (no. 10 and 11) for the RL (Fig. 3 and 4), although all these spots were observed in the RL and SL control gels (Fig. 4; Fig. S1 and S2). Five protein spots revealed quantitative differences between the analyzed lines, including four spots with significantly higher protein abundance in the more susceptible line (spots no. 2, 3, 7 and 9) and one spot with significantly higher abundance in the more resistant line (spot no. 1) (Fig. 4; Fig. S3). All the selected protein spots were subjected to mass spectrometry identification (Table 2). In some cases the selected proteins derived from wheat kernels were identified as homologues of proteins from related plant species (Table 2). For protein spot no. 1 no clear identification was found in the database. The spots no. 2, 4 and 5 were shown to be heterogeneous with more than one protein identified within them (Table 2). Thus, the abundance evaluation of particular proteins present in the spot was not possible and the total protein abundance was shown (Fig. S3). Approaches to deal with multi-protein spots would be required to determine relative abundance, for example, the spectral counting method described by Ishihama et al. [38]. The majority of spots were homogenous,

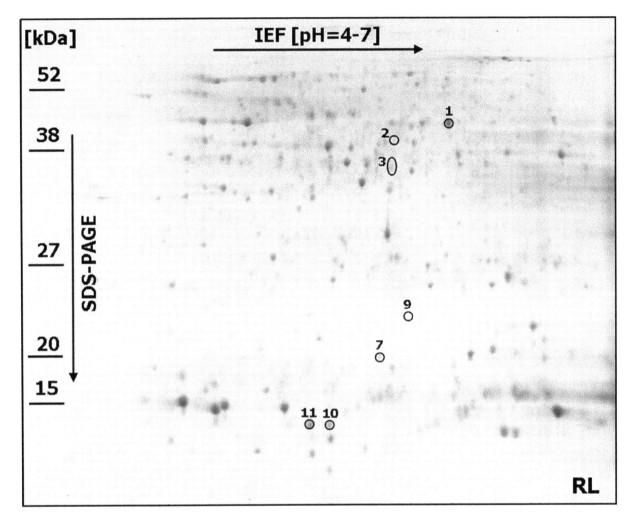

Figure 3. One representative 2-DE protein map of winter wheat (*Triticum aestivum*) kernel after *Fusarium culmorum* infection (*Fusarium*-damaged kernels) for the line more resistant (RL) to *Fusarium* head blight. The spots with differentially accumulated (p≤0.05) proteins (1–3, 7, 9–11) identified in the RL, are circled with a solid line. Molecular weight (MW) scale is shown.

containing one specific protein, and clear identifications were demonstrated for them.

The 0.19 dimeric alpha-amylase inhibitor which significantly accumulated in the SL samples was identified in two spots – spot no. 6 and spot no. 9 (Fig. 4). These proteins might vary in post-translation modifications, resulting in different isoelectric points and molecular weights affecting spot positions in the 2-D gels (Fig. 2). The higher molecular weights of the inhibitor proteins identified in spots nos. 6 and 9, observed on 2-D maps (experimental values), compared to the theoretical values revealed after MS protein identification (Table 2) further suggest the presence of post-translation modifications to these proteins. The 0.19 dimeric alpha-amylase inhibitor was also found to highly accumulate in the RL (spot no. 10) (Fig. 4) and this protein differed by 11 amino acids with the isoform identified in the SL. It is highly probable that these two isoforms could possess quite different properties (Fig. S4).

Significant differences were observed in the amino acid sequences of the monomeric alpha-amylase inhibitor (spot no.11) and the 0.19 dimeric alpha-amylase inhibitor (spot no. 10), both of which highly accumulated in the RL (Fig. 4). A total of 38 mismatched amino acids and seven gaps were found (Fig. S4). Interestingly, both inhibitors (spot no. 10 and 11) likely represent

the intact and functional proteins as almost no differences between their theoretical and experimentally evaluated molecular weights were observed (Table 2). A protein that slightly accumulated only in the SL was identified in spot no. 8 as the alpha-amylase/trypsin inhibitor CM3 (Fig. 4). Its higher theoretically evaluated molecular weight, compared to the experimentally calculated value, indicated partial degradation of this inhibitor (Table 2). Lastly, two further proteins identified as differentially accumulated between the RL and SL, and showing higher abundance in the more susceptible line were: serpin 1 (spot no. 3) and heat shock protein (spot no. 7) (Fig. 4).

Potential involvement of the identified proteins in resistance to FHB

The proteins with higher abundance after *Fusarium* infection in the SL revealed no differences in the accumulation level between the analyzed lines in the control conditions (without infection) (Fig. 4). These proteins are probably not all closely associated with the development of resistance to FHB. For example, in this study the up-regulated serpin 1 was identified in the susceptible line (spot no. 3). Eggert et al. [28] reported a 90–225% induction of this protein in wheat after *F. graminearum* infection, when compared

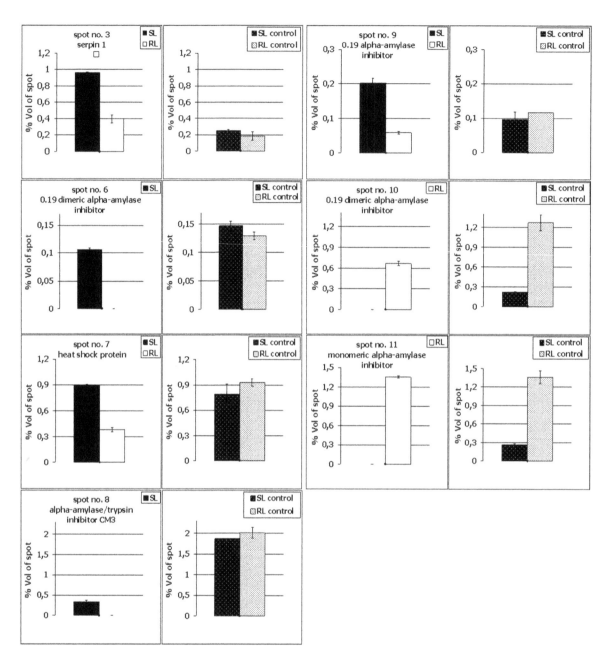

Figure 4. Comparison of selected kernel protein abundance after *Fusarium culmorum* infection and in the control conditions in the winter wheat (*Triticum aestivum*) SL (line more susceptible to *Fusarium* head blight) and the RL (line more resistant to *Fusarium* head blight). Spot numbering is the same as in Fig. 2, 3, S1 and S2. The standard deviation bars are shown. Only proteins identified from homogenous spots are shown.

to the control kernels. Serpins belong to a group of proteins which are involved in the inhibition of serine proteases. Eggert et al. [28] suggested that the pathogen infection enhanced serpin accumulation, which might prevent the digestion of seed storage proteins caused by a fungal pathogen. The involvement of the identified serpin in the process of resistance to FHB is, however, unclear. It was shown earlier that serpins could also function as storage proteins when they lost their inhibitory activity [39]. It thus cannot be excluded that proteins highly accumulated in the SL might not be active enough to perform their function efficiently or they possess a relatively low activity. On the other hand, two proteins absent in the SL, monomeric alpha-amylase inhibitor (spot no.11)

and 0.19 dimeric alpha-amylase inhibitor (spot no. 10), also had a lower accumulation level in this line during control conditions (Fig. 4 and Fig. S5), indicating lower resistance potential of the SL before infection.

Inhibition of alpha-amylase activity could be a crucial component of the resistance to FHB in winter wheat

Starch is the main reserve compound accumulated in the endosperm of kernels comprising approximately 70% of kernel weight of wheat and the other cereals [40]. In the study of Jackowiak et al. [41] and Packa et al. [42] severe damage of starch granules leading to their complete disappearance in wheat and

Table 2. The results of MS analysis performed on the spots that showed at least a 2.0 ratio (p≤0.05) in protein abundance between the more resistant and more susceptible winter wheat (*Triticum aestivum*) lines (spots no. 1–3, 7 and 9), and spots present only in one line (spots no. 4–6, 8, 10 and 11).

Spot no.[1]	Accession[2]	Identified protein[3]	Score[4]	Coverage (%)[5]	No. of peptide matched	Theor. MW [kDa]/pI[6]	Exp. MW [kDa]/pI[7]	Abundance[8]
1	NP001141324	uncharacterized protein [*Zea mays*]	1993	59	22	38.3/5.4	46/6.2	2.0 fold higher in the RL
2*	EMT04083	LL-diaminopimelate aminotransferase, chloroplastic [*Aegilops tauschii*]	995	41	11	46.3/5.5	42/6.0	2.3 fold higher in the SL
	CAO77315	putative acyl transferase 4 [*Triticum aestivum*]	890	37	13	46.8/5.7	42/6.0	2.3 fold higher in the SL
	AAB99745	HSP70 [*Triticum aestivum*]	835	20	10	71.4/5.1	42/6.0	2.3 fold higher in the SL
3	ACN59483	serpin 1 [*Triticum aestivum*]	896	29	10	43.3/5.4	37/5.9	2.4 fold higher in the SL
4*	BAK01819	predicted protein [*Hordeum vulgare* subsp. *vulgare*] blastp: Fructokinase-1 [*Oryza sativa*]	1152	50	14	34.9/5.7	36/6.0	present only in the SL
	EMS46550	putative NADP-dependent oxidoreductase P1 [*Triticum urartu*]	1014	37	12	38.4/5.5	36/6.0	present only in the SL
5*	BAK02140	predicted protein [*Hordeum vulgare* subsp. *vulgare*] blastp: peroxiredoxin-5 [*Zea mays*]	753	44	7	24.2/9.0	21/5.0	present only in the SL
	ABI54484	dimeric alpha-amylase inhibitor [*Triticum dicoccoides*]	654	69	6	13.7/6.5	21/5.0	present only in the SL
6	AAV39524	0.19 dimeric alpha-amylase inhibitor [*Aegilops tauschii*]	631	67	7	13.9/7.5	22/5.9	present only in the SL
7	BAJ94129	predicted protein [*Hordeum vulgare* subsp. *vulgare*] blastp: heat shock protein [*Triticum aestivum*]	692	52	6	16.9/5.8	19/5.8	2.4 fold higher in the SL
8	P17314	alpha-amylase/trypsin inhibitor CM3 [*Triticum aestivum*]	492	47	5	18.9/7.4	11/5.4	present only in the SL
9	P01085	0.19 dimeric alpha-amylase inhibitor [*Triticum aestivum*]	492	56	5	13.9/6.7	24/6.1	3.5 fold higher in the SL
10	AAV39517	0.19 dimeric alpha-amylase inhibitor [*Triticum aestivum*]	760	89	8	13.8/5.7	13/5.6	present only in the RL
11	ABO45988	monomeric alpha-amylase inhibitor [*Triticum aestivum*]	687	84	8	13.6/5.4	13/5.5	present only in the RL

[1] Spot numbering was the same as in Fig. 2, 3, S1 and S2.
[2] Database accession (according to NCBInr) of a homologous protein.
[3] Homologous protein and organism from which it originates.
[4] Mascot MudPIT (Multidimensional Protein Identification Technology) score.
[5] Amino acid sequence coverage for the identified proteins (primary identifications); amino acid sequences for proteins (primary identifications) derived from the homogenous spots were shown in Fig. S4.
[6] Theoretical molecular weight and isoelectric point calculated based on 2-D protein maps.
[7] Experimental molecular weight and isoelectric point revealed by Mascot software.
[8] Differences in accumulation level of proteins from the homogenous spots between the RL and SL after *Fusarium* infection. In case of heterogeneous spots the same value for all the indicated proteins present in the spot, were shown. *heterogeneous spots with more than only one protein; the most abundant proteins were shown.

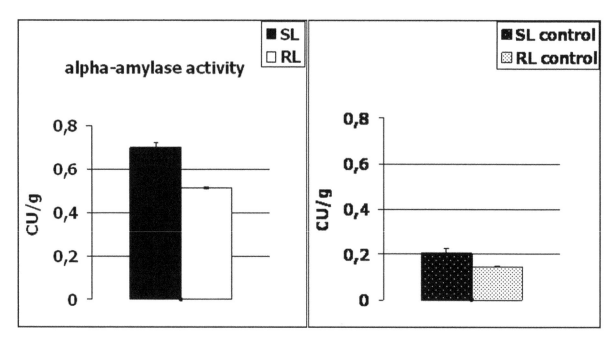

Figure 5. Comparison of alpha-amylase activity in the kernels of winter wheat (*Triticum aestivum*) SL (line more susceptible to *Fusarium* head blight) and RL (line more resistant to *Fusarium* head blight) after *Fusarium culmorum* infection (*Fusarium*-damaged kernels) and in control conditions. The enzyme activity was expressed in Ceralpha Units (CU) per gram of flour. The standard deviation bars are shown.

triticale kernels infected by *F. culmorum* were detected. Moreover, in the strongly infected kernels, the endosperm was replaced by mycelium. Alpha-amylase is a hydrolytic enzyme, which decomposes starch and makes carbohydrates available for germ development during sprouting, however, its activity is low in mature wheat kernels [43], [44]. *Fusarium* pathogens can use hydrolytic enzymes, including amylases to colonize kernels and acquire nitrogen and carbon from the endosperm [45]. The inoculation of wheat spikes with *F. culmorum* spores under field conditions showed increased alpha-amylase activity in kernels of different wheat species, including *T. monococum*, *T. dicoccum* and *T. aestivum* [46]. Thus, alpha-amylase inhibitors identified here in spots no. 10 and 11 could be involved in the development of resistance mechanisms and in the response to the synthesis of extracellular hydrolytic enzymes by infecting pathogens in the RL. The alpha-amylase inhibitors are thought to be important components of the active resistance of plants to necrotrophic pathogens [47].

To verify the above hypothesis, alpha-amylase activity was determined here in the kernels of the RL and SL, after *F. culmorum* infection (in FDK) and in the control conditions (Fig. 5). From our results, it was clearly visible that the enzyme activity was lower in the control conditions in the both analyzed wheat lines and it increased significantly after inoculation. Thus, this could suggest that amylase is involved in the propagation of pathogen infection and expansion of *Fusarium* biomass in the wheat heads. As indicated in Table 1 number and weight of FDK (% FDK) in the heads of the susceptible line were significantly higher, and a total kernel number and weight after inoculation, significantly lower, compared to the resistant line, in two applied locations of the experimental plots. Although other components of wheat resistance to FHB were not analyzed here in detail, it is quite possible that *Fusarium* expansion was higher in the SL within the infected kernels (FDK). The amylase activity revealed lower level ($p \leq 0.05$) in the RL, both in the control conditions and after

inoculation, and it is highly probable that it was due to the presence of monomeric alpha-amylase inhibitor (spot no.11) and dimeric alpha-amylase inhibitor (spot no. 10), both highly accumulated in the line with higher resistance to FHB.

The proteome network potentially involved in the resistance to FHB in cereals

The FHB resistant or susceptible cereal genotypes were characterized by many proteins related to carbon metabolism and photosynthesis which were down-regulated, while the up-regulated proteins were involved in antioxidant, jasmonic acid and ethylene signaling pathways, phenylpropanoid biosynthesis, antimicrobial compound synthesis, detoxification, cell wall fortification, defense-related responses, amino acid synthesis and nitrogen metabolism [48]. On the other hand, susceptible genotypes likely reflected the delayed activation of the salicylic acid defense pathway [49]. Eggert and Pawelzik [27] studied the effect of *F. graminearum* and *F. culmorum* infection on the proteome of naked barley grains. The proteins up-regulated in the infected samples in comparison to controls belong to a protein group involved in regulation of transcription (e.g. three Dof zinc finger proteins, DNA-direct RNA polymerase). Up-regulated proteins identified were also: one NBS-LRR disease-resistance protein and three serpins with protease-inhibitor and nutritional-reservoir functions. Down-regulated proteins were connected to starch synthesis processes, protein synthesis inhibition and fungal chitin hydrolysis. Zhou et al. [29] identified up-regulated proteins with antioxidant functions (superoxide dismutase, dehydroascorbate reductase and glutathione S-transferase). The PR-2 protein (β-1, 3 glucanase) was also shown to be up-regulated. Zhang et al. [50] studied two near isogenic wheat lines (NILs), NIL75 (*Fhb1*$^+$NIL) and NIL98 (*Fhb1*$^-$NIL), which were developed by backcrossing of 'Clark' (a highly FHB susceptible parent) to 'Ning 7840' (*Fhb1* donor) seven times (*Fhb1*$^+$NIL contains less than 0.5% of donor genome) and identified proteins which were accumulated in the

$Fhb1^{+}$NIL, but not in the $Fhb1^{-}$NIL, after *Fusarium* inoculation. These involved wheat proteins associated with defending fungal penetration (chitinases), photosynthesis (NAD(P)(+)-binding proteins, oxygen-evolving enhancer proteins) and energy metabolism (nucleoside diphosphate kinases).

Conclusions

We proved that FDK of both, the more resistant and more susceptible winter wheat lines differed in their proteome profiles after *F. culmorum* infection under field conditions and these kernels were shown to be a suitable plant material to identify the crucial proteins potentially involved in the resistance to FHB. The advantage of the research presented here is the selection of two proteins from hundreds of protein spots observed on 2-D maps, with the impact on the FHB resistance in the analyzed wheat lines. It was mainly due to: (i) pooled samples (bulks) used for the analysis and (ii) protein spot selection. Only the spots which showed at least a two-fold difference in protein abundance between two analyzed wheat lines were subjected to MS analyses. Moreover, we put special emphasis on the proteins which showed high accumulation levels in the control conditions and are potentially involved in the development of resistance before infection (amylase inhibitors highly accumulated in the RL). Alpha-amylase activity assays revealed the highest level of enzyme activity in the line more susceptible to FHB after *F. culmorum* infection. Finally, we suggest that the inhibition of pathogen amylase activity could be one of the most crucial mechanisms to prevent infection progress in the analyzed resistant wheat line. In our opinion, the presented results are an important contribution to the field comprising proteomic network associated with FHB resistance in wheat, however, further work to identify other components is still required.

Supporting Information

Figure S1 One representative 2-DE protein map of winter wheat (*Triticum aestivum*) kernel without *Fusarium culmorum* infection (control conditions) for the line more susceptible (SL) to *Fusarium* head blight. The spots with differentially accumulated (p≤0.05) proteins (1–11) identified in the SL and RL (line more resistant to *Fusarium* head blight) after infection, are circled with a solid line. Molecular weight (MW) scale is shown.
(TIF)

Figure S2 One representative 2-DE protein map of winter wheat (*Triticum aestivum*) kernel without *Fusarium culmorum* infection (control conditions) for the line more resistant (RL) to *Fusarium* head blight. The spots with differentially accumulated (p≤0.05) proteins (1–11) identified in the SL (line more susceptible to *Fusarium* head blight) and RL after infection, are circled with a solid line. Molecular weight (MW) scale is shown.
(TIF)

Figure S3 Comparison of selected kernel protein abundance after *Fusarium culmorum* infection and in the control conditions in the winter wheat (*Triticum aestivum*) SL (line more susceptible to *Fusarium* head blight) and the RL (line more resistant to *Fusarium* head blight). Spot numbering is the same as in the Fig. 2, 3, S1 and S2. The standard deviation bars are shown. Unidentified proteins and proteins derived from heterogeneous spots are shown.
(TIF)

Figure S4 A. Amino acid sequences for proteins (primary identifications) derived from the homogenous spots. In bold the amino acid sequence of peptides derived from winter wheat (*Triticum aestivum*), which were successfully matched to the protein sequences present in the database are indicated. Spot numbers, protein names and organism from which the protein originates are shown. B. Protein sequence alignment of alpha-amylase inhibitors identified in spots no. 6, 9, 10 and 11. C. Mascot search results for the identified proteins, including data for particular peptides.
(PDF)

Figure S5 Enlarged windows with spots no. 10 and 11 selected in 2-DE gels of winter wheat (*Triticum aestivum*) kernel without *Fusarium culmorum* infection (control conditions) and after inoculation, for the line more resistant (RL) and more susceptible (SL) to *Fusarium* head blight. The biological and technical replicates are shown.
(TIF)

Table S1 Meteorological conditions (sum of rainfalls and mean temperature) during the experiments performed in Cerekwica and Radzikow in 2013.
(PDF)

Author Contributions

Conceived and designed the experiments: DP HW AK. Performed the experiments: DP HW TG AK. Analyzed the data: DP HW TG MK MM AK. Contributed reagents/materials/analysis tools: DP HW TG AK. Contributed to the writing of the manuscript: DP HW TG MK MM AK.

References

1. Bottalico A, Perrone G (2002) Toxigenic *Fusarium* species and mycotoxins associated with head blight in small-grain cereals in Europe. Eur J Plant Pathol 108: 998–1003.
2. Champeil A, Doré T, Fourbet JF (2004) *Fusarium* head blight: epidemiological origin of the effects of cultural practices on head blight attacks and the production of mycotoxins by *Fusarium* in wheat grains. Plant Sci 166: 1389–1415.
3. Xu XM, Monger W, Ritieni A, Nicholson P (2008) Effect of temperature and duration of wetness during initial infection periods on disease development, fungal biomass and mycotoxin concentrations on wheat inoculated with single, or combinations of *Fusarium* species. Plant Pathol 56: 943–956.
4. Goliński P, Waśkiewicz A, Wiśniewska H, Kiecana I, Mielniczuk E, et al. (2010) Reaction of winter wheat (*Triticum aestivum* L.) cultivars to infection with *Fusarium* spp. – mycotoxin contamination in grain and chaff. Food Addit Contam Part A Chem Anal Control Expo Risk Assess 27: 1015–1024.
5. Mesterházy A (1995) Types and components of resistance against *Fusarium* head blight of wheat. Plant Breeding 114: 377–386.
6. Kiecana I, Mielniczuk E, Cegielko M (2004) Reaction of oats genotypes to *Fusarium avenaceum* (Fr.) Sacc., *Fusarium culmorum* (W.G.Sm.) Sacc. and *Fusarium graminearum* (Schwabe.) infection. Agronomijas Vestis. Latvian J.Agron. 7: 165–169.
7. Cowger C, Arrellano C (2010) Plump kernels with high deoxynivalenol linked to late *Gibberella zeae* infection and marginal disease conditions in winter wheat. Phytopathology 100: 719–728.
8. Bottalico A, Logrieco A (1998) Toxigenic *Alternaria* species of economic importance. In: Mycotoxins in Agriculture and Food Safety. Eds. Sinha KK, Bhatnager D, Dekker M, New York: 65–108.
9. Chełkowski J, Perkowski J, Grabarkiewicz-Szczęsna J, Kostecki M, Goliński P (2001) Toxigenic fungi and mycotoxins in cereal grains and feeds in Poland. In: Logrieco A. (ed.) Occurrence of Toxigenic Fungi and Mycotoxins in Plants, Food and Feeds in Europe, European Commission, COST Action 835, EUR 19695: 111–130.
10. Foroud NA, Eudes F (2009) Trichothecenes in cereal grains. Int J Mol Sci 10: 147–173.
11. Mardi M, Pazouki L, Delavar H, Kazemi MB, Ghareyazie B, et al. (2006) QTL analysis of resistance to *Fusarium* head blight in wheat using a 'Frontana' – derived population. Plant Breeding 125: 313–317.

12. Buerstmayr H, Ban T, Anderson JA (2009) QTL mapping and marker-assisted selection for *Fusarium* head blight resistance in wheat: a review. Plant Breeding 129: 1–26.

13. Mielniczuk E, Kiecana I, Perkowski J (2004) Susceptibility of oat genotypes to *Fusarium crookwellense* Burgess, Nelson and Toussoun infection and mycotoxin accumulation in kernels. Biologia Bratislava 59: 809–816.

14. Wiśniewska H, Perkowski J, Kaczmarek Z (2004) Scab response and deoxynivalenol accumulation in spring wheat kernels of different geographical origins following inoculation with *Fusarium culmorum*. J Phytopathol 152: 613–621.

15. Miedaner T, Reinbrecht C, Lauber U, Schollenberger M, Geiger HH (2001) Effects of genotype and genotype × environment interaction on deoxynivalenol accumulation and resistance to *Fusarium* head blight in rye, triticale, and wheat. Plant Breeding 120: 97–105.

16. Langevin F, Eudes F, Comeau A (2004) Effect of trichothecenes produced by *Fusarium graminearum* during *Fusarium* head blight development in six cereal species. Eur J Plant Pathol 110: 735–746.

17. Comeau A, Langevin F, Savard ME, Gilbert J, Dion Y, et al. (2008) Improving Fusarium head blight resistance in bread wheat and triticale for Canadian needs. Cereal Res Commun 36: 91–92.

18. Snijders CH, Perkowski J (1990) Effects of head blight caused by *Fusarium culmorum* on toxin content and weight of wheat kernels. Phytopathology 80: 566–570.

19. Paillard S, Schnurbusch T, Winzeler M, Messmer M, Sourdille P, et al. (2003) An integrative genetic linkage map of winter wheat (*Triticum aestivum* L.). Theor Appl Genet 107: 1235–1242.

20. Pirgozliev SR, Edwards SG, Hare MC, Jenkinson P (2003) Strategies for the control of *Fusarium* head blight in cereals. Eur J Plant Pathol 109: 731–742.

21. Edwards SG (2004) Influence of agricultural practices on *Fusarium* infection of cereals andsubsequent contamination of grain by trichothecene mycotoxins. Toxicol Lett 153: 29–35.

22. Veronese P, Ruiz MT, Coca MA, Hernandez-Lopez A, Lee H, et al. (2003) In defense against pathogens. Both plant sentinels and foot soldiers need to know the enemy. Plant Physiol 131: 1580–1590.

23. Pritsch C, Muehlbauer GJ, Bushnell WR, Somers DA, Vance CP (2000) Fungal development and induction of defense response genes during early infection of wheat spikes by *Fusarium graminearum*. Mol Plant Microbe Interact 13: 159–169.

24. Gottwald S, Samans B, Lück S, Friedt W (2012) Jasmonate and ethylene dependent defence gene expression and suppression of fungal virulence factors: two essential mechanisms of *Fusarium* head blight resistance in wheat? BMC Genomics 13: 369. doi:10.1186/1471-2164-13-369.

25. Yang F, Jensen JD, Spliid NH, Svensson B, Jacobsen S, et al. (2010a) Investigation of the effect of nitrogenon severity of *Fusarium* head blight in barley. J Proteomics 73: 743–752.

26. Yang F, Jensen JD, Svensson B, Jørgensen HJ, Collinge DB, et al. (2010b) Analysis of early events in the interaction between *Fusarium graminearum* and the susceptible barley (*Hordeum vulgare*) cultivar Scarlett. Proteomics 10: 3748–3755.

27. Eggert K, Pawelzik E (2011) Proteome analysis of *Fusarium* head blight in grains of naked barley (*Hordeum vulgare* subsp. *nudum*). Proteomics 11: 972–985.

28. Eggert K, Zörb C, Mühling KH, Pawelzik E (2011) Proteome analysis of *Fusarium* infection in emmer grains (*Triticum dicoccum*). Plant Pathol 60: 918–928.

29. Zhou W, Kolb FL, Riechers DE (2005) Identification of proteins induced or upregulated by *Fusarium* head blight infection in the spikes of hexaploid wheat (*Triticum aestivum*). Genome 48: 770–780.

30. Zhou W, Eudes F, Laroche A (2006) Identification of differentially regulated proteins in response to a compatible interaction between the pathogen *Fusarium graminearum* and its host, *Triticum aestivum*. Proteomics 6: 4599–4609.

31. Masojć P, Kosmala A (2012) Proteomic analysis of preharvest sprouting in rye using two dimensional electrophoresis and mass spectrometry. Mol Breed 30: 1355–1361.

32. Masojć P, Kosmala A, Perlikowski D (2013) Proteomic analysis of developing rye grain with contrasting resistance to preharvest sprouting. J Appl Genet 54: 11–19.

33. Hurkman WJ, Tanaka CK (1986) Solubilization of plant membrane proteins for analysis by two-dimensional gel electrophoresis. Plant Physiol 81: 802–806.

34. Neuhoff V, Arold N, Taube D, Ehrhardt W (1988) Improved staining of proteins in polyacrylamide gels including isoelectric focusing gels with clear background at nanogram sensitivity using Coomassie Brilliant Blue G-250 and R-250. Electrophoresis 9: 255–262.

35. Kosmala A, Perlikowski D, Pawłowicz I, Rapacz M (2012) Changes in the chloroplast proteome following water deficit and subsequent watering in a high- and a low-drought-tolerant genotype of *Festuca arundinacea*. J Exp Bot 63: 6161–6172.

36. Perlikowski D, Kosmala A, Rapacz M, Kościelniak J, Pawłowicz I, et al. (2014) Influence of short-term drought conditions and subsequent re-watering on the physiology and proteome of *Lolium multiflorum/Festuca arundinacea* introgression forms with contrasting levels of tolerance to long-term drought. Plant Biol 16: 385–394.

37. McCleary BV, McNally M, Monaghan D, Mugford DC (2002) Measurements of alpha-amylase activity in white wheat flour, milled malt and microbial enzyme preparations, using the ceralpha assay: collaborative study. J AOAC Int 85: 1096–1102.

38. Ishihama Y, Oda Y, Tabata T, Sato T, Nagasu T, et al. (2005) Exponentially modified protein abundance index (emPAI) for estimation of absolute protein amount in proteomics by the number of sequenced peptides per protein. Mol Cell Proteomics 4: 1265–1272.

39. Roberts TH, Hejgaard J (2008) Serpins in plants and green algae. Funct Integr Genomics 8: 1–27.

40. Thitisaksakul M, Jimenez RC, Arias MC, Beckles DM (2012) Effects of environmental factors on cereal starch biosynthesis and composition. J Cereal Sci 56: 67–80.

41. Jackowiak H, Packa D, Wiwart M, Perkowski J (2005) Scanning electron microscopy of *Fusarium* damaged kernels of spring wheat. Int J Food Microbiol 98: 113–123.

42. Packa D, Jackowiak H, Góral T, Wiwart M, Perkowski J (2008) Scanning electron microscopy of *Fusarium* infected kernels of winter triticale (×*Triticosecale* Wittmack.). Seed Sci Biotech 2: 27–31.

43. Lunn GD, Major BJ, Kettlewell PS, Scott RK (2001) Mechanism leading to excess *alpha*-amylase activity in wheat (*Triticum aestivum* L.) grain in U.K. 2001. J Cereal Sci 33: 313–329.

44. Mares D, Mrva K (2008) Late-maturity alpha-amylase: Low falling number in wheat in the absence of preharvest sprouting. J Cereal Sci 47: 6–17.

45. Wang J, Pawelzik E, Weinert J, Wolf GA (2005) Impact of *Fusarium culmorum* on the polysaccharides of wheat flour. J Agric Food Chem 53: 5818–5823.

46. Packa D, Graban Ł, Lajszner W, Załuski D, Hościk M (2013) Alpha-amylase activity in the kernels of hulled spring wheat after head inoculation with *Fusarium culmorum* (W.G.Smith) Sacc., Ejpau 16(4), #06.

47. Svensson B, Fukuda K, Nielsen PK, Bonsager BC (2004) Proteinaceous alpha-amylase inhibitors. Biochim Biophys Acta 1696: 145–156.

48. Yang F, Jacobsen S, Jorgensen HJL, Collinge DB, Svensson B, et al. (2013) *Fusarium graminearum* and its interactions with cereal heads: studies in the proteomics era. Front Plant Sci 4: 1–8.

49. Ding L, Xu HYH, Yang L, Kong Z, Zhang L, et al. (2011) Resistance to hemi-biotrophic F. *graminearum* infection is associated with coordinated and ordered expression of diverse defense signaling pathways. PLOS ONE 6: e19008. doi:10.1371/journal.pone.0019008.

50. Zhang X, Fu J, Hiromasa Y, Pan H, Bai G (2013) Differentially expressed proteins associated with *Fusarium* head blight resistance in wheat. PLOS ONE 8: e82079. doi:10.1371/journal.pone.0082079.

Crystal Structure of Human Protein N-Terminal Glutamine Amidohydrolase, an Initial Component of the N-End Rule Pathway

Mi Seul Park[1]ⓢ, Eduard Bitto[2]ⓢ, Kyung Rok Kim[1], Craig A. Bingman[3], Mitchell D. Miller[4], Hyun-Jung Kim[5], Byung Woo Han[1]*, George N. Phillips Jr[3,4]*

1 Research Institute of Pharmaceutical Sciences, College of Pharmacy, Seoul National University, Seoul, Korea, **2** Department of Chemistry and Biochemistry, Georgian Court University, Lakewood, New Jersey, United States of America, **3** Department of Biochemistry, Center for Eukaryotic Structural Genomics, University of Wisconsin Madison, Madison, Wisconsin, United States of America, **4** Department of Biochemistry and Cell Biology, Rice University, Houston, Texas, United States of America, **5** College of Pharmacy, Chung-Ang University, Seoul, Korea

Abstract

The N-end rule states that half-life of protein is determined by their N-terminal amino acid residue. N-terminal glutamine amidohydrolase (Ntaq) converts N-terminal glutamine to glutamate by eliminating the amine group and plays an essential role in the N-end rule pathway for protein degradation. Here, we report the crystal structure of human Ntaq1 bound with the N-terminus of a symmetry-related Ntaq1 molecule at 1.5 Å resolution. The structure reveals a monomeric globular protein with alpha-beta-alpha three-layer sandwich architecture. The catalytic triad located in the active site, Cys-His-Asp, is highly conserved among Ntaq family and transglutaminases from diverse organisms. The N-terminus of a symmetry-related Ntaq1 molecule bound in the substrate binding cleft and the active site suggest possible substrate binding mode of hNtaq1. Based on our crystal structure of hNtaq1 and docking study with all the tripeptides with N-terminal glutamine, we propose how the peptide backbone recognition patch of hNtaq1 forms nonspecific interactions with N-terminal peptides of substrate proteins. Upon binding of a substrate with N-terminal glutamine, active site catalytic triad mediates the deamination of the N-terminal residue to glutamate by a mechanism analogous to that of cysteine proteases.

Editor: Titus J. Boggon, Yale University School of Medicine, United States of America

Funding: This study was supported by GM064598, National Institutes of Health Protein Structure Initiative, (http://www.nigms.nih.gov/research/specificareas/PSI/pages/default.aspx), GNP GM074901, National Institutes of Health Protein Structure Initiative, (http://www.nigms.nih.gov/research/specificareas/PSI/pages/default.aspx, GNPGM094816), National Institutes of Health Protein Structure Initiative, (http://www.nigms.nih.gov/research/specificareas/PSI/pages/default.aspx), GNP1231306, The BioXFEL Science and Technology Center under National Science Foundation, (https://www.bioxfel.org/), MDM2013-043695, Ministry of Science, ICT and Future Planning of Korea, (www.msip.go.kr), BWH 2014-001848, Ministry of Science, ICT and Future Planning of Korea, (www.msip.go.kr), BWH A092006, Ministry of Health and Welfare of Korea, (www.mw.go.kr), BWH. The funders had no role in study design, data collection and analysis, decision to publish, or preparation of the manuscript.

Competing Interests: The authors have declared that no competing interests exist.

* Email: bwhan@snu.ac.kr (BWH); georgep@rice.edu (GNP)

ⓢ These authors contributed equally to this work.

Introduction

Aberrant polypeptides or proteins should be accurately removed in many physiological processes. Intracellular protein degradation is mainly conducted through the ubiquitin-proteasome pathway (UPP) or the lysosomal proteolysis [1]. The UPP is required for degradation of short-lived proteins in eukaryotic cells. In the UPP, ubiquitin first attaches to target proteins or polypeptides, which leads to their recognition by the 26S proteasome [2]. On the other hand, lysosomal proteolysis leads to breakdown of unnecessary proteins or polypeptides by lysosomes [3].

The N-end rule is related to the ubiquitin-dependent proteolytic system [4]. The N-end rule is one of common pathways for the degradation of polypeptides and proteins in prokaryotes and eukaryotes, which determines the stability of a protein by its N-terminal residue [1]. Val, Gly, and Pro are classified as stabilizing residues in mammals whereas Asp, Gln, Cys, and Arg are known as destabilizing residues [5]. In the N-end rule pathway, N-terminal glutamine and asparagine are tertiary destabilizing residues and these residues are converted into secondary destabilizing N-terminal glutamate and aspartate by deamidation [6,7]. Arginine is then conjugated to glutamate and aspartate residues by Arg-tRNA-protein transferase, converting target proteins into ones possessing a primary destabilizing residues [6–8]. The N-end rule mechanism is involved in degradation of misfolded proteins, regulation of DNA repair, apoptosis and meiosis [9–12]. The N-terminal amidohydrolases are classified into the N-terminal glutamine amidohydrolase (Ntaq) and the N-terminal asparagine amidohydrolase (Ntan), which share low amino acid sequence identity and mediate the deamidation of N-terminal glutamine and asparagine, respectively [6].

In this work, we present the crystal structure of hNtaq1 bound with the N-terminus of a symmetry-related Ntaq1 molecule at 1.5 Å resolution. The structure contains the catalytic triad (Cys-His-Asp) in the active site, which is well conserved among Ntaq proteins and transglutaminases from diverse organisms. Additionally, we conducted docking study with all the tripeptides containing N-terminal glutamine to elucidate how N-termini of proteins with N-terminal glutamine are recognized and positioned by hNtaq1 into catalytically conducive conformations. We also propose a catalytic mechanism of hNtaq1 based on the crystal structure of hNtaq1 and docking study.

Materials and Methods

Cloning, protein expression, and purification

The standard Center for Eukaryotic Structural Genomics pipeline protocols were used for cloning [13], protein expression [14], protein purification [15]. In summary, using Gateway cloning (Life Technologies, USA), hNtaq1 gene was cloned into pVP16 plasmid (Clontech, USA) containing N-terminal fusion (His)$_6$-Maltose Binding Protein (MBP) and a linker region with the TEV protease site for cleavage of target proteins. It results in the hNtaq1 construct mutated with Ser for the initial Met. The hNtaq1 construct was transformed into B834 *E. coli* cells (Novagen, USA) to express Se-Met labeled protein. Cells were cultured with auto-induction medium adapted from the work of Studier [16] and incubated in a shaker at 250 rpm, 25°C for 22~24 hours. Cells were harvested by centrifugation at 5000×g for 20 min and suspended in cell lysis buffer (20 mM sodium phosphate, pH 7.5, 500 mM sodium chloride, 20% ethylene glycol, 35 mM imidazole, 0.3 mM tris(2-carboxyethyl) phosphate (TCEP), and E64 protease inhibitor cocktail (Sigma Aldrich, USA). Cells were disrupted by sonication on ice and the cell lysate was centrifuged at 75,600×g for 30 min twice. The supernatant was collected and filtered through a 0.8 μm pore size filter.

The sample was loaded on the Ni-charged HiTrap chelating 5 ml HP column (GE Healthcare, UK) and washed with IMAC-washing buffer (20 mM sodium phosphate, pH 7.5, 500 mM sodium chloride, and 0.3 mM TCEP). The protein was eluted by applying a 30 column volume linear gradient from 10% to 80% IMAC-elution buffer (20 mM sodium phosphate, pH 7.5, 350 mM imidazole, 500 mM sodium chloride, and 0.3 mM TCEP) and buffer was exchanged to TEV proteolysis buffer (20 mM sodium phosphate, pH 7.5, 100 mM sodium chloride, and 0.3 mM TCEP) using a HiPrep 26/10 desalting column (GE Healthcare, UK). The (His)$_6$-MBP fusion hNtaq1 protein was treated with TEV protease (1:100 w/w) at 25°C for overnight. After the TEV protease treatment, the protein was loaded to HiTrap chelating 5 ml HP column (GE Healthcare, UK) and eluted with IMAC-washing buffer. The eluted sample was desalted with crystallization buffer (50 mM sodium chloride, 3 mM sodium azide, 0.3 mM TCEP, and 100 mM Bis-Tris, pH 7.0) using a HiPrep 26/10 desalting column (GE Healthcare, UK). For crystallization, the purified Se-Met hNtaq1 protein was concentrated to 10 mg/ml.

Crystallization, X-ray data collection, structure determination, and model evaluation

Crystals of the hNtaq1 were obtained by hanging-drop vapor diffusion method at 291 K by mixing the protein solution (10 mg/ml Se-Met protein, 50 mM sodium chloride, 3 mM sodium azide, 0.3 mM TCEP, and 100 mM Bis-Tris, pH 7.0) and the well solution (1% ethylene glycol, 1.8 M ammonium sulfate, 100 mM MES, pH 6.0) in 1:1 ratio. Crystals were cryoprotected in four

stages with well solution using containing 0 to 25% ethylene glycol and were flash-frozen in liquid nitrogen gas at 100 K. X-ray diffraction data were collected using synchrotron beam line 23-ID-D at the Advanced Photon Source of the Argonne National Laboratory. The crystal structure of the hNtaq1 was solved by SAD phasing at 1.5 Å resolution. SHARP [17] was used to solve experimental phase information, which was improved by density modification using DM [18]. Crystals of hNtaq1 belong to the space group $P2_12_12_1$ with unit cell parameters a = 34.3 Å, b = 64.0 Å, and c = 113.6 Å. Subsequent manual model building and refinement were carried out using *Coot* [19] and *REFMAC* [20] from CCP4 program suite [20]. All refinement steps were monitored using an R_{free} value based on 5.0% of the independent reflections. The stereochemical quality of the final model was assessed using PROCHECK [21] and *MolProbity* [22]. The data collection, phasing, and refinement statistics are summarized in Table 1.

Docking studies of tripeptides with N-terminal glutamine

AutoDock Vina program [23] was used for the docking studies of hNtaq1 with all the possible 400 tripeptides containing N-terminal glutamine (Gln-X-X, X is any amino acid residue, thus 20×20 candidates). Coordinatees for the 400 tripeptides were generated using Coot [19] and converted to pdbqt files using AutoDockTools4 [24]. The grid maps for docking studies were centered on C$_\alpha$ of Ser1 of the bound N-terminus from the symmetry-related hNtaq1 molecule in the substrate binding cleft and the maps comprised 30×30×30 points. AutoDock Vina program was run with four-way multithreading and the default settings were used for others computational parameters. Figures are generated using PyMol [25].

Data deposition

Atomic coordinates and structure factors have been deposited in the RCSB Protein Data Bank, accession code 4W79.

Results and Discussion

Overall structure of hNtaq1

The human C8orf32 gene encodes human N-terminal glutamine amidohydrolase isoform 1 (hNtaq1). Ntaq is an initial component of the N-end rule pathway and converts N-terminal glutamine to glutamate. In order to understand the relationship between the structure and function of hNtaq1, we determined the crystal structure of recombinant hNtaq1 bound with the adjacent N-terminus of a symmetry-related hNtaq1 molecule at 1.5 Å resolution. The hNtaq1 protein contains 205 amino acids, of which 202 have been successfully modeled in the presented hNtaq1 structure. Three C-terminal residues (Lys203, Asn204, and Cys205) were disordered in the crystal and could not be modeled. The R and R_{free} values of the final refined model were 14.4% and 17.0%, respectively. hNtaq1 is a monomeric globular protein with a novel structural fold of alpha-beta-alpha three-layer sandwich architecture (Figure 1A). To our surprise, the N-terminus of a symmetry-related hNtaq1 molecule was captured in the substrate binding cleft, even though Ser1 is the N-terminal residue, not an anticipated glutamine residue (Figure 1B). The core region of the protein shows antiparallel beta-sheets surrounded by helices. The catalytic triad (Cys28, His81, and Asp97) is highly conserved among Ntaq proteins, transglutaminases, and cysteine proteases of diverse organisms (Figure 1C). Ntaq and Ntan in human share only 13.1% sequence identity and the structure of Ntan has yet not been determined. Elucidation of subtle differences in the catalytic mechanism between Ntaq and

Table 1. Statistics for data collection, phasing, and model refinement.

Data collection and phasing[a]	
Space group	$P\,2_1\,2_1\,2_1$
Cell dimensions	
a, b, c (Å), α, β, γ (°)	34.32, 64.04, 113.66, 90, 90, 90
Data set	Se λ1 (peak)
X-ray wavelength (Å)	0.9794
Resolution range (Å)[b]	32.86–1.50 (1.53–1.50)
$<I/\sigma(I)>$	12.1 (2.5)
Multiplicity	12.2 (6.9)
Unique reflections	40,943 (2,566)
Completeness (%)	99.5 (95.9)
R_{merge} (%)[c]	0.5 (54.3)
Figure of merit[d] for SAD phasing: 0.44	
Refinement	
R_{work}[e]/R_{free}[f]	0.144/0.170
No. of protein atoms	1,666
No. of water atoms	332
No. of Non-water atoms	75
Mean B value (Å²)	18.9
Ramachandran plot analysis (for Chain A)	
Most favored regions	198 (96.6%)
Additional allowed regions	7 (3.4%)
Disallowed regions	0 (0%)
R.m.s. deviations from ideal geometry	
Bond lengths (Å)	0.017
Bond angles (°)	1.91

[a]Data collected at the Sector 23-ID-D of the Advanced Photon Source.
[b]Numbers in parentheses indicate the highest resolution shell of 20.
[c]$R_{merge} = \Sigma_h \Sigma_i |I(h)_i - <I(h)>|/\Sigma_h \Sigma_i I(h)_i$, where $I(h)$ is the observed intensity of reflection h, and $<I(h)>$ is the average intensity obtained from multiple measurements.
[d]Figure of merit $= <|\Sigma\,P(\alpha)e^{i\alpha}/\Sigma\,P(\alpha)|>$, where α is the phase angle and $P(\alpha)$ is the phase probability distribution.
[e]$R_{work} = \Sigma\,||F_o|-|F_c||/\Sigma\,|F_o|$, where $|F_o|$ is the observed structure factor amplitude and $|F_c|$ is the calculated structure factor amplitude.
[f]R_{free} = R-factor based on 5.0% of the data excluded from refinement.

Ntan in the N-end rule pathway awaits the structural information of Ntan.

To find out structural features of hNtaq1, we analyzed of structure similarity using the DALI server [26]. The result reveals that hNtaq1 shows low level of sequence conservation with other structurally similar proteins and the amino acid sequence identities are in the range of 9 to 15%. The result shows that protein-glutaminase from *Chryseobacterium proteolyticum* (PDB ID: 3A54) is the closest structural homolog of hNtaq1 (Z score 7.0 and RMSD distance 3.6 Å for 123 equivalent C$_\alpha$ positions out of 280 residues) and shares only 15% sequence identity. The protein-glutaminase from *Chryseobacterium proteolyticum* converts a glutamine residue to a glutamate and has catalytic triad comprising Cys156, His197, and Asp217 [27] (Figure 1D). The secreted effector protein SseI from *Sallonella typhimurium* (PDB ID: 4G2B) is the second best structural homolog of hNtaq1 (Z score 6.3 and RMSD distance 3.7 Å for 116 equivalent C$_\alpha$ positions out of 169 residues) and shares only 9% sequence identity. The secreted effector protein SseI from *Sallonella typhimurium* belongs to cysteine protease superfamily and also contains catalytic triad with Cys178, His216, and Asp 231 [28] (Figure 1E). LapG from *Legionella pneumophila* (PDB ID: 4FGP),

a bacterial transglutaminase-like cysteine protease (BTLCP) containing catalytic triad, also shares structural similarity to hNtaq1 (Z score 6.1 and RMSD distance 2.7 for 96 equivalent C$_\alpha$ positions out of 186 residues) [29] (Figure 1F). Even though structural homologs of hNtaq1 do not share high sequence identities, the catalytic triad is structurally well conserved, mediating similar type of reactions with different substrates. Uniqueness of substrate binding region except the catalytic triad could contribute to the specificity with respect to each substrates.

Active site with catalytic triad of hNtaq1

The structure of hNtaq1 shows the precise conformation of the active site and the catalytic triad, Cys28, His81, and Asp97. Previous researches showed that Cys and His residues play a vital role in the activity of Ntaq. In the case of mouse Ntaq Cys30Ala and His83Ala mutants, the mutations almost abolish activities as an N-terminal glutamine amidohydrolase [8]. In the structure of hNtaq1, Asp97 forms hydrogen bonds with His81, Tyr111 and a water molecule 1 (water403 in PDB) and His81 interacts with another water molecule 2 (water444 in PDB) via hydrogen bond. The distances from the sulfhydryl group of Cys28 to water2 and His81 are 3.4 Å and 3.8 Å, respectively. After proper conformational

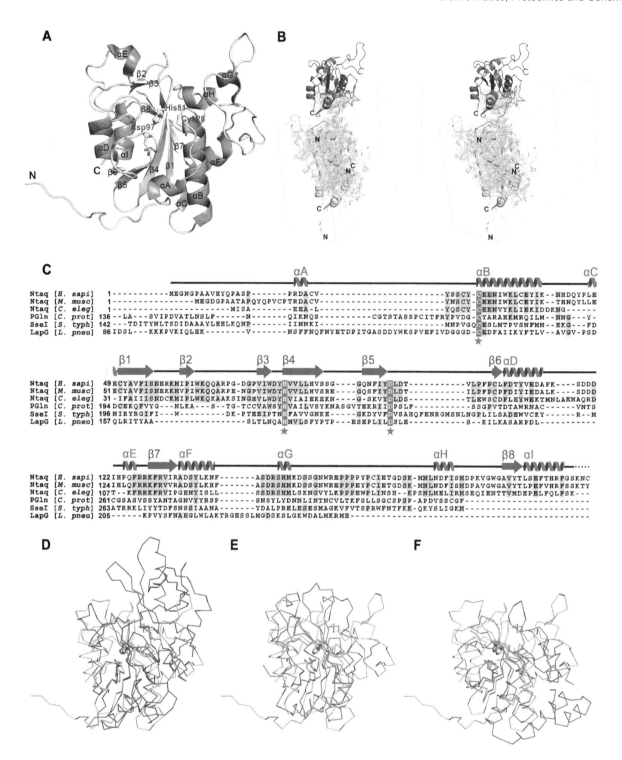

Figure 1. Overview of the crystal structure of hNtaq1. (A) Overall structure of hNtaq1. α-helices, β-strands, and loops are colored in orange, cyan, and white, respectively. The representative amino acid residues in the active site are shown as stick model (carbon, oxygen, nitrogen, and sulfur in yellow, red, blue, and gold color, respectively). (B) Stereo view of the crystallographic contact of hNtaq1 with a symmetry-related molecule. hNtaq1 and symmetry-related molecules are represented as green, cyan, and magenta cartoon, respectively. Unit cell is shown with green line and electron density map is shown as gray cloud. (C) Sequence alignment of hNtaq1 with Ntaq proteins from *Mus musculus, Caenorhabditis elegans*, protein-glutaminase from *Chryseobacterium proteolyticum*, secreted effector protein SseI from *Sallonella typhimurium*, and periplasmic protease LapG from *Legionella pneumophila*. Residues of the catalytic triad are represented by red asterisk below the amino acid sequences. Completely conserved, identical, moderately conserved residues are highlighted with red, green, and yellow shaded boxes, respectively. The secondary structure of hNtaq1 is shown on top of the sequence alignment. α-helix, β-sheet, and connecting region are represented by red spiral, blue arrow, and black line, respectively. Structural comparison of hNtaq1 with protein-glutaminase from *C. proteolyticum* (D), secreted effector protein SseI from *S. typhimurium* (E), and periplasmic protease LapG from *L. peumophila* (F). The catalytic triads are shown as stick models and main chains of hNtaq1, protein-glutaminase, SseI, and LapG are represented with ribbon diagram in green, magenta, orange, and blue, respectively.

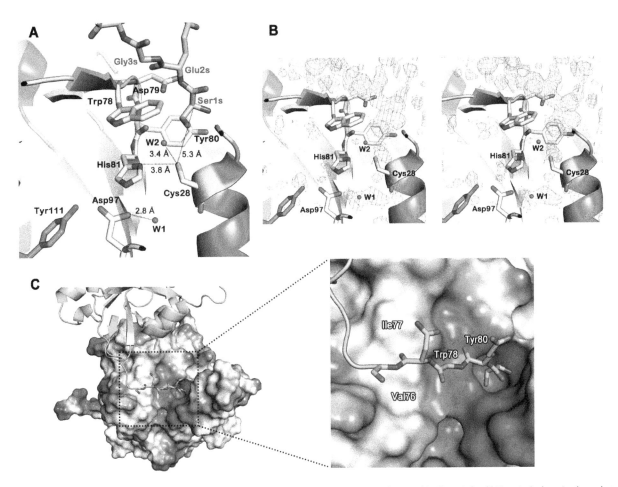

Figure 2. Active site and electrostatic potential surface charge of hNtaq1. (A) Substrate binding cleft of hNtaq1. Carbon in the substrate-mimicking peptide, catalytic triad, α-helices, β-strands, and loops are colored in green, yellow, orange, cyan, and white, respectively. Oxygen, nitrogen, and sulfur atoms are represented as red, blue, and gold, respectively. Two water molecules are shown as red sphere and labeled as W1 and W2. (B) Electron density map from an *Fo–Fc* omit map calculated without the bound substrate-mimicking peptide. Positive electron density are shown as a green mesh contoured at 2.0 σ, in a stereo view. (C) Electrostatic potential surface and substrate binding cleft region of hNtaq1. Negatively and positively charged surfaces are represented as red and blue shade, respectively. Residues interacting with the substrate-mimicking peptide molecule are labeled.

changes that shorten the distance between His81 and Cys28 during a catalytic conversion, the sulfhydryl group of Cys28 is deprived of proton and increases its nucleophilicity for a successive attack on substrates (Figure 2A).

Interestingly, the crystal structure of hNtaq1 revealed an unusual crystallographic contact between symmetry-related molecules. The N-terminus of a symmetry-related monomer is anchored into the active site of hNtaq1 with the side chain of the N-terminal residue (Ser1) in close proximity of the catalytic Cys28. The electron density of the bound N-terminus of a symmetry-related monomer could be clearly observed in an F_o–F_c omit map when calculated without the N-terminus residues as shown in Figure 2B. This phenomenon was similarly observed in the case of crystal structure of another N-end rule pathway protein, UBR box of ubiquitine ligases (PDB ID: 3NIS) [30]. The N-terminus of the symmetry-related monomer (from now on, the "substrate-mimicking" peptide) forms an antiparallel beta-strand segment that adheres to the surface exposed beta-strand of hNtaq1 (residues 76–79, β3 in Figure 1A). The strands are stabilized by several interactions: 1) three peptide backbone hydrogen bonds typical for beta-sheet structures (two between Ile77 and Asn4s, and one between Asp79 and Glu2s); 2) a hydrogen bond between side

chain of Asp79 and peptide backbone of the N-terminal residue Ser1s; 3) a hydrogen bond between Tyr80 and the amino group of the substrate-mimicking peptide; 4) a direct van der Waals contact of Trp78 with the substrate-mimicking peptide. The surface charge of the binding cleft is predominantly negative, and several hydrophobic and aromatic residues (Val76, Ile77, Tryp78, and Tyr80) are well oriented to recognize main chain backbone of the substrate-mimicking peptide and to stabilize aliphatic part of the bound substrate-mimicking peptide (Figure 2C). Overall, the active site bound with the substrate-mimicking peptide directly suggests possible interaction mode of hNtaq1 with substrates.

Docking study with anticipated substrate peptides

In order to further elucidate the interaction between hNtaq1 and anticipated substrate peptides with N-terminal glutamine, we conducted docking experiments with all the possible 400 tripeptides with N-terminal glutamine (Gln-X-X), using AutoDock Vina program [23]. Around the binding cleft, there exist three regions that are predominantly negatively-charged, positively-charged, and nonpolar. The docking study was performed to see how the three regions contributes to the recognition of its anticipated substrate. Control docking experiments performed

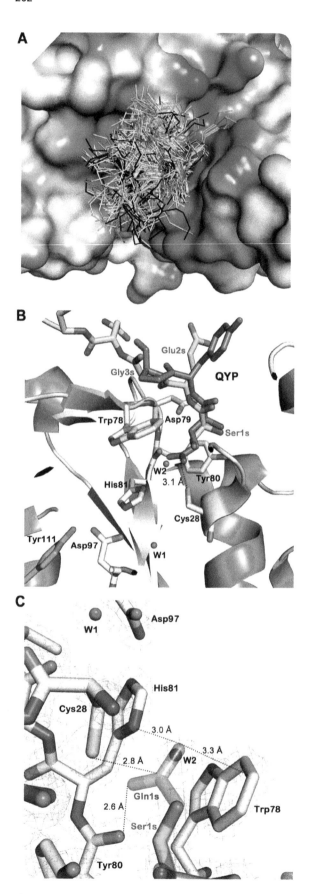

backbone of Ala, Gly, Ile, Leu, Met, Pro, Val tripeptides are colored in yellow, the backbone of Cys, Asn, Gln, Ser, Thr tripeptides are colored in green, the His, Lys, Arg tripeptides are colored in blue, the Phe, Trp, Tyr tripeptides are colored in black, and the Asp, Glu tripeptides are colored in orange. (B) The nearest tripeptide in docking study. The substrate-mimicking peptide is shown as green and predicted docking tripeptide Gln-Tyr-Pro is colored in magenta. (C) Binding mode of refined Ser1Gln hNtaq1 mutant on electron density map of hNtaq1. Carbon in substrate-mimicking peptide, catalytic triad, β-strands, loops, and Ser1Gln mutant are colored in green, yellow, cyan, white, and blue, respectively. Oxygen, nitrogen, and sulfur are represented as red, blue, and gold, respectively. Two water molecules are shown as red sphere and electron density map is represented as gray mesh contoured at 2.0 σ.

with the three N-terminal residues, Ser-Glu-Gly, showed similar binding poses to that from the crystal structure of hNtaq1 bound with the substrate-mimicking peptide. In the control docking study, stabilization energy of the best predicted binding mode was −4.8 kcal/mol. In the experimental docking study, the average stabilization energy of 400 docking results was 5.0 kcal/mol similar to that of the control docking. From the docking results, most stable tripeptides were QWF, QWW, QWQ, QHW, QWV, QQW, QHF, QGW, QEW, and QWS in the order of stabilization energy ranging from −5.8 to −5.3 kcal/mol. Backbones of all the 400 tripeptides seemed to be recognized by the predominantly negatively-charged patch around the substrate binding cleft, which confers nonspecific interaction with substrates regardless of side chains of second and thereafter residues or with very minor influences (Figure 3A). In the case of the closest tripeptide from the catalytic triad, Gln-Tyr-Pro, C_δ of glutamine residue is located 3.15 Å away from sulfhydryl group of Cys28 in the catalytic triad and stabilization energy was −4.6 kcal/mol (Figure 3B).

Interestingly, when we refined our structure with N-terminal Ser1Gln mutation to get a clue of reaction mechanism, modified first serine residue and water2 exactly overlapped with mutated glutamine in the refined structure, the actual substrate of hNtaq1, thus they are reminiscent of an actual substrate (Figure 3C). Sulfur of catalytic Cys28 is 2.8 Å away from the C_δ of Gln1s and amide nitrogen of Gln1s is 3.0 Å away from the spot on His81, optimal distance for protonation of a leaving amine group. Carbonyl oxygen of Gln1s seems to lock the N-terminal glutamine via a hydrogen bond to peptide carbonyl of Tyr80 (2.6 Å). We suggest that coordinates of Ser1 and water2 in our crystal structure mimic the pose of an N-terminal glutamine and this binding mode would represent the substrate binding step in hNtaq1 mechanism.

Proposed mechanism of hNtaq1

Based on our crystal structure of hNtaq1 and docking study with all the possible anticipated substrate tripeptides, we suggest a catalytic mechanism of hNtaq1 as shown in Figure 4. In the first step, nucleophilic sulfhydryl group of Cys28 approaches C_δ of the amide group of the N-terminal glutamine and becomes deprotonated by His81 as shown in Figure 4 step 1. The sulfhydryl group of Cys28 plays a crucial role in the nucleophilic attack on acyl group in the N-terminal glutamine side chain of substrates, which results in formation of a tetrahedral intermediate (Figure 4 step 2). Asp97 facilitates the process by forming a hydrogen bond and electrostatic interactions with His81. The ammonia is released upon productive collapse of the tetrahedral intermediate and a water molecule enters the active site cavity and attacks S-acyl intermediate to convert glutamine to a glutamate (Figure 4 step 3 and 4). As the final step, the glutamate side chain is cleaved from S-acyl of Cys28 (Figure 4 step 5 and 6). In these steps, His81 first

Figure 3. Docking study and suggested binding pose of N-terminal glutamine peptide in hNtaq1. (A) Tripeptides docking study of hNtaq1. Categorized by features of its second residue, the

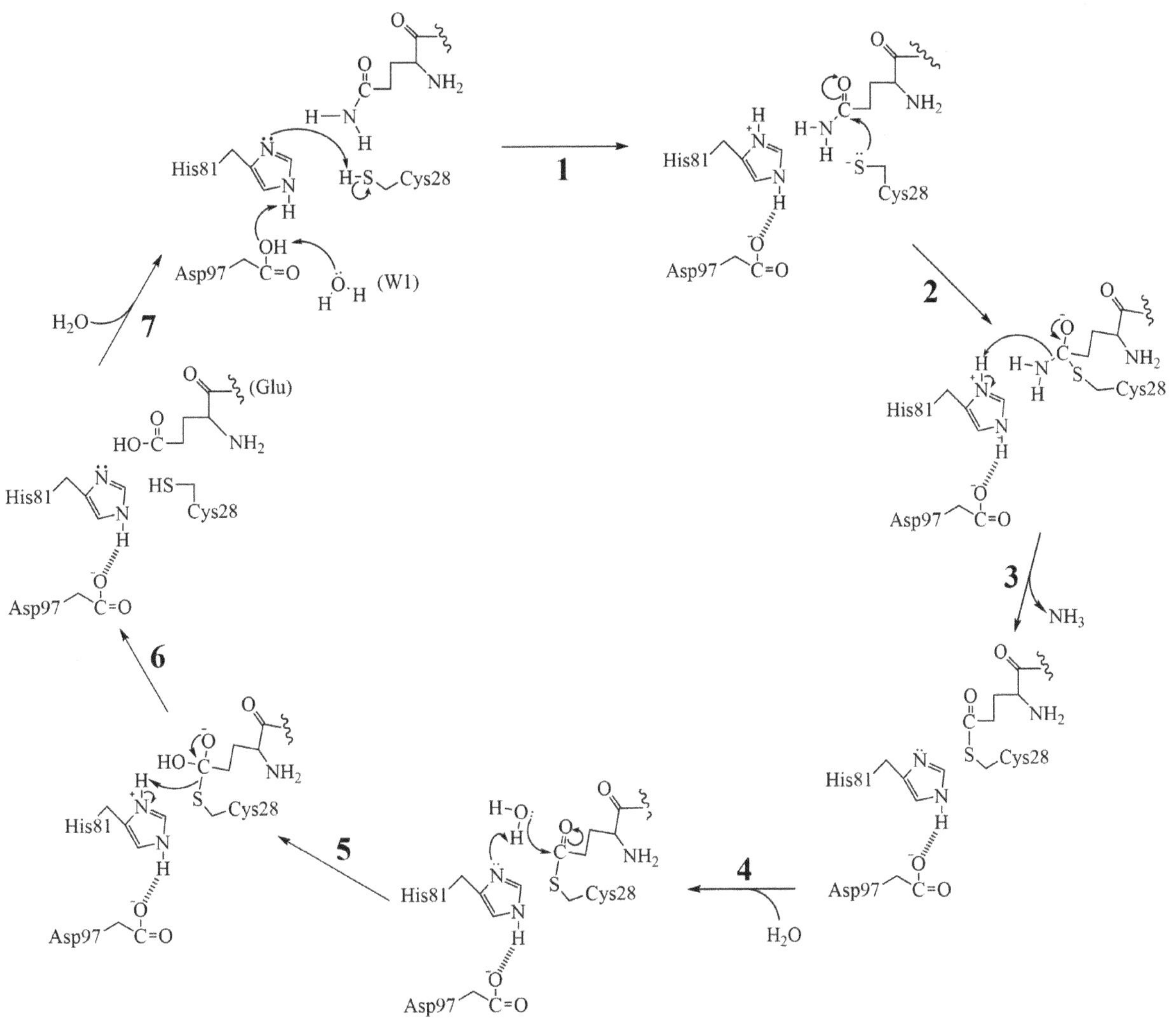

Figure 4. Proposed catalytic mechanism of hNtaq1.

acts as a general base activation water for a nucleophilic attack on the S-acyl intermediate, and then upon collapse of the tetrahedral intermediate acts a general acid to protonate the leaving group, i.e. the thiolate of Cys28. The substrate peptide with newly formed N-terminal glutamate is released from the binding cleft at this stage and the enzyme is ready for another round of catalysis (Figure 4 step 7). The proposed reaction mechanism of hNtaq1 and structural information from our study will provide valuable information for understanding the N-end rule pathway and the interaction between hNtaq1 and its protein substrate.

Acknowledgments

The authors thank all members of the Center for Eukaryotic Structural Genomics for various important contributions. GM/CA@APS has been funded in whole or in part with Federal funds from the National Cancer Institute (ACB-12002) and the National Institute of General Medical Sciences (AMG-12006). This research used resources of the Advanced Photon Source, a U.S. Department of Energy (DOE) Office of Science User Facility operated for the DOE Office of Science by Argonne National Laboratory under Contract No. DE-AC02-06CH11357.

Author Contributions

Conceived and designed the experiments: MSP EB CAB HK BWH GNP. Performed the experiments: MSP EB CAB KRK MDM. Analyzed the data: MSP EB CAB KRK MDM HK BWH GNP. Contributed reagents/materials/analysis tools: CAB MDM. Contributed to the writing of the manuscript: MSP EB CAB KRK MDM HK BWH GNP.

References

1. Mogk A, Schmidt R, Bukau B (2007) The N-end rule pathway for regulated proteolysis: prokaryotic and eukaryotic strategies. Trends Cell Biol 17: 165–172.
2. Lecker SH, Goldberg AL, Mitch WE (2006) Protein degradation by the ubiquitin-proteasome pathway in normal and disease states. J Am Soc Nephrol 17: 1807–1819.
3. Knop M, Schiffer HH, Rupp S, Wolf DH (1993) Vacuolar/lysosomal proteolysis: proteases, substrates, mechanisms. Curr Opin Cell Biol 5: 990–996.
4. Bachmair A, Finley D, Varshavsky A (1986) In vivo half-life of a protein is a function of its amino-terminal residue. Science 234: 179–186.
5. Tasaki T, Kwon YT (2007) The mammalian N-end rule pathway: new insights into its components and physiological roles. Trends Biochem Sci 32: 520–528.
6. Baker RT, Varshavsky A (1995) Yeast N-terminal amidase. A new enzyme and component of the N-end rule pathway. J Biol Chem 270: 12065–12074.
7. Kwon YT, Balogh SA, Davydov IV, Kashina AS, Yoon JK, et al. (2000) Altered activity, social behavior, and spatial memory in mice lacking the NTAN1p

amidase and the asparagine branch of the N-end rule pathway. Mol Cell Biol 20: 4135–4148.

8. Wang H, Piatkov KI, Brower CS, Varshavsky A (2009) Glutamine-specific N-terminal amidase, a component of the N-end rule pathway. Mol Cell 34: 686–695.

9. Hwang CS, Shemorry A, Varshavsky A (2009) Two proteolytic pathways regulate DNA repair by cotargeting the Mgt1 alkylguanine transferase. Proc Natl Acad Sci U S A 106: 2142–2147.

10. Eisele F, Wolf DH (2008) Degradation of misfolded protein in the cytoplasm is mediated by the ubiquitin ligase Ubr1. FEBS Lett 582: 4143–4146.

11. Kwon YT, Xia Z, An JY, Tasaki T, Davydov IV, et al. (2003) Female lethality and apoptosis of spermatocytes in mice lacking the UBR2 ubiquitin ligase of the N-end rule pathway. Mol Cell Biol 23: 8255–8271.

12. Ditzel M, Wilson R, Tenev T, Zachariou A, Paul A, et al. (2003) Degradation of DIAP1 by the N-end rule pathway is essential for regulating apoptosis. Nat Cell Biol 5: 467–473.

13. Thao S, Zhao Q, Kimball T, Steffen E, Blommel PG, et al. (2004) Results from high-throughput DNA cloning of Arabidopsis thaliana target genes using site-specific recombination. J Struct Funct Genomics 5: 267–276.

14. Sreenath HK, Bingman CA, Buchan BW, Seder KD, Burns BT, et al. (2005) Protocols for production of selenomethionine-labeled proteins in 2-L polyethylene terephthalate bottles using auto-induction medium. Protein Expr Purif 40: 256–267.

15. Jeon WB, Aceti DJ, Bingman CA, Vojtik FC, Olson AC, et al. (2005) High-throughput purification and quality assurance of Arabidopsis thaliana proteins for eukaryotic structural genomics. J Struct Funct Genomics 6: 143–147.

16. Studier FW (2005) Protein production by auto-induction in high density shaking cultures. Protein Expr Purif 41: 207–234.

17. Bricogne G, Vonrhein C, Flensburg C, Schiltz M, Paciorek W (2003) Generation, representation and flow of phase information in structure determination: recent developments in and around SHARP 2.0. Acta Crystallogr D Biol Crystallogr 59: 2023–2030.

18. Cowtan KD, Zhang KY (1999) Density modification for macromolecular phase improvement. Prog Biophys Mol Biol 72: 245–270.

19. Emsley P, Cowtan K (2004) Coot: model-building tools for molecular graphics. Acta Crystallogr D Biol Crystallogr 60: 2126–2132.

20. Murshudov GN, Vagin AA, Dodson EJ (1997) Refinement of macromolecular structures by the maximum-likelihood method. Acta Crystallogr D Biol Crystallogr 53: 240–255.

21. Laskowski RA, Moss DS, Thornton JM (1993) Main-chain bond lengths and bond angles in protein structures. J Mol Biol 231: 1049–1067.

22. Lovell SC, Davis IW, Arendall WB 3rd, de Bakker PI, Word JM, et al. (2003) Structure validation by Calpha geometry: phi, psi and Cbeta deviation. Proteins 50: 437–450.

23. Trott O, Olson AJ (2010) AutoDock Vina: improving the speed and accuracy of docking with a new scoring function, efficient optimization, and multithreading. J Comput Chem 31: 455–461.

24. Morris GM, Huey R, Lindstrom W, Sanner MF, Belew RK, et al. (2009) AutoDock4 and AutoDockTools4: Automated docking with selective receptor flexibility. J Comput Chem 30: 2785–2791.

25. The PyMOL Molecular Graphics System, Version 1.5.0.4 Schrödinger, LLC.

26. Holm L, Rosenstrom P (2010) Dali server: conservation mapping in 3D. Nucleic Acids Res 38: W545–549.

27. Hashizume R, Maki Y, Mizutani K, Takahashi N, Matsubara H, et al. (2011) Crystal structures of protein glutaminase and its pro forms converted into enzyme-substrate complex. J Biol Chem 286: 38691–38702.

28. Bhaskaran SS, Stebbins CE (2012) Structure of the catalytic domain of the Salmonella virulence factor SseI. Acta Crystallogr D Biol Crystallogr 68: 1613–1621.

29. Chatterjee D, Boyd CD, O'Toole GA, Sondermann H (2012) Structural characterization of a conserved, calcium-dependent periplasmic protease from Legionella pneumophila. J Bacteriol 194: 4415–4425.

30. Choi WS, Jeong BC, Joo YJ, Lee MR, Kim J, et al. (2010) Structural basis for the recognition of N-end rule substrates by the UBR box of ubiquitin ligases. Nat Struct Mol Biol 17: 1175–1181.

18

GroupRank: Rank Candidate Genes in PPI Network by Differentially Expressed Gene Groups

Qing Wang[1], Siyi Zhang[1], Shichao Pang[1], Menghuan Zhang[1], Bo Wang[1], Qi Liu[1,2,3*], Jing Li[1,4*]

1 Department of Bioinformatics & Biostatistics, School of Life Science and Biotechnology, Shanghai Jiao Tong University, Shanghai, China, 2 Department of Biomedical Informatics, Vanderbilt University School of Medicine, Nashville, Tennessee, United States of America, 3 Center for Quantitative Sciences, Vanderbilt University School of Medicine, Nashville, Tennessee, United States of America, 4 Shanghai Center for Bioinformation Technology, Shanghai, China

Abstract

Many cell activities are organized as a network, and genes are clustered into co-expressed groups if they have the same or closely related biological function or they are co-regulated. In this study, based on an assumption that a strong candidate disease gene is more likely close to gene groups in which all members coordinately differentially express than individual genes with differential expression, we developed a novel disease gene prioritization method GroupRank by integrating gene co-expression and differential expression information generated from microarray data as well as PPI network. A candidate gene is ranked high using GroupRank if it is differentially expressed in disease and control or is close to differentially co-expressed groups in PPI network. We tested our method on data sets of lung, kidney, leukemia and breast cancer. The results revealed GroupRank could efficiently prioritize disease genes with significantly improved AUC value in comparison to the previous method with no consideration of co-exprssed gene groups in PPI network. Moreover, the functional analyses of the major contributing gene group in gene prioritization of kidney cancer verified that our algorithm GroupRank not only ranks disease genes efficiently but also could help us identify and understand possible mechanisms in important physiological and pathological processes of disease.

Editor: Raya Khanin, Memorial Sloan Kettering Cancer Center, United States of America

Funding: The authors acknowledge the support by National Natural Science Foundation of China (31271416, 31000582, J1210047), National Key Basic Research Program (2011CB910204). Additional support from Pujiang Talent Program (12PJ1406600) and Program for "Chen Xing" Young Scholars, Shanghai Jiao Tong University. The funders had no role in study design, data collection and analysis, decision to publish, or preparation of the manuscript.

Competing Interests: The authors have declared that no competing interests exist.

* Email: qi.liu@vanderbilt.edu (QL); jing.li@sjtu.edu.cn (JL)

Background

It remains a big challenge to detect associations between diseases and genes although many disease candidate genes haven been reported through genetic studies such as linkage analysis [1] and association studies [2]. Prioritizing genes according to their likelihood of being disease genes using computational methods can help biologists find the most promising candidate genes for further downstream verification. Many tools have been developed, most of which use a guilt-by-association concept that ranks highest candidate genes similar to known disease genes.

Among them, Endeavour is a well-developed tool that ranks the candidates against the profile of the training set of genes known to be involved in a biological process or a disease of interest, combining 20 data sources such as functional annotations, expression data, regulatory information, literature, pathways, interactions, sequence, and disease probabilities [3,4]. In a variety of data sources, fast accumulating protein-protein interaction (PPI) data is a valuable resource for gene prioritization because the genes tend to be highly connected in the protein-protein interaction network when they are related to a specific biological function or similar disease phenotype [5]. Some tools have been developed to perform gene prioritization using this network and have performed well, including CGI [6], GeneWanderer [7], and DIR [8]. For example, comprising the interactions from HPRD [9], BIND [10], BioGrid [11], IntAct [12] and DIP [13],

GeneWanderer ranks candidate genes using a global network distance measure and random walk analysis for the definition of similarities to known disease genes in protein-protein interaction networks.

But the gene prioritization methods that measure the similarities to known disease genes by guilt-by-association or network distance cannot be applied accurately for a rare or even unknown disease gene. Recently, some efforts have been made to combine PPI network and global gene expression to conduct gene prioritization, the assumption of which is that nodes neighboring to differentially expressed genes are disease gene candidates [14,15]. The advantage of this kind of methods is that no prior knowledge about the biological process or disease genes is needed as a training set. However, we found that there is a risk of high false positive rates and low robustness when candidate genes are close to only a single gene with dramatic change in expression.

Genes usually show co-expression if they have the same or a closely related biological function or are co-regulated by the same transcript factor. In order to prioritize disease genes more precisely and robustly, we proposed a new algorithm called GroupRank to rank disease genes by integrating PPI network and gene groups clustered by coordinately differential expression. Our assumption is that, as well as differentially expressing in cases and controls, a strong disease gene candidate is more likely close to gene groups in which all members coordinately differentially express than to individual ones. To verify this assumption and evaluate the

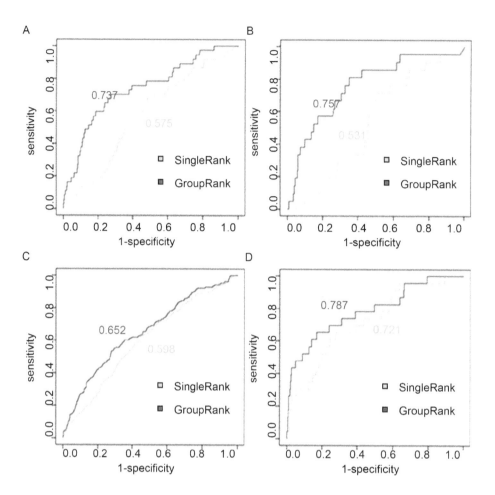

Figure 1. Mean rank ratio of GroupRank using different distance thresholds. The gene groups in GroupRank are partitioned based on a distance threshold with a gradient from 0.1 to 0.9. From A to D, the cancer types are lung cancer, kidney cancer, leukemia and breast cancer.

performance of our method, we applied GroupRank into the gene expression datasets of four cancer types including lung, kidney, leukemia and breast cancer.

Materials and Methods

Gene expression data collection

Four microarray gene expression datasets of humans in case-control design were downloaded from the NCBI Gene Expression Omnibus (GEO) [16] for lung cancer (GSE12428), kidney cancer (GSE6344), leukemia (GSE10631), and breast cancer (GSE29270). All these datasets were curated and reported in the GEO Datasets (GDS). More details about these datasets were summarized in Table S1.

Cancer gene list

We collected disease genes of lung, kidney, leukemia, and breast cancer respectively from OMIM [17] and Cancer Gene Census [18] (see Table S2). The OMIM database provides the connections between genes and lots of diseases. Cancer Gene Census is an ongoing effort to catalogue those genes for which mutations have been causally implicated in cancer.

PPI network

We used HINT as a protein-protein interaction network that is a database of high-quality protein-protein interactions in different

organisms (http://hint.yulab.org/) [19]. These PPI links have been compiled from different sources and then filtered both systematically and manually to remove erroneous and low-quality interactions. There are 27493 binary and 7629 co-complex interactions in HINT for *H.sapiens*.

Differential expression analysis

The statistical analysis of gene differential expression was computed by Student t-test and Bonferroni correction was applied. Only the genes having a corrected p-value less than 0.05 remained in the following gene grouping.

Gene grouping

In GroupRank, we first clustered the differentially expressed genes into the co-expressed groups. We defined the distance between two genes by $d = 1 - |Cor_{ij}|$. Here Cor_{ij} represents the Pearson correlation coefficient of the expression of gene i and gene j. Then, hierarchical clustering was applied to partition the differentially expressed genes into groups. The sizes and the number of groups are changed by adjusting the threshed of gene distance d within a group from 0 to 1.

Performance measurement

We measured the performance of ranking algorithms using the method described by Zhao *et al* (2011) [15]. Briefly, for a known disease gene in a candidate gene set of size N, if the predicted ranking position is r, then the rank ratio r/N may reflect how well this gene is ranked as a disease gene by our algorithm. Lower rank ratio represents better predictive performance. Optimized parameters could be determined through minimizing the average rank ratio of all known disease genes. In addition, we applied the receiver operating characteristics (ROC) analysis [18] [20] to evaluate the overall performance.

Algorithm of GroupRank

First, we defined the similarity matrix of genes by adopting *discrete diffusion kernel* from the Diffusion Rank algorithm reported by Yang et al. [21].

As described in Pinta [14], the transition probability matrix W of a random walk on a given graph G is defined as $W = D^{-1}A$. A is the adjacency matrix and D is the diagonal matrix of G. Consider $L = I - W$, and then we obtain the similarity matrix of genes

$$S = (I + \frac{-\alpha}{N}L)^N \tag{1}$$

where parameter α is the diffusion rate, and N is the number of iterations. In this paper, we set $\alpha = 0.5$ and $N = 3$ as the previous studies found that few iterations is sufficient to reach a considerably good performance [14,22].

Then, from the genes differentially expressed in cancer and normal control, we classify them into co-expressed gene groups by hierarchical clustering. When ranking a candidate gene using a gene group G_i in the PPI network, we define the rank score of a candidate gene obtained from group G_i as

$$r_i = e * s^{(1/log_2^{(1+n)})} \tag{2}$$

where s is the similarity score between the candidate gene and group G_i, which is measured with the geometric mean of the values in the similarity matrix S between the candidate gene and each member in group G_i. Parameter e represents the differential expression level of the gene group, which is computed by the geometric mean of log2 ratio (cancer/control) of each gene within the group G. n is the group size.

In the analysis of the active gene subnetwork of disease, highly connected nodes are often penalized and the size of the subnetwork is controlled [23]. To avoid bias and control possible false positives in the gene ranking that result from either the super group containing large numbers of gene members or the extremely high degree of the candidate gene itself as a hub in the PPI network, we adopted the method of Gaire et al [23] and added adjustable penalization parameters n_0 and k_0 into the following modified formula (3):

$$r_i = \frac{e * s^{(1/log_2^{(1+n)})}}{1 + log_{n_0}^n * log_{k_0}^k} \tag{3}$$

k is the degree of the candidate gene in network HINT. The smaller n_0 and k_0 are, the more stringent penalization is carried to the hub genes and the co-expressed group with super-size. Since the mean degree of network HINT is 6.7, we set $k_0 = 15$, and $n_0 = 20$ as default values in this paper.

Finally, the integrated ranking score of a candidate gene contributed by all gene groups is calculated as

$$r = \sum r_i \tag{4}$$

Results and Discussion

Performance evaluation

We tested GroupRank in four cancer related microarray datasets (lung, kidney, leukemia, and breast cancer) individually. Mean rank ratio (MRR) of known disease genes predicted by our algorithm was used to evaluate its overall performance. By adjusting the threshold of distance d from 0 to 1 with the gradient of 0.01 in defining a gene group, the best MRR was obtained when an optimized threshold was chosen (Table 1). In the results, the best thresholds of distance for different cancer types fell into 0.2–0.6 (Figure 1). A possible explanation is that using a more rigorous threshold, there are not enough effective groups that can be formed, while all genes are possibly classified into very few super groups with poor correlations if a more relaxed threshold is applied.

We compared the performance of GroupRank with the previous similar method that ranks candidate genes based on individual expressed genes in the PPI network [14]. To distinguish from GroupRank, we called this method SingleRank. As the results show in Figure2, the MRR of known disease genes predicted using GroupRank was lower than SingleRank in each testing dataset when a fixed distance threshold of 0.5 was used. We ran a paired Wilcoxon test and revealed that the improvement of MRR brought by GroupRank algorithm was significant (p-value< 0.001).

Additionally we plotted ROC curves to compare GroupRank and SingleRank. Figure 3 shows that GroupRank achieved AUC values from 0.65 to 0.80 in four cancers, which were higher than the values from SingleRank. The results suggest that GroupRank using co-expressed gene groups is a more efficient approach than the method simply using the individual genes in disease gene ranking. It implies that gene prioritization by gene groups could reduce noise and achieve better accuracy.

Table 1. MRR of GroupRank in four cancer types.

Cancer type	Lung cancer	Kidney cancer	Leukemia	Breast cancer
MRR	0.257	0.227	0.298	0.192
Optimized distance	0.42	0.60	0.52	0.24

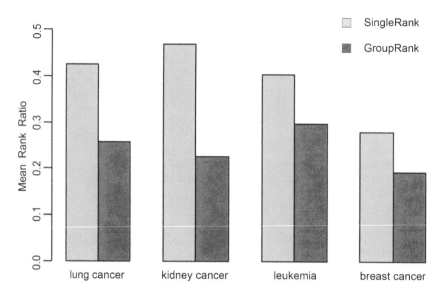

Figure 2. MRR comparisons of GroupRank and SingleRank. The colored bar chart shows the mean rank ratio (MRR) in disease gene ranking using GroupRank (red) and SingleRank (green). It indicates that GroupRank performs better with a lower MRR (p-value<0.001).

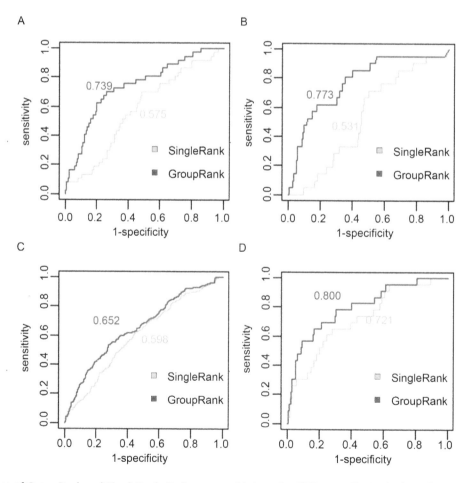

Figure 3. ROC curves of GroupRank and SingleRank. Performance validation using ROC curves. The AUC values of GroupRank and SingleRank achieved in each cancer type are labeled. From A to D, the cancer types are lung cancer, kidney cancer, leukemia and breast cancer.

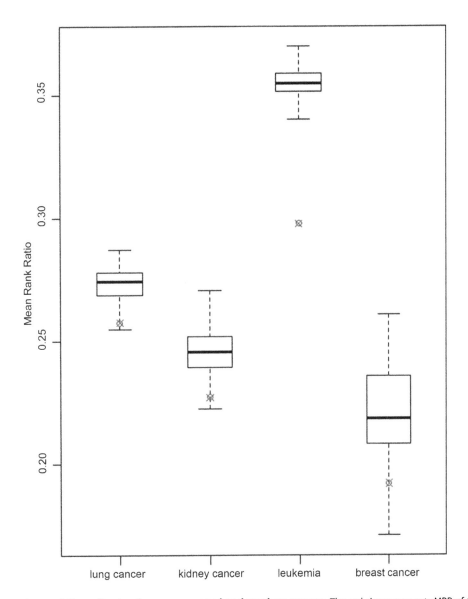

Figure 4. MRR Comparisons of GroupRank using co-expressed and random groups. The red sign represents MRR of GroupRank using co-expressed gene groups in four cancers. Boxplots show the distributions of MRRs using random groups of the same size. The random sampling was repeated 1000 times in each cancer type.

Grouping Efficiency by co-expression

In the GroupRank algorithm, we assumed that the differentially co-expressed gene groups are surrounding a good disease gene and thus are effective to rank disease gene candidates. In order to validate this assumption, we compared the ranking performance using co-expressed and random gene groups. The random groups having the same size were generated by randomly sampling from the PPI network. We repeated the sampling 1000 times. The results indicate that, in all four cancers we studied, the mean rank ratios using co-expressed groups are significantly better than using random gene groups (p-value<0.05) (see Figure 4). It suggested that the downstream genes of a strong disease gene tend to be co-expressed into a number of groups.

Major contributing groups in gene ranking

In the GroupRank algorithm, the co-expressed gene groups comprising the most significantly changed gene members in

cancers and normal controls must play major roles in cancer. Looking at it from another angle, further study on those major contributing groups can help us to explore and understand why a candidate gene is listed in the top rank and which pathway or biological process is influenced by this disease gene candidate in the disease condition. In this paper, kidney cancer was taken as an example, and we investigated the gene groups, especially the major contributing groups in the ranking of the top 20 gene candidates and 21 known kidney cancer genes. As illustrated in Figure 5, based on the accumulated contributions in ranking scores of known tumor genes using GroupRank, four gene groups emerged by explaining 64.7% of the ranking scores of all 21 known kidney cancer genes. We found that the top 20 ranked genes also had strong connections with those four groups. That indicates that these four gene groups are closely related with kidney cancer. We did GO enrichment analysis of these groups using WebGestalt [24] and found that these gene groups, which were differentially expressed in kidney cancer, are involved in cell proliferation,

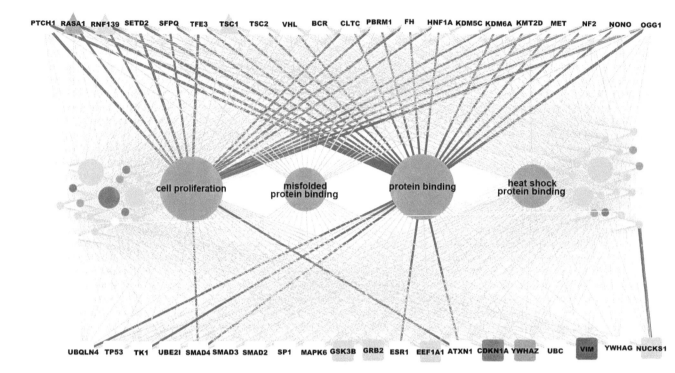

Figure 5. Schematic graph of gene ranking of kidney cancer using GroupRank. The graph illustrates gene ranking of kidney cancer using the algorithm GroupRank. The triangle nodes at the top represent known kidney cancer genes and the square nodes at the bottom represent the top 20 ranked genes of kidney cancer using GroupRank. The circle nodes in middle represent the co-expressed gene groups used to rank disease gene candidates. A known or putative cancer gene is connected with a gene group if it contributes more than 5% of the summed ranking score of this cancer gene. The width of the edge linked to a disease gene is proportional to the scoring contribution obtained from the corresponding gene group. The edges explaining more than 20% of the ranking score of the cancer gene candidate are highlighted in dark blue. The edge is colored in light blue if the scoring contribution of the gene group is from 15% to 20%. The darker node color indicates higher fold change at expression level in cancer and normal control. The size of the circle node representing gene group was proportional to its accumulated contribution in ranking scores of all known kidney cancer genes. The enriched functional annotation is labeled on each of the four major contributing gene groups.

protein binding, misfolded protein binding, and heat shock protein binding respectively (p-value<0.05, bonferroni multiple testing adjustment). It was reported by Short *et al.* (1993) that enhanced cell proliferation occurs at several stages of renal tumorigenesis [25]. Heat shock proteins (Hsps) are overexpressed in a wide range of human cancers and are implicated in tumor cell proliferation, differentiation, invasion, metastasis, death, and recognition by the immune system [26]. Misfolded proteins were also reported in the study of cancer, and targeted degradation of misfolded proteins has become one of the promising new therapeutic approaches in the treatment of cancer [27].

Conclusion

In this study, by combining PPI network and gene differential expression and co-expression data, we proposed a new algorithm GroupRank, in which disease candidate genes were ranked by the surrounding differentially co-expressed gene groups in PPI network. The results demonstrated that GroupRank could improve the accuracy of disease gene prioritization significantly. Furthermore, the further functional analysis of the major

contributing groups in ranking may not only help us predict disease gene candidates but also improve the biological interpretation of data.

Supporting Information

Table S1 The list of microarray gene expression datasets. (DOC)

Table S2 Cancer gene list. (DOC)

Acknowledgments

The authors wish to thank Margot Bjoring for editorial work on this Manuscript.

Author Contributions

Conceived and designed the experiments: QW QL JL. Performed the experiments: QW SZ SP BW. Analyzed the data: QW BW MZ. Wrote the paper: QW QL JL.

References

1. Kruglyak L, Daly MJ, Reeve-Daly MP, Lander ES (1996) Parametric and nonparametric linkage analysis: a unified multipoint approach. Am J Hum Genet 58: 1347–1363.

2. Klein RJ, Zeiss C, Chew EY, Tsai JY, Sackler RS, et al. (2005) Complement factor H polymorphism in age-related macular degeneration. Science 308: 385–389.

3. Aerts S, Lambrechts D, Maity S, Van Loo P, Coessens B, et al. (2006) Gene prioritization through genomic data fusion. Nat Biotechnol 24: 537–544.

4. Tranchevent LC, Barriot R, Yu S, Van Vooren S, Van Loo P, et al. (2008) ENDEAVOUR update: a web resource for gene prioritization in multiple species. Nucleic Acids Res 36: W377–384.

5. Gandhi TK, Zhong J, Mathivanan S, Karthick L, Chandrika KN, et al. (2006) Analysis of the human protein interactome and comparison with yeast, worm and fly interaction datasets. Nat Genet 38: 285–293.

6. Ma X, Lee H, Wang L, Sun F (2007) CGI: a new approach for prioritizing genes by combining gene expression and protein-protein interaction data. Bioinformatics 23: 215–221.

7. Kohler S, Bauer S, Horn D, Robinson PN (2008) Walking the interactome for prioritization of candidate disease genes. Am J Hum Genet 82: 949–958.

8. Chen Y, Wang W, Zhou Y, Shields R, Chanda SK, et al. (2011) In silico gene prioritization by integrating multiple data sources. PLoS One 6: e21137.

9. Peri S, Navarro JD, Kristiansen TZ, Amanchy R, Surendranath V, et al. (2004) Human protein reference database as a discovery resource for proteomics. Nucleic Acids Res 32: D497–501.

10. Bader GD, Betel D, Hogue CW (2003) BIND: the Biomolecular Interaction Network Database. Nucleic Acids Res 31: 248–250.

11. Stark C, Breitkreutz BJ, Reguly T, Boucher L, Breitkreutz A, et al. (2006) BioGRID: a general repository for interaction datasets. Nucleic Acids Res 34: D535–539.

12. Hermjakob H, Montecchi-Palazzi L, Lewington C, Mudali S, Kerrien S, et al. (2004) IntAct: an open source molecular interaction database. Nucleic Acids Res 32: D452–455.

13. Xenarios I, Rice DW, Salwinski L, Baron MK, Marcotte EM, et al. (2000) DIP: the database of interacting proteins. Nucleic Acids Res 28: 289–291.

14. Nitsch D, Tranchevent LC, Goncalves JP, Vogt JK, Madeira SC, et al. (2011) PINTA: a web server for network-based gene prioritization from expression data. Nucleic Acids Res 39: W334–338.

15. Zhao J, Yang TH, Huang Y, Holme P (2011) Ranking candidate disease genes from gene expression and protein interaction: a Katz-centrality based approach. PLoS One 6: e24306.

16. Edgar R, Domrachev M, Lash AE (2002) Gene Expression Omnibus: NCBI gene expression and hybridization array data repository. Nucleic Acids Res 30: 207–210.

17. Hamosh A, Scott AF, Amberger JS, Bocchini CA, McKusick VA (2005) Online Mendelian Inheritance in Man (OMIM), a knowledgebase of human genes and genetic disorders. Nucleic Acids Res 33: D514–517.

18. Futreal PA, Coin L, Marshall M, Down T, Hubbard T, et al. (2004) A census of human cancer genes. Nat Rev Cancer 4: 177–183.

19. Das J, Yu H (2012) HINT: High-quality protein interactomes and their applications in understanding human disease. BMC Syst Biol 6: 92.

20. Fawcett T (2006) An introduction to ROC analysis. Pattern Recognition Letters 27: 861–874.

21. Yang H, King I, Lyu MR (2007) DiffusionRank: a possible penicillin for web spamming. Proceedings of the 30th annual international ACM SIGIR conference on Research and development in information retrieval. Amsterdam, The Netherlands: ACM. 431–438.

22. Francisco AP, Goncalves JP, Madeira SC, Oliveira AL. Using personalized ranking to unravel relevant regulations in the saccharomyces cerevisiae regulatory network; 2009. 3–6.

23. Gaire RK, Smith L, Humbert P, Bailey J, Stuckery PJ, Haviv I (2013). Discovery and analysis of consistent active subnetworks in cancers. BMC Bioinformatics 2013, (Suppl 2): S7.

24. Zhang B, Kirov S, Snoddy J (2005) WebGestalt: an integrated system for exploring gene sets in various biological contexts. Nucleic Acids Res 33: W741–748.

25. Short BG (1993) Cell proliferation and renal carcinogenesis. Environ Health Perspect 101 Suppl 5: 115–120.

26. Ciocca DR, Calderwood SK (2005) Heat shock proteins in cancer: diagnostic, prognostic, predictive, and treatment implications. Cell Stress Chaperones 10: 86–103.

27. Kirkin V, McEwan DG, Novak I, Dikic I (2009) A role for ubiquitin in selective autophagy. Mol Cell 34: 259–269.

19

Significant Low Prevalence of Antibodies Reacting with Simian Virus 40 Mimotopes in Serum Samples from Patients Affected by Inflammatory Neurologic Diseases, Including Multiple Sclerosis

Elisa Mazzoni[1], Silvia Pietrobon[1], Irene Masini[1], John Charles Rotondo[1], Mauro Gentile[3], Enrico Fainardi[2], Ilaria Casetta[3], Massimiliano Castellazzi[3], Enrico Granieri[3], Maria Luisa Caniati[4], Maria Rosaria Tola[4], Giovanni Guerra[5], Fernanda Martini[1]*, Mauro Tognon[1]*

1 Department of Morphology, Surgery and Experimental Medicine, Section of Pathology, Oncology and Experimental Biology, University of Ferrara, Ferrara, Italy, 2 Unit of Neuroradiology, University Hospital of Ferrara, Ferrara, Italy, 3 Biomedical Sciences and Specialized Surgeries, Section of Neurology, School of Medicine, University of Ferrara, Ferrara, Italy, 4 Unit of Neurology, University Hospital of Ferrara, Ferrara, Italy, 5 Clinical Laboratory Analysis, University Hospital of Ferrara, Ferrara, Italy

Abstract

Many investigations were carried out on the association between viruses and multiple sclerosis (MS). Indeed, early studies reported the detections of neurotropic virus footprints in the CNS of patients with MS. In this study, sera from patients affected by MS, other inflammatory (OIND) and non-inflammatory neurologic diseases (NIND) were analyzed for antibodies against the polyomavirus, Simian Virus 40 (SV40). An indirect enzyme-linked immunosorbent assay (ELISA), with two synthetic peptides, which mimic SV40 antigens, was employed to detect specific antibodies in sera from patients affected by MS, OIND, NIND and healthy subjects (HS). Immunologic data indicate that in sera from MS patients antibodies against SV40 mimotopes are detectable with a low prevalence, 6%, whereas in HS of the same mean age, 40 yrs, the prevalence was 22%. The difference is statistically significant ($P = 0.001$). Significant is also the difference between MS vs. NIND patients (6% vs. 17%; $P = 0.0254$), whereas no significant difference was detected between MS vs OIND (6% vs 10%; $P > 0.05$). The prevalence of SV40 antibodies in MS patients is 70% lower than that revealed in HS.

Editor: Steven Jacobson, National Institutes of Health, United States of America

Funding: This study was sponsored, in part, by grants from The University of Ferrara, FAR Projects, University Hospital of Ferrara, Regione Emilia Romagna, ERMES, MS project to E.G., F.M. and M.T., Fondazione Cassa di Risparmio di Ferrara to E.G., and Fondazione Cassa di Risparmio di Cento to F.M. and M.T. Italy. The funders had no role in study design, data collection and analysis, decision to publish, or preparation of the manuscript.

Competing Interests: The authors have declared that no competing interests exist.

* Email: mrf@unife.it (FM); tgm@unife.it (MT)

Introduction

Multiple sclerosis (MS) is a chronic human demyelinating disease of the central nervous system (CNS) characterized by an autoimmune pathogenic process in genetically predisposed individuals [1,2]. Although the etiology of MS is unknown, genetic and environmental factors seem to play an important role.

Accumulating data, including animal study models, human models of virus inducing demyelination, epidemiologic and laboratory findings, have demonstrated that viruses and host genetic factors can interact to cause immune-mediated demyelination [3,4]. Infectious agents are environmental factors potentially involved in the MS onset. Specifically, ubiquitous viruses were found associated with the development or exacerbation of MS, including the *herpesviruses* (i) *Epstein-Barr virus* (EBV) [5,6,7,8], (ii) *human herpesvirus 6* (HHV-6) [9] and *human endogenous retrovirus* (HERV) families [10]. Although the current evidence supports a strong association between EBV and MS, the potential causality of this herpesvirus remains to be established [11].

EBV has been investigated for its putative role in the MS onset. Earlier studies found a higher prevalence of anti-EBV antibodies in MS patients compared to controls [7,8]. At present, it cannot be excluded that the abnormal response to EBV infection in MS patients is a consequence, rather than a cause. It has been reported that EBV cannot alone trigger the MS onset [7]. Further molecular evidences are needed to assess the real involvement of EBV in the MS onset.

HHV-6 strains A/B has been proposed as viral agents involved in several autoimmune disorders (AD), including MS. HHV-6A could participate in neuro-inflammation in the context of MS by promoting inflammatory processes through CD46 binding [9].

HERV-Fc1, which sequences map in chromosome X, has been associated with MS, mostly in Northern European populations. Association of the HERV-Fc1 polymorphism rs391745 with bout-onset MS susceptibility was also confirmed in Southern European cohorts [10].

Polyomaviruses, including Simian Virus 40 (SV40) [12,13] have been poorly investigated for their putative role in MS disease [14].

SV40, a monkey neurotropic polyomavirus, is responsible for the progressive multifocal leukoencephalopathy (PML) in immune-compromised macaques [15,16] while in humans its footprints have been detected in brain tumors and neurologic disorders [17,18,19].

Recently, the development of specific and sensitive serologic test for SV40 has been reported, which consists of an indirect ELISA employing synthetic peptides as mimotopes/antigens of SV40 viral capsid proteins (VPs). This immunologic assay was used to detect specific serum antibodies against SV40 VPs in normal individuals of different age [20,21,22]. Higher prevalence of SV40 antibodies was detected in oncologic patients affected by glioblastoma multiforme (GBM) [19], whereas SV40 sequences and large T antigen expression were detected in human brain tumors [23,24,25].

The complex interactions among the CNS, multiple infections with different infectious agents occurring in the periphery or within the CNS, and the immune response should be analyzed and elucidated in order to understand the etiology of MS.

The objective of the present study was to investigate whether serum samples from patients affected by MS, other OIND, NIND and HS carry SV40-antibodies. Sera were analyzed by an indirect ELISA employing synthetic peptides as mimotopes belonging to the viral capsid proteins (VPs).

Results

Low prevalence of antibodies reacting with SV40 mimotopes in serum samples from multiple sclerosis patients

Serum samples from MS patients were analyzed by indirect ELISA for the presence of IgG class antibodies against SV40 VP mimotopes/epitopes (Tables 1 and 2). The indirect ELISA was employed to test serum samples of MS affected patients (mean age = 37 yrs), which had been diluted at 1/20, for reactivity to SV40 epitopes from VP1, VP1 B peptide. Serum samples reacting with the SV40 VP1 B mimotopes reached an overall prevalence of 9%. Then, the same assay was addressed to detect IgG class serum antibodies against SV40 VP2/3 epitopes, which are known as VP2/3 C peptide. Serum samples reacted with the SV40 VP2/3 C peptide with a similar prevalence, 13%, as had been detected previously for the VP1 B peptide. Conversely, seronegative samples for the SV40 VP1 B peptide failed to react with SV40 VP2/3 C epitopes. The exceptions were negligible represented by a few serum samples, which were negative for VP1 B peptide, while testing positive for VP2/3 C peptide, and vice-versa. The difference was not statistically significant (P>0.05) (Table 1). The different prevalence of responses/OD reading of B and C peptides is probably due to the different immunogenicity of the two SV40 antigens. Indeed, the B peptide is present in the VP1 virion, in its pentameric form, 360 times, whereas the C peptide of VP2/3 is present in the virion 72 times [20].

As published before [19,20,21,22] in indirect ELISAs the human peptide hNPS which is unrelated to SV40, was employed as a negative control peptide to verify if, in our experimental conditions, a non-specific reaction may occur with SV40-positive and SV40-negative human and rabbit serum samples. Data indicate that this negative control peptide does not react with SV40-positive and SV40-negative sera. The OD value was usually in the range of 0.088-0.098, which is consistent with the OD background of both human and rabbit sera [19,20,21,22].

The two indirect ELISAs, with the two distinct VP B and C peptides gave overlapping results, thus confirming the presence of anti-SV40 VPs antibodies, although at low prevalence, in human sera from patients affected by MS (Tables 1 and 2).

In our investigation only those samples found positive for both B and C peptides were considered SV40-positive (Tables 1 and 2).

Altogether, our immunologic data indicate that combining the SV40-positive sera (6/93), both for the VP1 B and VP2/3 C peptides, the overall prevalence was 6% (Tables 2).

Control samples represented by sera from healthy subjects, with a similar mean age of MS patients, 40 yrs, were investigated with the same ELISA. Immunologic results indicated a prevalence 22% in these healthy subjects (HS1) (Table 2). These results are in agreement with previous immunologic data, obtained with other cohorts of HS, with the same mean age [20]. Indeed, ELISA data indicate that the prevalence of SV40 antibodies is similar in the cohorts of individuals with the same age and gender, despite coming from different regions in Italy. However, a decline of prevalence was detected in elderly [21].

MS and HS1 serologic profiles of serum antibody reactivity to SV40 mimotopes are shown in Figure 1.

SV40-positive sera, which were selected by the neutralization assay as reported elsewhere [19,20] tested by indirect ELISA diluted at 1/20 had a general cut-off, by spectrophotometric reading, in the range of 0.18 OD. This cut-off represents in our ELISA the value that discriminates SV40-negative (OD <0.18) from SV40-positive samples (OD>0.18).

ELISA data obtained with serum samples from OIND, NIND patients and HS

In the second step of this investigation, indirect ELISA was employed to test serum samples from OIND and NIND affected patients, and healthy subjects (HS2), with a similar mean age of these patients (51 yrs). Sera had been diluted at 1/20, for reactivity to SV40 epitopes from VP1, VP1 B peptide and then from VP 2/3, C peptide. Serum samples, which reacted with both SV40 VP1 B and VP2/3 mimotopes reached an overall prevalence of 10% in OIND, 17% in NIND, whereas the control samples represented by sera from HS2 the prevalence was 21%.

Serologic data obtained with the cohorts of HS1 and HS2 are in agreement with those reported before with other cohorts of HS, with the same mean age [20].

It is interesting to note that SV40 peptide C prevalence in these samples is very similar to the data obtained with the same samples for SV40 peptide B. OIND, NIND and HS2 serologic profiles of serum antibody reactivity to SV40 mimotopes are shown in Figure 1.

Discussion

SV40-antibody detection had been attempted in several studies using serologic methods with SV40 antigens, but due to the high homology among the three main Polyomaviruses (SV40, BKV and JCV), the results were always affected by some cross-reactivity [14,17,26,27,28,29].

Specific immunologic assays for the identification of SV40-seropositive individuals/patients and serum antibody reactivity to SV40 antigens are of paramount importance in revealing the prevalence of SV40 infection in patients and normal subjects.

In this study, serum samples from patients affected by neurologic diseases, including MS, together with those from healthy individuals employed as controls, were analyzed for exposure to SV40 infection. The indirect ELISA employed in this study used specific SV40 antigens/mimotopes. Indeed, the synthetic peptides from SV40 VP 1–3 employed mimic the corresponding epitopes of the viral capsid peptides [19,20,21,22].

Table 1. MS, OIND and NIND patients and healthy subjects.

Serum Subject (N)	Female (%)	Mean yrs ± SD (range)	Subtype	Number Of Subtype	SV40-positive sample/ sample analyzed (%)
MS	66	37±11.2	RR	78	6/78 (7)
(93)		(13–68)	SP	10	0/10
			PP	5	0/5
OIND	50	53±14.2	Inflammatory Demyelinating	34	2/34(6)
(77)		(18–83)	Meningitis/Encephalitis/Myelitis	25	3/25(12)
			Mononeuritis	4	0/4
			Comnnectivitis/Vasculitis	13	3/13(23)
			PML	1	0/1
NIND	50	52±16.8	ALS	8	2(25)
(81)		(21–85)	Dementia	6	0/6
			MSA	1	0/1
			Arteriovenous malformation	3	1/3(33)
			Migraine	10	5/10(50)
			Toxic encephalopathy	2	0/2
			Epilepsy	4	0/4
			Hereditary ataxia	3	1/3(33)
			Brain Tumor	8	0/8
			Hydrocephalus	2	0/2
			Spondylotic myelopathy	4	0/4
			Peripheral neuropathy	9	2/9(22)
			Pseudotumor cerebri	2	1/2(50)
			Funicular myelopathy	1	0/1
			Stroke/TIA	18	2/18(11)
HS1	67	40±11.7	Healthy subjects	180	40/180(22)
(180)		(17–67)			
HS2	50	51±11.7	Healthy subjects	160	34/160(21)
(160)		(18–81)			

MS (Multiple Sclerosis); OIND (Other Inflammatory Neurologic Disease); NIND (Non Inflammatory Neurologic Disease); HS (Healthy Subjects); RR (Relapsing-Remitting); SP (Secondary-Progressive); PP (Primary-Progressive); PML (Progressive Multifocal Leucoencephalopathy); ALS (Amyotrophic Lateral Sclerosis); MSA (Multiple System Atrophy); TIA (Transient Ischemic Attack). Among SV40-positive patients, 2 OIND patients were found to be affected by Inflammatory Demyelinating, 3 patients were affected by Meningitis/Ancephalitis/Myelitis and 3 by Comnnectivitis/Vasculitis. Among NIND patients SV40-positive were 2 patients affected by Amyotrophic Lateral Sclerosis (ALS), 1 by Arteriovenous Malformation, 1 by Hereditary Ataxia, 2 by Peripheral Neuropathy, 1 by Pseudotumor Cerebri and 2 by Transient Ischemic Attacks (TIA). The three MS patients were affected by relapsing remitting MS forms.

Table 2. Prevalence of immunoglobulin G antibodies reacting with Simian Virus 40 (SV40) viral protein (VP) mimotopes in serum samples of patients affected by MS, OIND, NIND and HS^.

Serum^ sample	Number of patients/subject	Female %	Number of positive samples (%)		
			VP B	VP C	VP B+C
MS	93	66	8 (9)	12 (13)	6 (6)*
HS1	180	67	63 (35)	52 (29)	40 (22)
OIND	77	50	20 (26)	11 (14)	8 (10)**
NIND	81	50	16 (20)	16 (20)	14 (17)
HS2	160	52	45 (28)	40 (25)	34 (21)

^Human sera were from patients affected by multiple sclerosis (MS), other inflammatory neurologic diseases (OIND), non-inflammatory neurologic diseases (NIND) and healthy subjects (HS1), (HS2).
*The prevalence of SV40 antibodies in MS patients is statistically significant lower that those detected in NIND patients ($P = 0.0254$) and in HS1 ($P = 0.001$), whereas no significant was the different prevalence detected in MS and OIND patients ($P > 0.05$).
** The prevalence of SV40 antibodies in OIND patients is statistically significant lower that those detected in HS2 ($P = 0.0403$). The different prevalence of SV40 antibodies between the cohorts of NIND patients was not significant compared with the HS2 ($P > 0.05$). Statistical analysis was performed using the χ^2 test.

A

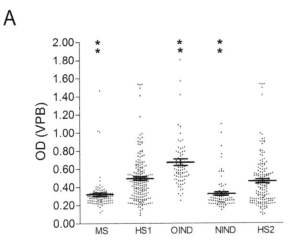

B

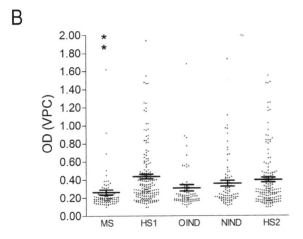

C

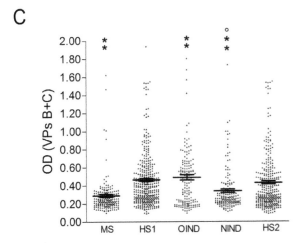

Figure 1. Serologic profile of serum antibody reactivity to SV40 mimotopes VP1 B (A) and VP2/3 C (B) and VPs B+C (C). Immunologic data are from serum samples from Multiple Sclerosis patients (MS), Healthy Subjects (HS1), Other Inflammatory Neurologic Diseases (OIND), Non-Inflammatory Neurologic Diseases (NIND) and Healthy subjects (HS2). Results are presented as values of optical density (OD) readings at λ 405 nm, of serum samples diluted at 1:20, detected in indirect ELISA. In scatter dot plotting, each plot represents

the dispersion of OD values to a mean level indicated by the line inside the scatter with Standard Error Mean (SEM) for each group of subjects analyzed. A) The mean OD of sera (VP B ± Std Error) in MS (0.32±0.02) were lower than that in HS1 (0.50±0.02) and in OIND (0.67±0.04). Moreover the mean OD of sera in OIND were higher than that in NIND (0.32±0.02) and HS2 (0.46±0.02). The mean OD of sera in NIND were lower than that in HS2. B) The mean OD of sera (VP C ± Std Error) in MS (0.26±0.03) were lower than that in HS1 (0.43±0.02). C) The mean OD of sera (VPs B+C ± Std Error) in MS (0.29±0.02) were lower than that in HS1 (0.46±0.02) and in OIND (0.49±0.03). Moreover, the mean OD of sera in OIND were higher than that in NIND (0.34±0.02). The mean OD in NIND were lower than that in HS2 (0.43±0.02) and in OIND sera. Statistical analysis was performed using Anova and Newman-Keuls Comparison test. (**P<0.001; °P<0.01).

Our immunologic data suggest that specific SV40 antibodies are detectable in human serum samples from neurologic patients and healthy individuals.

The prevalence of SV40 antibodies observed in Italian patients affected by MS and OIND, revealed a lower prevalence of SV40 antibodies in these patients respect to controls represented by NIND patients and HS. Indeed, our serologic assays indicated that the prevalence of SV40 antibodies in serum samples of patients affected by MS (6%) is lower than that determined in HS1 (22%) (P = 0.001) and in NIND patients (17%) (P = 0.0254). Similarly, the prevalence of SV40 antibodies determined in OIND patients (10%) is lower than that determined in HS2 (21%) (P = 0.04).

It should be noted that MS patients included in our study, at the time of the serum collection, were not subjected to any immuno-modulatory therapy. This data indicates that the low prevalence of SV40 antibodies is not due to the immune-modulatory therapy.

It has been proposed that viruses may play a role in MS pathogenesis acting as environmental triggers. However, it remains to be elucidated whether viruses are causal agents of the MS onset.

Our data represent the first study indicating a significant low prevalence of serum antibodies against SV40 in patients affected by inflammatory neurologic diseases, including MS. One may speculate that these patients, with specific impairments of their immune system, are poor responders to SV40 infection because unable to present these polyomaviral antigens. In this peculiar condition of the host, SV40 escaping from the immune surveillance could exerts its pathologic effect in human oligoden-drocytes. Indeed, SV40 is responsible in immunodepressed/suppressed macaques of the PML, another inflammatory neuro-logic disease related to the myelin degradation [15,16]. It should be noted that SV40 in HIV-positive or AIDS patients does not increase its viral load indicating that human cells are only semi-permissive to its multiplication, whereas the inflammatory molecules do not efficiently reactivate this polyomavirus [30,31,32].

A recent investigation reported that the IgG levels in MS patients against distinct herpesviruses do not differ from that of controls, whereas a higher IgG prevalence against EBV was detected [5]. Our immunologic data suggest that a specific dysregulation in the IgG response to SV40 in multiple sclerosis and OIND patients may occur. At present, the nature of this immunologic defect is not known. It has been reported that the cronic inflammation, which characterize these patients, is respon-sible of several immune impairments [33]. The low level of SV40 antibodies detected in serum samples of MS patients could be related to the cronic inflammatory conditions of these patients that may affect the response/stability of SV40 antibody over time.

Alternatively, the low prevalence of SV40 antibodies detected in these patients indicates that they are less prone to the SV40 infection, or a closely related yet unknown human polyomavirus.

Additional studies are needed to verify if SV40 plays a role in the MS onset, such as the evaluation of the viral DNA load, virus isolation, HLA characterization of the host and IgG isotype analysis. These investigations are feasible and they will be part of our next investigation.

Materials and Methods

Human Samples

A total of 591 serum samples were collected at the University Hospital of Ferrara, Department of Neurology and Clinical Laboratory Analysis. Human sera were from discarded clinical laboratory analysis samples, anonymously collected, coded with indications of age, gender and pathology, if any. Sera from MS patients (n = 93), as well as from other inflammatory neurologic (OIND) (n = 77), and non-inflammatory neurologic diseases (NIND) (n = 81) affected patients and healthy subjects (HS) with a different mean age, HS1, HS2 (n = 180, mean age = 40 yrs; n = 160, mean age = 51 yrs, respectively) were analyzed for IgG antibodies reacting to SV40 viral capsid protein (VP) mimotopes, represented by synthetic peptides, which mimic VP antigens (Table 1). To this purpose indirect ELISA was set up employing synthetic peptides, which correspond to SV40 VP 1-2-3 mimotopes, together with an unrelated human peptide, hNPS, employed as a negative control of the serum reactivity [34]. Informed written consent was obtained by patients/individuals. The project was approved by the County Ethical Committee, Ferrara, Italy.

Synthetic Peptides

Computer assisted analyses allowed us to select 2 specific SV40 peptides, from the late viral region by comparing the three capsid proteins, VP 1-2-3 from SV40, with the amino acids of the human BK (BKV) and JC (JCV) polyomaviruses which are highly homologous to SV40, as well as with other, less homologous polyomaviruses [20,22]. Previous ELISA results indicated that the two SV40 peptides did not cross-react with the BKV and JCV hyperimmune sera that were employed as controls [20,22]. The two peptides belong to the VP1/VP2/VP3 viral capsid proteins (VP1 ID: 1489598; VP3 ID: 9486895; web site, http://www.ncbi. nlm.nih.gov/nuccore). The amino acid sequences of the two peptides, known as VP1 B and VP2/3 C, respectively, are as follows:

VP1 B: NH2- NPDEHQKGLSKSLAAEKQFTDDSP- COOH
VP2/3 C: NH2- IQNDIPRLTSQELERRTQRYLRD- COOH

VP1 B and VP2/3 C mimotopes were selected as they react specifically in indirect ELISA with the rabbit hyperimmune serum that had been experimentally immunized with SV40 (positive control serum). SV40-positive and SV40-negative human sera were also employed as controls. These SV40 control sera, selected by the neutralization assay, were from our collections [19,20,21,22] (see below the other technical details). BKV and JCV hyperimmune rabbit sera did not react with VP1 B or VP2/3 C peptides (negative control sera). The amino acid residues of the two specific SV40 VP peptides show low homology with the BKV, JCV and other polyomaviruses VPs [20,22]. The characterization of these two peptides were published before [20,22]. The synthetic peptides were synthesized using standard procedures and were purchased from UFPeptides s.r.l., Ferrara, Italy [20].

Indirect Enzyme-Linked Immunosorbent Assay (ELISA)

Indirect ELISA was developed and standardized to detect specific antibodies against SV40 in human sera using VP1 B VP 2/3 C synthetic peptides [20,22]. *Peptide coating.* Plates were coated with 5 µg of the selected peptide for each well and diluted in 100 µl of Coating Buffer, pH 9.6 (Candor Bioscience, Germany). *Peptide blocking.* Blocking was made with 200 µl/well of the Blocking Solution, containing the casein (Candor Bioscience, Germany) at 37°C for 90 min. *Primary antibody adding.* Different wells were covered with 100 µl containing the following sera: positive-control, represented by immune rabbit serum containing anti-SV40 antibodies, negative controls represented by immune sera anti-BKV and anti-JCV, and three human serum samples, which were found to be SV40 negative in our previous investigations [20,22]. Sera under analysis were diluted at 1:20 in Low Cross-Buffer, pH 7.2 (Candor Bioscience, Germany). *Secondary antibody adding.* The solution contained a goat anti-human IgG heavy and light chain specific peroxidase-conjugate (Calbiochem-Merck, Germany) diluted 1:10,000 in Low Cross-Buffer. *Dye treatment and spectrophotometric reading.* Samples were treated with 100 µl of 2,2'-azino-bis 3-ethylbenzthiazoline-6-sulfonic acid (ABTS) solution (Sigma-Aldrich, Milan) and then read on the spectrophotometer (Thermo Electron Corporation, model Multiskan EX, Finland) at a wavelength (λ) of 405 nm. This approach detects the color intensity in wells where the immuno-complexes were formed by optical density (OD). *Cut-off determination.* The cut-off point was determined in each assay by an OD reading of three negative controls, that were added to the standard deviation and multiplied three times (+3SD). The choice of using the mean plus 3 times the standard deviation is to eliminate false positive samples and to increase the specificity of the ELISA test.

Sera with antibodies against SV40 were considered VP-positive upon reacting to both peptides of the late region and when sera that had been analyzed three times, with independent experiments, by indirect ELISA testing gave the same positive result.

SV40 specificity of the indirect ELISA employing synthetic peptides which mimic the VPs antigens

In previous investigations, comparative computer assisted analyses by BLAST program were carried out with the SV40 VP peptides B and C and the corresponding amino acid (a.a.) sequences of the new human polyomaviruses (HPyV) and hundreds of different BKV and JCV serotypes [20,22]. Results indicate a low homology for the BKV and JCV prototypes and other polyomaviruses. [20,22]. Indirect ELISA data indicate that the two SV40 peptides B and C did not cross-react with the BKV and JCV hyperimmune sera (negative controls), as described before [20,22]. Briefly, hyperimmune sera against SV40 and BKV were obtained in rabbits that had been inoculated with purified viral stocks as previously reported [20]. The serum against JCV was kindly provided by Dr. Major, NIH, Bethesda (MD), U.S.A [20,22]. The immune serum anti-BKV was titered using a hemagglutination inhibition (H.A.I.) test employing human erythrocytes group 0, Rh+. Anti SV40 serum was titered by neutralization assay [20].

SV40 VP1 B and VP2/3 C mimotopes were selected as they react specifically in indirect ELISA with the rabbit hyperimmune serum that had been experimentally immunized with SV40 (positive control) [20,22], and with SV40-positive human sera. These SV40-positive human sera, from our collection, were analyzed before by SV40 neutralization assay [19,20]. The human peptide hNPS, a.a. sequence SFRNGVGTGMKKTSFQRAKS [34] was employed as a negative control peptide [20,22].

Serum samples tested by indirect ELISA diluted at 1/20 were considered SV40-positive when above OD = 0.17–0.19, according to the spectrophotometric reading. Indeed, this cut-off point represents the value that discriminates SV40-negative (sample with OD below 0.17–0.19) from SV40-positive samples (OD above 0.17–0.19). The positive controls, represented by the SV40 hyperimmune rabbit serum, had an OD of up to 1.8, while the two JCV and BKV hyperimmune rabbit sera, which were employed as negative controls, had an OD of less than 0.1. In each ELISA plate 3 SV40-positive and 3 SV40-negative human sera, from our collections [19,20,21,22] were also added. The selection of these human sera was based on their SV40-neutralization activity [19,20].

Statistical analyses

All analyses were performed by Prism 4.0 (GraphPad software). For all tests, we considered P values < 0.05 to be statistically significant. To determine significances between two groups we used two-sided chi-square test. The serologic profile of serum antibody reactivity to SV40 mimotopes was statistically analyzed using Anova and Newman-Keuls Comparison test.

Acknowledgments

The authors thank Dr. Eugene O. Major, the Laboratory of Molecular Medicine and Neuroscience, the National Institute of Neurological Disorders and Stroke, Bethesda, MD, for the hyperimmmune serum against JCV, and Dr. Kamel Khalili, Department of Neuroscience, Center for Neurovirology, Temple University School of Medicine, Philadelphia, PA, for the JCV viral stock.

The members of ERMeS groups are, in the Region Emilia Romagna Multiple Sclerosis: E Granieri (coordinator), I. Casetta, M. Castellazzi, V. Govoni, R. De Gennaro, E. Groppo, M. Gentile, L. Piccolo, M. Padroni, M. Pastore; M. R. Tola, L. Caniatti, E. Baldi, E. Fainardi (Ferrara); D. Guidetti, P. De Mitri, P. Immovilli, E. Terlizzi (Piacenza); E. Montanari, I. Pesci, B. Allegri, MP (Fidenza); G. Terzano, F. Granella, M. I. Antonelli (Parma); N. Marcello, L. Motti (Reggio Emilia); M. Santangelo, C. M. Stucchi, L. Vaghi (Carpi); P. Nichelli, P. Sola, D. Ferraro, A. M. Simone (Modena); P. Avoni, A. Baldrati, A. Baruzzi, R. D'Alessandro, Michelucci, L. Sabattini, R. F. Salvi, T. Saquegna, C. Scandellari, S. Stecchi (Bologna); V. Mussuto (Imola); W. Neri, P. Prati, S. Strumia (Forlì); F. Rasi, C. Callegarini, M. Galeotti, L. Fiorani, M. Spadoni, (Ravenna, Faenza, Lugo); S. Malagù (Cesena); A. Ravasio, J.Andruccioli (Rimini) and S. Guttmann, C. Monaldini (Repubblica di San Marino).

Funding: Dr. Elisa Mazzoni is a post doctoral fellow of the Fondazione Veronesi, Milan, Italy. This work was supported, in parts, by grants from The University of Ferrara, FAR Projects and Regione Emilia Romagna, ERMES, MS project to E.G., F.M. and M.T., Fondazione Cassa di Risparmio di Ferrara to E.G., and Fondazione Cassa di Risparmio di Cento to F.M. and M.T. Italy.

Author Contributions

N/A. Conceived and designed the experiments: MT FM EG. Performed the experiments: EM SP IM JCR. Analyzed the data: EM SP IM JCR IC MG GG EG EF. Contributed reagents/materials/analysis tools: MC MG EF EG MRT MLC IC GG. Wrote the paper: EM SP MG EG FM MT.

References

1. Granieri E (1997) The epidemiology study of exogenous factors in the etiology of multiple sclerosis. Introduction. Neurology 49: S2:S4.

2. Compston A, Coles A (2002) Multiple sclerosis. Lancet 359: 1221–1231.

3. Giovannoni G, Cutter GR, Lunemann J, Martin R, Munz C, et al. (2006) Infectious causes of multiple sclerosis. Lancet neurology 5: 887–894.

4. Giovannoni G, Ebers G (2007) Multiple sclerosis: the environment and causation. Current opinion in neurology 20: 261–268.

5. Myhr KM, Riise T, Barrett-Connor E, Myrmel H, Vedeler C, et al. (1998) Altered antibody pattern to Epstein-Barr virus but not to other herpesviruses in multiple sclerosis: a population based case-control study from western Norway. Journal of neurology, neurosurgery, and psychiatry 64: 539–542.

6. Mameli G, Cossu D, Cocco E, Masala S, Frau J, et al. (2013) EBNA-1 IgG titers in Sardinian multiple sclerosis patients and controls. Journal of neuroimmunology 264: 120–122.

7. Pakpoor J, Giovannoni G, Ramagopalan SV (2013) Epstein-Barr virus and multiple sclerosis: association or causation? Expert review of neurotherapeutics 13: 287–297.

8. Kvistad S, Myhr KM, Holmoy T, Bakke S, Beiske AG, et al. (2014) Antibodies to Epstein-Barr virus and MRI disease activity in multiple sclerosis. Multiple sclerosis.

9. Broccolo F, Fusetti L, Ceccherini-Nelli L (2013) Possible role of human herpesvirus 6 as a trigger of autoimmune disease. TheScientificWorldJournal 2013: 867389.

10. de la Hera B, Varade J, Garcia-Montojo M, Alcina A, Fedetz M, et al. (2014) Human endogenous retrovirus HERV-Fc1 association with multiple sclerosis susceptibility: a meta-analysis. PloS one 9: e90182.

11. Libbey JE, Cusick MF, Fujinami RS (2013) Role of Pathogens in Multiple Sclerosis. International reviews of immunology.

12. Barbanti-Brodano G, Martini F, De Mattei M, Lazzarin L, Corallini A, et al. (1998) BK and JC human polyomaviruses and simian virus 40: natural history of infection in humans, experimental oncogenicity, and association with human tumors. Adv Virus Res 50: 69–99.

13. Barbanti-Brodano G, Sabbioni S, Martini F, Negrini M, Corallini A, et al. (2006) BK virus, JC virus and Simian Virus 40 infection in humans, and association with human tumors. Adv Exp Med Biol 577: 319–341.

14. Ribeiro T, Fleury MJ, Granieri E, Castellazzi M, Martini F, et al. (2010) Investigation of the prevalence of antibodies against neurotropic polyomaviruses BK, JC and SV40 in sera from patients affected by multiple sclerosis. Neurol Sci 31: 517–521.

15. Horvath CJ, Simon MA, Bergsagel DJ, Pauley DR, King NW, et al. (1992) Simian virus 40-induced disease in rhesus monkeys with simian acquired immunodeficiency syndrome. The American journal of pathology 140: 1431–1440.

16. Kaliyaperumal S, Dang X, Wuethrich C, Knight HL, Pearson C, et al. (2013) Frequent infection of neurons by SV40 virus in SIV-infected macaque monkeys with progressive multifocal leukoencephalopathy and meningoencephalitis. The American journal of pathology 183: 1910–1917.

17. Barbanti-Brodano G, Sabbioni S, Martini F, Negrini M, Corallini A, et al. (2004) Simian virus 40 infection in humans and association with human diseases: results and hypotheses. Virology 318: 1–9.

18. Martini F, Corallini A, Balatti V, Sabbioni S, Pancaldi C, et al. (2007) Simian virus 40 in humans. Infect Agent Cancer 2: 13.

19. Mazzoni E, Gerosa M, Lupidi F, Corallini A, Taronna AP, et al. (2014) Significant prevalence of antibodies reacting with simian virus 40 mimotopes in sera from patients affected by glioblastoma multiforme. Neuro-oncology 16: 513–519.

20. Corallini A, Mazzoni E, Taronna A, Manfrini M, Carandina G, et al. (2012) Specific antibodies reacting with simian virus 40 capsid protein mimotopes in serum samples from healthy blood donors. Hum Immunol 73: 502–510.

21. Mazzoni E, Tognon M, Martini F, Taronna A, Corallini A, et al. (2013) Simian virus 40 (SV40) antibodies in elderly subjects. J Infect 67: 356–358.

22. Taronna A, Mazzoni E, Corallini A, Bononi I, Pietrobon S, et al. (2013) Serological evidence of an early seroconversion to Simian virus 40 in healthy children and adolescents. PLoS One 8: e61182.

23. Martini F, De Mattei M, Iaccheri L, Lazzarin L, Barbanti-Brodano G, et al. (1995) Human brain tumors and simian virus 40. J Natl Cancer Inst 87: 1331.

24. Martini F, Iaccheri L, Lazzarin L, Carinci P, Corallini A, et al. (1996) SV40 early region and large T antigen in human brain tumors, peripheral blood cells, and sperm fluids from healthy individuals. Cancer Res 56: 4820–4825.

25. Tognon M, Casalone R, Martini F, De Mattei M, Granata P, et al. (1996) Large T antigen coding sequences of two DNA tumor viruses, BK and SV40, and nonrandom chromosome changes in two glioblastoma cell lines. Cancer Genet Cytogenet 90: 17–23.

26. Viscidi RP, Rollison DE, Viscidi E, Clayman B, Rubalcaba E, et al. (2003) Serological cross-reactivities between antibodies to simian virus 40, BK virus, and JC virus assessed by virus-like-particle-based enzyme immunoassays. Clin Diagn Lab Immunol 10: 278–285.

27. Lundstig A, Eliasson L, Lehtinen M, Sasnauskas K, Koskela P, et al. (2005) Prevalence and stability of human serum antibodies to simian virus 40 VP1 virus-like particles. J Gen Virol 86: 1703–1708.

28. Kjaerheim K, Roe OD, Waterboer T, Sehr P, Rizk R, et al. (2007) Absence of SV40 antibodies or DNA fragments in prediagnostic mesothelioma serum samples. Int J Cancer 120: 2459–2465.

29. Kean JM, Rao S, Wang M, Garcea RL (2009) Seroepidemiology of human polyomaviruses. PLoS Pathog 5: e1000363.

30. Martini F, Dolcetti R, Gloghini A, Iaccheri L, Carbone A, et al. (1998) Simian-virus-40 footprints in human lymphoproliferative disorders of HIV- and HIV+ patients. Int J Cancer 78: 669–674.

31. Comar M, Zanotta N, Bonotti A, Tognon M, Negro C, et al. (2014) Increased Levels of C-C Chemokine RANTES in Asbestos Exposed Workers and in

Malignant Mesothelioma Patients from an Hyperendemic Area. PloS one 9: e104848.

32. Comar M, Rizzardi C, de Zotti R, Melato M, Bovenzi M, et al. (2007) SV40 multiple tissue infection and asbestos exposure in a hyperendemic area for malignant mesothelioma. Cancer research 67: 8456–8459.

33. Hoffmann F, Meinl E (2014) B cells in multiple sclerosis: good or bad guys? An article for 28 May 2014 - World MS Day 2014. European journal of immunology 44: 1247–1250.

34. Guerrini R, Salvadori S, Rizzi A, Regoli D, Calo G (2010) Neurobiology, pharmacology, and medicinal chemistry of neuropeptide S and its receptor. Med Res Rev 30: 751–777.

Proteome Folding Kinetics Is Limited by Protein Halflife

Taisong Zou[1], Nickolas Williams[2], S. Banu Ozkan[1], Kingshuk Ghosh[2]*

1 Center for Biological Physics, Department of Physics, Arizona State University, Tempe, Arizona, United States of America, **2** Department of Physics and Astronomy, University of Denver, Denver, Colorado, United States of America

Abstract

How heterogeneous are proteome folding timescales and what physical principles, if any, dictate its limits? We answer this by predicting copy number weighted folding speed distribution – using the native topology – for E.coli and Yeast proteome. E.coli and Yeast proteomes yield very similar distributions with average folding times of 100 milliseconds and 170 milliseconds, respectively. The topology-based folding time distribution is well described by a diffusion-drift mutation model on a flat-fitness landscape in free energy barrier between two boundaries: i) the lowest barrier height determined by the upper limit of folding speed and ii) the highest barrier height governed by the lower speed limit of folding. While the fastest time scale of the distribution is near the experimentally measured speed limit of 1 microsecond (typical of barrier-less folders), we find the slowest folding time to be around seconds (≈ 8 seconds for Yeast distribution), approximately an order of magnitude less than the fastest halflife (approximately 2 minutes) in the Yeast proteome. This separation of timescale implies even the fastest degrading protein will have moderately high (96%) probability of folding before degradation. The overall agreement with the flat-fitness landscape model further hints that proteome folding times did not undergo additional major selection pressures – to make proteins fold faster – other than the primary requirement to "sufficiently beat the clock" against its lifetime. Direct comparison between the predicted folding time and experimentally measured halflife further shows 99% of the proteome have a folding time less than their corresponding lifetime. These two findings together suggest that proteome folding kinetics may be bounded by protein halflife.

Editor: Emanuele Paci, University of Leeds, United Kingdom

Funding: KG acknowledges support from NSF (award number 1149992), and TZ and SBO acknowledge ASU-CLAS funding. The funders had no role in study design, data collection and analysis, decision to publish, or preparation of the manuscript.

Competing Interests: The authors have declared that no competing interests exist.

* Email: kghosh@du.edu

Introduction

Diverse pool of protein sequences give rise to an astonishing degree of heterogeneity in the biophysical properties across the proteome. This raises a fundamental question: how heterogeneous is the proteome? Recent work showed biophysical properties have broad distributions across the proteome and their consequences at the phenotypic level [1–5]. While sequence variation alone would lead to such diverse biophysical properties, there are other features of the cellular environment – for example protein abundance, role of chaperones, co-translational folding – that can further influence these distributions. Protein copy number – although neglected in the earlier calculations of distributions – in particular can play a crucial role due to a possible correlation with biophysical properties such as folding stability [6]. It has been well established that highly abundant proteins are slowly mutating [7,8]. The reason behind this negative correlation is believed to be the selection pressure against cytotoxicity of misfolded proteins arising due to lower stability. Rules of protein biophysics has been used to quantitatively establish the relation between abundance and stability [6,8]. On the other hand, it is believed that there may be a possible correlation between stability and folding speed [9–11]. Thus, it is tempting to hypothesize that protein abundance

and folding speed may be related as well. A natural question arises – how does protein abundance alter, if at all, the folding time distribution? Without *a priori* knowledge of the effect of protein abundance on the folding time distribution, it is imperative that any attempt to predict the folding time distribution of a proteome should consider the effect of abundance as well.

Learning about the extent of heterogeneity in biophysical properties across the proteome in itself is a fundamental question – leading further inquires on the details of the distribution. For example in case of folding time distribution, what are the lower and upper speed limits? What physical principle dictates these limits? What is the peak value, if any, of the distribution? Is there a limiting behavior due to competition with other time scales such as diffusion, protein synthesis, degradation? If kinetic stability [12] – introducing higher barrier height while keeping the same value for the free energy difference between the folded and the unfolded state – is a strategy cells use to minimize exposure to unfolded states to avoid lethal effects of aggregation or degradation [13], do we expect proteomes to be biased towards higher folding times? And if so, how do these timescales compare with protein halflife, in other words is the proteome folding timescale still able to beat the degradation clock with an increased barrier height? While outpacing degradation appears to be important, are there any other selection pressures that may have influenced proteome folding kinetics? Furthermore, how do these distributions vary across different kingdoms of life – for example between Escherichia coli (E.coli) and Yeast – or is there an universality in the shape of the distribution? In this article, we attempt to determine proteome folding kinetics distribution and address some of these fundamental questions.

Materials and Methods

Determining the folding speed of a protein

Plaxco, Simons, Baker [14] made the observation that relative contact order (CO), a metric based on the native topology of the protein, correlates well with the folding speed measured *in vitro*. CO is defined as the average residue separation – normalized by the chain length – of atomic contacts present in the native structure of the protein [14]. Since the pioneering work of Plaxco, Simons, Baker there have been numerous efforts to understand its implication [15] and establish the role of other native-centric metric [16–21] and their relative performances to predict the folding speed of proteins using native structure [18,20,21]. One such effort has shown absolute contact order (ACO) – defined as the product of CO and the chain length – predicts folding speeds more accurately than CO for bigger set of proteins [16]. In a nutshell, all these different metrics provide a prescription to predict the folding speed of a protein with the knowledge of the native structure alone. We utilize this powerful idea to predict the folding time distribution for proteins in the proteome for which the exact (or highly homologous) native structures are known. Recent work by Rustad and Ghosh [21] has provided a first principle explanation – employing polymer physics arguments – for the observed correlation between absolute contact order (ACO) [16] and folding speed. Furthermore, within a perturbative scheme, the work has proposed an extension of the metric (ACO) that captures the effect of different loop topologies [21]. This new metric, minor variation of ACO, provides slight improvement over ACO when benchmarked against the largest set (116 proteins) of *in vitro* folding speed data. We use this new modified metric, instead of ACO, to predict the folding speed from the native structure of the protein. For a given protein, we predict folding speeds for different domains, assuming each domain folds independently. Since the domain with the slowest folding speed is rate limiting, we use the folding speed of the slowest folding domain to be the folding speed of the protein.

Curating the fraction of proteome that have both the structure and abundance data available

In order to predict folding speed, as described above, we need the information about the native structures of proteins in the proteome. We collect proteins from the Yeast and E.coli proteome for which the structures of proteins are available. For the Yeast proteome we use domain assignment from Yeast resource center (YRC) database [22]. Next we perform a BLAST search of the corresponding sequences to identify the best possible match for their structures. We list only those proteins that simultaneously satisfy a minimum of 80% sequence coverage and 50% identity match. In order to predict copy number weighted folding time distribution, we gather proteins for which both the structure and abundance information are available. We cross reference the curated list of proteins with available structure, described above, against the integrated list from PaxDB database [23]. The integrated list is the most comprehensive list of protein abundance values. We choose this list to ensure maximum coverage of proteins from the proteome. This method yields a total of 755

Yeast proteins. For E.coli proteome, we follow a similar approach but use the dataset collected by O'Brien et al. [24]. The original dataset reported in O'Brien et al. categorizes proteins (and their domains) based on a single abundance scale. We cross reference the combined list against the integrated list of abundance from PaxDb [23] yielding a total of 848 E.coli proteins. In summary, our datasets (Table S1 and S2) provide the largest fraction of proteomes (in E.coli and Yeast) for which both the abundance and structural informations are now available.

Results and Discussion

Folding time distribution is heterogeneous

Copy number weighted folding speed (lnk_f, k_f being the folding speed) distributions in E.coli and Yeast show a broad range of folding speeds, from several microseconds^{-1} to minutes^{-1} (Figure 1). The fastest folding time is in the neighborhood of microseconds. This is consistent with studies on ultrafast folding proteins defining the speed limit of protein folding [21,25,26]. It is interesting to note the lower speed limit is of the order of seconds to minutes, in proximity to the scale of halflives of short-lived proteins [27]. The implication of this observation will be discussed in detail in the section below. The average folding time (τ_f) for copy number weighted distribution is calculated as

$$\ln \tau_f \approx -\langle \ln k_f \rangle = -\frac{\sum_i \ln k_{fi} N_i}{\sum_i N_i} \qquad (1)$$

where, k_{fi} and N_i are the folding speed and the copy number, respectively, of the i th protein. Average folding time without accounting for differential protein abundance levels can be obtained by simply setting $N_i = 1$. For E.coli, we find the average is approximately 100 milliseconds for copy number weighted distribution. The average remains almost unaltered when the distribution is not weighted by the protein expression level (i.e. setting $N_i = 1$, distribution not shown here). The average folding time for Yeast proteome is 170 milliseconds and 60 milliseconds for copy number weighted and unweighted distributions, respectively.

Recent work – grounded in the hypothesis of global selection against toxic effect of misfolding explaining observed correlation between abundance and evolution rate [8] – predicts highly abundant proteins are more stable [6]. Given this link between stability-abundance and *possible* interdependence between stability and folding kinetics [9–11], it is natural to expect a possible relation between abundance and folding speed as well. However, based on the results stated above, we do not see any noticeable effect of abundance on folding kinetics in E.coli. A possible explanation, among many other alternative ones, could be that the proteome can not afford to under-express slow folding proteins due to functional reasons. Furthermore, we notice a marginal slowing down of the proteome folding speed in Yeast upon weighting by protein abundance. Given the inherent uncertainties in predicting folding speed from native topology, a three-fold slowing down of the proteome is probably a very weak effect. However, if slowing down of the proteome due to copy number weighting is indeed beyond uncertainty, it may imply slow folding proteins are over-expressed for strong functional reasons despite the threat of misfolding. It may also imply the proteome is equipped with mechanisms such as chaperone-assisted folding, complex chaperone-substrate network [28] to mitigate possible deleterious effects of misfolding due to lower folding speed. As will be seen in later sections, three fold lowering of the speed around 60 millisecond timescale still allows proteins enough time to fold

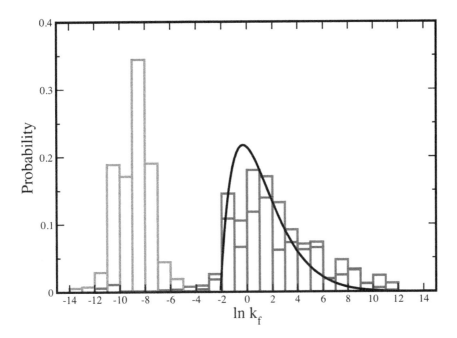

Figure 1. Folding speed (lnk_f) distribution – calculated using native topology – of E.coli (in red) and Yeast (in blue) weighted by protein copy number. The distribution of average lifetime for proteins in Yeast [27] is shown in green. The predicted folding time distribution using a diffusion-drift model (equation 5) with the boundary condition of the maximum folding time of 8 seconds is shown in black. Maximum folding time of 8 seconds was determined by best fitting Yeast distribution.

before degradation. It is interesting to note folding speed distributions in E.coli and Yeast – baring minor variations mentioned above – are very similar, indicating a universal behavior in the folding kinetics.

One caveat of our analysis is that the folding speed is predicted using models that have been benchmarked against *in vitro* folding data. However recent work, although limited, does not show significant differences between folding times measured *in vivo* and *in vitro* [29]. It is also important to note major conclusions remain the same if other metric such as ACO is used to predict the folding speed.

Diffusion-drift model of mutations on a flat-fitness landscape explains the predicted distribution of folding speed

Apart from minor differences in details, the overall shape and the range of the distributions for E.coli and Yeast are roughly similar. The universal distribution (Figure 1) of the folding speed, irrespective of the details of the species, is well explained by a diffusion-drift model of mutations altering folding free energy barrier ($\Delta G^{\dagger}$). Shakhnovich *et al.* [1] used a similar model to describe a universal distribution of stability (ΔG). Due to close analogy between the two models, we briefly describe the stability model first. Further details of the model can be found in the work of Shakhnovich *et al.* [1]. Their model uses diffusion - arising from random mutations - with a drift to explain the stability distribution $P(\Delta G)$. The model also imposes two boundary conditions $P(\Delta G_{min}) = P(\Delta G_{max}) = 0$ at the maximum (ΔG_{max}) and minimum (ΔG_{min}) values of allowed stability. These two constraints can be explained as follows (Figure 2A): from design perspective, it is impossible to make proteins indefinitely stable, thus imposing an upper limit on the stability, hence $P(\Delta G_{max}) = 0$. The boundary condition on the lower limit of stability, on the other hand, arises

from the requirement of minimal stability to avoid misfolding that can be lethal to the phenotype of the organism. The model assumes a flat-fitness landscape for all values of stability greater than the minimum, i.e. $\Delta G > \Delta G_{min}$. The fitness is severely compromised if stability drops below the threshold i.e. $\Delta G < \Delta G_{min}$, imposing the constraint $P(\Delta G_{min}) = 0$. Thus, the fitness landscape is 'step-like' near the threshold (see Figure 2A). The time evolution of the probability distribution of stability in this mutational model with the flat 'step-like' landscape is given by [1]

$$\frac{\partial P}{\partial t} = cP - mh\frac{\partial P}{\partial \Delta G} + \frac{m}{2}(h^2 + D)\frac{\partial^2 P}{\partial(\Delta G)^2};\tag{2}$$

$$p(\Delta G_{max}) = p(\Delta G_{min}) = 0$$

where, c is a constant related to the birth rate of the population, m is the mutation rate per gene (or protein), h and D are the average and variance, respectively, of the distribution of stability changes upon mutation. Formally, $h = \langle \Delta\Delta G \rangle$ and $h^2 + D = \langle(\Delta\Delta G)^2\rangle$, where $\langle...\rangle$ denotes the average over all possible mutations and $\Delta\Delta G = \Delta G_{mutant} - \Delta G_{wt}$. The second derivative in equation 2 describes diffusion, while drift is captured by the first derivative (in the right hand side of the equation). Using the long-time limit solution $P(\Delta G,t) = \exp(\lambda t)P(\Delta G)$ [1], we require the steady state solution to be the eigenfunction of the differential equation

$$-mh\frac{\partial P}{\partial \Delta G} + \frac{m}{2}(h^2 + D)\frac{\partial^2 P}{\partial(\Delta G)^2}\tag{3}$$

subject to the boundary conditions. Thus, the steady state solution – within a normalization constant A – is given by

$$P(\Delta G) = A \exp\left(\frac{h\Delta G}{h^2 + D}\right) \sin\left(\pi \frac{\Delta G - \Delta G_{min}}{\Delta G_{max} - \Delta G_{min}}\right) \quad (4)$$

Noticing one-to-one relation between folding speed (k_f) and barrier height ($\Delta G^\dagger$), we employ similar idea to model the distribution of barrier height to ultimately predict the folding speed distribution. We use the same diffusion-drift model where mutations alter the free energy barrier of folding instead of folding stability. Analogous to the stability model, we impose two boundary conditions, $P(\Delta G^\dagger_{min}) = P(\Delta G^\dagger_{max}) = 0$, at the two extremities of the free energy barrier, $\Delta G^\dagger_{min}$ and $\Delta G^\dagger_{max}$ (see Figure 2B). On one hand it is simply impossible to make proteins that fold faster than the speed limit of folding, setting the lower limit of the barrier $\Delta G^\dagger_{min}$. On the other hand, extremely slow folding proteins – if not folded at birth – even if highly stable will not be able to fold in time before degradation. Stated differently, for functional reasons, proteins would require to fold before their lifetime (inside the cell) expires. Also, slow folding proteins would be a potential hazard due to unfolded-state induced aggregation propensity. This sets a selection pressure against slow folding proteins with extremely high barriers ($\Delta G^\dagger_{max}$). Similar to the stability model, we assume a flat-fitness landscape for $\Delta G^\dagger < \Delta G^\dagger_{max}$, with a severe drop in fitness for $\Delta G^\dagger > \Delta G^\dagger_{max}$ (Figure 2B). In reality, fitness can gradually decrease around the threshold value of $\Delta G^\dagger_{max}$. However, in order to keep the calculation simple and analogous to the work of Shakhnovich et al., we make the simplifying assumption of a 'step-like' fitness function. Thus the model assumes all proteins are subjected to a single global constraint of lifetime implying a single value of $\Delta G^\dagger_{max}$. Noticing the exact analogy between the model for the stability and the barrier height, the predicted distribution for the free energy barrier can be easily obtained by replacing the stability (ΔG) by the barrier height $\Delta G^\dagger$ in equation 4. Thus,

$$P(\Delta G^\dagger) = A \exp\left(\frac{h\Delta G^\dagger}{h^2 + D}\right) \sin\left(\pi \frac{\Delta G^\dagger - \Delta G^\dagger_{min}}{\Delta G^\dagger_{max} - \Delta G^\dagger_{min}}\right) \quad (5)$$

where, A is a normalization constant, $h = \langle \Delta\Delta G^\dagger \rangle$, $h^2 + D = \langle (\Delta\Delta G^\dagger)^2 \rangle$; $\Delta\Delta G^\dagger = \Delta G^\dagger_{mutant} - \Delta G^\dagger_{wt}$, and $\langle ... \rangle$ denotes the average over all possible mutations of barrier height. Three parameters of the model h, D, and $\Delta G^\dagger_{min}$, can be estimated from the literature. From the dataset of 858 mutations across

24 different proteins [30], we find $h = 0.6(k_b T)$ and $h^2 + D = 1.12(k_b T)^2$; k_b is the Boltzmann constant and T is the room temperature.

The lower limit of the barrier is assumed to be zero, $(\Delta G^\dagger_{min} = 0)$, consistent with barrier-less folding proteins that define the speed limit of folding [25,26].

Now we focus on the determination of $\Delta G^\dagger_{max}$. We hypothesize the lower speed limit i.e. the maximum folding time ($t_{f,max}$) – setting the upper limit of folding barrier ($\Delta G^\dagger_{max}$) – has to be less than the protein halflife ($t_{1/2}$). Experimentally reported halflife measures the time scale over which the copy number of a given protein, upon inhibition of synthesis, decreases by half [27]. This timescale does not distinguish between unfolded or folded state degradation, instead simply provides an estimate of the lifetime of a protein inside a cell. Based on this definition of halflife, it is natural to expect that proteins would be required to fold in a timescale lower than their halflife. Assuming lifetime distribution to be Poisson, average lifetime (t_l) and halflife ($t_{1/2}$) are related $t_l = t_{1/2} / \ln 2$. If the average folding time of a given protein is t_f, the probability of *folding before degradation* (P_{fbd}) is

$$P_{fbd} = \frac{1}{1 + t_f / t_l}. \quad (6)$$

Clearly, if $t_f >> t_l$ most of the proteins will be degraded before folding. At the other extreme if $t_l >> t_f$, almost all of the proteins will be folded before degradation. It is also important to note, even if $t_l \approx t_f$, nearly 50% of the proteins will be degraded before folding which is not very efficient either. Thus we do not assume the boundary condition due to the maximum folding time to be exactly equal to the average lifetime of the fastest degrading protein. Instead, we fit topology-based folding speed distribution to determine the maximum allowed folding time for the diffusion-drift model. We find the best fit value of $\Delta G^\dagger_{max}$ to be $16 k_b T$, yielding the maximum folding time $t_{f,max} \approx 8$ seconds (for Yeast distribution). In the above we used the speed-barrier height relation $k_f = k_0 \exp(-\Delta G^\dagger / k_b T)$ and $k_0 \approx 1 \text{ microsecond}^{-1}$. The numerical value of k_0 is consistent with several estimates of folding speed limit [21,25,26,31,32].

Figure 1 shows the best fit distribution is in reasonable agreement with the Yeast distribution. The implication of this is threefold: i) the diffusion-drift model provides an independent test of our topology-based model prediction for the distribution of folding kinetics; ii) $t_f = t_{f,max} = 8$ seconds and $t_l = 2/.69 = 3$ min

Figure 2. A) Accessible range in stability (ΔG increasing towards right) is shown between blue and red lines. Black line shows the flat-fitness landscape for all values of stability greater than the minimum; i.e. $\Delta G > \Delta G_{min}$, with the red line showing the drop in fitness when stability is lower than the minimum due to cytotoxic effects from aggregation/misfolding. Blue line shows the upper limit of stability (ΔG_{max}) due to design challenge. B) Accessible range in the folding free energy barrier height ($\Delta G^\dagger$ increasing to the right) between blue and red lines. Black line shows the flat-fitness landscape for all values of barrier heights less than the maximum allowed i.e. $\Delta G^\dagger < \Delta G^\dagger_{max}$, with the red line showing the compromised fitness when the barrier height is greater than the maximum leading to slow folding proteins, prone to aggregation and degradation. Blue line shows it is not possible to create proteins faster than the speed limit of folding set by barrier-less folders.

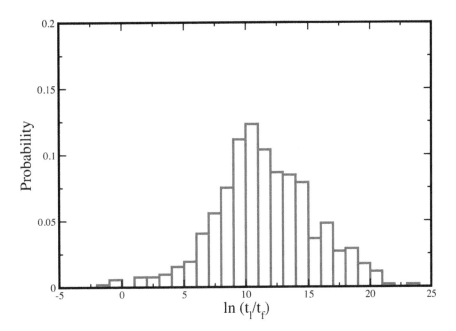

Figure 3. Distribution of the ratio of protein lifetime and protein folding time.

(for the fastest degrading protein in Yeast) argues even the fastest degrading protein in Yeast has roughly 96% probability of folding before the expiration of its lifetime. This supports the hypothesis that the slowest folding processes may be constrained by protein lifetime allowing sufficient chance for proteins to fold before degradation; iii) the assumption of flat-fitness landscape is reasonable. This implies proteome folding kinetics is not subjected to any major selection criteria to make it faster other than the primary requirement of staying sufficiently below the maximum allowed timescale set by protein halflife. However, it can not be ruled out that there are other secondary pressures to alter folding kinetics that can further improve the agreement between the diffusion-drift and topology-based model of folding kinetics. We have also fitted E.coli speed distribution with the diffusion-drift model, yielding $t_{f,max} = 2$ seconds (data not shown). However we do not provide details since a corresponding comparison with lifetime is not possible due to lack of lifetime information for E.coli proteome.

Diffusion-drift mutation model makes further prediction on the upper limit of the number of mutations per portion of the genome encoding essential genes per replication. As mentioned above, long time limit solution is given by $P(\Delta G^\dagger, t) = \exp(\lambda t) P(\Delta G^\dagger)$. In order for the population to survive, we require $\lambda \geq 0$. This requirement sets an upper limit on the number of mutations per portion of the genome encoding essential genes per replication. This limit can be obtained in terms of h, D, $\Delta G^\dagger_{max} - \Delta G^\dagger_{min}$ (see equation 8 from [1] for details). Using the values for the parameters noted above, our estimate for the upper limit is ≈ 5.5. This is indeed close to $5.7 (\approx 6)$ predicted by Shakhnovich et al. from the consideration of the stability distribution and matches well with experiments [1].

Proteome folding time is lower than the lifetime

The analysis above provides indirect support to the hypothesis that proteome lifetime may limit folding kinetics. We further test this hypothesis by directly plotting the distribution of average lifetime (t_l converted from experimentally measured halflife) values

[27] for Yeast proteome (Figure 1 in green). It is evident that the folding time and lifetime distributions are well separated. However, we also notice slight overlap between the two time scales at the boundary. This observation, at first, may indicate existence of some proteins for which the folding time may be higher than the lifetime, implying a possible contradiction to our hypothesis that protein folding is faster than degradation. In order to further test the validity of our hypothesis, we directly compare these measured lifetime values [27] and predicted folding times for each individual proteins. We select proteins from our list – used to predict the folding time in the Yeast proteome – for which lifetimes are known [27]. We compute the ratio of the lifetime and folding time for each protein in our dataset (Table S3). Figure 3 shows the distribution of the ratios of these two time scales. We find less than 1% of the proteome (4 out of 520 proteins in our list) has a folding time higher than their lifetime. The overwhelming number of proteins with a lower folding time than their lifetime, further supports the hypothesis that the lower limit of protein folding speed is indeed bounded by protein lifetime.

Although 1% is a minor fraction, one can further reason these possible exceptions. First, chaperones can play an important role to facilitate folding [5,33–35]. Chaperones can favorably alter the ratio of lifetime and folding time to help proteins escape the selection against degradation. Second, it is possible that the kinetics of the slowest folding domains are altered due to possible interdependence between multiple domains [36], an aspect not included in our model. Third, it should also be noted that the reported halflife in the work of O'Shea et al. [27] has an inherent uncertainty of a factor of two. In order to determine if any of the reasons mentioned above may be responsible, we further studied in detail the four proteins (corresponding open reading frames of YER070W, YFL041W, YJL200C and YLR304C) for which the predicted folding time is higher than the lifetime. We find three of these proteins (YFL041W, YJL200C and YLR304C) have folding time within twice their average lifetime, within the measurement uncertainty [27]. The only protein that has significantly higher folding time (fourfold higher than the lifetime) is YER070W with 80% probability of degradation before folding. However, it is interesting to note that this protein is also one of the highly abundant (top 5%) protein in the Yeast proteome [23]. The high abundance is likely due to its important biological function of facilitating synthesis of DNA. Furthermore, high abundance may offset the effect of slow folding ensuring enough copies (in absolute numbers) of the protein are present inside the cell despite the low probability of folding before degradation. Moreover, this protein has eighteen chaperone interaction partners as reported in ChaperoneDB database [28]. While the exact role of such unusually high number of chaperones to folding speed is not known at this time, it may be possible that some specific chaperones from this list or the entire chaperone network – in concert – facilitate folding of this protein in reasonable time scale to lower the burden of degradation.

Conclusions

In summary, we predict the folding time distributions for E.coli and Yeast proteome weighted by protein expression levels. We make four key observations. First, we notice E.coli and Yeast have broad distributions of folding speed with roughly similar features and ranges of the distribution. Second, the underlying distribution is reasonably explained by an independent model of diffusion-drift of mutations in free energy barrier on a "flat-fitness landscape" with two boundary conditions. While the boundary at the upper speed limit (minimum folding time) is determined by barrierless folding proteins, we find the maximum folding time to be $t_{f,max} \approx 8$ seconds (for Yeast proteome). Comparing this with the average lifetime of the fastest degrading protein ($t_l = 3$ min), we find even the fastest degrading protein in Yeast has roughly 96% probability of folding before the expiration of its lifetime. This supports the hypothesis that the slowest folding time may be bounded by protein lifetime allowing sufficient chance for proteins to fold before degradation. Third, direct comparison between measured lifetime and predicted folding time shows 99% of the proteome has a folding time less than the corresponding lifetime. Finally, the reasonable agreement between the topology-based speed distribution and the diffusion-drift model on "flat-fitness

landscape" further justifies the assumption of flat-fitness landscape. This implies the primary selection pressure for proteome folding kinetics is perhaps to outrun degradation only.

Supporting Information

Table S1 Dataset of folding time and abundance for E.coli proteome. First column reports protein name as reported in O'Brien et al. [24]; second column reports $\ln k_f$ where k_f is the folding speed (in the units of s^{-1}) for the slowest folding domain; third column reports abundance value (in ppm) from PaxDB Integrated list [23].
(PDF)

Table S2 Dataset of folding time and abundance for Yeast proteome. First column reports Open Reading Frame as reported in YRC [22]; second column reports $\ln k_f$ where k_f is the folding speed for the slowest folding domain in the units of s^{-1}; third column reports abundance value (in ppm) from PaxDB Integrated list [23].
(PDF)

Table S3 Dataset of folding time and halflife for Yeast proteome. First column reports Open Reading Frame as reported in YRC [22]; second column reports halflife (in minutes) from O'Shea et al. [27]; third column reports $\ln k_f$ where k_f is the folding speed for the slowest folding domain in the units of s^{-1}.
(PDF)

Acknowledgments

TZ and SBA acknowledge XSede for CPU time. We dedicate the paper in the memory of Nicklas Williams.

Author Contributions

Conceived and designed the experiments: SBO KG. Performed the experiments: TZ NW KG. Analyzed the data: TZ SBO KG. Contributed reagents/materials/analysis tools: TZ. Wrote the paper: SBO KG.

References

1. Zeldovich K, Chen P, Shakhnovich E (2007) Protein stability imposes limits on organism complexity and speed of molecular evolution. Proc Natl Acad Sci 104: 16152–16157.

2. Ghosh K, Dill K (2010) Cellular proteomes have broad distributions of protein stability. Biophys J 99: 3996–4002.

3. Sawle L, Ghosh K (2011) How do thermophilic proteins and proteomes withstand high temperature? Biophys J 101: 217–227.

4. Dill K, Ghosh K, Schmit J (2011) Physical limits of cells and proteomes. Biophys J 108: 17876.

5. Rollins G, Dill K (2014) General mechanism of two-state protein folding kinetics. J Am Chem Soc 136: 11420–11427.

6. Serohijos A, Lee S, Shakhnovich E (2013) Highly abundant proteins favor more stable 3d structures in yeast. Biophys J 104: L1–3.

7. Drummond D, Bloom J, Adami C, Wilke C, Arnord F (2005) Why highly expressed proteins evolve slowly. Proc Natl Acad Sci 102: 14338–14343.

8. Serohijos A, Rimas Z, Shakhnovich E (2012) Protein biophysics explains why highly abundant proteins evolve slowly. Cell Reports 2: 249–256.

9. Clarke J, Cota E, Fowler S, Hamill S (1999) Folding studies of immunoglobulin-like beta-sandwich proteins suggest that they share a common folding pathway. Structure 7: 1145–1153.

10. Dinner A, Karplus M (2001) The roles of stability and contact order in determining protein folding rates. Nature Structural Biology 8: 21–22.

11. Wang T, Zhu Y, Gai F (2004) Folding of a three-helix bundle at the folding speed limit. J Phys Chem B 108: 3694–3697.

12. Baker D, Agard D (1994) Kinetics versus thermodynamics in protein folding. Biochemistry 33: 7505–7509.

13. Braselmann E, Chaney J, Clark P (2013) Folding the proteome. Trends in Biochemical Sciences 38: 337–344.

14. Plaxco K, Simons K, Baker D (1998) Contact order, transition state placement and the refolding rates of single domain proteins. J Mol Biol 277: 985.

15. Chan H (1998) Protein folding: Matching speed and locality. Nature 392: 761–763.

16. Ivankov D, Garbuzynskiy S, Alm E, Plaxco K, Baker D, et al. (2003) Contact order revisited: Influence of protein size on the folding rate. Protein Sci 12: 2057–2062.

17. Gromiha M, Selvaraj S (2001) Comparison between long-range interactions and contact order in determining the folding rate of two-state proteins: application of long-range order to folding rate prediction. J Mol Biol 310: 27–32.

18. Ouyang Z, Liang J (2008) Predicting protein folding rates from geometric contact and amino acid sequence. Protein Sci 17: 1256.

19. De Sancho D, Munoz V (2011) Integrated prediction of protein folding and unfolding rates from only size and structural class. Phys Chem Chem Phys 13: 17030–17043.

20. Zou T, Ozkan S (2011) Local and non-local native topologies reveal the underlying folding landscape of proteins. Physical Biology 8: 066011.

21. Rustad M, Ghosh K (2012) Why and how does native topology dictate the folding speed of a protein? J Chem Phys 137: 205104.

22. Drew K, Winters P, Butterfoss G, Berstis V, Uplinger K, et al. (2011) The proteome folding project: proteome-scale prediction of structure and function. Genome Res 21: 1981–1994.

23. Wang M, Weiss M, Simonovic M, Haertinger G, Schrimpf S, et al. (2012) Paxdb, a database of protein abundance averages across all three domains of life. Mol Cell Proteomics 11: 492–500.

24. Ciryam P, Morimoto R, Vendruscolo M, Dobson C, O'Brien E (2013) In vivo translation rates can substantially delay the cotranslational folding of the e. coli cytosolic proteome. Proc Natl Acad Sci 110: E132–140.

25. Hagen S, Hofrichter J, Szabo A, Eaton W (1996) Diffusion-limited contact formation in unfolded cytochrome c: estimating the maximum rate of protein folding. Prot Natl Acad Sci 93: 11615–17.

26. Ghosh K, Ozkan S, Dill K (2007) The ultimate speed limit to protein folding is conformational searching. J Am Chem Soc 129: 11920–11927.

27. Belle A, Tanay A, Bitincka L, Shamir R, O'Shea E (2006) Quantification of protein half-lives in the budding yeast proteome. Proc Natl Acad Sci 103: 13004–13009.

28. Gong Y, Kakihara Y, Krogan N, Greenblatt J, Emili A, et al. (2009) An atlas of chaperoneprotein interactions in saccharomyces cerevisiae: implications to protein folding pathways in the cell. Molecular Systems Biology 5: 275.

29. Guo M, Xu Y, Gruebele M (2012) Temperature dependence of protein folding kinetics in living cells. Proc Natl Acad Sci 109: 17863–17867.

30. Naganathan A, Munoz V (2010) Insights into protein folding mechanisms from large scale analysis of mutational effects. Proc Natl Acad Sci 107: 8611–8616.

31. Yang W, Gruebele M (2003) Folding at the speed limit. Nature 423: 193–197.

32. Changbong H, Thirumalai D (2012) Chain length determines the folding rates of RNA. Biophys J 102: L11–L13.

33. Mashaghi A, Kramer G, Bechtluft P, Zachmann-Brand B, Driessen A, et al. (2013) Reshaping of the conformational search of a protein by the chaperone trigger factor. Nature 500: 98–101.

34. Brinker A, Pfeifer G, Kerner M, Naylor D, Hartl F, et al. (2001) Dual function of protein confinement in chaperonin-assisted protein folding. Cell 107: 223–233.

35. Cuyle J, Texter F, Ashcroft A, Masselos D, Robinson C, et al. (1999) Groel accelerates the refolding of hen lysozyme without changing its folding mechanism. Nature 6: 683–690.

36. Batey S, Clarke J (2006) Apparent cooperativity in the folding of multidomain proteins depends on the relative rates of folding of the constituent domains. Proc Natl Acad Sci 103: 18113–8.

Permissions

The contributors of this book come from diverse backgrounds, making this book a truly international effort. This book will bring forth new frontiers with its revolutionizing research information and detailed analysis of the nascent developments around the world.

We would like to thank all the contributing authors for lending their expertise to make the book truly unique. They have played a crucial role in the development of this book. Without their invaluable contributions this book wouldn't have been possible. They have made vital efforts to compile up to date information on the varied aspects of this subject to make this book a valuable addition to the collection of many professionals and students.

This book was conceptualized with the vision of imparting up-to-date information and advanced data in this field. To ensure the same, a matchless editorial board was set up. Every individual on the board went through rigorous rounds of assessment to prove their worth. After which they invested a large part of their time researching and compiling the most relevant data for our readers.

The editorial board has been involved in producing this book since its inception. They have spent rigorous hours researching and exploring the diverse topics which have resulted in the successful publishing of this book. They have passed on their knowledge of decades through this book. To expedite this challenging task, the publisher supported the team at every step. A small team of assistant editors was also appointed to further simplify the editing procedure and attain best results for the readers.

Apart from the editorial board, the designing team has also invested a significant amount of their time in understanding the subject and creating the most relevant covers. They scrutinized every image to scout for the most suitable representation of the subject and create an appropriate cover for the book.

The publishing team has been an ardent support to the editorial, designing and production team. Their endless efforts to recruit the best for this project, has resulted in the accomplishment of this book. They are a veteran in the field of academics and their pool of knowledge is as vast as their experience in printing. Their expertise and guidance has proved useful at every step. Their uncompromising quality standards have made this book an exceptional effort. Their encouragement from time to time has been an inspiration for everyone.

The publisher and the editorial board hope that this book will prove to be a valuable piece of knowledge for researchers, students, practitioners and scholars across the globe.

List of Contributors

Erik Helgeland, Lars Ertesvåg Breivik, Marc Vaudel, Frode Steingrimsen Berven, Anne Kristine Jonassen and Hilde Garberg
Department of Biomedicine, Faculty of Medicine and Dentistry, University of Bergen, Bergen, Norway

Øyvind Sverre Svendsen
Department of Anaesthesia and Surgical Services, Haukeland University Hospital, Bergen, Norway

Jan Erik Nordrehaug
Department of Clinical Science, Faculty of Medicine and Dentistry, University of Bergen, Bergen, Norway

Scott Fields., Benben Song., Bareza Rasoul, Julie Fong, Melissa G. Works, Kenneth Shew, Ying Yiu, Jon Mirsalis and Annalisa D'Andrea
Biosciences Division, SRI International, Menlo Park, California, United States of America

Jiabin Wang, Jian Yang, Song Mao, Xiaoqiang Chai, Yuling Hu, Xugang Hou, Yiheng Tang, Cheng Bi and Xiao Li
College of Life Sciences, Sichuan University, Ministry of Education Key Laboratory for Bio-resource and Eco-environment, Sichuan Key Laboratory of Molecular Biology and Biotechnology, Chengdu, People's Republic of China

Mehdi Bagheri Hamaneh and Yi-Kuo Yu
National Center for Biotechnology Information, National Library of Medicine, National Institutes of Health, Bethesda, MD, United States of America

Ranjan Kumar Barman
Biomedical Informatics Centre, National Institute of Cholera and Enteric Diseases, Kolkata, West Bengal, India

Santasabuj Das
Biomedical Informatics Centre, National Institute of Cholera and Enteric Diseases, Kolkata, West Bengal, India
Division of Clinical Medicine, National Institute of Cholera and Enteric Diseases, Kolkata, West Bengal, India

Sudipto Saha
Bioinformatics Centre, Bose Institute, Kolkata, West Bengal, India

Min Oh and Youngmi Yoon
Department of Computer Engineering, Gachon University, Seongnam, Korea

Jaegyoon Ahn
Department of Integrative Biology and Physiology, University of California Los Angeles, Los Angeles, California, United States of America

Juan Yang, Lingyu Zhao, Liying Liu, Yannan Qin, Xiaofei Wang,Tusheng Song and Chen Huang
Key Laboratory of Environment and Genes Related to Diseases of the Education Ministry, Department of Genetics and Molecular Biology, Medical School of Xi9an Jiaotong University, Xi9an, China

Jin Yang and Yan Gao
Department of Medical Oncology, First Affiliated Hospital of Medical School of Xi'an Jiaotong University, Xi9an, China

André Weiss, Hanna Joerss and Jens Brockmeyer
Institute of Food Chemistry, Westfälische Wilhelms-Universität Münster, Münster, Germany

Mohamed Diwan M. AbdulHameed, Gregory J. Tawa, Kamal Kumar and Anders Wallqvist
Department of Defense Biotechnology High Performance Computing Software Applications Institute, Telemedicine and Advanced Technology Research Center, U.S. Army Medical Research and Materiel Command, Fort Detrick, Maryland, United States of America

Danielle L. Ippolito, John A. Lewis and Jonathan D. Stallings
U.S. Army Center for Environmental Health Research, Fort Detrick, MD, United States of America

Hong Yu
Department of Quantitative Health Sciences, University of Massachusetts Medical School, Worcester, Massachusetts, United States of America
VA Central Massachusetts, Leeds, Massachusetts, United States of America

Qing Zhang
Department of Quantitative Health Sciences, University of Massachusetts Medical School, Worcester, Massachusetts, United States of America

Cinzia Franchin and Giorgio Arrigoni
Proteomics Center of Padova University, Padova, Italy

Luca Cesaro and Mauro Salvi
Department of Biomedical Sciences, University of Padova, Padova, Italy

Lorenzo A. Pinna
Department of Biomedical Sciences, University of Padova, Padova, Italy
CNR Institute of Neurosciences, Padova, Italy

Naif Zaman and Edwin Wang
Biotechnology Research Institute National Research Council, Montréal, QC, Canada

Paresa N. Giannopoulos
Lady Davis Institute for Medical Research, Segal Cancer Centre - Jewish General Hospital, Montréal, QC, Canada

Shafinaz Chowdhury
Biotechnology Research Institute National Research Council, Montréal, QC, Canada
Lady Davis Institute for Medical Research, Segal Cancer Centre - Jewish General Hospital, Montréal, QC, Canada

Eric Bonneil and Pierre Thibault
Institut de recherche en immunologie et en cance´rologie, Université de Montréal, Montréal, QC, Canada

Mark Trifiro and Miltiadis Paliouras
Lady Davis Institute for Medical Research, Segal Cancer Centre - Jewish General Hospital, Montréal, QC, Canada
Department of Medicine, Department of Oncology and Division of Experimental Medicine, McGill University, Montréal, QC, Canada

Ngaam J. Cheung and Hong-Bin Shen
Institute of Image Processing and Pattern Recognition, Shanghai Jiao Tong University, and Key Laboratory of System Control and Information Processing, Ministry of Education of China, Shanghai, China

Xue-Ming Ding
School of Optical-Electrical and Computer Engineering, University of Shanghai for Science and Technology, Shanghai, China

Catarina Correia and Astrid M. Vicente
Departamento de Promoção da Saúde e Doenças não Transmissíveis, Instituto Nacional de Saúde Doutor Ricardo Jorge, 1649-016 Lisboa, Portugal
Center for Biodiversity, Functional & Integrative Genomics, Faculty of Sciences, University of Lisbon, 1749-016 Lisboa, Portugal
Instituto Gulbenkian de Ciência, 2780-156 Oeiras, Portugal

Guiomar Oliveira
Unidade Neurodesenvolvimento e Autismo, Centro de Desenvolvimento, Hospital Pediátrico (HP) do Centro Hospitalar e Universitário de Coimbra (CHUC), 3000-602 Coimbra, Portugal
Centro de Investigação e Formação Clinica do HP-CHUC, 3000-602 Coimbra, Portugal
Faculdade de Medicina da Universidade de Coimbra, 3000-548 Coimbra, Portugal

Laura Bennett, Songsong Liu and Lazaros G. Papageorgiou
Centre for Process Systems Engineering, Department of Chemical Engineering, UCL (University College London), Torrington Place, WC1E 7JE, London, United Kingdom

Aristotelis Kittas and Sophia Tsoka
Department of Informatics, King's College London, Strand, WC2R 2LS, London, United Kingdom

Dawid Perlikowski, Halina Wiśniewska, Michał Kwiatek, Maciej Majka and Arkadiusz Kosmala
Institute of Plant Genetics of the Polish Academy of Sciences, Poznan, Poland

Tomasz Góral
Plant Breeding and Acclimatization Institute – National Research Institute, Radzikow, Blonie, Poland

Hyun-Jung Kim
College of Pharmacy, Chung-Ang University, Seoul, Korea

Mi Seul Park, Kyung Rok Kim and Byung Woo Han
Research Institute of Pharmaceutical Sciences, College of Pharmacy, Seoul National University, Seoul, Korea

Eduard Bitto
Department of Chemistry and Biochemistry, Georgian Court University, Lakewood, New Jersey, United States of America

Craig A. Bingman
Department of Biochemistry, Center for Eukaryotic Structural Genomics, University of Wisconsin Madison, Madison, Wisconsin, United States of America

Mitchell D. Miller
Department of Biochemistry and Cell Biology, Rice University, Houston, Texas, United States of America

George N. Phillips Jr
Department of Biochemistry, Center for Eukaryotic Structural Genomics, University of Wisconsin Madison, Madison, Wisconsin, United States of America
Department of Biochemistry and Cell Biology, Rice University, Houston, Texas, United States of America

Qing Wang, Siyi Zhang, Shichao Pang, Menghuan Zhang and Bo Wang
Department of Bioinformatics & Biostatistics, School of Life Science and Biotechnology, Shanghai Jiao Tong University, Shanghai, China

Qi Liu
Department of Bioinformatics & Biostatistics, School of Life Science and Biotechnology, Shanghai Jiao Tong University, Shanghai, China
Department of Biomedical Informatics, Vanderbilt University School of Medicine, Nashville, Tennessee, United States of America
Center for Quantitative Sciences, Vanderbilt University School of Medicine, Nashville, Tennessee, United States of America

Jing Li
Department of Bioinformatics & Biostatistics, School of Life Science and Biotechnology, Shanghai Jiao Tong University, Shanghai, China
Shanghai Center for Bioinformation Technology, Shanghai, China

Elisa Mazzoni, Silvia Pietrobon, Irene Masini, John Charles Rotondo Fernanda Martini and Mauro Tognon
Department of Morphology, Surgery and Experimental Medicine, Section of Pathology, Oncology and Experimental Biology, University of Ferrara, Ferrara, Italy

Mauro Gentile, Ilaria Casetta, Massimiliano Castellazzi and Enrico Granieri
Biomedical Sciences and Specialized Surgeries, Section of Neurology, School of Medicine, University of Ferrara, Ferrara, Italy

Enrico Fainardi
Unit of Neuroradiology, University Hospital of Ferrara, Ferrara, Italy

Maria Luisa Caniati and Maria Rosaria Tola
Unit of Neurology, University Hospital of Ferrara, Ferrara, Italy

Giovanni Guerra
Clinical Laboratory Analysis, University Hospital of Ferrara, Ferrara, Italy

Taisong Zou and S. Banu Ozkan
Center for Biological Physics, Department of Physics, Arizona State University, Tempe, Arizona, United States of America

Nickolas Williams and Kingshuk Ghosh
Department of Physics and Astronomy, University of Denver, Denver, Colorado, United States of America

Index

A

Absolute Contact Order (aco), 220

Adaptive Firefly Algorithm, 148-149, 158

Amino Acid Residue, 197-198

Androgen Receptor (ar), 134, 146

Angiotensinogen, 84-85, 93-95

Antibodies, 4, 11-12, 82, 212-217

Autism Genetic Resource Exchange (agre), 169

Autism Risk Genes, 160, 165

Autism Spectrum Disorder (asd), 160

B

Benzalkonium Chloride (bzk), 10-11

Biological Function, 15, 20, 23, 205, 224

Biological Pathways, 37, 39-40, 44-45, 47, 49, 51, 70, 108, 134

Biomarkers, 1, 5, 8, 10-11, 15-18, 20, 37, 75-76, 78-80, 82-83, 96, 98, 108-109, 145, 147, 170, 184

Biomedical Domain, 111

C

Cardioprotection, 1-2, 5, 8-9

Cellular Signaling, 123, 145

Central Nervous System (cns), 212

Chaotic Mapping Operator, 148

Chemical Exposures, 96, 98, 100-101, 105-106

Clear Cell Renal Cell Carcinoma, 75, 82-83

Clinical Prostate Cancer, 134

Community Structure Detection, 171-172, 181

Comparative Toxicogenomics

Database (ctd), 64, 99, 102

Computational Approaches, 23, 110

Contact Order (co), 220

Crystal Structure, 95, 136, 146, 170, 197-198, 200-202

D

Dimeric Alpha-amylase Inhibitors, 186

Disjoint Communities, 171-173, 183

Drug-centric Approach, 63

Drug-disease Associations, 63-65, 67, 69, 72

E

Enrichment Analysis, 39, 41-44, 47, 49, 51, 53, 58-61, 71, 97-99, 101, 104, 109, 133, 136, 139, 164-166, 169-170, 209

Enterohemorrhagic Escherichia Coli (ehec), 84

Eukaryotic Host, 53

F

Female Genital Tract, 10, 21

Functional Characterization, 84

Functional-linkage Network, 22

Fusarium Head Blight, 186, 188-192, 194-196

G

Gene Associations, 39-40, 42, 45-46, 49-50

Gene Grouping, 206

Gene Groups, 40, 205-207, 209-210

Gene Prioritization, 23, 30, 32, 38, 165, 205, 207, 210-211

Genome-wide Association Studies (gwas), 160

H

Host Proteins, 53, 55-56, 60, 84, 92

Human Plasma Proteome, 1

Human Serine Protease Inhibitors, 84

Humoral Mediators, 1

I

Inflammatory Host Response, 84, 94

Inflammatory Neurologic Diseases, 212, 214-216

Information Flow, 39-40, 51

Interactome, 22, 37, 62, 134-137, 141, 143, 170, 185, 211

K

K-nearest Neighbor Model (knn), 112

Kernel Proteins, 186, 188

L

Lambda Phosphatase, 123, 125

Light Absorption Coefficient, 148-150

Liver Fibrosis, 96-109

M

Machine Learning Methods, 53

Mammalian Mitochondria, 22, 25

Mathematical Programming, 171-173, 181, 183-184

Matrix-assisted Laser Desorption Ionization (maldi), 75

Microbicide Toxicity, 10, 15, 18

Mitochondrial Diseases, 22-23, 30, 33

Molecular Actions, 63

Monomeric Alpha-amylase, 186, 191-194

Multiple Sclerosis, 161, 165, 169, 212-215, 217-218

N

N-end Rule Pathway, 197-199, 201, 203-204

N-terminal Glutamine

Amidohydrolase, 197, 199

Network-based Classification Model, 63

Non-hispanic, 134, 140-145

O

Overlapping Modules, 171-172, 184

P

Parameter Analysis, 148

Peptide Backbone Recognition

Patch, 197

Phenotypic Level, 219

Phosphopeptidome, 123-124, 127, 130-131

Polo-like Kinase 2 (plk2), 123

Potential Serum, 75-76, 78

Protein Halflife, 219, 222-223

Protein Interaction Network, 23, 37, 39-41, 45, 62, 111, 136-138, 145, 174, 184, 206

Protein Interaction Networks, 38, 40, 138, 160, 171, 184-185

Protein-protein Interaction, 24, 26, 28-29, 37, 39-40, 51, 53, 62, 96, 98, 102, 104, 108, 111, 137, 160-161, 169, 171, 174, 184, 205-206, 211

Proteome, 1-2, 5, 7-8, 14, 20, 22-23, 25-26, 29-32, 37-38, 123, 125-128, 132, 137, 145, 184, 187-190, 194-196, 219-220, 223-225

Proteome Folding Kinetics, 219-220, 223

Proteomexchange, 1-2, 4, 8

Proteomic Biomarkers, 75

Proteomic-coupled-network, 134

Q

Quantitative Proteomics, 10, 38, 83, 123, 132

R

Randomization Parameter, 148-150, 154

Rank Candidate Genes, 205

Remote Ischemic Preconditioning, 1, 8

S

Simian Virus, 54, 212, 214, 217

Somatic Mutations, 70, 134, 138, 143, 145-146

Support Vector Machines (svm), 53-55

Systems Level Analysis, 96, 108

T

Topological Structures, 110-111

Toxic Liver Injury, 96

Toxicogenomics Database, 51, 64, 73, 96, 98-99, 102, 109

Triticum Aestivum, 186-187, 189-196

V

Vaginal Epithelial Cell Line, 10, 17

W

Winter Wheat, 186-196

Z

Zinc Finger Protein 3 (zfp3), 75, 80

Printed in the USA
CPSIA information can be obtained
at www.ICGtesting.com
JSHW051417221024
72173JS00006B/1372